Ecological Responses at Mount St. Helens: Revisited 35 years after the 1980 Eruption

Charles M. Crisafulli • Virginia H. Dale
Editors

Ecological Responses at Mount St. Helens: Revisited 35 years after the 1980 Eruption

Springer

Editors
Charles M. Crisafulli
USDA Forest Service
Pacific Northwest Research Station
Mount St. Helens National Volcanic Monument
Amboy, WA, USA

Virginia H. Dale
Department of Ecology & Evolutionary Biology
The University of Tennessee
Knoxville, TN, USA

ISBN 978-1-4939-7449-8 ISBN 978-1-4939-7451-1 (eBook)
https://doi.org/10.1007/978-1-4939-7451-1

Library of Congress Control Number: 2017959605

Cover Illustration: Cover photo of indian paintbrush (*Castilleja miniata*) on the Pumice Plain with Mount St. Helens in the background, July 2009. (Photo by Charles M. Crisafulli).

Printed on acid-free paper

This Springer imprint is published by Springer Nature
The registered company is Springer Science+Business Media, LLC
The registered company address is: 233 Spring St, New York, NY 10013, U.S.A.

Preface

The spectacular 18 May 1980 eruption of Mount St. Helens was the largest and most deadly volcanic eruption in historical times in the conterminous United States. The catastrophic eruption and its aftermath captured the attention of people worldwide who witnessed the event as it was broadcast by numerous media. Social, economic, and ecological systems over a wide swath of the Pacific Northwest were disrupted, and the hazards of living in a volcanically active region were quickly realized. At the same time, the eruption created unprecedented scientific opportunities for investigating initial and long-term responses of ecosystems to a range of volcanic disturbance processes. Given its close proximity to population centers, transportation networks, and academic institutions, scores of scientists arrived on the scene within months of the eruption, where they established hundreds of studies that focused on numerous biological groups (microbes to mammals) and ecological processes across a range of terrestrial and aquatic system types. A number of these studies are still active today (2017), and, with nearly 40 years of accumulated data, they provide a wealth of information on the resilience of the Mount St. Helens landscape and waters. Many lessons have been distilled from this body of research, which has strongly influenced and even reshaped our understanding of biological community assembly, successional processes, and disturbance ecology. This book tells the story of ecological responses around Mount St. Helens spanning almost four decades since the eruption and synthesizes much of what has been learned. Mount St. Helens offers an exemplary arena to investigate both the forces that cause dramatic change in ecosystems and the resilience of nature. A diversity of disturbance mechanisms operated at the volcano, and they created high gradients of disturbance intensity. The variability in response trajectories was large, with rapid ecological responses in some places and slow transformation in others. Where they occurred, biological legacies, both living and dead, strongly influenced development of ecological communities. Chance events and contingencies played a large role in reestablishment processes, particularly where no biological legacies persisted and where populations experienced boom and bust cycles. Spatial heterogeneity in both disturbance and the pace and pattern of ecological response is a hallmark of the post-eruption area and has led to high biodiversity, rich with complex early-seral communities that are of both local and regional significance. In addition, secondary disturbances, ongoing geological processes, and human activities were important in some locations.

The study of ecological responses in the diverse environments at Mount St. Helens has been conducted by a team of dedicated scientists. For several scientists, their research and careers were entirely redirected as they focused intensively on the area's ecology. Although the ecologists working at Mount St. Helens come from a variety of institutions and locations, their interests have been held together by a participation in a regular gathering called a "pulse" at which scientists monitor the status of Mount St. Helens ecosystems. During these week-long field camps, researchers visit many of field sites, collect data, forge new collaborations, plan synthesis efforts, and recruit the next generation of early- to mid-career scientists. The first pulses occurred in the summers of 1980 and 1981 and were cosponsored by the USDA Forest Service's Pacific Northwest Research Station and National Science Foundation. During the recent decades, they have been held at 5-year intervals and funded by the Forest Service.

This volume serves to link both the science and the many scientists conducting research at Mount St. Helens. In reviewing each other's work, we have learned about the commonalities and differences of ecological responses across a range of volcanic disturbance processes.

Over the years, we have shared what has been learned at Mount St. Helens via publications and speaking at professional meetings. But the most enthusiastic audiences are often students and youth. Even preschool children are fascinated by volcanoes and learning about ecological interactions. As a way to engage others in the geological and ecological learning from Mount St. Helens, each May 18, Virginia makes a volcano cake and shares it with whatever group she is speaking to on that anniversary of the eruption. This three-dimensional model provides a vivid means to explain volcanic and ecological processes. One fifth-grade teacher said that her students had never been more interested in discussing a cake than eating it.

Photograph This volcano cake illustrates many of the geological and ecological features of Mount St. Helens. The unseen strawberry cake layer depicts magma, which is hidden below the Earth's surface. The next layer of chocolate cake represents soil lying under the volcanic material. Chocolate icing symbolizes the pyroclastic flows, debris avalanche, and lahars that inundated the surface. White icing stands for the glaciers that remain on the mountain top. The crater still emits steam, as demonstrated by dry ice. Although there was little lava emitted during the eruption of Mount St. Helens (at least until later dome-building episodes), enjoyment of strawberries enticed us to add lots of them. Pretzels represent the downed trees, and thyme and rosemary symbolize the vegetation that is reestablishing. Scattered gummy butterflies herald the return of insects to the ecosystem. Delectable chocolate-covered raisins represent volcanic bombs. Hidden in the cake are gummy fish that memorialize the many animals killed by the eruption, and gummy worms representing animals that survived in some locations by being sheltered underground during the eruption. A Mount St. Helens cake is a special treat every May 18.

The ecological studies at Mount St. Helens could not have been accomplished without diverse financial support. The Forest Service and National Science Foundation funded initial studies, and their support has continued to the present. In addition, individual projects were funded by grants from the National Geographic Society, Earthwatch Institute, the Washington Department of Fish and Wildlife, and several foundations. Much work has been accomplished by personal initiative, volunteers, and building upon related projects. The Forest Service has provided continuous support for the work by Crisafulli and others at Mount St. Helens and for collecting, documenting, and archiving data sets and physical and biological specimens from long-term ecological studies in the area.

This book is the direct result of contributions of many people in addition to the authors. Kathryn Ronnenberg (USDA Forest Service, Pacific Northwest Research Station) worked tirelessly to standardize all tables, create many of the disturbance zone maps, edit and generate final figures, and performed an excellent job as technical editor for the entire book. Kelly Christiansen (USDA Forest Service, Pacific Northwest Research Station) assisted with cartography and GIS projects. Special thanks are owed to the numerous scientists who reviewed early versions of the chapters. For these important reviews, we wish to thank Scott Abella, Ivan Arismendi, Pete Bisson, John Campbell, Cynthia Chang, Roger del Moral, Karen Dillman, Luciana Elizalde, Jack Feminella, Mark Fonstad, Jerry Franklin, Doria Gordon, Karen Gran, Raymond Iglay, Sarah Jovan, Raffaele Lafortezza, Doug Larson, Kelly Maloney, Connie Millar, Erich Mueller, Jim O'Connor, Bob Parmenter, Daniel Schindler, Bjarni Sigurdsson, Siobhan Staunton, Fred Swanson, Jonathan Titus, Shiro Tsuyuzaki, Lars Walker, Frank Wilhelm, Michael Willig, Ellen Wohl, Don Zobel, and several reviewers who wish to remain anonymous.

The institutional homes of the editors provided essential support for their work at Mount St. Helens, including writing and editing this book. Charlie gratefully acknowledges the support of the Pacific Northwest Research Station and particularly the sustained interest of Cindy Miner and Deanna Olson. Virginia appreciates the support from the Environmental Sciences Division at Oak Ridge National Laboratory (ORNL), which is managed by UT–Battelle for the US Department of Energy. The editors thank the Gifford Pinchot National Forest, the Mount St. Helens National Volcanic Monument, and the Mount St. Helens Institute and their staffs for logistical support and access to records, maps, and research sites. All royalties from the purchase of this book will go to the Mount St. Helens Institute.

On a personal note, during the past decades, we have spent much time in the volcanic landscape learning a great deal about disturbance ecology and becoming quite familiar with the area. Friends, colleagues, and family members have all supported our work, and many have been with us as we learned about this intriguing landscape. Virginia wishes to especially thank her family and friends, who enjoy assisting in locating plots, digging soil samples, monitoring plots, and conducting other fieldwork and did not mind the hours she spent on this book during weekends and early mornings. In particular, her commitment to finish this book was encouraged by the interests of her grandchildren—especially 5-year old Lily, who is a budding volcanologist. Charlie thanks and acknowledges his daughters, Erica and Teal, for their interest, enthusiasm, and tolerance for his passion and obsession with Mount St. Helens. He thanks his friend Abigail Groskopf for countless conversations, outings into Cascadia, and meals together during the preparation of this book. Thanks and appreciation are given to Fred Swanson for his interest, collaboration, and shared adventures into the fascinating world of volcano ecology. Finally, appreciation and insights to Mount St. Helens ecology have been sharpened by annual trips to investigate contemporary eruption sites in Chile with collaborators Mauro E. González, Antonio Lara, and their students at Universidad Austral de Chile and also with Barbara Corrales and Pablo Saumann at Parque Valle Los Ulmos. The editors and authors owe special gratitude to Jerry Franklin, Jim MacMahon, and the late Jim Sedell for their commitments to science at Mount St. Helens.

We thank authors and publishers for the permission to reprint the three poems. John Daniel's poem "To Mount St. Helens" was first published in his book *All Things Touched by Wind* (Salmon Run Press, 1994) and subsequently published in *Of Earth: New and Selected Poems* (Lost Horse

Press, 2012). Gary Snyder's poem "Pearly Everlasting" first appeared in his book *Danger on Peaks* (Counterpoint, 2004, 2014). These two poems and Ursula K. Le Guin's "Another Weather: Mount St. Helens" appeared in *In the Blast Zone* (Oregon State University Press, 2008).

As we write this Preface, small earthquakes are common occurrences, and magma is recharging beneath Mount St. Helens, which signals the ongoing geological unrest and potential future ecological consequences for the region. Although such events have occurred several times in recent years and are not a cause for concern, they do remind us that this is an active volcano. The ecological systems are in flux as well. The ecosystems are moving toward, but still many long years from, conditions more characteristic of the forested landscapes of the Pacific Northwest than the gray, barren, ash-covered valleys and ridges that were prevalent immediately after the eruption. There is still much to be learned from this dynamic system. We hope that science reported in this volume and our prior book, *Ecological Responses to the 1980 Eruption of Mount St. Helens*, makes unique and important contributions to the study of succession and disturbance ecology. Collectively, this body of research is supported by well-documented, publicly accessible descriptions of long-term field plots and associated data. Most of all, we hope that the thrill of learning about ecological systems experienced by all of us working at Mount St. Helens can be conveyed by reading about the complex interaction between organisms and their environment at Mount St. Helens.

Charles M. Crisafulli
Virginia H. Dale
July 2017

Contents

Contributors

Michael F. Allen Departments of Plant Pathology and Microbiology, Center for Conservation Biology, University of California, Riverside, CA, USA

Joseph A. Antos Department of Biology, University of Victoria, Victoria, BC, Canada

Tara E. Blackman U.S. Department of Agriculture, Forest Service, Pacific Northwest Research Station, Mount St. Helens National Volcanic Monument, Amboy, WA, USA

Shannon M. Claeson U.S. Department of Agriculture, Forest Service, Pacific Northwest Research Station, Wenatchee Forestry Sciences Laboratory, Wenatchee, WA, USA

Charles M. Crisafulli U.S. Department of Agriculture, Forest Service, Pacific Northwest Research Station, Mount St. Helens National Volcanic Monument, Amboy, WA, USA

Virginia H. Dale Department of Ecology & Evolutionary Biology, The University of Tennessee, Knoxville, TN, USA

Roger del Moral Department of Biology, University of Washington, Seattle, WA, USA

Elsie M. Denton U.S. Department of Agriculture, Agricultural Research Service, Eastern Oregon Agricultural Research Center, Burns, OR, USA

James E. Gawel Environmental Sciences and Studies, School of Interdisciplinary Arts and Sciences, University of Washington, Tacoma, Tacoma, WA, USA

Linda H. Geiser U.S. Department of Agriculture, Forest Service, Pacific Northwest Region, Air Resources Management Program, Corvallis, OR, USA

U.S. Department of Agriculture, Forest Service, Water, Fish, Wildlife, and Rare Plants, Washington Office - Yates Building, Washington, DC, USA

William J. Gerth Department of Fisheries and Wildlife, Oregon State University, Corvallis, OR, USA

Jeffrey J. Gerwing Environmental Science and Management Department, Portland State University, Portland, OR, USA

Jonathan J. Halvorson U.S. Department of Agriculture, Agricultural Research Service, Northern Great Plains Research Laboratory, Mandan, ND, USA

Jasper H. Hardison III U.S. Fish and Wildlife Service, Water Resources Branch, Anchorage, AK, USA

Lindsey L. Karr Environmental Sciences and Management Department, Portland State University, Portland, OR, USA

Thomas E. Lisle U.S. Department of Agriculture, Forest Service, Pacific Southwest Research Station, Arcata, CA, USA

James A. MacMahon Department of Biology and the Ecology Center, Utah State University, Logan, UT, USA

Jon J. Major U.S. Geological Survey, Cascades Volcano Observatory, Vancouver, WA, USA

Bruce McCune Department of Botany and Plant Pathology, Oregon State University, Corvallis, OR, USA

Rich Miller Environmental Science and Management Department, Portland State University, Portland, OR, USA

Adam R. Mosbrucker U.S. Geological Survey, Cascades Volcano Observatory, Vancouver, WA, USA

Peter R. Nelson Arts and Sciences Division, University of Maine-Fort Kent, Fort Kent, ME, USA

Kristine A. Nichols Rodale Institute, Kutztown, PA, USA

Matthew R. O'Neill Department of Biology, University of California-Riverside, Riverside, CA, USA

Robert R. Parmenter National Park Service, Valles Caldera National Preserve, Jemez Springs, NM, USA

Cheryl A. Parmenter Museum of Southwestern Biology, University of New Mexico, Albuquerque, NM, USA

Gary L. Parsons Department of Entomology, A.J. Cook Arthropod Research Collection, Michigan State University, East Lansing, MI, USA

Danny Shpeley Department of Biological Sciences, University of Alberta, Edmonton, AB, Canada

Kurt Spicer U.S. Geological Survey, Cascades Volcano Observatory, Vancouver, WA, USA

Frederick J. Swanson U.S. Department of Agriculture, Forest Service, Pacific Northwest Research Station, Corvallis Forestry Sciences Laboratory, Corvallis, OR, USA

Jonathan H. Titus Department of Biology, State University of New York at Fredonia, Fredonia, NY, USA

Tim Wheeler Arlee, MT, USA

Donald B. Zobel Department of Botany and Plant Pathology, Oregon State University, Corvallis, OR, USA

About the Editors

Charles M. Crisafulli is a research ecologist with the USDA Forest Service, Pacific Northwest Research Station. His primary research focuses on initial and longer-term ecological responses of organisms in aquatic and terrestrial ecosystems following explosive volcanism. From 1980 to present (2017), he has worked at Mount St. Helens, and since 2008, he has annually conducted field research at several other contemporary eruption sites in Patagonia, Chile. His primary research themes are processes of succession and assembly of biological communities and chiefly focus on small mammals, birds, amphibians, and arthropods. He is currently engaged in expanding and testing the generality of lessons drawn from Mount St. Helens to volcanoes in other regions of the world such as South America, Asia, and Alaska. More broadly, he is building a global volcano ecology database to assess the current state of the field and to develop science strategies for the future related to investigations of volcanically disturbed ecosystems.

U.S. Department of Agriculture, Forest Service, Pacific Northwest Research Station, Mount St. Helens National Volcanic Monument, Amboy, WA 98601, USA
Phone: 360-449-7834
email: ccrisafulli@fs.fed.us

Virginia H. Dale is a corporate fellow at Oak Ridge National Laboratory (ORNL) and an adjunct professor in the Department of Ecology & Evolutionary Biology at the University of Tennessee. She was on the first team of biologists who entered the red zone at Mount St. Helens after the 1980 eruption and has been studying vegetation reestablishment there ever since. Her primary research interests are disturbance ecology, plant succession, land use change, sustainability, and environmental decision making. She has served on national scientific advisory boards for five agencies of the United States and several committees of the National Academy of Sciences. She was among the members of the international science community who contributed to the Intergovernmental Panel on Climate Change Scientific Assessment that in 2007 received the Nobel Peace Prize. In 2013, she was recognized with the Distinguished Landscape Ecologist award by the US Regional Association of the International Association for Landscape Ecology. Her current interest is in understanding risk and resilience in a changing world.

Department of Ecology & Evolutionary Biology, The University of Tennessee, Knoxville, TN 37996-1610, USA
Phone: 865-974-3065
email: vdale@utk.edu

Prologue. "To Mount St. Helens"

John Daniel

You were the perfect one,
the saint of symmetry.
We glanced at your benign
bright face, and you shined back
your blessing, you smiled
peacefully upon us.
We didn't much believe
Your smoke and stir, we thought
your restiveness would pass—
and then you shuddered hard
and blasted yourself across
four states, engulfed a lake,
gorged rivers with gray mud,
flattened entire forests
and whatever lives they held
in your searing smother.
Your evenness and grace
exploded twelve miles high,
then showered down as grit
on our trim lawns and gardens—
and there you slouch, smudged
and gaping, spewing smoke,
resting in your rubble.
You did it, Mount St. Helens.
As all of us looked on
you stormed in solitude,
you shrugged and shook aside
what we called beautiful
as if none of us were here,
no animals, no trees,
no life at all outside
your ancient fiery joy—
I admired you, mountain,
but I never loved you until now.

(from *In the Blast Zone*, Oregon State University Press. 2008, p. 52–53)

Ecological Responses to the 1980 Eruption of Mount St. Helens: Key Lessons and Remaining Questions

1

Virginia H. Dale and Charles M. Crisafulli

1.1 Introduction

The role of disturbance in influencing ecosystems and shaping landscapes is clearly recognized (White 1979; Pickett and White 1985; Turner 1987, 2010; Turner and Dale 1998). The heterogeneous and complex effects of large infrequent disturbances on ecosystems contribute to the operation of natural systems, including patterns and processes of ecological response following such events (Turner et al. 1997). As such, disturbances are not viewed by ecologists as catastrophic agents of destruction but rather as normal, perhaps even integral, parts of long-term system dynamics (Weatherhead 1986). Volcanic eruptions are powerful agents of large infrequent disturbances and create exemplary living laboratories that can be used to investigate both initial disturbance effects on ecosystems and longer-term successional processes (Dammerman 1948; Thornton et al. 1988; del Moral and Grishin 1999; Dale et al. 2005a; Crisafulli et al. 2015).

Mount St. Helens and the other 16 major volcanic landforms in the Cascade Range of northern California, Oregon, and Washington state are part of the "ring of fire" that borders the Pacific Ocean; the Pacific Rim has about three-fourths of the world's active and dormant terrestrial volcanoes (Harris 1976). Mount St. Helens is the most active volcano in the Pacific Northwest in recent centuries and has experienced at least 20 eruptions in the past 4000 years (Crandell and Mullineaux 1978).

The 1980 eruption of Mount St. Helens was dramatic both in terms of the variety of volcanic processes and the spatial extent of their disturbance effects on ecosystems (Lipman and Mullineaux 1981; Dale et al. 2005a). Complex geophysical forces created exceptional opportunities to study initial ecological responses and longer-term successional processes across a diversity of ecosystems (lowland to subalpine forests, meadows, lakes, streams, and rivers), and the presence of forests across the landscape gave geologists many opportunities to interpret geophysical processes from observations of damage to vegetation. These conditions make Mount St. Helens an extremely rich environment for learning about the ecology of volcanic areas and, more generally, about ecosystem response to many types of disturbance, including land management.

Dozens of ecologists took advantage of the research opportunities created by the 1980 eruption, establishing hundreds of studies that included numerous taxa, ecological processes, and system types across an array of volcanic disturbance types and intensities. These efforts resulted in hundreds of research articles and several syntheses (e.g., Keller 1982, 1986; Bilderback 1987; Dale et al. 2005a; Swanson and Crisafulli, Chap. 16, this volume). For example, nearly 40% of the world's published studies on plant and animal responses to volcanic eruptions have taken place at Mount St. Helens (Dale et al. 2005b; Edwards 2005; Crisafulli et al. 2015; Swanson and Crisafulli, Chap. 16, this volume). The 2005 volume *Ecological Responses to the 1980 Eruption of Mount St. Helens* (Dale et al. 2005a) provided a 25-year perspective on ecological change distilled from several important studies conducted at the volcano. Now, after the passage of another 12 years of research, it is timely to undertake a further synthesis of knowledge of ecosystem responses to the 1980 eruption. This companion volume builds on early foundational work by exploring and reporting on ecological patterns and

V.H. Dale (✉)
Department of Ecology & Evolutionary Biology, The University of Tennessee, Knoxville, TN 37996, USA
e-mail: vdale@utk.edu

C.M. Crisafulli
U.S. Department of Agriculture, Forest Service, Pacific Northwest Research Station, Mount St. Helens National Volcanic Monument, 42218 NE Yale Bridge Rd, Amboy, WA 98601, USA
e-mail: ccrisafulli@fs.fed.us

C.M. Crisafulli, V.H. Dale (eds.), *Ecological Responses at Mount St. Helens: Revisited 35 years after the 1980 Eruption*,
https://doi.org/10.1007/978-1-4939-7451-1_1

processes at Mount St. Helens up to 35 years since the eruption.

Mount St. Helens is a unique living laboratory that can be used to address questions that require long-term observations and experiments that can take advantage of its diverse conditions. Mount St. Helens National Volcanic Monument was established, in part, to foster scientific investigations. Thus, researchers have some confidence that long-term studies and collaborations among disciplines will continue.

1.2 The 1980 Eruption of Mount St. Helens

The cataclysmic eruption of Mount St. Helens on the morning of 18 May 1980 involved a variety of volcanic and hydrological processes: a massive *debris avalanche*, a laterally directed pyroclastic density current (commonly referred to as a lateral blast and herein called the *blast PDC*), *lahars* (volcanic mudflows), *pyroclastic flows*, and extensive ash and pumice deposition (*tephrafall*) (Lipman and Mullineaux 1981; Swanson and Major 2005) (Table 1.1). Subsequent minor eruptions generated pyroclastic flows, tephrafalls, and lahars and created several lava domes in the new crater (Lipman and Mullineaux 1981; Sherrod et al. 2008). These geological processes dramatically affected ecosystems over ~570-km^2 area in a broad arc north of the volcano and a far larger area (10s of thousands of km^2) to the east with much lesser intensity through wind-dispersed deposition of tephra.

Pre-eruption forests consisted of naturally regenerated conifer stands, including numerous old-growth forests, as well as tree plantations that were established following clearcut logging (Swanson et al. 2005). The highly dissected landscape was drained by abundant fast-flowing streams, and high-elevation areas were dotted with small subalpine lakes. Meadows were common on the flanks of the volcano, which had a very low tree line, well below the climatic limit, reflecting volcanic activity during recent centuries (Lawrence 1938). Meadows were also common in the Mount Margaret Range, north of the volcano.

In the aftermath of the eruption, a highly heterogeneous environment resulted from numerous factors. The eruption involved a complex set of geophysical forces, varying in their intensity and acting upon an array of different pre-eruption biological communities, successional stages, and site conditions, distributed across complex montane topography (Table 1.2). These factors led to a landscape mosaic of broadly defined disturbance zones that were characterized by

Table 1.1 Definitions of volcanic processes used in this book.

Volcanic process	Definition
Pyroclastic density current (PDC)[a, b]	Rapid flow of a dry mixture of hot (commonly >700 °C) solid particles, gases, and air, which can range in character from a dense, ground-hugging flow (*pyroclastic flow*) to a turbulent, low-density cloud of mostly fine ash and superheated gases and air (*pyroclastic surge*). A single PDC commonly involves both flow types as a result of gravitational segregation. Flows are generally gravity driven but may be accelerated initially by impulsive lateral forces of directed volcanic explosions. Flows typically move at high velocity (up to several hundred km h^{-1}).
Blast pyroclastic density current[c]	A form of pyroclastic density current initiated by rapid decompression of lava domes or cryptodomes (magma bodies cooled high within a volcanic edifice) owing to sudden collapse. Rapid decompression results in a directed explosion that initially impels the current laterally before it becomes a gravity-driven flow. In the case of the 1980 Mount St. Helens eruption, failure of the volcano's north flank unroofed pressurized magma and superheated water. Rapid exsolution of magmatic gases and conversion of superheated water to steam produced a laterally directed blast, which formed a density current that flowed across rugged topography. The current contained fragmented rock debris as well as shattered forest material.
Debris avalanche[a]	A rapid granular flow of an unsaturated or partly saturated mixture of volcanic rock particles (± ice) and water, initiated by the gravitational collapse and disintegration of part of a volcanic edifice. Debris avalanches differ from debris flows in that they are not water-saturated. Although debris avalanches commonly occur in association with eruptions, they can also occur during periods when a volcano is inactive.
Lahar[a]	An Indonesian term for a rapid granular flow of a fully saturated mixture of volcanic rock particles (± ice), water, and commonly woody debris. A lahar that has ≥50% solids by volume is termed a *debris flow*; one that has roughly 10–50% solids by volume is termed a *hyperconcentrated flow*. Flow type can evolve with time and distance along a flow path as sediment is entrained or deposited.
Tephrafall[a]	Volcanic particles raining to the ground following their ejection into the atmosphere by an explosive eruption. Tephra is a collective term for particles of any size, shape, or composition ejected in an explosive eruption.

[a]Terms for size classes of pyroclastic particles are as follows: Ash, particles <2 mm in diameter; lapilli, particles 2–64 mm in diameter; blocks or bombs, particles >64 mm in diameter. Once pyroclastic particles are deposited and later remobilized, they are referred to as *secondary volcaniclastic deposits*, and their particles are described using standard sedimentology size terminology (e.g., clay, silt, sand, pebbles, cobbles, and boulders; Folk 1980; Pierson and Major 2014).

[b]A generalized definition based on definitions of PDCs provided in Sigurdsson et al. (2015) and Pierson and Major (2014).

[c]Sigurdsson et al. (2015) and Pierson and Major (2014).

Table 1.2 Characteristics of deposits from the 18 May 1980 eruption of Mount St. Helens.

Event	Volume of uncompacted deposit (km^3)[a]	Area affected (km^2)[a]	Thickness of deposit (m)[a]	Survival of biota
Debris avalanche	2.5	60	10–195	Almost none[b]
Blast PDC	0.20	550	0.01–1	Pockets of survival in limited protected areas[c]
Lahar	0.05	50	0.1–3	Some trees survived[d]
Pyroclastic flows	0.3	15	0.25–40	None[c]
Proximal tephrafall	0.1	1100	>0.01	High[e]

[a]Data from Lipman and Mullineaux (1981).
[b]Dale et al. (2005c).
[c]Crisafulli et al. (2005a).
[d]Frenzen et al. (2005).
[e]Antos and Zobel (2005).

the amount, type, and distribution of residuals (biological legacies—organisms and organic materials that remain after a disturbance) from the pre-eruption ecosystems (Fig. 1.1). Collectively, these zones can be viewed as a gradient of disturbance severity ranging from complete annihilation of all life immediately in front of the crater in the pyroclastic-flow zone, to areas of high survivorship of most taxa 30 km to the north in the tephrafall zone (Fig. 1.2). Areas occurring between these two endpoints (debris-avalanche deposit, lahar deposits, blast scour, blowdown and standing dead forest zones) had intermediate levels of biological legacies. Certainly, the eruption led to extensive mortality of plants and animals across a broad tract of Cascadian terrain and also to the tragic loss of 57 human lives. However, a most surprising fact in the early post-eruption years was the realization that nearly all plant and animal species inhabiting the landscape before the eruption had at least some individuals that survived by many mechanisms in small and isolated safe havens throughout the majority of the landscape and waterways (del Moral 1983; Halpern and Harmon 1983; Andersen and MacMahon 1985; Franklin et al. 1985; Crawford 1986; Zobel and Antos 1986, 1992; Adams et al. 1987; Hawkins and Sedell 1990) (Fig. 1.3). Subsequent ecological change resulted from growth and spread of surviving populations, joined by numerous colonists that immigrated from distant source populations; community and ecosystem development was influenced by physical changes to the environment such as weathering of substrates, nutrient enrichment, and secondary disturbance (Figs. 1.4 and 1.5). These processes both retarded and accelerated plant establishment and growth and, by extension, animal community assembly, depending on local circumstances (Fig. 1.6).

Secondary physical disturbances increased the heterogeneity of developing biological communities and landscapes. These disturbances included natural events such as shallow, rapid landslides, lahars, and erosion, as well as land management activities such as salvage logging, road construction, and aerial seeding of plant species with the intent to reduce erosion. Each activity was location specific, and none approached the magnitude or extent of the 1980 eruption. Furthermore, as the ecosystem became more complex, biotic interactions such as mutualism, herbivory, predation, decomposition, and competition gained importance. More than three decades after the cataclysmic eruption, the intertwining of geophysical and ecological changes is critical to understanding the developing Mount St. Helens system.

1.3 Lessons Learned from Ecological Research at Mount St. Helens

What has already been learned from research on ecological responses at Mount St. Helens is influencing the understanding of succession, animal community assembly, disturbance ecology, ecosystem management, evolution and the origin of life, trophic interactions, and landscape ecology (Table 1.3). Together, these studies reveal how much knowledge about ecological succession and disturbance ecology has expanded as a result of the numerous long-term studies at Mount St. Helens (Fig. 1.7). The work on the great diversity of taxa examined in the aftermath of the 1980 eruption provides numerous insights on ecosystem responses to a large, intense natural disturbance, including patterns and processes of survival, colonization, and biotic assembly (Table 1.4). The following key findings are discussed in depth in Crisafulli et al. (2005b) and Dale et al. (2005d, e).

- *Disturbances are complex.* Like most explosive eruptions, the 1980 event at MSH involved several types of volcanic disturbance processes that occurred concurrently or consecutively within a 12-h period.
- *Mechanisms of disturbance vary in type and intensity and thus severity of effects.* Mechanisms of disturbance included the force of impact, heat, abrasion, burial, plant-canopy loading, suspended particles in air or water, and chemical toxicity, and their intensities influenced the severity of effects on the biota. These mechanisms varied substantially among the types of disturbance processes that operated at Mount St. Helens in 1980; generally, their intensity decreased with distance from the crater. Several gradients of disturbance intensity occurred across the different disturbance processes involved in the 1980 eruption. The gradients included impact force, heat, and deposit thickness. For example, the large, 570-km^2 area that was disturbed by the blast PDC varied from areas

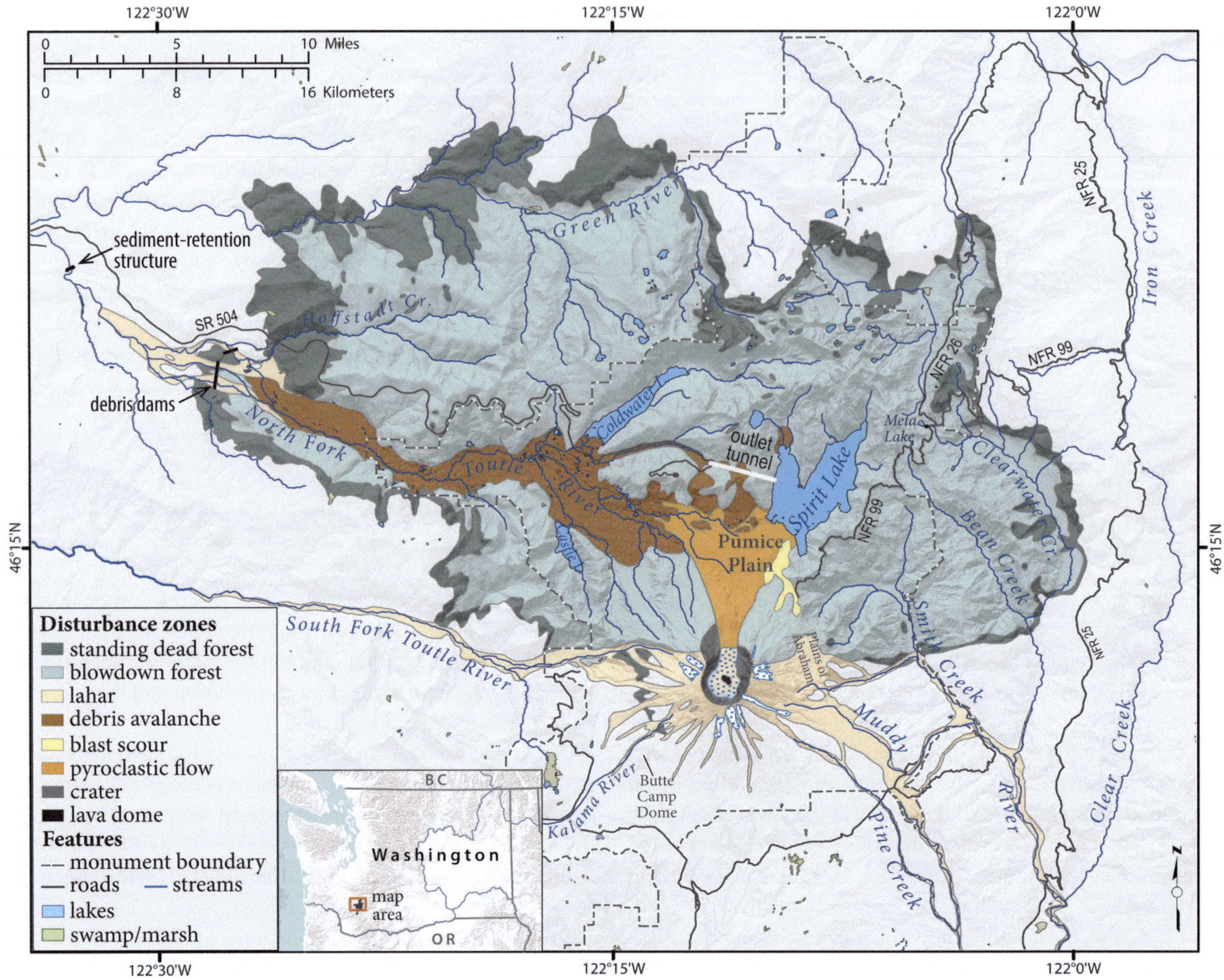

Fig. 1.1 Disturbance zones created during the 18 May 1980 eruption of Mount St. Helens. (Cartography by Kathryn Ronnenberg, USDA Forest Service, Pacific Northwest Research Station. Original data source for disturbance zones was Lipman and Mullineaux 1981).

closest to the crater where trees were entirely removed (shattered, incinerated, or carried away), to more distant locations where the force leveled trees, to the outer limits of the blast area where the force was not powerful enough to knock trees down but heat seared the foliage resulting in a standing dead forest. Tephrafall thickness also demonstrated an intensity gradient that ranged from several tens of cm deposited 12 km northeast of the crater to areas 500 km from the volcano in eastern Washington where just a few millimeters of tephra were deposited.

- *Biotic and abiotic legacies from the disturbance influence patterns and rates of ecosystem development.* In most of the post-eruption landscape and waterways that were influenced by the powerful blast PDC, biotic and abiotic legacies were prevalent and persisted. However, some organisms that survived the initial disturbance did not live long enough to reproduce but still contributed to ecosystem development. The types, amounts, and spatial distribution of these legacies had profound implications for the pace and pattern of ecological response in both the short and long term.
- *Ecological response is rapid in some systems but is spatially and temporally heterogeneous.* The ecological response has been surprisingly rapid in many disturbance zones, but the rate and trajectory of these responses have been variable across the landscape as well as across system types (e.g., upland versus riparian, stream versus lake). New habitats were created by both physical and biotic processes (e.g., several new lakes and ~120 ponds formed after water filled surface depressions or at the mouths of tributary streams dammed by the debris-avalanche deposit).
- *Spatial heterogeneity was a common feature of this large, disturbed landscape.* This heterogeneity resulted from the

Fig. 1.2 Photographs of the Mount St. Helens landscape showing conditions produced by the 18 May 1980 eruption: (**a**) tephrafall zone—late-seral Pacific silver fir forest (*Abies amabilis*) ~25 km NNE of the volcano; (**b**) standing dead forest zone, upper Clearwater Creek valley; (**c**) blowdown zone with extensive toppled trees, upper Green River valley; (**d**) debris-avalanche deposit, North Fork Toutle River valley looking east; (**e**) lahar, Muddy River looking downstream—note helicopter (*white circle*, at *right*) for scale; (**f**) pyroclastic-flow deposits on the Pumice Plain (light-colored features that flowed from *left* to *right*). (Figure modified from Figure 3.1 of Dale et al. 2005a).

Fig. 1.3 Organisms that survived in the blast area or tephrafall zone of the 18 May 1980 eruption of Mount St. Helens: (**a**) fireweed (*Chamerion angustifolium*) emerges from blast-PDC and tephrafall deposits (J. Franklin, July 1980); (**b**) a diverse assemblage of herbs emerges from erosion gullies where ~20 cm of pumiceous tephra were deposited (Joe Antos, 1980); (**c**) northwestern salamanders (*Ambystoma gracile*) survived in burrows and frozen lakes and ponds; (**d**) boreal toads (*Anaxyrus boreas*) were still in hibernation underground at the time of the eruption; (**e**) northern pocket gophers (*Thomomys talpoides*) experienced widespread survival because of their fossorial habitat association; (**f**) ants (*Camponotus* spp.) survived in logs and underground. (Modified from Figure 73.3 of Crisafulli et al. 2015. Photos **c**–**f** by Charles M. Crisafulli).

Fig. 1.4 Examples of plant species, representing several life forms, that colonized the blast area in the years following the 1980 eruption of Mount St. Helens: (**a**) prairie lupine (*Lupinus lepidus*), a nitrogen-fixing, keystone species, was an early and persistent colonizer of primary-successional surfaces at the volcano; (**b**) Cardwell's penstemon (*Penstemon cardwellii*), a ground-hugging shrub, was a common colonizer of upper-elevation slopes on the volcano (July 2009); (**c**) giant red Indian paintbrush (*Castilleja miniata*) became a conspicuous component of the herbaceous flora of the Pumice Plain beginning in the 1990s (July 2009); (**d**) *Racomitrium* moss (*Racomitrium canescens*), a dominant species on primary-successional surfaces (May 2017); (**e**) Sitka willow (*Salix sitchensis*) and green alder (*Alnus viridis*) have formed dense shrub thickets throughout much of the blast-PDC area (2015); and (**f**) Douglas-fir (*Pseudotsuga menziesii*) is among the most common conifer species to colonize the blast area (2006). (Photos by Charles M. Crisafulli).

Fig. 1.5 Examples of animals that colonized the blast-PDC area in the years following the 1980 eruption of Mount St. Helens: (**a**) elk (*Cervus elaphus*); **b** deer mouse (*Peromyscus maniculatus*); (**c**) white-crowned sparrow (*Zonotrichia leucophrys*); (**d**) common nighthawk (*Chordeiles minor*); (**e**) two-stripped grasshopper (*Melanoplus bivitattus*); (**f**) a carabid beetle (*Calosoma tepidum*). (Adapted from Figure 73.9 in Crisafulli et al. 2015. Photos by Charles M. Crisafulli).

Fig. 1.6 Photo chronosequence showing changes in stream-channel conditions and riparian plant community development at Herrington Creek on the debris-avalanche deposit, 1981–2012. Herrington Creek did not exist before the 1980 eruption and formed during the post-eruption evolution of the watershed. (Photo credits: Brian Franzen and Pete Bisson).

Table 1.3 Selected examples of the types of research about ecological responses that have occurred at Mount St. Helens.

Topic of research	References
Succession (including plants, mycorrhizae, and lichens)	Turner et al. (1998), Walker and del Moral (2003), del Moral and Chang (2015), Meiners et al. (2015), Allen et al. (Chap. 11, this volume), del Moral and Titus (Chap. 7, this volume), Nelson et al. Chap. 10, this volume), Zobel and Antos, (Chap. 6, this volume)
Animal community assembly (including insects, mammals, and fish)	MacMahon et al. (1989), Crisafulli et al. (2005a, c, Chap. 15, this volume), Parmenter et al. (2005, Chap. 12, this volume), Blackman et al. (Chap. 14, this volume), Claeson et al. (Chap. 13, this volume)
Disturbance ecology	Turner et al. (1997), Turner and Dale (1998), del Moral and Grishin (1999), Dale and Denton, (Chap. 8, this volume), Karr et al. (Chap. 9, this volume)
Hydroecology	Gawel et al. (Chap. 4, this volume)
Soil ecology	Halvorson et al. (Chap. 5, this volume)
Ecosystem management	Swanson and Franklin (1992), Dale et al. (1998, 2000), Franklin et al. (2002)
Evolution and the origin of life	Baross and Hoffman (1985)
Trophic interactions	Fagan and Bishop (2000), Bishop (2002), Bishop et al. (2005)
Landscape ecology	Foster et al. (1998), Lawrence and Ripple (2000), Raab et al. (2012)

Additional papers can be found in the literature-cited sections of the listed publications.

complexities of the primary disturbance processes themselves and ecosystem responses to these processes. For example, successful seed dispersal and establishment of plants were affected by local site conditions as much as by distance from seed sources. Animal populations and community structure were strongly influenced by biological legacies and plant community development.

- *Secondary disturbances and ongoing geophysical processes influence ecological response to the 1980 eruption.* Secondary disturbances, often caused by a sequence of geophysical processes (such as erosion), occur both chronically and episodically on the landscape and can be important in regulating ecological responses to the primary disturbance. For example, ongoing erosion along the newly formed streams creates a dynamic environment in which plants are not able to become permanently established. These processes led to increased spatial heterogeneity and development of early-seral communities that promoted high biotic diversity across the landscape.
- *Random events are important.* Chance and contingency have been central in shaping ecological responses; for example, the timing of the event (i.e., in spring, when patches of the landscape were still snow-covered, and in daytime, when nocturnal burrowing animals would have been protected underground) greatly affected the severity of disturbance, and thus the types, amounts, and distribution of organisms that survived in refugia, ranging from single individuals to entire lake ecosystems (Figs. 1.8 and 1.9). Chance establishment of certain species influenced substrate nutrient status and community development. Such factors complicated the ability to predict successional rate and pattern at any one time or place.
- *Ecological succession is complex.* Succession processes were multifaceted rather than simple as some theories have claimed. Ecological succession during the first three post-eruption decades was not a unidirectional path toward convergence but rather involved many nuances that accelerated, retarded, and sometimes redirected systems. In areas undergoing primary succession, sites close together often supported dissimilar communities, suggesting stochastic processes were paramount. Yet other sites showed more similar patterns, suggesting that a primary factor influencing community assembly is environmental filtering, which refers to abiotic factors that prevent the establishment or persistence of species in a particular location (Kraft et al. 2015). With myriad different points of initiation in the post-eruption landscape, many successional pathways were underway.
- *Human activities influence patterns and rates of ecological response.* Human intervention in post-eruption landscapes and waterways was sometimes implemented to ensure public safety or was desired if restoration of some resource or ecological function was deemed important. Decisions to attempt to influence processes or organisms in a disturbed area impeded, promoted, or had little effect on the rate of ecological response, yet often affected research opportunities. Attention to natural processes could have improved the potential for positive outcomes of management efforts.
- *Resource managers should plan for disturbances.* Management actions such as aerial seeding and stream engineering (e.g., constructed outlet channels for lakes created by volcanic deposits that blocked streams) were most effective when tailored to particular disturbance characteristics and management goals and based on ecological principals that promote desired response pattern and process. It is critical that management plans recognize the potential for disturbances to occur and design responses in tune with the particular situation.

Possibly the most critical lesson gained from more than three decades of studies following the 1980 eruption is the importance of persistence, both of these ecological systems and in research efforts. Ecological persistence refers to the ability of the ecological systems to survive or return to equilibrium after a perturbation, possibly to different stable states rather than to prior conditions. The plants and animals that survived and those that later colonized the different

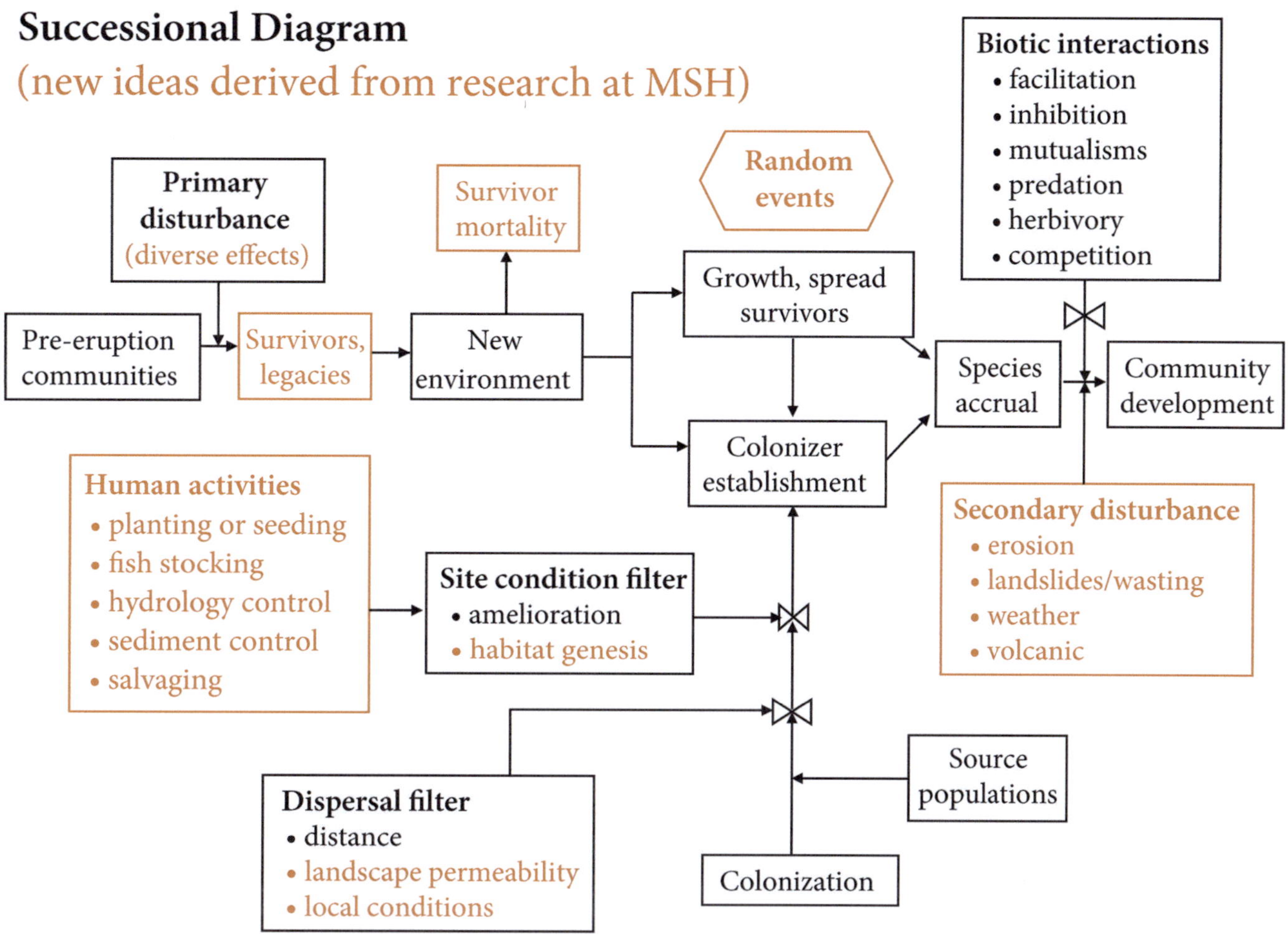

Fig. 1.7 Generalized depiction of key biological, human, and physical factors and their flow paths influencing ecological succession at Mount St. Helens after the 1980 eruption. New ideas that were fostered by lessons from Mount St. Helens shown in color. (Modified after Crisafulli et al. 2005c).

Table 1.4 Taxa studied following the 1980 eruption of Mount St. Helens, by disturbance zone.

Taxon	Tephrafall	Standing dead forest	Blowdown	Lahar	Debris avalanche	Pyroclastic flow	Study period[a]
Fungi (mycorrhizae)	x	x	x			x	1982–2015
Lichens	x	x	x	x	x	x	2013
Plants	x	x	x	x	x	x	1980–2017
Phytoplankton			x		x	x	1980–2017
Zooplankton			x		x	x	1980–2017
Stream macroinvertebrates	x		x	x		x	1980–2015
Pond macroinvertebrates					x		2003–2017
Beetles	x		x			x	1982–2015
Earwigs	x		x			x	1982–2015
Grasshoppers and allies	x		x			x	1982–2015
Ants	x		x			x	1982–2015
Spiders	x		x			x	1982–2015
Stream fishes	x	x	x	x	x	x	1980–2014
Lake fishes	x	x	x		x	x	1980–2017
Stream amphibians	x	x	x			x	1980–2016
Lake amphibians	x	x	x		x	x	1980–2017
Birds	x	x	x	x	x	x	1981–2017
Small mammals	x	x	x	x	x	x	1980–2017
Large mammals		x	x	x	x	x	1980–2017

[a]Studies were not necessarily conducted every year during the sampling period.

Fig. 1.8 Photo chronosequence showing changes in plant community development from 1983 (post-eruption year 3) through 2004 in the blowdown zone at Meta Lake basin, a site that was snow-covered at the time of the 18 May 1980 eruption, which led to the survival of numerous plants, including understory conifer saplings. Compare this sequence with Fig. 1.9, a site lacking snow at the time of the eruption. (Photos by Peter Frenzen, USDA Forest Service).

disturbance zones have attributes allowing them to persist in the face of what often appeared to be difficult challenges. These attributes include the ability of some plants to regenerate from root or stem fragments, and of others to survive by having stems and leaves that rose above burial by tephra or lahars. Mutualistic relationships that developed between plants and bacteria (N_2 fixation) and plants and fungi (mycorrhizae) led to the successful establishment and growth of vegetation, which in turn facilitated colonization by numerous other plants and animals. Both our earlier volume (Dale et al. 2005a) and this book convey many examples of ecological persistence via survival and reestablishment of the fauna and flora in the vicinity of Mount St. Helens.

The importance of persistence of the scientists who have worked at Mount St. Helens is another key lesson from the research conducted at the volcano. Those of us

Fig. 1.9 Photo chronosequence showing changes in plant community development from 1983 (post-eruption year 3) through 2016 in the blowdown zone at upper Smith Creek, a site that was not snow-covered at the time of the 18 May 1980 eruption. Compare this sequence with Fig. 1.8, a site where vegetation was snow-protected at the time of the eruption. (Photos by Peter Frenzen, USDA Forest Service).

biologists who took the first official helicopter ride into the Red Zone to survey the damage and begin looking for survivors were enthralled by the opportunity to document the initial stages of survival and reestablishment after such a large disturbance. Now, more than three decades later, we are still intrigued on each return trip to Mount St. Helens to observe and record key patterns and processes in this dynamic and fascinating landscape. And that enthusiasm is infectious, as new faces have joined the research force, while others have moved on to different activities. The group of scientists engaged in the ongoing research is diverse in terms of disciplines, expertise, and institutions. Many of the scientists from the original cohort have reached retirement age, and regrettably, a few have died, but others are continuing the research by taking on the responsibilities of data collection and analysis.

The ability to continue these long-term studies is largely due to the persistence of a few dedicated scientists and to the long-term support of the USDA Forest Service via the Pacific Northwest Research Station and the Mount St. Helens National Volcanic Monument. Although support for this work has come from a diversity of public and private funds, the Forest Service's Pacific Northwest Research Station has continuously supported a science program since the eruption, as well as research pulses every 5 years, when the Mount St. Helens ecological-research community convenes to share key lessons, develop collaborations, and to collect new information. In addition, the National Science Foundation has provided both the initial and sustained funding to various researchers at the volcano from 1980 through the present (2017). Mount St. Helens as a research site has benefitted for 37 years from the efforts of a large community of investigators and has served as a training ground for hundreds of undergraduate and graduate students. The research community working at Mount St. Helens includes a plethora of specialists (botanists, ecologists, entomologists, zoologists, lichenologists, mycologists, soil scientists, hydrologists, and geologists) who come from a variety of institutions (universities, government agencies, and non-governmental organizations).

Over time, the effects of the 1980 eruption of Mount St. Helens have become apparent in the regional biodiversity and ecosystem (Swanson et al. 2011). The high habitat diversity that developed in the National Volcanic Monument, where the ecological systems are allowed to proceed on their own course, provides resources for numerous early-successional species and assemblages as well as ecosystem processes. The most notable animals are the abundant populations and diverse assemblages of amphibians, mammals, passerine birds, beetles, and lepidopterans (butterflies and moths). The large, disturbed area experienced different patterns and rates of succession than other parts of the Pacific Northwest landscape where fire suppression and intensive forest management have promoted development of a conifer forest canopy. As the most recently active of the 17 major volcanic features in the Cascade Range (Harris 1976), Mount St. Helens continues to add diversity to the landscape.

Worldwide, a variety of ecosystems appear to have similar patterns of survival and assembly of biotic communities in response to volcanic disturbance (Dale et al. 2005b; Edwards 2005; Swanson and Crisafulli, Chap. 16, this volume). The type and intensity of eruption processes and characteristics of the erupted material (e.g., tephra) can have widely varying effects on biota. Establishment of plants and animals is influenced by the types, amounts, and spatial distribution of biological legacies; distance from source populations; substrate conditions; climate and weather; biotic interactions; and time since disturbance. Site-specific predictions of establishment patterns are challenging because each location experiences unique effects during an eruption, has site-specific initial vegetation and soil and microclimate conditions as well as unique legacies, and is influenced by stochastic processes.

1.4 Questions Still Remaining

There is still much to learn from the ongoing succession and assembly processes at Mount St. Helens. The variety and large spatial extent of the primary disturbance and subsequent ecological and geophysical processes that continue to unfold create rich opportunities for addressing numerous ecological questions at the volcano.

- *What is the role of spatial heterogeneity in response patterns and processes?* The spatial variability in resource islands, nutrient pools, seed sources, and substrate conditions in the aftermath of the eruption creates circumstances in which no one successional pathway or process occurs in all places. How different habitat patches interact with each other can influence the pattern, rate, and dominant processes as ecosystems change over time. The extent to which habitat patches are connected to each other and to regional source populations may strongly influence ecological stability, persistence, and change.
- *How can information be appropriately extrapolated from plot-based studies to landscapes and from years to decades?* Extrapolation of ecological patterns across spatial and temporal scales is an important research challenge at Mount St. Helens and elsewhere. The diversity of patterns and rates of ecological response across disturbance zones offers a means to compare and contrast scaling approaches under several different sets of conditions. The interplay of disturbance types with scaling can help build understanding of how underlying homogeneity or heterogeneity affects the ability to extrapolate information and discern what kinds of processes are most important at certain scales. Broad-scale and long-term analyses can reveal patterns of ecosystem assembly not possible at finer scales, such as plot-based work conducted for only a few years.
- *How do human activities influence ecological succession?* Humans have affected ecosystems around Mount St. Helens in numerous ways, including spread of non-native plant species, fish stocking, interruption of fish passage, game harvesting and protection, road and trail building, dam construction, active timber management (outside of the National Volcanic Monument), and control of erosion, sediment transport, and water flow. Hence, Mount St. Helens offers opportunities to examine

long-term effects of human activities on ecological and geological processes. Findings from such studies can directly influence resource management designs that address past and future natural and anthropogenic disturbances in the Pacific Northwest and elsewhere.

- *What is the role of species interactions in early succession of diverse environments?* Interspecific relationships are now better understood as a result of ongoing studies at Mount St. Helens. For example, much has been learned about the effects that a diverse set of herbivores has on plant community structure in this system, including numerous insect and several mammal species [e.g., elk (*Cervus elaphus*), American beaver (*Castor canadensis*), voles (*Microtus* spp.), and northern pocket gophers (*Thomomys talpoides*)]. Considerable research at Mount St. Helens has focused on mutualistic relationships among plants and fungi (mycorrhizae; Allen et al., Chap. 11, this volume) and plants and bacteria (nitrogen fixation) and how they have influenced soil microbial biomass and nutrient pools. As succession proceeds, future research will provide insights into how effects of interspecific relationships vary across the heterogeneous landscapes at Mount St. Helens, how interactions between species influence the pace and pattern of succession and the assembly of animal communities, and how roles change over time.
- *How does succession occur in heterogeneous environments?* Often succession is studied in places that are fairly uniform in terms of disturbance type and intensity. But the disturbances created by the 1980 eruption of Mount St. Helens were diverse, and effects varied depending on the disturbance type, distance from the crater, site conditions at the time of the eruption (e.g., presence of snow), elevation, and other factors. Hence the Mount St. Helens landscape offers an ideal environment to study how successional patterns and processes occur over the long run and from different starting points.
- *What are the key interactions between physical and biological processes?* Any preconception that physical and biological processes were distinct was quickly erased for the team of researchers working at Mount St. Helens. Understanding interactions between physical and biological processes is essential to discerning key patterns of biotic survival and establishment. The characteristics of those interactions vary with disturbance type and intensity. The Mount St. Helens environment provides an excellent place to flesh out the implications of a variety of interactions.
- *How do interactions among aquatic and terrestrial ecosystems affect those systems?* Ecosystem productivity shifted dramatically, comparing levels observed before the eruption with those after the eruption. Highly productive terrestrial ecosystems were impoverished through loss of vegetation and burial of soil beneath volcanic deposits, whereas aquatic systems, particularly lakes, went from low levels of nutrients to high concentrations. As time since the eruption continues to elapse, some of these systems are transitioning back toward their pre-eruption productivity state, and investigation of the rate and pattern of this transition offers intriguing learning opportunities. Similarly, investigations of the movement of organisms and resources between terrestrial and aquatic ecosystems can improve understanding of factors that influence and constrain development of complex ecosystems. For example, to what extent do highly productive post-eruption lakes subsidize adjacent nutrient-impoverished terrestrial habitats through energy exports such as emergence of aquatic macroinvertebrates and amphibians that disperse to terrestrial habitats?

Although this book addresses some of these questions, others remain to be considered.

1.5 Overview of the Book

This book contributes to the ecological literature and provides a compendium of original research that builds on existing work exploring succession, disturbance ecology, and the interface between geophysical and biological systems at Mount St. Helens. This introductory chapter sets forth relevant details of the 1980 eruption along with key concepts and lessons learned over the first decades of study at Mount St. Helens. Chapters 2 and 3 review changes in hydrology and geomorphology of two major watersheds draining the volcano. Chapter 4 focuses on the physical, chemical, and biological characteristics of Spirit Lake. Chapter 5 reports on the development of nitrogen and carbon pools and evidence for formation of key building blocks for ecosystem development in nutrient-impoverished pyroclastic-flow deposits. Chapter 6 provides an overview of patterns of forest understory development following burial by tephra, specifically focusing on measures of inertia and resilience, whereas Chapter 7 looks at the rates, determinism, and alternative states of plant communities during early primary succession on pyroclastic-flow and lahar deposits. Chapter 8 reports on the patterns of plant succession on the debris-avalanche deposit and assesses differences in plant community structure, comparing plots that were aerially seeded with mostly non-native plants for erosion control against plots undergoing natural succession. Chapter 9 examines the influence of human activities (i.e., roads and silviculture) on the presence, species richness, and abundance of non-native plant species in the pyroclastic, blowdown, and tephrafall zones. Chapter 10 provides the first published treatment of lichen species composition and distribution across several disturbance zones at the volcano. Chapter 11 builds on existing literature addressing the role of mutualistic symbioses between plants and fungi across a range of volcanic disturbance types and intensities. Chapters 12, 13, 14, and 15 focus on animal responses to the 1980 eruption. Chapter 12 describes community assembly of ground-dwelling beetles in the pyroclastic-flow zone. Chapter 13 reports on the development of aquatic macroinvertebrate communities in dozens of ponds and wetlands that were created within days to months of the eruption. Chapter 14 focuses on dynamics of a new rainbow trout population in Spirit Lake and its tributary streams. Chapter 15 takes a detailed look at the rates and patterns of arrival and establishment of a diverse small-mammal assemblage on the Pumice Plain, an area undergoing primary succession. Chapter 16 places the work conducted at Mount St. Helens in the context of the broader field of volcano ecology. Together, these chapters provide an in-depth but still incomplete analysis of ecological patterns of response after the 1980 eruption of Mount St. Helens.

Throughout the book, conventional terminology for volcanic disturbance processes is used (Table 1.1). Where needed, chapters include a glossary for defining technical terms (emphasized in the text in ***bold italic face***), and the locations of study sites are shown on a map that matches the geographic scale and details referenced in the text. The major taxonomic source for species mentioned in the book is the Integrated Taxonomic Information System (http://www.itis.usda.gov). Additional information about the area and research results is available at http://www.fsl.orst.edu/msh/ and http://www.fs.fed.us/pnw/mtsthelens/index.shtml.

This book is both an update and extension of our earlier book, *Ecological Responses to the 1980 Eruption of Mount St. Helens*, published 25 years after the eruption. Now, 37 years after the eruption, ecologists working in the vicinity of the volcano bring their long-term studies up to date and report findings that extend early understanding about some organisms and report for the first time on a suite of organisms that have not previously been covered. Collectively, the chapters in this book address general themes of resistance, resilience, succession, and community assembly and lay the foundation for future research. This book synthesizes understanding of ecological change in the complex and changing environment around the Mount St. Helens volcano, which is unique in the study of volcanic eruptions and subsequent ecological recovery.

Acknowledgments We appreciate the insightful review of an earlier version of the chapter by Fred Swanson. We thank Kathryn Ronnenberg for generating final figures and tables and for copy editing the text. Part of the work by VHD was supported by Oak Ridge National Laboratory, which is managed by UT-Battelle, LLC, for the US Department of Energy under contract DE-AC05-00OR22725. Support for CMC has been provided by the USDA Forest Service, Pacific Northwest Research Station and the National Science Foundation (LTREB Program DEB-0614538). The Gifford Pinchot National Forest granted access to study sites and provided logistical support.

References

Adams, A.B., V.H. Dale, A.R. Kruckeberg, and E. Smith. 1987. Plant survival, growth form and regeneration following the May 18, 1980, eruption of Mount St. Helens, Washington. *Northwest Science* 61: 160–170.

Andersen, D.C., and J.A. MacMahon. 1985. Plant succession following the Mount St. Helens volcanic eruption: Facilitation by a burrowing rodent, *Thomomys talpoides*. *American Midland Naturalist* 114: 62–69.

Antos, J.A., and D.B. Zobel. 2005. Plant responses in forests of the tephra-fall zone. In *Ecological responses to the 1980 eruption of Mount St. Helens*, ed. V.H. Dale, F.J. Swanson, and C.M. Crisafulli, 47–58. New York: Helens. Springer.

Baross, J.A., and S.E. Hoffman. 1985. Submarine hydrothermal vents and associated gradient environments as sites for the origin and evolution of life. *Origins of Life and Evolution of the Biosphere* 15: 327–345.

Bilderback, D.E., ed. 1987. *Mount St. Helens 1980: Botanical consequences of the explosive eruptions*. Los Angeles: University of California Press.

Bishop, J.G. 2002. Early primary succession on Mount St. Helens: Impact of insect herbivores on colonizing lupines. *Ecology* 83: 191–202.

Bishop, J.G., W.F. Fagan, J.D. Schade, and C.M. Crisafulli. 2005. Causes and consequences of herbivory on prairie lupine (*Lupinus lepidus*) in early primary succession. In *Ecological responses to the 1980 eruption of Mount St. Helens*, ed. V.H. Dale, F.J. Swanson, and C.M. Crisafulli, 151–161. New York: Springer.

Crandell, D.R., and D.R. Mullineaux. 1978. *Potential hazards from future eruptions of Mount St. Helens volcano, Washington*, Bulletin 1383-C. Washington, DC: U.S. Geological Survey.

Crawford, B.A. 1986. *Recovery of game fish populations impacted by the May 18, 1980 eruption of Mount St. Helens*, Fishery Management Report 85-9B. Olympia: Washington Department of Game.

Crisafulli, C.M., J.A. MacMahon, and R.R. Parmenter. 2005a. Small mammal survival and colonization on the Mount St. Helens Volcano: 1980–2002. In *Ecological responses to the 1980 eruption of Mount St. Helens*, ed. V.H. Dale, F.J. Swanson, and C.M. Crisafulli, 199–218. New York: Springer.

Crisafulli, C.M., F.J. Swanson, and V.H. Dale. 2005b. Overview of ecological responses to the eruption of Mount St. Helens—1980–2005. In *Ecological responses to the 1980 eruption of Mount St. Helens*, ed. V.H. Dale, F.J. Swanson, and C.M. Crisafulli, 287–299. New York: Springer.

Crisafulli, C.M., L.S. Trippe, C.P. Hawkins, and J.A. MacMahon. 2005c. Amphibian responses to the 1980 eruption of Mount St. Helens. In *Ecological responses to the 1980 eruption of Mount St. Helens*, ed. V.H. Dale, F.J. Swanson, and C.M. Crisafulli, 183–197. New York: Springer.

Crisafulli, C.M., F.J. Swanson, J.J. Halvorson, and B. Clarkson. 2015. Volcano ecology: Disturbance characteristics and assembly of biological communities. In *Encyclopedia of volcanoes*, 2nd ed. H. Sigurdsson, B. Houghton, S.R. McNutt, H. Rymer, and J. Stix, 1265–1284. Amsterdam: Elsevier.

Dale, V.H., A. Lugo, J. MacMahon, and S. Pickett. 1998. Ecosystem management in the context of large, infrequent disturbances. *Ecosystems* 1: 546–557.

Dale, V.H., S. Brown, R.A. Haeuber, N.T. Hobbs, N. Huntly, R.J. Naiman, W.E. Riebsame, M.G. Turner, and T.J. Valone. 2000. Ecological principles and guidelines for managing the use of land. *Ecological Applications* 10: 639–670.

Dale, V.H., D.R. Campbell, W.M. Adams, C.M. Crisafulli, V. Dains, P.M. Frenzen, and R. Holland. 2005a. Plant succession on the Mount St. Helen debris-avalanche deposit. In *Ecological responses to the 1980 eruption of Mount St. Helens*, ed. V.H. Dale, F.J. Swanson, and C.M. Crisafulli, 59–73. New York: Springer.

Dale, V.H., C.M. Crisafulli, and F.J. Swanson. 2005b. 25 years of ecological change at Mount St. Helens. *Science* 308: 961–962.

Dale, V.H., J. Delgado-Acevedo, and J. MacMahon. 2005c. Effects of modern volcanic eruptions on vegetation. In *Volcanoes and the environment*, ed. J. Marti and G.G.J. Ernst, 227–249. Cambridge: Cambridge University Press.

Dale, V.H., F.J. Swanson, and C.M. Crisafulli, eds. 2005d. *Ecological responses to the 1980 eruption of Mount St. Helens*. New York: Springer.

———. 2005e. Ecological perspectives on management of the Mount St. Helens landscape. In *Ecological responses to the 1980 eruption of Mount St. Helens*, ed. V.H. Dale, F.J. Swanson, and C.M. Crisafulli, 277–286. New York: Springer.

Dammerman, K.W. 1948. The fauna of Krakatau, 1883–1933. Verhandelingen Koniklijke Nederlansche Akademie van Wetenschappen, Afdeling Natuurkunde II 44:1–594.

del Moral, R. 1983. Initial recovery of subalpine vegetation on Mount St. Helens. *American Midland Naturalist* 109: 72–80.

del Moral, R., and C.C. Chang. 2015. Multiple assessments of succession rates on Mount St. Helens. *Plant Ecology* 216: 165–176.

del Moral, R., and S.Y. Grishin. 1999. Volcanic disturbances and ecosystem recovery. In *Ecosystems of disturbed ground*, Ecosystems of the World, ed. L.R. Walker, vol. 16, 137–160. New York: Elsevier.

Edwards, J. 2005. Animals and volcanoes: Survival and revival. In *Volcanoes and the environment*, ed. J. Marti and G. Ernst, 250–272. Cambridge, UK: Cambridge University Press.

Fagan, W.F., and J.G. Bishop. 2000. Trophic interactions during primary succession: Herbivores slow a plant reinvasion at Mount St. Helens. *American Naturalist* 155: 238–251.

Folk, R.L. 1980. *Petrology of sedimentary rocks*. Austin: Hemphill.

Foster, D.R., D.H. Knight, and J.F. Franklin. 1998. Landscape patterns and legacies resulting from large, infrequent forest disturbances. *Ecosystems* 1: 497–510.

Franklin, J.F., J.A. MacMahon, F.J. Swanson, and J.R. Sedell. 1985. Ecosystem responses to the eruption of Mount St. Helens. *National Geographic Research* 1: 198–216.

Franklin, J.F., T.A. Spies, R. Van Pelt, A.B. Carey, D.A. Thornburgh, D.R. Berg, D.B. Lindenmayer, M.E. Harmon, W.S. Keeton, D.C. Shaw, K. Bible, and J.Q. Chen. 2002. Disturbances and structural development of natural forest ecosystems with silvicultural implications, using Douglas-fir forests as an example. *Forest Ecology and Management* 155: 399–423.

Frenzen, P.M., K.S. Hadley, J.J. Major, M.H. Weber, J.F. Franklin, J.H. Hardison III, and S.M. Stanton. 2005. Geomorphic change and vegetation development on the Muddy River mudflow deposit. In *Ecological responses to the 1980 eruption of Mount St. Helens*, ed. V.H. Dale, F.J. Swanson, and C.M. Crisafulli, 75–91. New York: Springer.

Halpern, C.B., and M.E. Harmon. 1983. Early plant succession on the Muddy River mudflow, Mount St. Helens. *American Midland Naturalist* 110: 97–106.

Harris, S.L. 1976. *Fire and ice: The Cascade volcanoes*. Seattle: The Mountaineers, Pacific Search Press.

Hawkins, C.P., and J.R. Sedell. 1990. The role of refugia in the recolonization of streams devastated by the 1980 eruption of Mount St. Helens. *Northwest Science* 64: 271–274.

Keller, S.A.C., ed. 1982. *Mount St. Helens: One year later*. Cheney: Eastern Washington University Press.

———., ed. 1986. *Mount St. Helens: Five years later*. Cheney: Eastern Washington University Press.

Kraft, N.J.B., P.B. Adler, O. Godoy, E.C. James, S. Fuller, and J.M. Levine. 2015. Community assembly, coexistence and the environmental filtering metaphor. *Functional Ecology* 29: 592–599. https://doi.org/10.1111/1365-2435.12345.

Lawrence, D.B. 1938. Trees on the march: Notes on the recent volcanic and vegetational history of Mount St. Helens. *Mazama* 20: 49–54.

Lawrence, R.L., and W.J. Ripple. 2000. Fifteen years of revegetation of Mount St. Helens: A landscape-scale analysis. *Ecology* 81: 2742–2752.

Lipman, P.W., and D.R. Mullineaux, eds. 1981. *The 1980 eruptions of Mount St. Helens, Washington*, Professional Paper 1250. Washington, DC: U.S. Geological Survey.

MacMahon, J.A., R.R. Parmenter, K.A. Johnson, and C.M. Crisafulli. 1989. Small mammal recolonization on the Mount St. Helens volcano: 1980–1987. *American Midland Naturalist* 122: 365–387.

Meiners, S.J., M.W. Cadotte, J.D. Fridley, S.T.A. Pickett, and L.R. Walker. 2015. Is successional research nearing its climax? New approaches for understanding dynamic communities. *Functional Ecology* 29: 154–164.

Parmenter, R.R., C.M. Crisafulli, N. Korbe, G. Parsons, M. Edgar, and J.A. MacMahon. 2005. Posteruption arthropod succession on the Mount St. Helens volcano: The ground-dwelling beetle fauna (Coleoptera). In *Ecological responses to the 1980 eruption of Mount St. Helens*, ed. V.H. Dale, F.J. Swanson, and C.M. Crisafulli, 139–150. New York: Springer.

Pickett, S.T.A., and P.W. White, eds. 1985. *The ecology of natural disturbance and patch dynamics*. New York: Academic Press.

Pierson, T.C., and J.J. Major. 2014. Hydrogeomorphic effects of explosive volcanic eruptions on drainage basins. *Annual Review of Earth and Planetary Sciences* 42: 469–507.

Raab, T., J. Kruemmelbein, A. Schneider, W. Gerwin, T. Maurer, and M.A. Naeth. 2012. Initial ecosystem processes as key factors of landscape development—a review. *Physical Geography* 33: 305–343.

Sherrod, D.R., W.E. Scott, and P.H. Stauffer, eds. 2008. *A volcano rekindled: The renewed eruption of Mount St. Helens, 2004–2006*, Professional Paper 1750. Washington, DC: U.S. Geological Survey.

Sigurdsson, H., B. Houghton, S.R. McNutt, H. Rymer, and J. Stix, eds. 2015. *The encyclopedia of volcanoes*. Amsterdam: Academic/Elsevier.

Swanson, F.J., and J.F. Franklin. 1992. New forestry principles of ecosystem analysis of Pacific Northwest forests. *Ecological Applications* 2: 262–274.

Swanson, F.J., and J.J. Major. 2005. Physical events, environments, and geological-ecological interactions at Mount St. Helens, March 1980–2004. In *Ecological responses to the 1980 eruption of Mount St. Helens*, ed. V.H. Dale, F.J. Swanson, and C.M. Crisafulli, 27–44. New York: Springer.

Swanson, F.J., C.C. Crisafulli, and D.K. Yamaguchi. 2005. Geological and ecological settings of Mount St. Helens before May 18, 1980. In *Ecological responses to the 1980 eruption of Mount St. Helens*, ed. V.H. Dale, F.J. Swanson, and C.M. Crisafulli, 13–26. New York: Springer.

Swanson, M.E., J.F. Franklin, R.L. Beschta, C.M. Crisafulli, D.A. DellaSala, R.L. Hutto, D.B. Lindenmayer, and F.J. Swanson. 2011. The forgotten stage of forest succession: Early successional ecosystems on forest sites. *Frontiers in Ecology and the Environment* 9: 117–125.

Thornton, I.W.B., R.A. Zann, P.A. Rawlinson, C.R. Tidemann, A.S. Adikerana, and A.H.T. Widjor. 1988. Colonization of the Krakatau Islands by vertebrates: Equilibrium, succession, and possible delayed extinction. *Proceedings of the National Academy of Sciences USA* 88: 515–518.

Turner, M.G., ed. 1987. *Landscape heterogeneity and disturbance*. New York: Springer-Verlag.

———. 2010. Disturbance and landscape dynamics in a changing world. *Ecology* 91: 2833–2849.

Turner, M.G., and V.H. Dale. 1998. What have we learned from large, infrequent disturbances? *Ecosystems* 1: 493–496.

Turner, M.G., V.H. Dale, and E.H. Everham. 1997. Crown fires, hurricanes and volcanoes: A comparison among large-scale disturbances. *Bioscience* 47: 758–768.

Turner, M.G., W.I. Baker, C.J. Peterson, and R.K. Peet. 1998. Factors influencing succession: Lessons from large, infrequent natural disturbances. *Ecosystems* 1: 511–523.

Walker, L.R., and R. del Moral. 2003. *Primary succession and ecosystem rehabilitation*. Cambridge: Cambridge University Press.

Weatherhead, P.J. 1986. How unusual are unusual events? *American Naturalist* 128: 150–154.

White, P.S. 1979. Patterns, processes and natural disturbances in vegetation. *Botanical Review* 45: 229–299.

Zobel, D.B., and J.A. Antos. 1986. Survival of prolonged burial by subalpine forest understory plants. *American Midland Naturalist* 115: 282–287.

———. 1992. Survival of plants buried for eight growing seasons by volcanic tephra. *Ecology* 73: 698–701.

Sediment Erosion and Delivery from Toutle River Basin After the 1980 Eruption of Mount St. Helens: A 30-Year Perspective

2

Jon J. Major, Adam R. Mosbrucker, and Kurt R. Spicer

2.1 Introduction

Explosive volcanic eruptions can alter broad swaths of landscape by damaging or destroying vegetation, draping hillsides with volcanic ash, and filling valleys with volcanic sediment (e.g., Lipman and Mullineaux 1981; Sigurdsson et al. 1984; Newhall and Punongbayan 1996; Surono et al. 2012; Major and Lara 2013; Pierson and Major 2014). As a result, production and routing of runoff from rainfall and snowmelt can be greatly altered, erosional regimes modified, and fluxes of water and sediment greatly increased. In response to changes in water and sediment fluxes, geomorphic processes that modulate erosion, delivery, and storage of sediment can also change. Variations in geomorphic response and stability following volcanic eruptions in turn affect the nature and pace of ecological responses to eruptions (Dale et al. 2005a; Swanson and Major 2005).

Sediment-response trajectories after volcanic disturbance commonly exhibit a two-stage evolution (Gran et al. 2011) that is a function of topography. Explosive eruptions deposit fresh sediment in topographically diverse environments—on hillsides and in river valleys. The relative proportions of hillside and channel deposits vary with the nature of the eruption, basin orientation, and proximity to the volcano. Initial erosion of sediment from both hillsides and channels is typically very rapid and dramatic, which generates rates of sediment delivery that commonly are many tens to hundreds of times above typical pre-eruption levels (Pierson and Major 2014). Absent sediment recharge by frequent eruptions, this initial phase of response usually lasts only a few years before erosion and sediment delivery decline markedly. Commonly, only this immediate and acute phase of response affects catchments characterized predominantly by disturbed hillsides or those in which channels are relatively little disturbed. In such catchments, geomorphic processes usually return to those typical of pre-eruption conditions within a few years and sediment delivery swiftly returns to a level that is within the range of pre-eruption variation (e.g., Collins and Dunne 1986; Meyer and Martinson 1989; Major et al. 2000; Gran and Montgomery 2005; Yamakoshi et al. 2005). Although responses in such catchments commonly are immediate and short lived, some geomorphic responses and sediment delivery may be delayed or protracted owing to accumulation and rearrangement of large woody debris delivered to channels by an eruption (e.g., Lisle et al., Chap. 3, this volume), to landslides that may be more prevalent several years after an eruption (e.g., Swanson and Major 2005), or to releases of lakes created or enlarged by blockages of volcanic sediment (e.g., White et al. 1997). In contrast, rates and magnitudes of sediment release typically are greater and more prolonged from catchments in which valleys are thickly filled with volcanic sediment. In those catchments, initial geomorphic response and magnitude of sediment delivery owe largely to substantial incision and widening of channels as they adjust to altered flow regimes, changes in sediment character and supply, and changes in morphology (e.g., Janda et al. 1984; Manville 2002; Gran and Montgomery 2005; Pierson and Major 2014).

After channel forms and networks are reestablished, the most easily erodible sediment is depleted, and the hydraulic resistance of channels increases (e.g., Simon 1992), geomorphic response enters a second phase. In this response phase, sediment delivery is greatly diminished from peak levels but remains higher than pre-eruption levels because of persistent channel change. This phase of diminished, but still elevated, sediment delivery can persist for decades, perhaps centuries, depending on the magnitude of disturbance (Major 2004;

Glossary terms appear in ***bold italic face***.

J.J. Major (✉) • A.R. Mosbrucker • K.R. Spicer
U.S. Geological Survey, Cascades Volcano Observatory, 1300 SE Cardinal Court, Bldg 10, Suite 100, Vancouver, WA 98683, USA
e-mail: jjmajor@usgs.gov; amosbrucker@usgs.gov; krspicer@usgs.gov

C.M. Crisafulli, V.H. Dale (eds.), *Ecological Responses at Mount St. Helens: Revisited 35 years after the 1980 Eruption*,
https://doi.org/10.1007/978-1-4939-7451-1_2

Manville et al. 2009; Gran et al. 2011; Meadows 2014). In some instances, exceptional sediment delivery can be rejuvenated if lakes impounded by volcanic sediment breach and reinvigorate channel erosion (e.g., White et al. 1997; Manville et al. 2007).

Catchments affected by severe volcanic disturbance may never recover to pre-eruption geomorphic conditions—at least not over typical human timeframes. Instead, they may attain a degree of stability and ecological function under different equilibrium conditions (e.g., Gran and Montgomery 2005; O'Connor et al. 2013; Pierson and Major 2014), and this can have prolonged societal consequences. Here, we provide a multi-decade perspective of the hydrologic and geomorphic (hydrogeomorphic) responses of Toutle River basin at Mount St. Helens in the context of the two-stage conceptual-response framework. Toutle River basin, which drains the volcano's north and west flanks as well as terrain to the north of the mountain, was affected by the most diverse and severe disturbances caused by the 18 May 1980 eruption. As a result, its catchments have shown the most varied and prolonged hydrogeomorphic responses.

The 18 May 1980 eruption of Mount St. Helens consisted of an ensemble of volcanic processes (Fig. 2.1) that dramatically altered landscapes and ecosystems of several catchments surrounding the volcano (Lipman and Mullineaux 1981; Dale et al. 2005b). That ensemble—a rockslide–***debris avalanche***; a laterally directed ***pyroclastic density current*** (PDC), commonly called the lateral blast, but herein called the ***blast PDC***; ***lahars*** (volcanic debris flows); ***pyroclastic flows***; and fall of volcanic ash (***tephrafall***)—caused gradients of landscape disturbances (e.g., Lipman and Mullineaux 1981; Swanson and Major 2005; Lisle et al., Chap. 3, this volume) which affected the fluxes of water and sediment through catchments to varying degrees and for substantially different durations (Major et al. 2000; Major and Mark 2006).

The nature and duration of hydrogeomorphic responses at Mount St. Helens reflect the nature and severity of landscape disturbance. In some catchments, such as Green River and Clearwater Creek (Fig. 2.1), hydrogeomorphic responses were modest and short-lived, even though large proportions of their areas were affected, because only hillsides were disturbed by the primary events (Collins and Dunne 1986; Meyer and Martinson 1989; Lisle et al., Chap. 3, this volume). Fluxes of water and sediment in those catchments returned to typical pre-eruption levels within a few years (Major et al. 2000; Major and Mark 2006; Lisle et al., Chap. 3, this volume). In other catchments where the eruption buried substantial proportions of valley bottoms with thick volcanic deposits, such as North Fork Toutle River (Fig. 2.1), hydrogeomorphic responses were of greater magnitude and have been much longer lived. In those catchments, the greatest magnitudes of channel change occurred in the first few years after the eruption, when channels were furthest out of equilibrium (e.g., Rosenfeld and Beach 1983; Martinson et al. 1984; Meyer et al. 1986; Meyer and Martinson 1989; Simon 1999; Zheng et al. 2014). After an initial adjustment period, both the magnitudes and rates of channel changes in those catchments slowed considerably (Martinson et al. 1986; Meyer and Dodge 1988; Simon 1999; Hardison 2000; Zheng et al. 2014), and consequent sediment delivery declined sharply (Major et al. 2000). In North Fork Toutle River catchment, however, sediment fluxes remain significantly above pre-eruption levels even after more than 30 years since the eruption.

In this chapter, we summarize erosion and long-term sediment delivery from Toutle River basin. We examine some of the underlying causes for its persistently high sediment fluxes, particularly from North Fork Toutle River catchment. We first provide a synopsis of the disturbances caused by the cataclysmic 1980 eruption and their hydrogeomorphic impacts. We then present an update of the suspended-sediment flux from Toutle River basin. Analyses of volumetric change in erosion and deposition along upper North Fork Toutle River valley during the second and third decades after eruption follow. We use analyses of volumetric change and patterns of erosion to consider why sediment flux from upper North Fork Toutle River has remained so persistently elevated.

2.2 Volcanic Disturbance Processes and Deposits in Toutle River Basin

Toutle River basin was affected by four major volcanic processes (Fig. 2.1): a rockslide–debris avalanche, a blast PDC and associated rain of fine ash (particles $\leq$0.25 mm) from its turbulent cloud, large lahars, and pyroclastic flows. Detailed discussions and summaries of these processes and the sequencing of processes on 18 May are given in Lipman and Mullineaux (1981), Criswell (1987), Fairchild (1987), Scott (1988), Glicken (1996), Swanson and Major (2005), Belousov et al. (2007), Esposti Ongaro et al. (2012), and Waitt (2015).

The rockslide–debris avalanche resulting from collapse of the volcano's north flank buried upper North Fork Toutle River valley. It filled about 60 km^2 of valley with 2.5 km^3 of mostly gravelly sand to an average depth of 45 m. Locally, the deposit is as thick as 200 m (Voight et al. 1981). The debris avalanche obliterated mature forest on the valley floor and slammed against Johnston Ridge opposite the volcano. Although most of it turned and flowed down the valley after hitting Johnston Ridge, part of it surged through Spirit Lake, and part crossed Johnston Ridge and flowed down South Coldwater Creek valley before rejoining North Fork Toutle River valley (Figs. 2.2 and 2.3). Its deposit blocked several

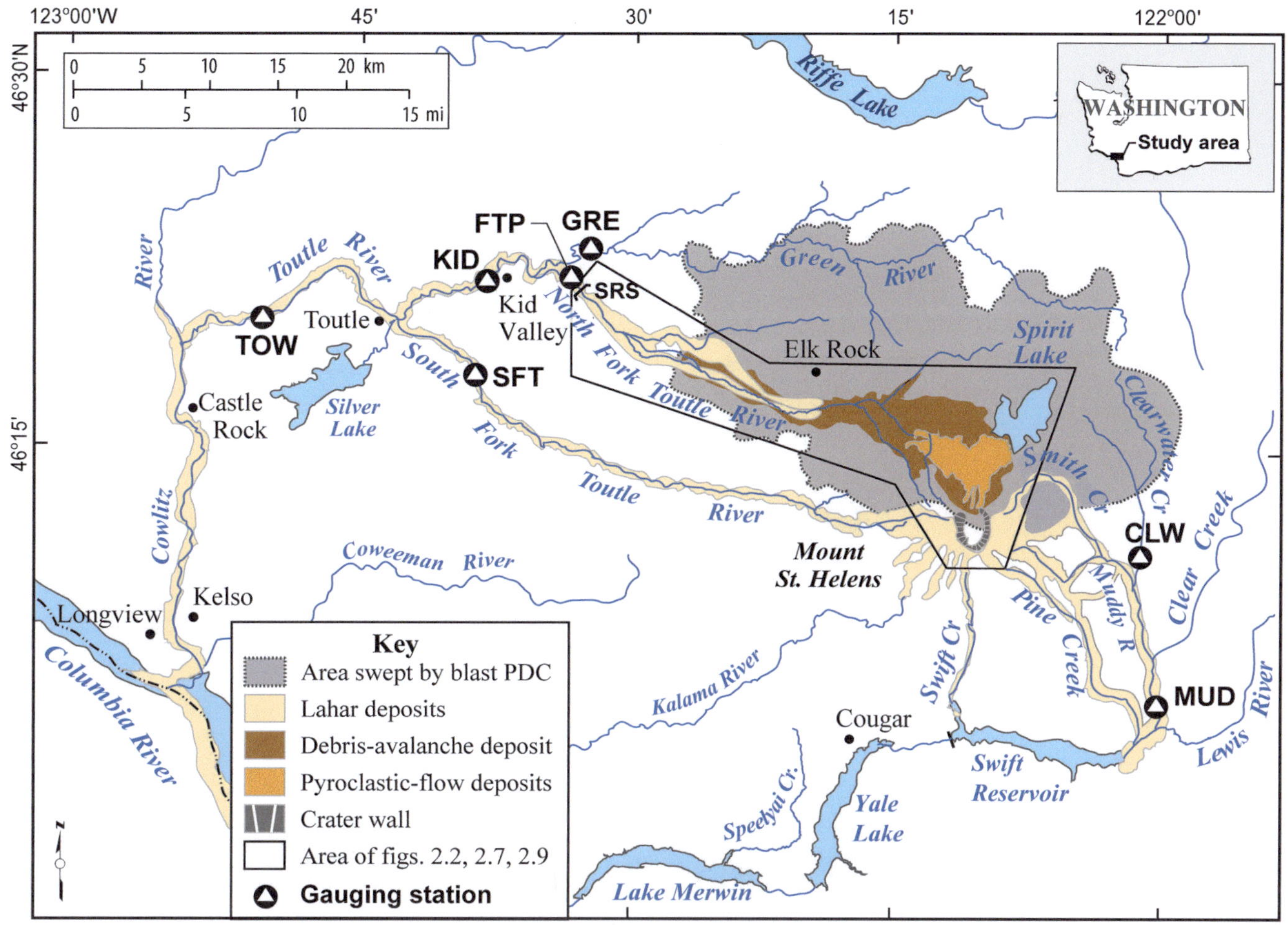

Fig. 2.1 Distribution of volcanic disturbance zones of 1980 Mount St. Helens eruption and locations of gauging stations. *SRS* sediment-retention structure. Gauging stations: *GRE* Green River, *FTP* additional North Fork Toutle River station, *KID* Kid Valley on North Fork Toutle River, *SFT* South Fork Toutle River, *TOW* lower Toutle River, *CLW* Clearwater Creek, *MUD* Muddy River. (Modified from Major et al. 2000).

tributary channels and impounded new lakes of varying size. Several small lakes breached their impoundments within weeks to months (Janda et al. 1984; Simon 1999). The two largest new lakes (Castle and Coldwater) still retain substantial volumes of water (20 and 90 million m^3, respectively) behind impoundments having reinforced outlets (Figs. 2.2 and 2.3). Spirit Lake was displaced upward and its outlet buried. To prevent catastrophic breaching of the blockage, water was pumped temporarily from the lake over the blockage crest and onto a stable part of the deposit, forming Truman channel (Paine et al. 1987; see Fig. 2.2), while an outlet tunnel was bored through bedrock (Fig. 2.2). Since 1985, water from Spirit Lake has drained westward through the tunnel into South Coldwater Creek and eventually back to North Fork Toutle River.

Failure of the volcano's north flank unroofed magma that had intruded high into the edifice and unleashed a violent, laterally directed, gas-charged explosion (Hoblitt et al. 1981; Waitt 1981; Voight 1981). That explosion triggered a widespread, rapidly moving, hot PDC—the blast PDC—that swept across 570 km^2 of rugged landscape to the north and removed, toppled, or scorched mature forest. The blast PDC deposited as much as 1.5 m of graded gravel to fine sand capped by as much as 6 cm of fine silt (Hoblitt et al. 1981; Waitt 1981). Within about 10 km of the volcano, the blast PDC stripped the landscape of soil and vegetation; from there to its distal limits (20 km north and about 30 km east to west extent), it toppled mature forest (Moore and Sisson 1981; Swanson and Major 2005). The zone of toppled trees is fringed by standing, scorched forest (Winner and Casadevall 1981). In Green River catchment (Fig. 2.1), the blast PDC was the only volcanic disturbance process.

Rapid melting of snowpack by the blast PDC in the headwaters of South Fork Toutle River catchment on the volcano's west flank and dewatering of debris-avalanche deposit in North Fork Toutle River valley triggered two large (14–140 million m^3) lahars in Toutle River basin on 18 May (Janda et al. 1981; Fairchild 1987; Scott 1988; Waitt 1989).

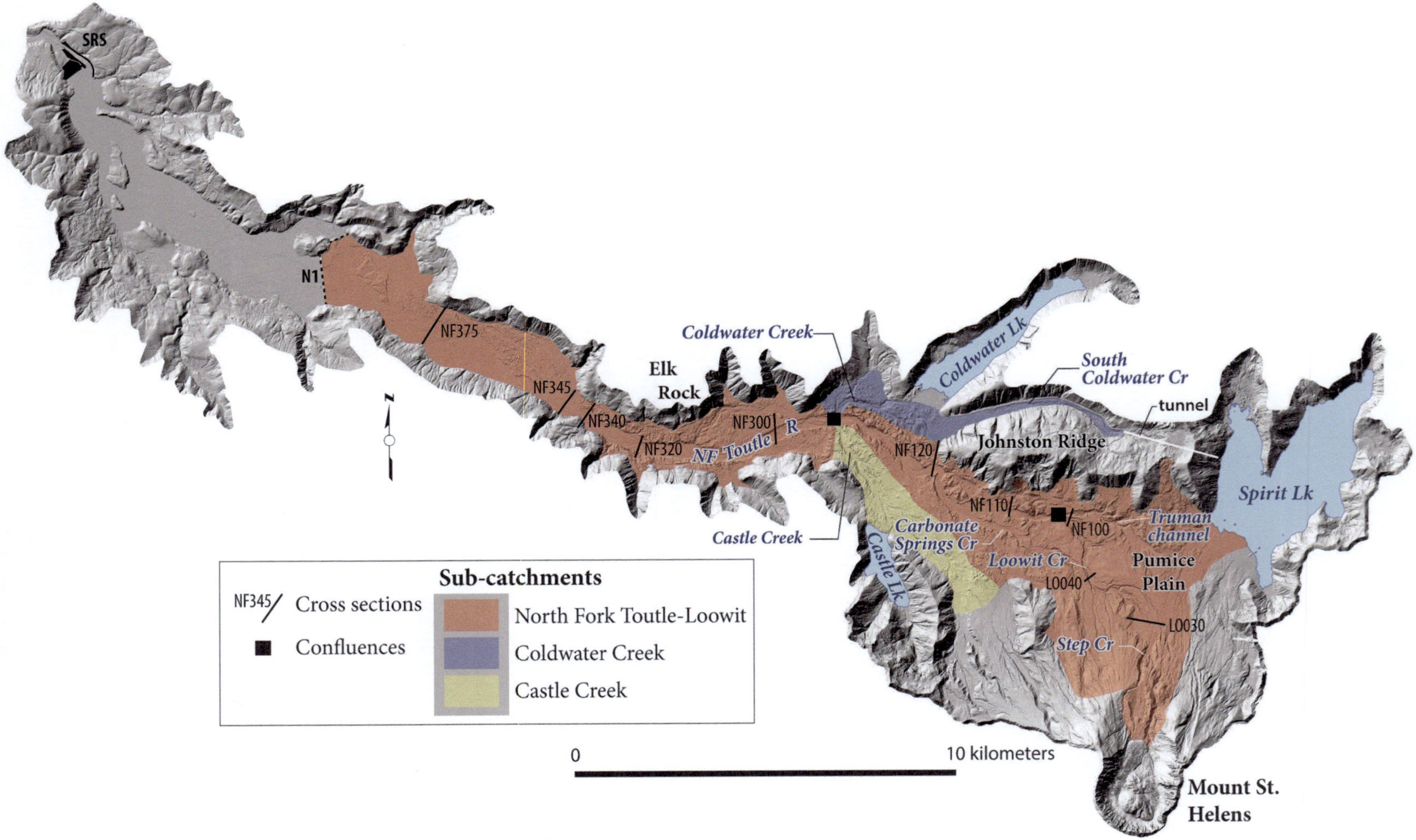

Fig. 2.2 Shaded-relief digital elevation model of upper North Fork Toutle River catchment showing 2009 topography (Mosbrucker 2014), locations cited in text, and select cross sections of channel geometry surveys (e.g., NF320). *Red* shading shows approximate upper North Fork Toutle–Loowit channel sub-catchment in which decadal-scale volumes of erosion and deposition across debris-avalanche deposit were estimated (outlined downstream only to site of former N1 retention structure); *yellow* shading shows Castle Creek sub-catchment; *blue* shading shows Coldwater Creek sub-catchment. Major channels across the debris-avalanche deposit are labeled. *Black boxes* mark confluences of North Fork Toutle River, Castle Creek, and Coldwater Creek (west) and North Fork Toutle River, Truman channel, and Loowit Creek (east). These confluences, and cross sections NF340 and LO030, denote reach boundaries shown in Table 2.1. *SRS* is the sediment-retention structure.

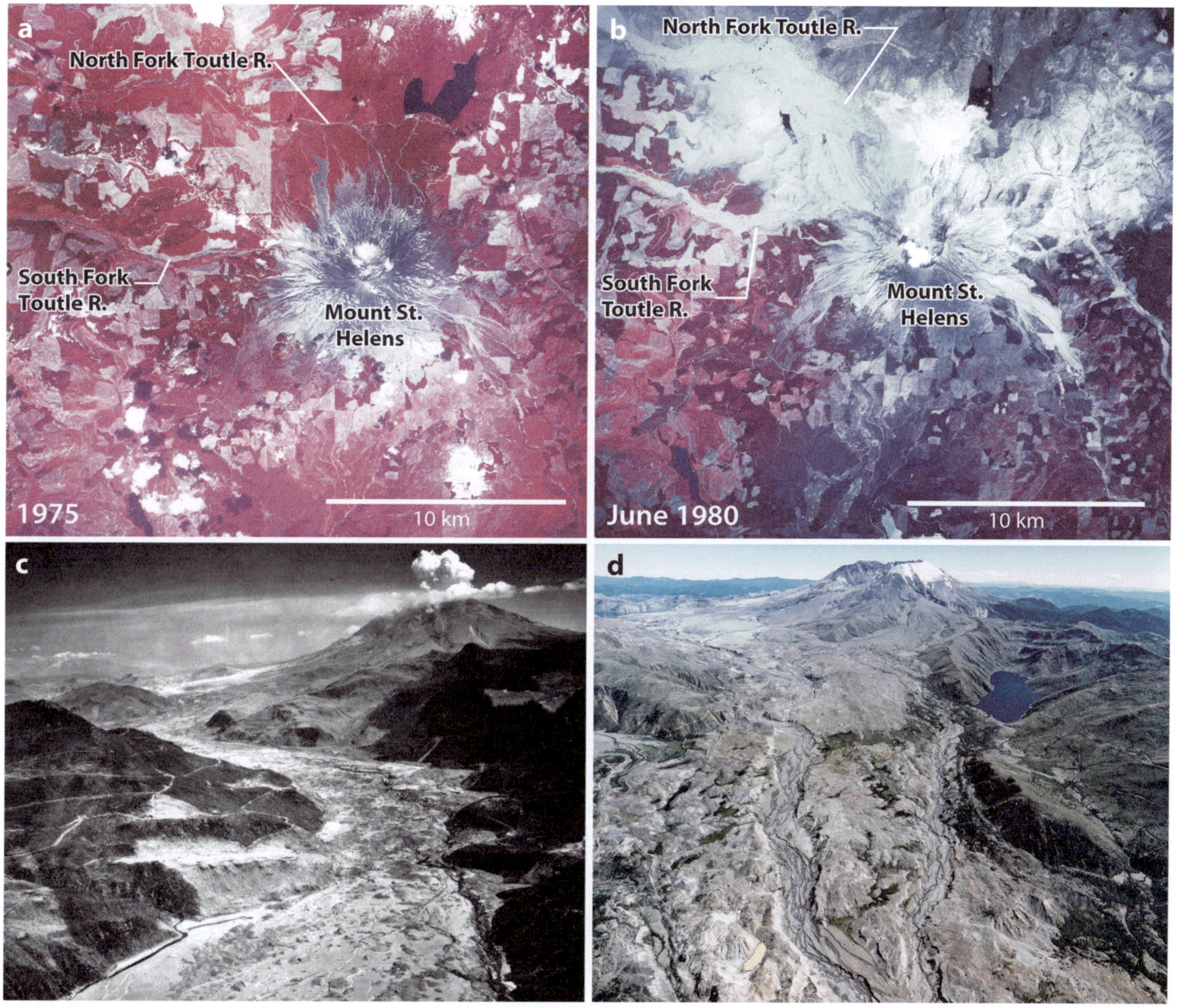

Fig. 2.3 Images of volcanic impact and erosion in upper North Fork Toutle River valley. **a**, **b** are high-altitude (false-color infrared) vertical aerial photographs of Mount St. Helens taken in 1975 and on 19 June 1980, respectively. Spirit Lake is in the *upper right* quadrant of each image. Compare zones of volcanic impact in (**b**) with disturbance zones in Fig. 2.1. **c** Oblique aerial photograph of debris-avalanche deposit in upper North Fork Toutle River valley. Valley constriction is the Elk Rock reach shown in Fig. 2.2. USGS photograph by A. Post and R.M. Krimmel, 30 June 1980. **d** Oblique aerial photograph of upper North Fork Toutle River valley showing erosion of debris-avalanche deposit. Castle Lake is in the *upper right* quadrant of photograph; lower edge of South Coldwater Creek valley is in the *left center* of photograph. (Image courtesy of Washington State Department of Transportation approximately 1993).

The smaller, flash-flood-like South Fork Toutle River lahar deposited gravelly sand across valley floor and floodplain, ranging in thickness from a few centimeters on the volcano's flank to a few meters farther downstream (Janda et al. 1981; Scott 1988; Waitt 1989). The North Fork Toutle River lahar was a much larger, longer-lasting event that started later in the day. Its exceptional volume (140 million m^3 just downstream of the debris-avalanche deposit and 120 million m^3 at the mouth of Toutle River, 70 km from source), minimal attenuation, and long duration (11 h at the mouth of Toutle River) damaged houses, lumber facilities, and infrastructure along the North Fork Toutle, Toutle, and Cowlitz River valleys (Janda et al. 1981; Fairchild 1987). The North Fork Toutle River lahar deposited meters-thick gravelly to muddy sand along the channel bed, valley floor, and floodplain, especially in lower Toutle, Cowlitz, and Columbia Rivers (e.g., Lombard et al. 1981; Schuster 1981).

Within tens of minutes of the debris avalanche and blast PDC, a sustained vertical eruption column rose more than 20 km (Christiansen and Peterson 1981). Ash (particles $\leq$2 mm) from that eruption column drifted downwind and deposited tephrafall over tens of thousands of square kilometers

(Sarna-Wojcicki et al. 1981). This eruptive phase lasted for more than 9 h. During that time, partial collapses of the eruption column generated pumiceous pyroclastic flows that rushed from the new crater and deposited sheets of ash and pumice lapilli (particles 2–64 mm) atop the debris-avalanche deposit in the area now known as the Pumice Plain (Criswell 1987; Brand et al. 2014; Figs. 2.1 and 2.2). Five subsequent eruptions in summer 1980 also produced pumiceous pyroclastic flows that added to the fill atop the debris-avalanche deposit (Rowley et al. 1981). Deposits from these flows cover about 15 km^2; thicknesses of individual flow units range from 0.25 to 10 m, and the collective deposit thickness from the succession of flows is as much as 40 m (Rowley et al. 1981).

2.3 Hydrogeomorphic Impacts of the 1980 Eruptions in Toutle River Basin

Landscape changes caused by the cataclysmic eruption of Mount St. Helens profoundly altered landscape hydrology and geomorphology. Volcanic processes damaged or destroyed mature forest over hundreds of square kilometers, draped hillsides with sediment having a nearly impervious surface, and filled or altered the character of major channels draining the volcano (Fig. 2.3). Later eruptions in summer 1980 added to, but did not cause additional, landscape disturbances. Aside from upper North Fork Toutle River and South Coldwater Creek valleys, which were deeply buried by debris-avalanche deposit, sediment deposition caused little change to fundamental pre-eruption landforms. Even so, prevailing geomorphic processes were altered. Because disturbances caused by the 1980 eruptions occurred before the onset of the wet season, they can be considered as a single event with regard to post-eruption hydrogeomorphic response. Therefore, in the following discussions, we refer to them in aggregate as the 1980 eruption.

2.3.1 Hillsides

Forested Cascade Range hillsides typically have permeable, organic-rich soils showered by low- to moderate-intensity rainfall (14–30 mm h^{-1}; Miller et al. 1973; see also Western Regional Climate Center 2015). As a result, infiltration capacities are high, about 75–100 mm h^{-1} or more (Johnson and Beschta 1980; Leavesley et al. 1989; McGuire et al. 2005), and flow moves through the landscape largely in the subsurface (Jones 2000; McDonnell 2003; McGuire et al. 2005). Consequently, annual sediment yields are low, a few hundred tonnes (t) km^{-2} (e.g., Major et al. 2000; Roering et al. 2010; Czuba et al. 2011).

At Mount St. Helens, forest damage and deposition of fine ash in 1980 altered fluxes of water from hillsides to channels. Destruction of a broad swath of forest significantly reduced foliar interception and altered seasonal evapotranspiration. Loss of foliar interception allowed more precipitation to reach the ground surface, and changes in evapotranspiration kept soils wetter than usual. Deposition of fine ash markedly reduced the infiltration of rainfall and snowmelt across the landscape near Mount St. Helens. Rainfall experiments conducted in summer 1980 on hillsides affected by the blast PDC showed spatially averaged infiltration capacities had been reduced to as little as 2 mm h^{-1} (Leavesley et al. 1989), and they remained <10 mm h^{-1} 1 year later (Swanson et al. 1983; Leavesley et al. 1989). After nearly 20 years, plot-specific infiltration capacities remained three to five times lower than pre-eruption capacities (Major and Yamakoshi 2005). Loss of foliar interception, changes in evapotranspiration, and reduction of surface infiltration changed the mode of water transfer through the hydrological system from one dominated by subsurface flow to one dominated by overland flow (Leavesley et al. 1989; Pierson and Major 2014). The dominance of overland flow persisted for more than 1 year, and even after 20 years high-intensity (>25 mm h^{-1}) rainstorms remained capable of triggering overland flow (Major and Yamakoshi 2005).

This fundamental change in hillside hydrology altered sediment erosion and transport processes. Hillside erosion and transport in the pre-eruption landscape were dominated by tree throw, bioturbation, soil creep, and mass wasting (e.g., Gabet et al. 2003; Roering et al. 2010). Dominance of overland flow after the 1980 eruption triggered extensive slope erosion through development of rill and gully networks across ash-covered slopes (Swanson et al. 1983; Collins and Dunne 1986).

2.3.2 Channels

Channel changes caused by the eruption had variable hydrological ramifications. The channels of South Fork Toutle, lower North Fork Toutle, and Toutle Rivers were straightened and smoothed and changed from gravel-bedded, pool-riffle systems to simplified sand-bedded corridors stripped of riparian vegetation (Janda et al. 1984). These changes enhanced the hydraulic efficiency of channels (Major and Mark 2006; Pierson and Major 2014). In contrast, deposition of the debris avalanche in upper North Fork Toutle River valley hydrologically disconnected the upper valley from the lower, which temporarily reduced water delivery downstream. The debris-avalanche deposit blocked tributary channels, and its irregular surface, formed of mounds and closed depressions, disrupted through-going flow. Drainage development across the debris-avalanche deposit began on the afternoon of 18 May when springs poured forth from melting ice and trapped groundwater, and ponds that formed in depressions filled and breached. Sediment entrained during this dewatering and breaching process across the debris-avalanche deposit eventually generated the North Fork Toutle River lahar (Janda et al. 1981, 1984). Breakouts of small lakes impounded at mouths of tributary channels, controlled releases of water from the largest impounded lakes, pumping of water from Spirit Lake across the deposit surface during tunnel construction, subsequent meltwater lahars and floods in later eruptions (e.g., Waitt et al. 1983), and rainfall runoff augmented drainage development. It took nearly 3 years to fully integrate a new drainage network across the deposit surface (Rosenfeld and Beach 1983; Meyer 1995; Simon 1999).

These hydrological changes resulted in discharge peaks in Toutle River basin that were larger than pre-eruption peaks for a given amount of rainfall (Major and Mark 2006). Hydrological responses were greater in the South Fork Toutle River and North Fork Toutle River catchments, subject to both hillside and channel disturbances, than in the Green River catchment, subject only to hillside disturbance. Peak flows generated by runoff from autumn storms increased tens of percent along all major channels of Toutle River basin through 1984 and increased to a lesser extent, as measured along lower Toutle River, from 1985 to 1989. Pre- and post-eruption peak discharges are statistically indistinguishable after 1989.

Channel-network development across the debris-avalanche deposit, in conjunction with increased transport capacity, triggered substantial erosion and sediment transport (Janda et al. 1984; Dinehart 1998; Simon 1999; Major 2004). Even low- to moderate-magnitude flows (discharges ≤2-year flood) transported large quantities of sediment (Major 2004). Repeat surveys of an extensive network of channel cross sections in Toutle River basin show the greatest channel changes were on North Fork Toutle River where it crossed the debris-avalanche deposit (Meyer and Martinson 1989). There, early channel changes commonly followed a multistage evolutionary process (Meyer and Martinson 1989): (1) channel initiation, (2) incision, (3) aggradation and widening, and (4) episodic scour and aggradation with little net change in bed elevation. Persistent and progressive incision and widening dominated channel development upstream of a valley constriction at Elk Rock (upstream of NF320; Figs. 2.2 and 2.4). Downstream of that valley constriction, aggradation and widening dominated until the mid-1980s, after which secondary incision accompanied continued widening (Fig. 2.4). North Fork Toutle River channel and those of its headwater tributaries incised by tens of meters and widened by hundreds of meters upstream of Elk Rock (e.g., NF100, NF300; Fig. 2.4). Downstream, the channel aggraded as much as 15 m and widened by hundreds of meters (e.g., NF345; Fig. 2.4). The greatest and most rapid channel changes occurred within the first few years after the 1980 eruption, but the principal channel and its main tributaries were subject to progressive, long-term geometric adjustments (Fig. 2.4).

Channel gradient affected initial changes of South Fork Toutle River channel as well as lahar-affected channels east of the volcano (Meyer and Martinson 1989; Hardison 2000). In steep upstream reaches, lahar-affected channels generally incised by several meters and widened by a few tens of meters during the first year following eruption but then aggraded by a few meters and underwent episodic scour and aggradation in subsequent years. In contrast, shallow-gradient reaches downstream aggraded by a few meters and widened by a few tens of meters during the first year after eruption and then incised by a few meters in following years. Similar to channel changes across the debris-avalanche deposit, those in lahar-affected channels were most rapid during the first few years after the eruption.

In Green River catchment, affected only by the blast PDC, the channel aggraded by as much as a meter during the first few winter storms following the 1980 eruption as sediment flushed from hillsides accumulated. However, during later storms that same winter, the channel incised (Meyer and Martinson 1989). Subsequently, little except local channel change occurred. Widening of Green River channel, as well as of channels in other catchments affected only by the blast PDC (e.g., Clearwater Creek; Fig. 2.1), was generally constrained by riparian vegetation that rapidly resprouted and maintained bank cohesion and was strongly influenced by large woody debris toppled into the channel by the PDC (Lisle 1995; Lisle et al., Chap. 3, this volume).

Approximately 5 years after the 1980 eruption, rates and magnitudes of geometric changes in channels affected by lahars and those in catchments affected by the blast PDC had largely diminished (Meyer and Martinson 1989; Simon 1999; Hardison 2000; Lisle et al., Chap. 3, this volume). By then, relatively minor changes in bed elevations and channel widths were likely within the range of variations characteristic of pre-eruption conditions, especially in catchments affected only by the blast PDC. It is worth noting, however, that regional climate was relatively dry during the decade following the eruption and few large floods occurred (Major 2004; Swanson and Major 2005). Large storm events in the 1990s and 2000s generated a few larger floods (20-year to ~100-year return interval flows), which caused channel changes that not only exceeded the likely range of pre-eruption variations, but in some channels caused changes that exceeded those documented during the first decade after the eruption (Simon 1999; Lisle et al., Chap. 3, this volume; Mosbrucker et al. 2015).

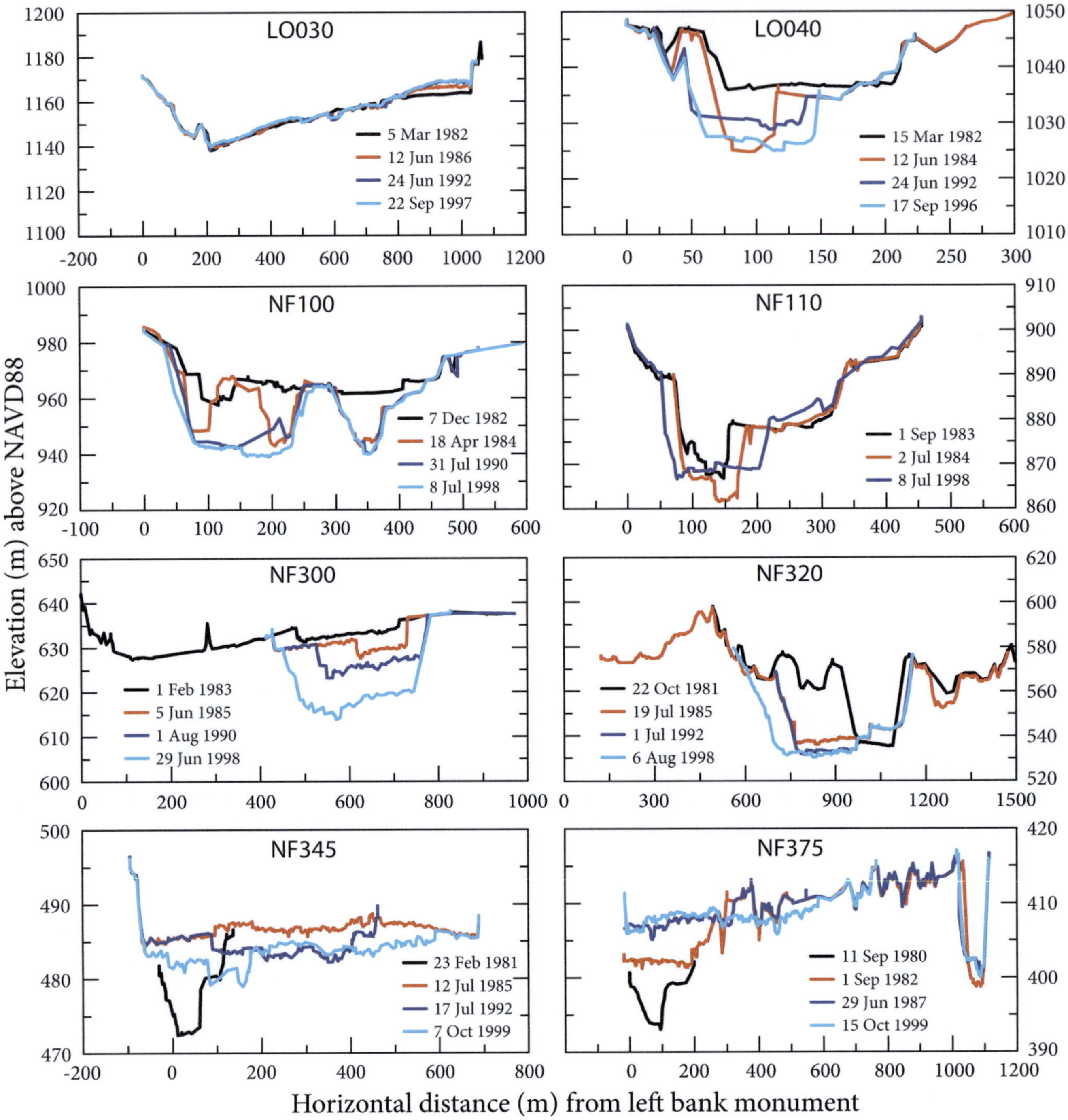

Fig. 2.4 Cross-section profiles from select locations on debris-avalanche deposit from 1980 to 1999. (See Figs. 2.2 and 2.7 for section locations).

2.4 Long-Term Sediment Transport from Toutle River Basin

Post-eruption sediment redistribution at Mount St. Helens is mostly by fluvial transport (Major 2004). The dominance of post-eruption fluvial transport contrasts with predominant lahar transport at many other volcanoes, especially those in tropical climates (Pierson and Major 2014). The scarcity of post-1980 lahars at Mount St. Helens is largely a consequence of the region's low to moderate rainfall intensities. However, it also reflects engineering measures undertaken to prevent catastrophic breaching of impounded lakes (Willingham 2005) and the general character of post-1980 eruptions at the volcano—quiescent dome building rather than violent explosions. The few explosions and minor dome collapses that occurred while the volcano was clad in snow triggered the most notable post-1980 lahars (e.g., Waitt et al. 1983; Pringle and Cameron 1999).

Gauging stations operating at the time of or established after the 1980 eruption provide measures of discharges of water and suspended sediment. Stations in Toutle River basin (Fig. 2.1) are located downstream of the debris-avalanche deposit (KID on lower North Fork Toutle River; TOW[1] on lower Toutle River), along a lahar-affected channel (SFT on lower South Fork Toutle River), and in a basin affected solely by the blast PDC (GRE on Green River). KID measured combined discharges from the North Fork Toutle and Green River catchments, and TOW integrates discharges from the entire Toutle River basin. All four stations operated continuously from 1982 to 1994; KID and GRE were decommissioned in 1994, and SFT was decommissioned in 2014. As of 2015, TOW remained in operation. An additional station (FTP) was established on North Fork Toutle River just above the Green River confluence and has measured water discharge intermittently since 1990 and suspended-sediment discharge intermittently since 2001.

Suspended-sediment transport in Toutle River basin increased significantly following the 1980 eruption (Figs. 2.5 and 2.6). The magnitude and duration of excessive sediment loads vary as a function of the type of volcanic disturbance. Elevated levels of suspended sediment have been greatest and of longest duration from upper North Fork Toutle River catchment, buried deeply by debris-avalanche deposit (Dinehart 1998; Major et al. 2000; Major 2004). By contrast, elevated levels of suspended sediment were least and of shortest duration from Green River catchment, deforested and draped with tephra by the blast PDC. Fluctuations in annual loads (Fig. 2.5) owe mainly to variations in annual climate and infrequent, high-discharge floods.

Normalized suspended-sediment yields (load normalized by upstream basin area) measured at KID and TOW were initially several hundred times greater than probable pre-eruption yields (Fig. 2.6). Those high yields, caused mainly by sediment flushed from hillsides in Green River catchment and from channels eroded on the debris-avalanche deposit, declined sharply over the first 5 years following the 1980 eruption. They declined further in the late 1980s following construction by the US Army Corps of Engineers of a large sediment-retention structure (SRS) on North Fork Toutle River above the Green River confluence (Willingham 2005) (Fig. 2.1). Despite the decline from peak yields in the early 1980s, yield from the debris-avalanche deposit (measured below the SRS) continues at a level ~10–50 times greater than probable pre-eruption level even after more than 30 years (Fig. 2.6). By contrast, yield from Green River catchment was initially 100 times less than that from North Fork Toutle River catchment and only about 10 times greater than probable pre-eruption level (Fig. 2.6). Initially high rates were caused by development of rills and gullies on hillsides. Because erosion of tephra from hillsides stabilized quickly (Swanson et al. 1983; Collins and Dunne 1986), sediment yield from Green River catchment declined rapidly and returned to pre-eruption level within about 5 years (Fig. 2.6). Yield from South Fork Toutle River, a catchment affected mainly by a large lahar, was initially about 20 times greater than probable pre-eruption level (Fig. 2.6) as both newly deposited and centuries-old sediment were scoured from the system. But it also decreased sharply in the first few years after the eruption once the most easily entrained sediment was evacuated. However, sediment flux from South Fork Toutle River catchment subsequently fluctuated considerably (Fig. 2.5), with loads in the 1990s and 2000s sometimes equaling those of the mid-1980s. Although its yield may still be above probable pre-eruption level (Fig. 2.6), since 2000, its average yield has been within the range of variation of rivers in the western Cascade Range.

Substantial post-eruption sediment transport and resultant channel aggradation are common around Cascade Range volcanoes. Thick, post-eruptive alluvial fills have been described along other channels draining Mount St. Helens (Crandell 1987; Major and Scott 1988) as well as along channels draining Mount Hood (Pierson et al. 2011), Mount Rainier (Zehfuss et al. 2003), Mount Baker (Pringle and Scott 2001), and Glacier Peak (Beget 1982). Persistent sediment release from the Mount St. Helens debris-avalanche deposit has caused significant socioeconomic impact (Willingham 2005) and continues to present mitigation challenges. But why has such elevated sediment release persisted, and what are the predominant sediment sources? In Sect. 5, we address these questions by analyzing topographic changes among repeated collections of topographic data.

2.5 Topographic Change Across the Debris-Avalanche Deposit from 1987 to 2009: DEM Analyses

With the exception of Green River catchment, high sediment yields from Toutle River basin were caused mainly by post-eruption channel erosion rather than by hillside erosion. Both within and among disturbance zones, rates and

[1]Prior to 1980, the station measuring discharge on Toutle River was located near Silver Lake, but that station was destroyed by the North Fork Toutle lahar. A temporary station (THW) was established 9 km downstream of the location of station TOW (Fig. 2.1) shortly after the eruption. Station TOW was established in March 1981. THW and TOW operated simultaneously until the end of water year 1982. Because there are no significant tributaries that enter Toutle River between TOW and THW, their records are combined to provide the sediment discharge for water year 1981 (see Dinehart 1998).

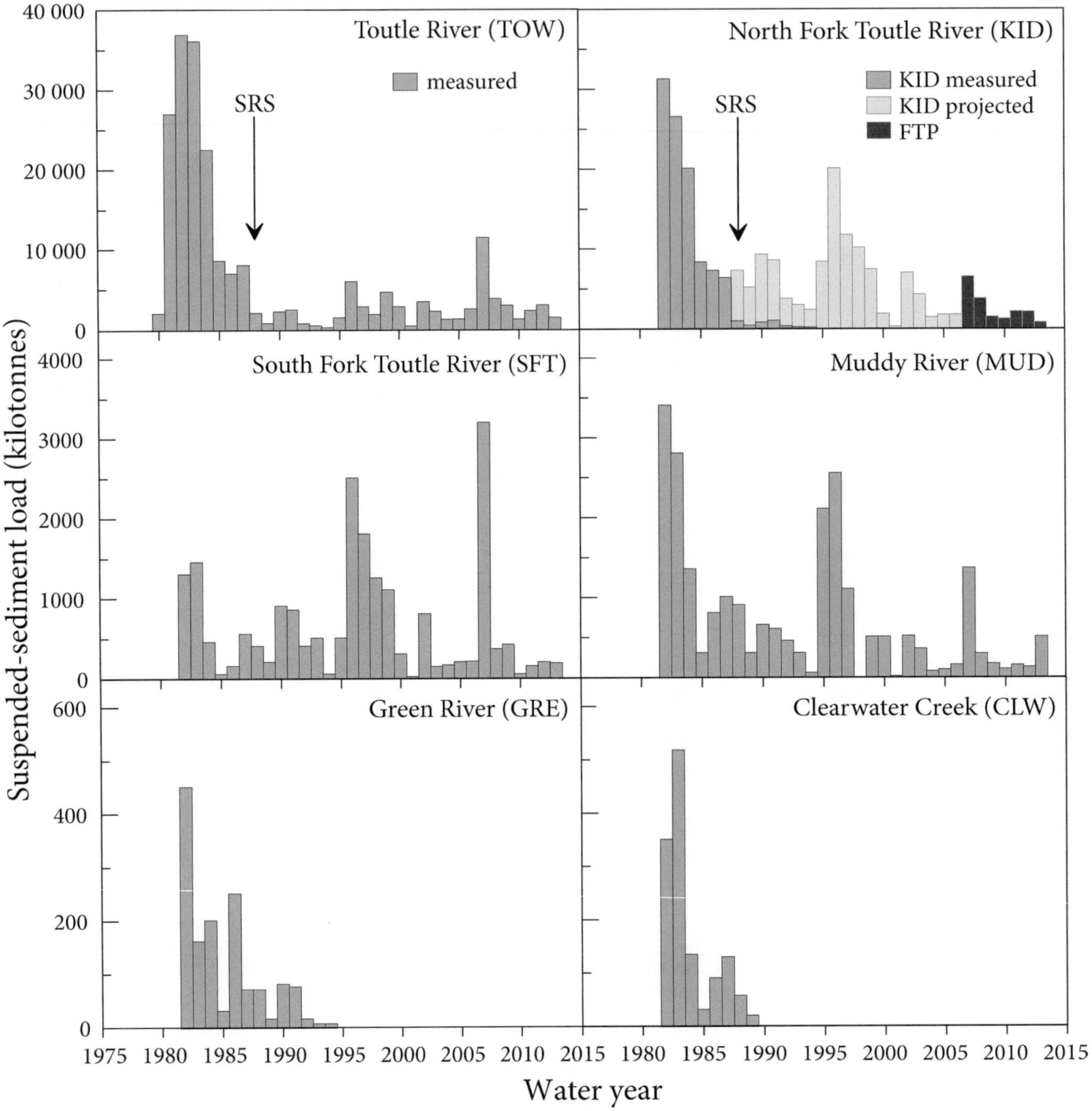

Fig. 2.5 Time series of suspended-sediment loads measured at various gauging stations at Mount St. Helens. The *lightest* colored bars for North Fork Toutle River show projected load that would have been measured if the sediment-retention structure (SRS) had not been constructed; the *darkest* bars show loads measured at station FTP just downstream of the SRS. Note the approximately order-of-magnitude change in vertical scale for different disturbance zones. See Fig. 2.1 for basin disturbance and gauging station locations. (Modified from Major 2004).

magnitudes of channel changes and degrees of channel stabilization varied greatly. Changes in rates and magnitudes of sediment discharge from catchments of Toutle River basin can be tied closely with changes in rates and magnitudes of channel and hillslope erosion. For example, the relative magnitudes and swift decline of sediment yield from Green River catchment compared with other catchments in the basin (Major et al. 2000) can be tied to modest deposit thicknesses and rapid stabilization of rills and gullies on hillsides in the catchment (Collins and Dunne 1986). Measurements of sediment discharge among basin catchments show clearly that the primary source of persistently elevated sediment delivery

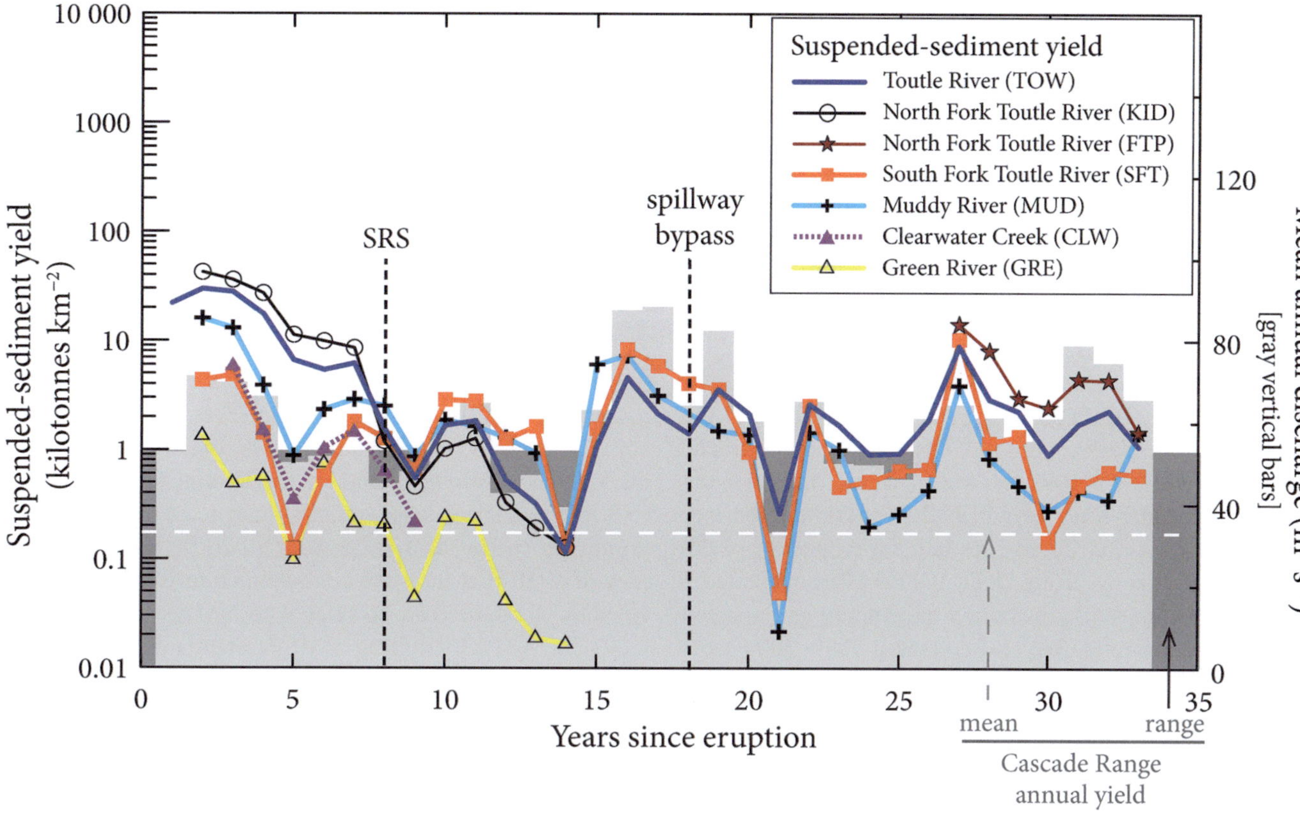

Fig. 2.6 Time series of suspended-sediment specific yields (load per unit basin area) at Mount St. Helens. See Fig. 2.1 for basin disturbances and gauge locations. *Background shaded region* depicts range and *dashed horizontal line* depicts approximate mean value of mean annual yields of several Cascade Range rivers. *Vertical bars* show mean annual discharge measured at TOW on Toutle River. *SRS line* marks completion date of the sediment-retention structure on North Fork Toutle River in water year 1988. *Spillway bypass line* indicates when water and sediment began passing over the SRS spillway in 1998. (Modified from Major et al. 2000).

is the debris-avalanche deposit in upper North Fork Toutle River valley. Here we examine spatial patterns of erosion and deposition across the debris-avalanche deposit during two time periods that follow the initial peak and sharp decline of sediment delivery from upper North Fork Toutle valley: (1) the approximate decade from 1987 to 1999 and (2) the decade from 1999 to 2009. These decadal periods are bracketed by digital elevation models (DEMs) that can be used to produce derivative models showing topographic change across the deposit. These broad snapshots of spatial change provide a more holistic view of evolving patterns of geomorphic change across the deposit than can be obtained from periodic resurveys of spatially discrete channel cross sections. We use these decadal-scale holistic views of patterns of geomorphic change to elucidate sources and causes of persistently elevated levels of sediment delivery from upper North Fork Toutle valley.

Channel cross-section surveys on the debris-avalanche deposit (e.g., Fig. 2.4) show rates and magnitudes of change in channel-bed elevation have diminished greatly since the mid-1980s (Meyer and Martinson 1989; Simon 1999; Zheng et al. 2014). Persistently elevated sediment delivery despite these diminished changes in channel-bed elevation implies that lateral channel adjustment is the principal cause of ongoing fluvial instability. Lateral channel adjustment can take many forms. A river can widen a channel by eroding channel banks and valley walls; it can migrate across the valley floor and mine channel-bed and floodplain sediment without causing significant incision; or it can entrain sediment delivered to the channel from external sources, such as from tributary channels or extrafluvial mass failures. All of these types of adjustments are evident at Mount St. Helens.

Since construction of the SRS, several DEMs of all or part of the debris-avalanche deposit have been created from aerial photography and lidar surveys to assess topographic change. Some of these DEMs have been used to produce DEMs of difference (DoDs) (Wheaton et al. 2010) in order to assess spatial distributions of erosion and deposition and to quantify volumetric topographic changes and compute morphological sediment budgets (Bradley et al. 2001). DoDs are produced by subtracting the elevations of pixels in one DEM from those of corresponding pixels in a later DEM. Such differencing provides the net vertical change between matched pixels. A net negative vertical change (erosion) does not necessarily represent vertical incision. Lateral erosion between times of DEMs can also produce net negative vertical change. This is particularly evident where large magnitudes of net erosion along a channel or valley margin are juxtaposed with substantially lesser magnitudes of net erosion (or even deposition) on a valley floor. Here, we compare and contrast a DoD representing topographic changes during the second decade after the 1980 eruption (1987–1999 DoD) with one that represents changes during the third decade after eruption (1999–2009 DoD).

Bradley et al. (2001) produced a DoD from DEMs of the debris-avalanche surface derived from aerial photographs obtained in 1987 and 1999. However, neither they nor the more comprehensive report which their work summarizes (West Consultants 2002) provides vertical accuracies of the DEMs or uncertainty of the DoD. We therefore recreated a 1987–1999 DoD having quantified uncertainty and a stated change-detection limit. We also created a 1999–2009 DoD from the 1999 DEM and a DEM based on a 2009 lidar survey (Mosbrucker 2014).

2.5.1 Estimates of Error in the DEMs and Resulting DoDs

One of the most useful aspects of a DoD is its ability to quantify volumetric change. However, quantitative estimates of erosion and deposition are subject to errors within each DEM and to the propagation of those individual errors to the DoD. The metadata that accompanied the original 1987 and 1999 DEMs provide scales of photographs acquired, intended scales of topographic maps to be generated from the photography, target contour intervals, and photogrammetric points from which the topographic maps and DEMs were derived. Those metadata do not provide estimates of vertical accuracies. The 1987 and 1999 photographs were acquired to support 1:2400-scale contour maps having a 1.22-m (4-ft) contour interval. Photogrammetric point densities support DEM resolutions of about 20 m, but West Consultants (2002) generated DEMs using triangulated irregular networks (TINs) interpolated to 3-m-grid resolution.

Because the vertical accuracies of the 1987 and 1999 DEMs are not specified, they must be estimated. We can estimate approximate vertical accuracy of the DEMs if we assume that the underlying photography met appropriate specifications and that photogrammetric techniques were properly executed when the consequent points, maps, and DEMs were compiled. The National Map Accuracy Standard (NMAS) (Maune et al. 2001), as applied to contour maps, specifies that elevation differences of no more than 10% of tested, well-defined points on a map may exceed one-half the contour interval. Thus, the intended map scales of the photography and target contour intervals suggest vertical accuracies of at least 0.61 m at the 90% confidence level. The American Society for Photogrammetry and Remote Sensing (ASPRS) further refined vertical accuracy standards for contour maps. The ASPRS standard (ASPRS 1990; Maune et al. 2001) for a class 1 contour map (the most accurate) states that the limiting root-mean-square vertical error ($RMSE_z$) for well-defined features between contours is one-third the contour interval. If the photography and consequent map compilations met that standard, then the intended map scales and target contour intervals suggest a limiting $RMSE_z$ of 0.41 m. However, these standards apply to contour maps, and neither is directly applicable to DEMs. The National Standard for Spatial Data Accuracy (NSSDA) of 1998 is specifically designed to be applied to digital geospatial data (Maune et al. 2001). Vertical accuracy of a DEM at the 95% confidence interval according to NSSDA standards can be related to NMAS vertical accuracy and ASPRS limiting $RMSE_z$ standards for contour maps by (Maune et al. 2001):

$$\text{Accuracy}_z(\text{NSSDA}) = 1.1916 \times \text{NMAS vertical accuracy} \quad (2.1)$$

$$\text{Accuracy}_z(\text{NSSDA}) = 1.9600 \times \text{RMSE}_z \quad (2.2)$$

Equation (2.1) can be related to map contour interval by:

$$\text{Accuracy}_z(\text{NSSDA}) = 1.1916 \times \begin{pmatrix} 0.5 \times \text{contour} \\ \text{interval} \end{pmatrix} \quad (2.3)$$

$$\text{Accuracy}_z(\text{NSSDA}) = 0.5958 \times \text{contour interval} \quad (2.4)$$

If we assume that vertical errors are normally distributed and that the photography and mapping that support the DEMs met their intended scales and contour intervals, then the estimated vertical accuracies of the 1987 and 1999 DEMs at the 95% confidence level range from 0.73 to 0.80 m (Eqs. 2.2 and 2.4). From 2010 to 2012, we spot-checked elevations in the DEMs along North Fork Toutle valley in the vicinity of Coldwater Lake in places that had not changed since the photographs were acquired. Those limited field checks confirm that our estimate of vertical accuracies of those DEMs is reasonable but probably a bit high.

The 2009 lidar-derived DEM has higher resolution and greater vertical accuracy than those derived from aerial photographs. The report accompanying the 2009 lidar survey states that the DEM has a 1-m resolution, and vertical accuracy obtained from spot checks along a bare road surface is 0.071 m at the 95% confidence level (Mosbrucker 2014). Spot checks of 103 elevations on the surface of the debris-avalanche deposit, however, show a lower accuracy on natural terrain; there, the DEM has a vertical accuracy ranging from 0.10 to 0.50 m at the 95% confidence level, with a mean accuracy of 0.35 m (Mosbrucker 2014). To match the interpolated resolution of the 1987 and 1999 DEMs, we degraded the 2009 DEM to 3-m-grid resolution.

Estimates of volumetric change in sediment erosion and deposition are computed by subtracting surface elevations of one DEM from those of another. However, measurements of change between two DEMs incorporate errors of each DEM. Therefore, potential vertical error of the DoD must be estimated in order to determine the minimum level of change that distinguishes detectable change from noise.

Simple error propagation (Bevington 1969; Brasington et al. 2003) is used to determine the minimum level of change detection in the DoD. Errors from individual DEMs are propagated into the DoD as:

$$\delta_{\mathrm{DoD}} = \sqrt{\delta_{\mathrm{old}}^2 + \delta_{\mathrm{new}}^2} \tag{2.5}$$

where δ_{DOD} is the propagated vertical error in the DoD, δ_{old} is the vertical error of the earlier DEM, and δ_{new} is the vertical error of the later DEM. This estimate of propagated error assumes that errors are normally distributed and those for each cell of each DEM are random and independent.

We compute the estimated error of the recreated 1987–1999 DoD using an approximate value of the estimated vertical accuracy of each DEM. Substituting a midrange value (0.75 m) of the estimated vertical accuracy for the 1987 and 1999 DEMs into Eq. 2.5 yields an estimated δ_{DOD} of 1.06 m. Hence, propagation of estimated vertical accuracies of the individual DEMs indicates our 1987–1999 DoD has a threshold level of change detection of about 1 m. Using the midrange value of estimated vertical accuracy for the 1999 DEM and the range of vertical accuracy values at the 95% confidence level for the 2009 DEM indicates the threshold level of change detection in the 1999–2009 DoD ranges from 0.75 to 0.90 m, a level of detection that is influenced largely by the assumed vertical accuracy of the 1999 DEM. Because the propagated vertical accuracy in the 1999–2009 DoD is influenced so strongly by assumed accuracy of the 1999 DEM, and because vertical accuracy of the 1999 DEM is based on many assumptions, we also use 1 m as the threshold level of change detection for the 1999–2009 DoD. Therefore, in both the 1987–1999 and 1999–2009 DoDs we produced, vertical elevation changes of less than 1 m are considered noise. That change-detection threshold appears reasonable for changes on the valley floor and in areas of high point density. However, it may be too low in areas of low point density (≤1 point per m^2) such as on valley slopes greater than 30° or under locally heavy vegetation. We thus restrict our analyses of change to the valley bottom. This restriction loses little information of relevance because steep valley slopes that bound the debris-avalanche deposit contribute little sediment to the river.

2.5.2 Topographic Changes from 1987 to 1999

Like Bradley et al. (2001), we find that there was net erosion from 1987 to 1999 along the entire North Fork Toutle River channel system upstream of the former N1 structure—a small sediment-retention structure built upstream of the SRS in the early 1980s (Willingham 2005; Fig. 2.7). Documented net erosion includes an 8-km-long reach downstream of Elk Rock (Fig. 2.7) where the valley widens considerably—a reach where Simon (1999) and Zheng et al. (2014) documented predominantly channel-bed aggradation (e.g., NF375; Fig. 2.4). By 1999, channel area had developed across 20% of the debris-avalanche deposit. However, the boundary of the DoD does not extend all the way to the lip of the crater because the 1987 photography did not extend that far. Thus, erosion along Loowit channel from the mid-Pumice Plain (approximately cross section LO030) to the crater lip (Figs. 2.2 and 2.7) is unaccounted in the analysis. The documented erosion in the DoD is thus a minimum volume within the valley.

Much of the erosion documented from 1987 to 1999 was focused locally (Table 2.1; Fig. 2.7). The greatest erosion occurred along North Fork Toutle River channel from near Elk Rock to about its confluence with Castle and Coldwater Creeks (Table 2.1 and Fig. 2.7). Other significant erosion occurred along Castle Creek and Coldwater Creek channels and locally along North Fork Toutle River channel upstream of the Castle Creek/Coldwater Creek confluence (Fig. 2.7 and Table 2.1). Channel erosion between Elk Rock and the Castle Creek/Coldwater Creek confluence accounted for about 40% of the volume eroded during the second decade after the 1980 eruption, even though this reach accounted for only about 20% of the active channel area (~4% of the area of the avalanche deposit). Castle Creek and Coldwater Creek channels combined accounted for nearly as much active channel area (17%) as did the reach from Elk Rock to the Castle/Coldwater confluence, but they contributed only about half as much sediment volume. By contrast, channel erosion above the Castle Creek/Coldwater Creek confluence—about 30% of the active channel area within the boundary of the DoD (~7% of the area of the avalanche deposit)—accounted for 28% of the volume eroded (Fig. 2.7 and Table 2.1). About 10% of documented erosion was downstream of Elk Rock (Table 2.1), and a very minor amount of erosion (about 1%) was from re-entrainment of sediment deposited behind the SRS.

Comparisons of net vertical changes detected by differencing the DEMs with measured changes in channel cross sections (e.g., Fig. 2.4) show that much of the channel change from 1987 to 1999 resulted from lateral bank erosion. The most extensive bank erosion occurred at the outsides of channel bends (Fig. 2.7). Locally, tens to hundreds of meters of lateral bank erosion caused as much as 57 m of negative (erosional)

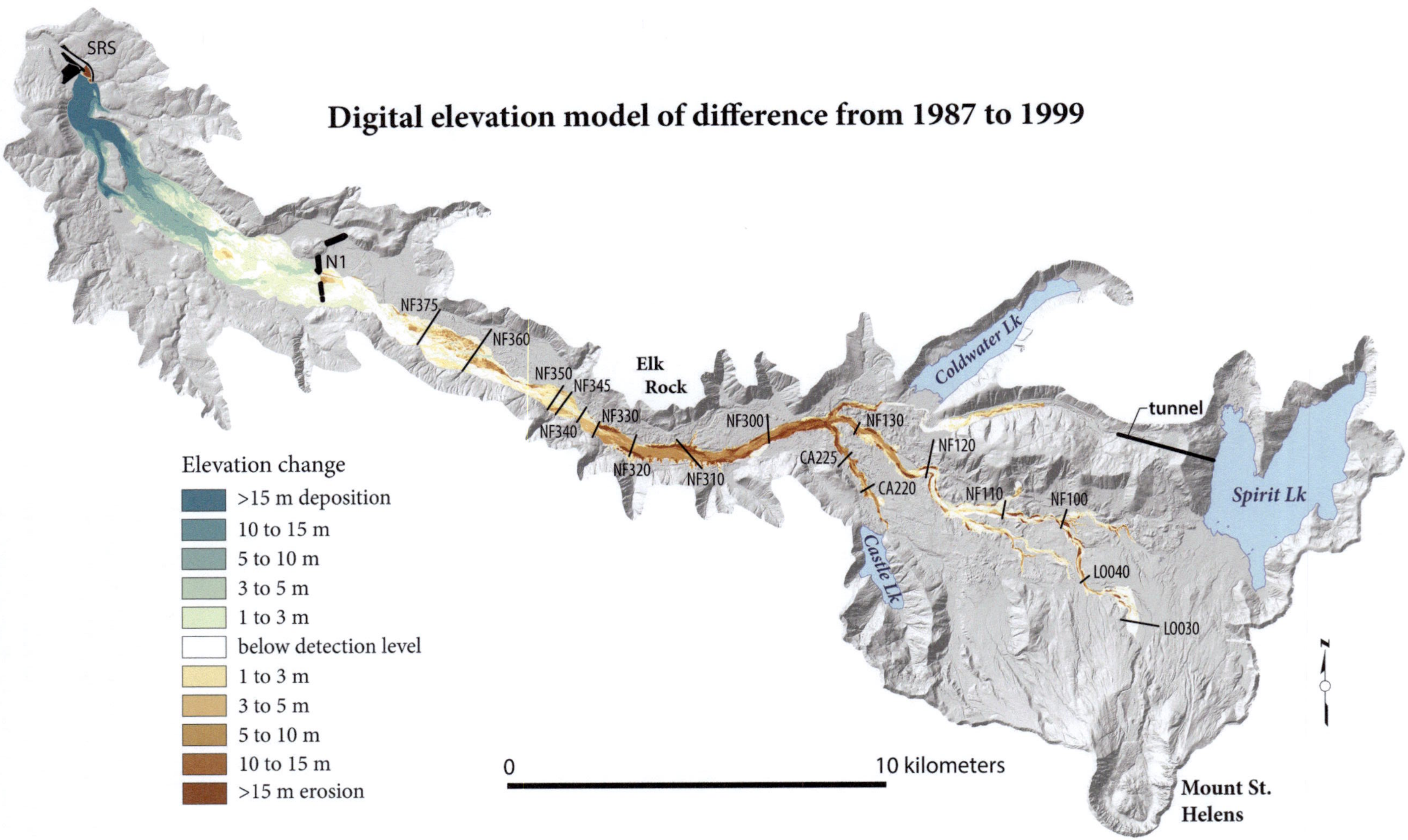

Fig. 2.7 Digital elevation model of topographic difference (DoD) created by differencing digital elevation models derived from aerial photography in 1987 and 1999 (see also Bradley et al. 2001). The DoD has been draped over a hill-shaded topographic model derived from a lidar survey in 2009. Locations of select cross sections shown (see Fig. 2.4). *SRS* is the location of sediment-retention structure completed in 1988. *N1* is the location of former small retention structure constructed in the early 1980s (see Willingham 2005). The deep erosion at SRS spillway is an artifact of construction activity; material had not yet been excavated at the time the 1987 photos were acquired.

Table 2.1 Comparison of results from DEMs of difference from 1987 to 1999 and 1999 to 2009.

	1987–1999				1999–2009			
Reach[a]	Volume (million m^3)[b]	Area (million m^2)[c]	Average erosion yield[d] (m^3 m^{-2})	% of total erosion	Volume (million m^3)	Area (million m^2)	Average erosion yield (m^3 m^{-2})	% of total erosion
Erosion								
Total above SRS	57.6 ± 1.3	23.8	2.4		38.9 ± 2.6	24.8	1.6	
Total erosion dilated by 16%[e]	66.8 ± 1.5				45.1 ± 3.0			
SRS–N1[f]	0.4 ± 0.2	9.8	0.04	1	0.9 ± 0.8	10.3	0.1	2
N1–NF340	6.9 ± 0.5	5.0 (3.5 on avalanche deposit)	1.4	12	6.2 ± 1.1	5.0 (3.5 on avalanche deposit)	1.2	16
NF340–NF/CA/CW confluence	22.4 ± 0.1	2.6	8.6	39	8.7 ± 0.1	2.0	4.4	22
NF/CA/CW–NF/TR/LO confluence[g]	12.4 ± 0.1	2.8	4.4	22	6.3 ± 0.2	2.8	2.3	16
Above NF/TR/LO confluence	3.6 ± 0.1	1.2	3.0	6	5.0 ± 0.04	1.2	4.2	13
Above LO030	–[h]	–	–	–	8.6 ± 0.04	1.7	5.1	22
Castle	7.4 ± 0.02	1.0	7.4	13	1.8 ± 0.1	0.6	3.0	5
Coldwater	4.5 ± 0.2	1.4	3.2	8	1.4 ± 0.3	1.2	1.2	4
Deposition								
SRS–N1	66.2 ± 0.6	9.8			14.7 ± 1.2	10.3		

[a]NF/CA/CW is the confluence of North Fork Toutle River, Castle Creek, and Coldwater Creek. NF/TR/LO is the confluence of North Fork Toutle River, Truman channel, and Loowit channel (see Fig. 2.2).
[b]Estimated from reanalysis of DEM of difference shown in Fig. 2.7. Values differ from those of Bradley et al. (2001) because we use a 1-m level-of-detection threshold.
[c]These values (and those for 1999–2009) represent areas of channel reworked or eroded by incision and migration. The area for SRS to N1 represents the total area of the sediment plain formed by deposition. Because channels there are braided, we did not compute specific areas of active channel erosion.
[d]Spatially averaged volume of sediment eroded per planimetric area of channel over the entire time period.
[e]To compare net erosion to values to net deposition, differences in deposit density must be taken into account. The average bulk density of the debris-avalanche deposit is 1.85 t m^{-3} (Glicken 1996). In general, bulk densities of fluvial deposits range from about 1.4 to 1.7 t m^{-3}. For computational purposes, Bradley et al. (2001) assumed a fluvial bulk density of 1.55 t m^{-3}—a value 16% less than the average density of the debris-avalanche deposit. Therefore, net erosion volumes are dilated 16% to compare with net deposition volumes.
[f]Although this reach is predominantly depositional, sediment was locally re-entrained.
[g]Computed volumes and channel areas include the Carbonate Springs reach (see Fig. 2.2).
[h]This reach is beyond the boundary of the 1987 DEM. Therefore, volumes of erosion cannot be computed.

vertical elevation change during this period as the river clawed into tall banks across the debris-avalanche deposit.

Even though channel widening was the dominant fluvial process during the second decade following eruption, thalweg elevation generally continued to change. Within and above the Elk Rock reach, thalweg elevation generally lowered but by locally variable amounts (Fig. 2.4; Zheng et al. 2014). By contrast to general channel degradation within and above the Elk Rock reach from 1987 to 1999, the sediment plain behind the SRS aggraded by as much as 36 m (Fig. 2.7). Despite the magnitudes of channel erosion, the overall longitudinal profile of North Fork Toutle River changed little between 1987 and 1999 except in the sediment plain behind the SRS (Bradley et al. 2001; Zheng et al. 2014). By this time, rates and magnitudes of channel incision had diminished considerably compared with the early 1980s, and most channel change was lateral. Locally persistent incision of a few meters or less had little effect on the overall channel profile (Simon 1999; Zheng et al. 2014; Mosbrucker et al. 2015). The documented dominance of lateral erosion, in conjunction with bed coarsening (Simon and Thorne 1996), rather than vertical incision during this period is consistent with Meyer and Martinson's (1989) conclusion that channel changes across much of the debris-avalanche deposit 5 years after the eruption were dominated mainly by channel widening accompanied by fluctuating, low-magnitude, vertical adjustments.

2.5.3 Morphological Sediment Budget Versus Measured Sediment Flux from 1987 to 1999

A morphological sediment budget obtained from estimates of net volume change in the DoD and measurements of sediment transport father downstream show the SRS trapped most sediment eroded from 1987 to 1999. The DoD shows 57.6 (±1.3) million m^3 of sediment eroded from upper North Fork Toutle valley and 66.2 (±0.6) million m^3 deposited in the sediment plain behind the SRS. This apparent discrepancy between erosion and deposition arises because the debris-avalanche sediment was deposited in a state denser than that typical of redeposited fluvial sediment. From several field measurements, Glicken (1996) found the mean bulk density of the debris-avalanche deposit is 1.85 t m^{-3}. By contrast, typical fluvial deposits have bulk densities that range from about 1.4 to 1.7 t m^{-3} (e.g., Wilcox et al. 2014; Warrick et al. 2015). Thus, sediment deposited behind the SRS likely has a bulk density that is probably 10–25% lower than that of the debris-avalanche deposit. To directly compare net erosion to net deposition, Bradley et al. (2001) used a 16% density difference (1.55 t m^{-3} or 88 lbs ft^{-3}, a mid-range density value commonly used for engineering purposes) to adjust net erosion volume, and we do likewise (Table 2.1). Thus, the density-adjusted net erosion was 66.8 (±1.5) million m^3. The morphological sediment budget therefore balances erosion and deposition. However, the DoD does not extend to the headwaters of the valley and therefore excludes erosion in the uppermost active channels. Also, erosion likely occurred below our change-detection limit. Furthermore, we know that a substantial amount of suspended sediment was transported downstream of the SRS from 1987 to 1999 (Major et al. 2000; Figs. 2.5 and 2.6). Therefore the volume of sediment eroded from the valley was greater than measured in the DoD.

Suspended-sediment discharge downstream of the SRS was measured on North Fork Toutle River (KID; Fig. 2.1) from 1982 to 1994, on South Fork Toutle River (SFT; Fig. 2.1) from 1982 to 2014, on Green River (GRE; Fig. 2.1) from 1982 to 1994, and along lower Toutle River (TOW) from 1981 to present (2015). From 1987 to 1994, 9.75 million t of suspended sediment passed KID, and 0.34 million t passed GRE, showing that North Fork Toutle River delivered 9.4 million t of suspended sediment below the SRS (Major et al. 2000). If we assume that after 1994 Green River delivered negligible sediment load, and that erosion between Green River confluence and TOW was negligible, then we can assume that the difference between suspended-sediment fluxes measured at TOW and SFT can be attributed to sediment delivery from North Fork Toutle River. The difference in fluxes between TOW and SFT from 1995 to 1999 is 9.2 million t (Major et al. 2000). Therefore, North Fork Toutle River delivered an estimated 18.6 million t of suspended sediment downstream of the SRS from 1987 to 1999. This estimate is probably accurate to within about 20% (Major 2004). If we assume bulk sediment density ranges from 1.4 to 1.7 t m^{-3}, then North Fork Toutle River delivered about 11–13 million m^3 of suspended sediment downstream of the SRS. Undetected erosion below our threshold limit and erosion upstream of the DoD boundary could easily balance this volumetric discrepancy. We therefore infer that about 80 million m^3 (density adjusted) of sediment eroded from upper North Fork Toutle River valley from 1987 to 1999 and that about 85% of that sediment accumulated behind the SRS.

2.5.4 Topographic Changes from 1999 to 2009

Cross sections of North Fork Toutle River along the debris-avalanche deposit (e.g., Figs. 2.2 and 2.8) show variable types and magnitudes of topographic change between 1999 and 2009. Although topographic changes in these sections are dominated by channel widening, channel incision continued to modify channel geometry. In general, magnitudes of incision were greatest and widening least upstream of the Elk Rock constriction, whereas magnitudes of widening were greatest and incision least downstream (generally above and below section NF345; Figs. 2.2 and 2.8). Channel incision outpaced lateral erosion at some sections along Loowit channel (LO040) and along the toe of the Pumice Plain (e.g., NF100 and NF110) (Figs. 2.2 and 2.8). Although channel incision is evident between sections NF120 and NF300 (Figs. 2.2 and 2.8), widening outpaced incision as that reach generally incised by only about 2 m but widened from a few tens to more than 100 m (e.g., NF300; Fig. 2.8). The greatest amounts of widening in this reach were between the Castle Creek/Coldwater Creek confluence and section NF300. Where the valley constricts near Elk Rock, the channel incised as much as 6 m but widened little (e.g., NF320; Fig. 2.8). Incision between NF300 and NF330 left abandoned terraces that narrowed the active channel. (Note the difference in active channel areas from 1987 to 1999 and from 1999 to 2009 above NF340 in Table 2.1. Also compare Figs. 2.7 and 2.9.) Where the valley widens downstream of the constriction, submeter- to meter-scale vertical fluctuations of the channel bed, caused by scour and fill, accompanied as much as 300 m of channel widening (e.g., NF345; Fig. 2.8). Here, the river transitions from a steep channel gradient past Elk Rock (about 20 m km^{-1}) to a much flatter gradient in the sediment plain behind the SRS (about 2.8 m km^{-1} from a few km downstream of former N1 retention structure to the SRS). Owing to this significant change in gradient, the river transitions from a single thread to a braided channel. As a result, hundreds of meters of lateral erosion of

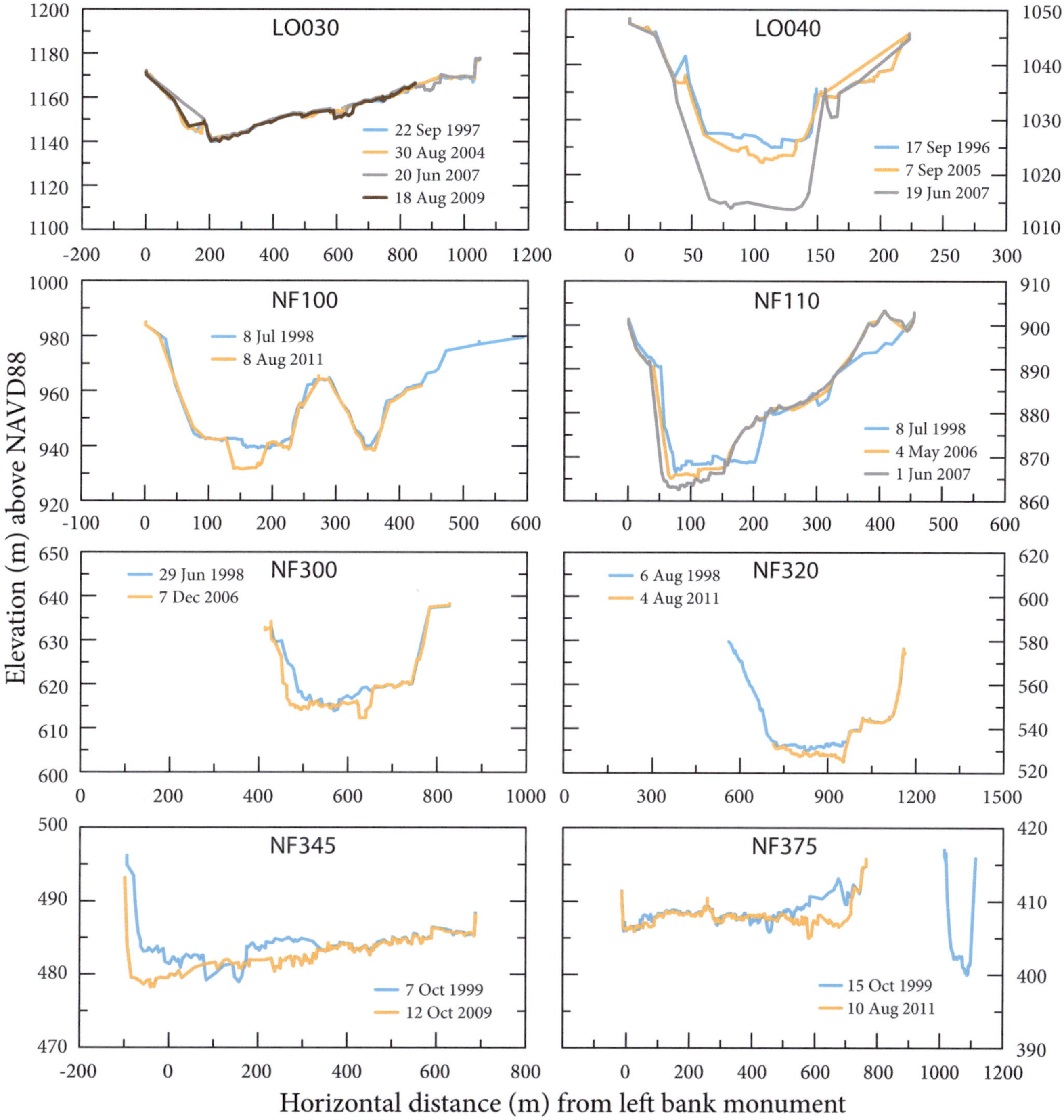

Fig. 2.8 Cross-section profiles from select locations on debris-avalanche deposit bounding the period 1999–2009. (See Figs. 2.2 and 2.9 for section locations. Also see Fig. 2.4 for channel geometry profiles prior to 1999).

relatively short (≤2–5-m-tall) banks and terraces (e.g., NF375; Fig. 2.8) dominated topographic changes.

Measurements of channel geometries along the debris-avalanche deposit show channel-bed elevation appears to have adjusted into a relatively stable profile by the third decade after the 1980 eruption (Zheng et al. 2014). Yet, sediment yield from North Fork Toutle catchment remains above probable pre-eruption level three decades after eruption (Fig. 2.6). The rapid reduction of sediment delivery from Green River catchment (Collins and Dunne 1986; Major et al. 2000; Major 2004) demonstrates that major catchments tributary to North Fork Toutle River are not the sources of persistently elevated sediment yield. Thus, persistently high sediment delivery from North Fork Toutle River must result from in-channel processes or from reworking of sediment delivered to the channel by local mass failures. To obtain broader insights on topographic change in North Fork Toutle River valley above the SRS during the third decade after eruption, we compared the 1999 DEM to a 2009 lidar-derived DEM of the upper valley (Mosbrucker 2014).

The 1999–2009 DoD (Fig. 2.9) shows sediment erosion from the debris-avalanche deposit in the third decade after eruption was less voluminous and more spatially distributed than during the second decade (Figs. 2.7 and 2.9; Table 2.1). It also shows that the uppermost reaches of North Fork Toutle River headwater tributaries (Loowit Creek), reaches that were beyond the boundary of the 1987–1999 DoD, were erosional hotspots (Fig. 2.9). Most erosion between 1999 and 2009 (73%) was along North Fork Toutle River channel above Elk Rock and along Loowit channel (Figs. 2.2, 2.9 and Table 2.1). This reach of drainage accounted for about 50% of the active channel area across the avalanche deposit (~13% of the deposit area). The channels of Castle Creek and Coldwater Creek were much less active in the third decade after the eruption compared with the second decade. About 10% of erosion was along Castle Creek and Coldwater Creek (including South Coldwater Creek) channels combined, which account for 12% of active channel area. The remaining erosion was along North Fork Toutle River channel below Elk Rock. There, sediment was eroded not only from the channel above the former N1 retention structure (16%) but was also entrained from the sediment plain behind the SRS (2%; Table 2.1 and Fig. 2.9). After deposited sediment filled to the level of the SRS spillway in 1999, the sediment plain became a minor source for downstream sediment delivery.

Although erosion from 1999 to 2009 was more spatially distributed along the active channel system compared with erosion from 1987 to 1999, locally focused erosion was nevertheless important. Particular areas of focused erosion were along the volcano's lower north flank (upstream of cross section LO030), across the Pumice Plain (between cross sections LO030 and NF100), near Coldwater Lake between cross section NF120 and the Castle Creek/Coldwater Creek confluence, and near Elk Rock between cross sections NF310 and NF330 (Fig. 2.9). Downstream of NF120, erosion was focused particularly at channel bends. On the volcano's lower north flank, a few tens of meters of lateral bank erosion produced as much as 90 m of vertical elevation change because flow, strong winds, and mass failures gnawed away very tall banks, especially along upper Loowit Creek and Step Creek (Figs. 2.2 and 2.9). Oblique aerial photographs from late November 2006 indicate some, and perhaps much, of the documented erosion in upper Loowit Creek channel system happened during a large storm in early November 2006. Across the Pumice Plain, a few tens of meters of channel widening caused large changes in vertical elevation. Below the toe of the Pumice Plain (near cross section NF110; Fig. 2.9), a large mass failure of debris-avalanche sediment off Johnston Ridge in January 2006 deposited about 600 000 m^3 of sediment in the North Fork Toutle River channel. That deposit forced the river to erode laterally by tens of meters, which caused as much as 15–20 m of vertical elevation change. Near Elk Rock, channel shifting caused as much as 100–300 m of lateral bank erosion locally, and farther downstream it caused variable amounts of erosion. Magnitudes of valley-floor erosion diminished considerably downstream of cross section NF375 (Fig. 2.9).

Reaches of focused erosion along upper North Fork Toutle River channel were punctuated by reaches of substantive deposition (Fig. 2.9). At the base of the volcano's north flank, but above cross section LO030, as much as 4 m of deposit accumulated on the valley floor. Downstream of the fan of landslide debris that filled the channel near cross section NF110, fluvial and debris-flow deposits from the large storm in November 2006 left a 6-m-thick fill on the valley floor. Sediment transported during that storm also accumulated on the valley floor below cross section NF130 and displaced the confluence of North Fork Toutle River and Coldwater Creek several hundred meters downstream from its previous position. Below the former N1 retention structure, sediment deposition dominated changes in valley-floor topography although re-entrainment of previously deposited sediment is evident (Fig. 2.9). To minimize sediment re-entrainment and induce additional sedimentation behind the SRS, the US Army Corps of Engineers installed engineered log jams in 2010 (Townsend 2013) and raised the lip of the SRS spillway by 2.1 m in 2013.

Significantly less sediment eroded from upper North Fork Toutle River valley in the third decade after the 1980 eruption compared with the second. Total erosion measured upstream of the SRS between 1999 and 2009 (45 million m^3 density-adjusted) is about 60% of that eroded between 1987 and 1999 (67 million m^3 density-adjusted above the SRS plus about 12 million m^3 passed downstream) (Table 2.1). Furthermore, the proportion of eroded sediment that accumulated behind the SRS from 1999 to 2009 was substantially less than the proportion that accumulated from 1987 to 1999; only about 30% of eroded sediment volume accumulated behind the SRS from 1999 to 2009 (Table 2.1), whereas we infer about 85% accumulated from 1987 to 1999. That difference in relative proportion of sediment accumulation reflects the decrease in trap efficiency of the SRS, which by 1999 permitted most sediment delivered to the sediment plain to bypass the structure. Sediment that accumulated from 1999 to 2009 aggraded the sediment plain as much as 7 m, compared with 36 m of maximum aggradation from 1987 to 1999 when the SRS largely stored delivered sediment.

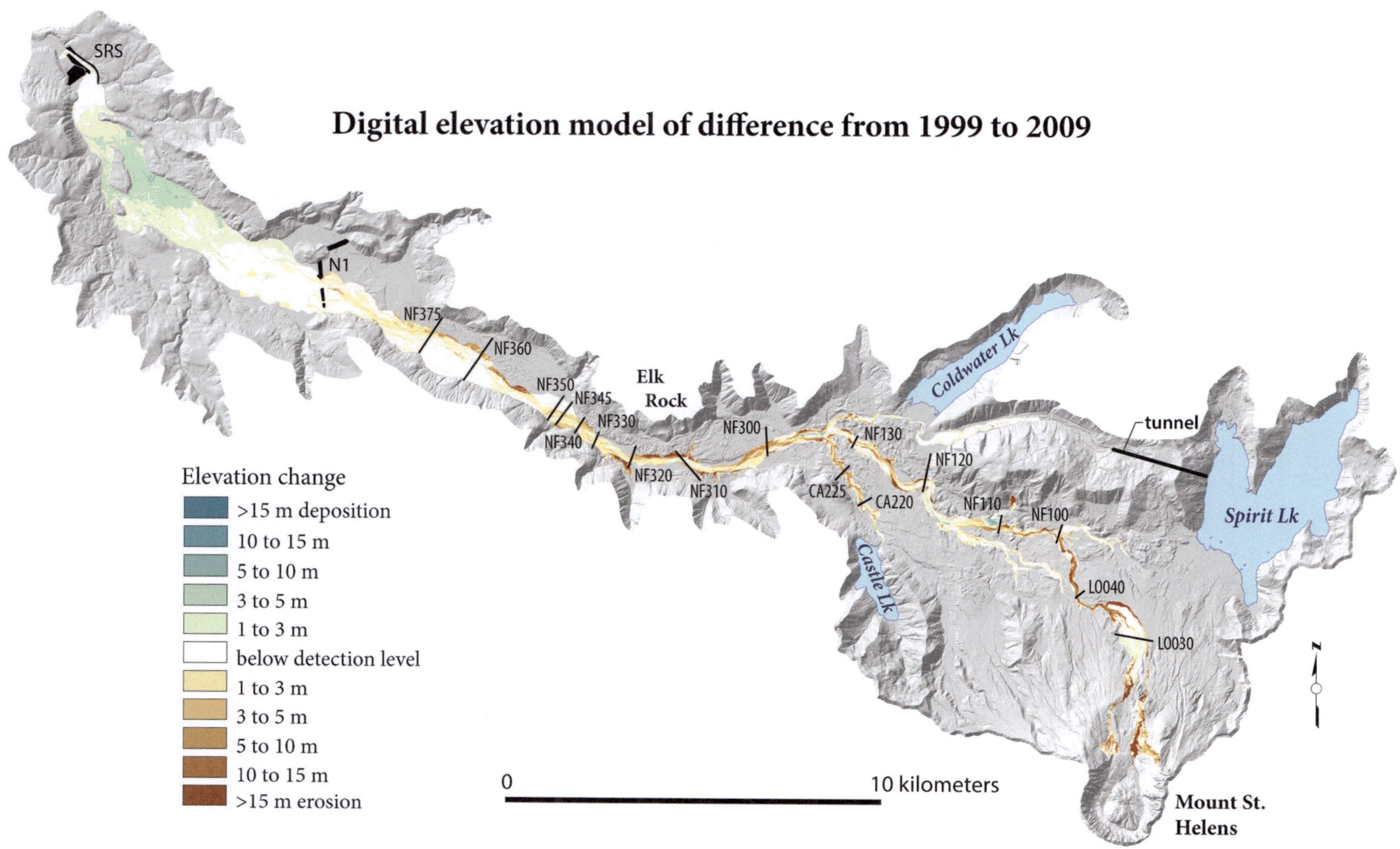

Fig. 2.9 Digital elevation model of topographic difference (DoD) created by differencing digital elevation models derived from aerial photography in 1999 (Bradley et al. 2001) and an airborne lidar survey in 2009 (Mosbrucker 2014). The DoD has been draped over a hill-shaded topographic model derived from the 2009 lidar survey. Locations of select cross sections shown (see Fig. 2.8). *SRS* is the location of sediment-retention structure completed in 1988. *N1* is the location of former small retention structure constructed in the early 1980s (see Willingham 2005).

2.5.5 Morphological Sediment Budget Versus Measured Sediment Flux from 1999 to 2009

The 1999–2009 DoD shows 38.9 (±2.6) million m^3 (45.1 ± 3.0 million m^3 density adjusted) net erosion from upper North Fork Toutle valley and 14.7 (±1.2) million m^3 net deposition behind the SRS (Table 2.1). Hence, the DoD shows about 30 (±1.8) million m^3 of sediment bypassed the SRS from 1999 to 2009, about three times the suspended-sediment load measured downstream and attributed to North Fork Toutle River from 1987 to 1999. By 1999, accumulated sediment had filled to the level of the SRS spillway, and substantial quantities of sediment, likely including sandy bedload, began passing the structure.

Fluxes of suspended sediment delivered by North Fork Toutle River downstream of the SRS after 1994 must be estimated by proxy because KID gauging station was decommissioned. Using the same assumptions as in Sect. 5.3, we estimate that North Fork Toutle River delivered about 15 million t of suspended sediment downstream of the SRS from 1999 to 2006. From 2007 to 2009, station FTP just downstream of the SRS (Fig. 2.1) measured 11.3 million t of suspended-sediment discharge (Fig. 2.5). Thus, from 1999 to 2009, North Fork Toutle River discharged about 26 million t, or 15–20 million m^3, of suspended sediment downstream of the SRS. That suspended-sediment load accounts for about 50–70% of the net sediment loss from upper North Fork Toutle River valley estimated from DoD analysis. The imbalance is most likely related to two main factors: (1) unmeasured bedload downstream of the SRS, and (2) fluvial storage of sediment between the SRS and TOW gauging station on lower Toutle River. Although the amount of bedload that may have passed the SRS is unknown, bedload typically comprises about 20% or more of total sediment load in steep mountain rivers, but proportions are highly variable (e.g., Wohl 2000; Major et al. 2012; Magirl et al. 2015). If we assume 20% of the load that passed the SRS was bedload, then net suspended-sediment loss from the upper valley is about 22–25 million m^3. If 20% of that potential suspended-sediment load was lost to channel, floodplain, and overbank storage between the SRS and TOW gauging station, then the amount of suspended-sediment load measured at TOW from 1999 to 2009 and attributed to North Fork Toutle River roughly balances net loss of suspended sediment estimated from the morphological sediment budget. Although these estimates of the fraction of sediment bypassing the SRS attributed to bedload and to suspended sediment lost to possible channel storage downstream are unconstrained, those processes provide a plausible explanation for the imbalance between the morphological sediment budget and suspended-sediment load measured downstream.

2.6 Discussion

Mount St. Helens' 1980 eruption severely disrupted the hydrogeomorphic conditions of Toutle River basin, in particular those of upper North Fork Toutle River valley. There, a massive debris-avalanche deposit reset valley topography. Its thick fill (average thickness of 45 m across 60 km^2), having a chaotic surface that lacked through-going flow, truncated confluences of tributary channels and hydrologically disconnected most of the upper valley (above the Green River confluence) from the channel downstream. An initial phase of rapid channel change across the debris-avalanche deposit ensued. Following channel initiation, incision and widening dominated channel change upstream of a valley constriction at Elk Rock (20 km down valley from the volcano), whereas aggradation and widening dominated downstream. During the early phase of channel evolution, North Fork Toutle River delivered sediment downstream at a rate several hundred times above pre-eruption levels. Concurrently, tephra erosion from hillsides in Green River catchment and lahar-deposit erosion along South Fork Toutle River channel augmented sediment delivery to lower Toutle River. Within 2 years of the eruption, each disturbed catchment achieved peak sediment delivery. Within 3–4 years, rates and magnitudes of hillside and channel erosion declined, and magnitudes of sediment delivery dropped sharply (Figs. 2.5 and 2.6). Magnitudes of sediment delivery were much less and rates of decline were more precipitous, from Green River catchment, where rills and gullies rapidly stabilized and hillside erosion swiftly diminished (Collins and Dunne 1986), and from South Fork Toutle catchment, where the river rapidly evacuated the most easily erodible lahar sediment (Meyer and Dodge 1988), than from North Fork Toutle catchment. Within 5 years of eruption, sediment delivery from Green River catchment returned to pre-eruption levels. Sediment delivery from South Fork Toutle catchment has fluctuated considerably over three decades since eruption (Fig. 2.6), caused largely by sensitive responses to storms that produced large floods. Although its level of sediment delivery may still be above pre-eruption levels, its average annual delivery since 2000 falls within the range of variation of sediment delivery from western Cascade Range rivers.

After 30 years, sediment delivery from upper North Fork Toutle River is far below its peak level of early 1980s but still many times above probable pre-eruption level. The great reduction of sediment delivery is due largely to diminished vertical incision, coarsening of the channel bed, and stabilization of the channel longitudinal profile, as well as to mitigation measures including construction of a massive sediment-retention structure (SRS). Analysis of thalweg elevations upstream of the SRS (Zheng et al. 2014) confirms that changes in bed elevation were swift in the early 1980s but by 1987 had slowed considerably. From 1987 to 2009, the longitudinal profile of the channel changed little except where sediment accumulated behind the SRS. On the basis of observations and modeling, Zheng et al. (2014) inferred that relaxation of channel slope in conjunction with channel widening and bed coarsening reduced local stream power, which over time caused rates of vertical channel change to decrease and the profile of North Fork Toutle River to stabilize. Decades-long persistence of high sediment yield thus owes largely to ongoing channel widening, episodic extra-fluvial input of mass failures, and reworking of the valley floor and floodplain through episodic avulsion and channel migration. Decades of repeated surveys of channel cross sections (e.g., Simon and Thorne 1996; Simon 1999; Mosbrucker et al. 2015) show that lateral erosion has been a dominant process sculpting the riverine landscape in upper North Fork Toutle River valley since the mid-1980s. Such prolonged channel instability, which maintains sediment transport at rates above pre-eruption levels but below peak rates immediately after eruptions, is a typical long-term fluvial geomorphic response to large explosive eruptions (Manville et al. 2009; Gran et al. 2011; Gran 2012; Pierson and Major 2014).

Patterns of topographic change across the debris-avalanche deposit were examined for two approximately decadal-scale periods—1987–1999 and 1999–2009—to assess spatial locations of ongoing channel instability and to quantify morphological sediment budgets. These periods are bracketed by digital elevation models that can be used to generate models of topographic difference (DoDs). Evident topographic changes show strong patterns of erosion and deposition. From 1987 to 1999, net channel erosion was highly focused along the channel of North Fork Toutle River in a reach 10–20 km downstream of the volcano as well as along Coldwater Creek and Castle Creek (Fig. 2.7). That reach of North Fork Toutle River coincides with some of the greatest changes in channel gradient relative to pre-eruption gradient (Simon 1999), and thus is an area where initial vertical changes of bed elevation were greatest. Rapid incision through that reach promoted bank instability, extensive lateral erosion, and channel braiding (Simon 1999). As the channel bed coarsened through the 1980s and early 1990s (Simon and Thorne 1996; Simon 1999) and rates of change of the channel longitudinal profile greatly diminished (Zheng et al. 2014), vertical incision through this steep-gradient reach became more difficult and lateral erosion more dominant. About 85% of the sediment eroded from 1987 to 1999 accumulated as thick fill immediately behind the SRS. By the third decade after the 1980 eruption (1999–2009), the channel profile along the debris-avalanche deposit had largely stabilized, hydraulic geometries of the channel system had largely adjusted to imposed sediment loads (Pitlick et al. 2006), and net channel changes caused by both vertical and lateral adjustments were of lesser magnitude and more spatially distributed (Fig. 2.9). Thus, by the third decade after eruption, the Elk Rock reach of North Fork Toutle River had become a less dominant sediment source. In contrast to the second decade after eruption, only about 30% of eroded sediment accumulated behind the SRS because fill had aggraded to the level of the structure's spillway. Between 1999 and 2009, trap efficiency of the SRS greatly diminished.

Net erosion of the fluvial system upstream of the SRS was about 40% less from 1999 to 2009 than from 1987 to 1999. This decline in net erosion resulted largely because channels eroded across the debris-avalanche deposit are wide ($\geq$100 m wide) relative to flow footprints (bankfull flow about 10–20 m wide). As a result, the river and its headwater tributaries have less contact with steep channel banks and walls of valleys carved across the deposit than they did from 1987 to 1999 when channels were still actively evolving and adjusting width-to-depth ratios. Hydrological differences pose another possible cause of the large difference in net erosion between these periods. The period 1987–1999 contained a cycle of wetter-than-average conditions from 1995 to 1999 as well as floods of record in Toutle River basin in 1996 (Major 2004). In contrast, 1999–2009 was a cycle of generally drier-than-average conditions. But except for the 1996 flood, magnitudes and distributions of annual peak flows from 1987 to 1999 and from 1999 to 2009 were distributed similarly. Furthermore, average mean annual flow from 1999 to 2009 was only 5% less than that from 1987 to 1999 (57.2 m^3 s^{-1} versus 60.4 m^3 s^{-1} measured at TOW; see Fig. 2.6). These modest hydrological differences likely do not account for the large difference in net erosion between these two periods.

The rapid peak and sharp decline of sediment delivery from Toutle River basin followed by a decades-long phase of lower-level but still elevated sediment delivery (Figs. 2.5 and 2.6) is consistent with a posited conceptual model of geomorphic response following large explosive eruptions (Gran et al. 2011). In that model, the prolonged phase of geomorphic adjustments that causes persistently elevated sediment delivery is attributed to ongoing fluvial instability dominated by valley widening and reworking of the valley floor. Gran (2012) noted that seasonal influx and accumulation of sand on channel beds at Mount Pinatubo led to enhanced bed-sediment mobility and persistent low-level channel incision. From measured cross sections and observations at Mount St. Helens, Janda et al. (1984) and Simon (1999) noted that bank-toe erosion and mass wasting of bank materials caused large increases in channel width during early phases of channel evolution. Decadal-scale snapshots of spatial topographic changes at Mount St. Helens during the second and third decades after the 1980 eruption show the phase of persistent fluvial instability along channels on the debris-avalanche deposit is dominated by bank failures leading to channel widening. Those snapshots show large vertical changes in elevation (many meters to tens of meters) accompanied only modest vertical degradation and even aggradation of adjacent channel floor—changes also documented by cross-section surveys. Such large net vertical changes in channel topography in a fluvial system experiencing diminishing rates of vertical bed-elevation change (Zheng et al. 2014) can be caused only by mass wasting of tall vertical banks and valley walls on the avalanche deposit.

Elevated sediment flux and lateral channel erosion provide a positive feedback loop that prolongs channel instability. Elevated sediment flux causes local channel-bed aggradation or sediment accumulation on channel bars, which distorts the flow field and establishes cross-stream gradients that drive flow toward opposing channel banks (Dunne et al. 2010; Wickert et al. 2013). Such flow-field distortion, along with conditions that promote mass failure of channel banks (e.g., toe erosion and elevated water tables), causes lateral bank erosion, which in turn introduces additional sediment to the river and increases sediment flux. As a result, the river channel remains dynamic and can maintain lateral migration (Dunne et al. 2010). Eventually, the developed floodplain becomes sufficiently wide and the active channel sufficiently narrow that the river only rarely flows against valley walls. By then, vegetation can stabilize channel banks, and the river mainly reworks its bed and floodplain within a zone of fairly consistent channel width. Gradually, channel mobility and sediment flux decrease until the system achieves a state of dynamic equilibrium (Wickert et al. 2013). In the case of volcanically disturbed river systems, and as is evident at Mount St. Helens, this state of equilibrium can take decades, perhaps centuries, to achieve (Gran et al. 2011; Pierson et al. 2011; Meadows 2014; Pierson and Major 2014).

Although the debris-avalanche deposit provides an enormous source of erodible sediment to North Fork Toutle River and its headwater tributaries, most of that sediment is sequestered from erosion. As of 2010, only about 20% of the planimetric area of the avalanche deposit had been modified by riverine and local mass-failure processes (Table 2.1), and only about 13% of the deposit volume had been eroded. Most likely, sediment outside the narrowly focused corridor that North Fork Toutle River and its headwater tributaries occupy will remain in place as channel mobility diminishes, floodplains develop, and forest and riparian vegetation recolonize the deposit and floodplain surfaces. Comparison of the 1987–1999 and 1999–2009 DoDs shows the overall position and shape of the river corridor has changed little in the second and third decades following the 1980 eruption.

2.7 Conclusions

More than 30 years after the 1980 eruption of Mount St. Helens, high sediment load continues to be released from Toutle River basin. Nearly all elevated load comes from persistent channel erosion across the debris-avalanche deposit that fills upper North Fork Toutle River valley. Sediment discharge by Green River, from a catchment having hillsides deforested and thinly draped with volcanic sediment by the blast PDC, returned to pre-eruption levels within 5 years after the eruption. Sediment discharge by South Fork Toutle River, reamed by a large (14 million m^3) lahar, has fluctuated greatly over three decades since the eruption. Although its sediment discharge may still be above pre-eruption levels, its average annual discharge has been within the range of variation of western Cascade Range rivers since 2000. Measurements of suspended-sediment discharge along North Fork Toutle River below a massive sediment-retention structure (SRS) and along lower Toutle River show that average sediment yield from North Fork Toutle River catchment remains more than ten times greater than probable pre-eruption level even with the presence of upstream mitigation measures.

Since the mid-1980s, channel change across the debris-avalanche deposit has been dominated by lateral channel adjustments. Although channel-bed incision has persisted within and upstream of a valley constriction near Elk Rock (20 km downstream of the volcano), the rate of bed-elevation change has diminished, and magnitudes of lateral erosion have generally outpaced those of incision.

Comparisons of digital elevation models from 1987, 1999, and 2009 reveal patterns of topographic changes across the debris-avalanche deposit in the second and third decades after the 1980 eruption. A digital elevation model of topographic difference (DoD) shows that channel erosion during the second decade after eruption (1987–1999) was strongly focused along a 6-km-long channel reach at and upstream of the valley constriction near Elk Rock and in the channels of Castle Creek and Coldwater Creek. During the third decade after eruption (1999–2009), a DoD shows erosion was more spatially distributed throughout the channel system. In conjunction with channel cross sections that show diminishing channel incision above Elk Rock and channel-bed aggradation below, those DoDs show large fractions of topographic changes resulted from lateral erosion owing to mass failure of tall banks.

Net erosion from the debris-avalanche deposit from 1999 to 2009 was about 40% less than that from 1987 to 1999. However, unlike the second decade after the eruption, when only about 15% of eroded sediment passed the SRS, nearly 70% of eroded sediment passed the SRS from 1999 to 2009, because by 1999, sediment fill behind the SRS had reached its spillway. Comparison of a morphological sediment budget developed from the 1999–2009 DoD with measured suspended-sediment discharge downstream of the SRS indicates that sediment bypassing the SRS (largely sand) may have moved downstream as unmeasured bedload and some suspended sediment may have been stored along the channel or on the floodplain between the SRS and lower Toutle River valley.

The active river corridor has sculpted only about 20% of the surface of the debris-avalanche deposit. Because of the vertical depth of that corridor, large areas of the avalanche deposit will remain sequestered from fluvial processes. Diminishing topographic changes in areas affected by fluvial processes indicate that parts of the drainage network across the deposit are stabilizing.

Acknowledgments We thank the US Army Corps of Engineers Portland District Office for generously providing us digital elevation data for 1987, 1999, and 2009. Dennis Saunders and Tami Christianson helped us compile suspended-sediment and cross-section data. Tim Meadows, Jim O'Connor, Charles Crisafulli, and two anonymous reviewers provided comments that improved this chapter.

Glossary

Blast pyroclastic density current A form of pyroclastic density current initiated by rapid decompression of lava domes or cryptodomes (magma bodies cooled high within a volcanic edifice) owing to sudden collapse. Rapid decompression results in a directed explosion that initially impels the current laterally before it becomes a gravity-driven flow [Sources: a generalized definition based on definitions of PDCs provided in Pierson and Major 2014, Sigurdsson et al. (2015)]. In the case of the Mount St. Helens 1980 eruption, failure of the volcano's north flank unroofed pressurized magma and superheated water. Rapid exsolution of magmatic gases and conversion of superheated water to steam produced a laterally directed blast, which formed a density current that flowed across rugged topography. The current contained fragmented rock debris as well as shattered forest material (Lipman and Mullineaux 1981).

Debris avalanche A rapid granular flow of an unsaturated or partly saturated mixture of volcanic rock particles (± ice) and water, initiated by the gravitational collapse and disintegration of part of a volcanic edifice. Debris avalanches differ from debris flows in that they are not water saturated. Although debris avalanches commonly occur in association with eruptions, they can also occur during periods when a volcano is dormant. Sources: Pierson and Major (2014), Sigurdsson et al. (2015).

Lahar An Indonesian term for a rapid granular flow of a fully saturated mixture of volcanic rock particles (± ice), water, and commonly woody debris. A lahar that has ≥50% solids by volume is termed a *debris flow*; one that has roughly 10–50% solids by volume is termed a *hyperconcentrated flow*. Flow type can evolve with time and distance along a flow path as sediment is entrained or deposited. Sources: Pierson and Major (2014), Sigurdsson et al. (2015).

Pyroclastic density current (PDC) Rapid flow of a dry mixture of hot (commonly >700 °C) solid particles, gases, and air, which can range in character from a dense, ground-hugging flow (*pyroclastic flow*) to a turbulent, low-density cloud of mostly fine ash and superheated air (*pyroclastic surge*). A single PDC commonly involves both flow types as a result of gravitational segregation. Flows are generally gravity driven but may be accelerated initially by impulsive lateral forces of directed volcanic explosions. Flows typically move at high velocity (up to several hundred km h^{-1}). Source: Pierson and Major (2014).

Pyroclastic flow See pyroclastic density current (PDC).

Tephrafall A rain of volcanic particles to the ground following ejection into the atmosphere by an explosive eruption. Tephra is a collective term for particles of any size, shape, or composition ejected in an explosive eruption. Sources: Pierson and Major (2014), Sigurdsson et al. (2015).

References

ASPRS. 1990. ASPRS accuracy standards for large-scale maps. *Photogrammetric Engineering and Remote Sensing* 56: 1068–1070.

Beget, J.E. 1982. *Postglacial volcanic deposits at Glacier Peak, Washington, and potential hazards from future eruptions*, Open-File Report 82–830. Washington, DC: U.S. Geological Survey.

Belousov, A., B. Voight, and M. Belousova. 2007. Directed blasts and blast-generated pyroclastic density currents: A comparison of the Bezymianny 1956, Mount St. Helens 1980, and Soufrière Hills, Montserrat 1997 eruptions and deposits. *Bulletin of Volcanology* 69: 701–740.

Bevington, P.R. 1969. *Data reduction and error analysis for the physical sciences*. New York: McGraw-Hill.

Bradley, J.B., T.R. Grindeland, and H.R. Hadley. 2001. *Sediment supply from Mount St. Helens—20 years later*. Proceedings of the Seventh Federal Interagency Sedimentation Conference, March 25–29, 2001, Reno, Nevada, USA, volume 2: x-9–x-16.

Brand, B.D., C. Mackaman-Lofland, N.M. Pollock, S. Bendaña, B. Dawson, and P. Wichgers. 2014. Dynamics of pyroclastic density currents: Conditions that promote substrate erosion and self-channelization—Mount St. Helens, Washington (USA). *Journal of Volcanology and Geothermal Research* 276: 189–214.

Brasington, J., J. Langham, and B. Rumsby. 2003. Methodological sensitivity of morphometric estimates of coarse fluvial sediment transport. *Geomorphology* 53: 299–316.

Christiansen, R.L., and D.W. Peterson. 1981. The 1980 eruptions of Mount St. Helens. In *The 1980 eruptions of Mount St. Helens, Washington*, Professional Paper 1250, ed. P.W. Lipman and D.R. Mullineaux, 15–51. Washington, DC: U.S. Geological Survey.

Collins, B.D., and T. Dunne. 1986. Erosion of tephra from the 1980 eruption of Mount St. Helens. *Geological Society of America Bulletin* 97: 896–905.

Crandell, D.R. 1987. *Deposits of pre-1980 pyroclastic flows and lahars from Mount St. Helens volcano, Washington*, Professional Paper 1444. Washington, DC: U.S. Geological Survey.

Criswell, C.W. 1987. Chronology and pyroclastic stratigraphy of the May 18, 1980 eruption of Mount St. Helens, Washington. *Journal of Geophysical Research* 92: 10,237–10,266.

Czuba, J.A., C.S. Magirl, C.R. Czuba, E.E. Grossman, C.A. Curran, A.S. Gendaszek, and R.S. Dinacola. 2011. *Sediment load from major rivers into Puget Sound and its adjacent waters*, Fact Sheet 2011-3083. Washington, DC: U.S. Geological Survey.

Dale, V.H., F.J. Swanson, and C.M. Crisafulli. 2005a. Disturbance, survival, and succession: Understanding ecological responses to the 1980 eruption of Mount St. Helens. In *Ecological responses to the 1980 eruption of Mount St. Helens*, ed. V.H. Dale, F.J. Swanson, and C.M. Crisafulli, 3–11. New York: Springer.

———, eds. 2005b. *Ecological responses to the 1980 eruption of Mount St. Helens*. New York: Springer.

Dinehart, R.L. 1998. *Sediment transport at gaging stations near Mount St. Helens, Washington, 1980–1990: Data collection and analysis*, Professional Paper 1573. Washington, DC: U.S. Geological Survey.

Dunne, T., J.A. Constantine, and M.B. Singer. 2010. The role of sediment transport and sediment supply in the evolution of river channel and floodplain complexity. *Transactions of the Japanese Geomorphological Union* 31 (2): 155–170.

Esposti Ongaro, T., A.B. Clarke, B. Voight, A. Neri, and C. Widiwijayanti. 2012. Multiphase flow dynamics of pyroclastic density currents during the May 18, 1980 lateral blast of Mount St. Helens. *Journal of Geophysical Research* 117: B06208. https://doi.org/10.1029/2011JB009081.

Fairchild, L.H. 1987. The importance of lahar initiation processes. In *Debris flows/avalanches: Process, recognition, and mitigation*, Reviews in Engineering Geology, ed. J.E. Costa and G.F. Wieczorek, vol. VII, 51–61. Boulder: Geological Society of America.

Gabet, E.J., O.J. Reichman, and E.W. Seabloom. 2003. The effects of bioturbation on soil processes and sediment transport. *Annual Review of Earth and Planetary Sciences* 31: 249–273.

Glicken, H.X. 1996. *Rockslide-debris avalanche of May 18, 1980, Mount St. Helens Volcano, Washington*, Open-File Report 96–677. Washington, DC: U.S. Geological Survey.

Gran, K.B. 2012. Strong seasonality in sand loading and resulting feedbacks on sediment transport, bed texture, and channel planform at Mount Pinatubo, Philippines. *Earth Surface Processes and Landforms* 37: 1012–1022.

Gran, K.B., and D.R. Montgomery. 2005. Spatial and temporal patterns in fluvial recovery following volcanic eruptions—Channel response to basin-wide sediment loading at Mount Pinatubo, Philippines. *Geological Society of America Bulletin* 117: 195–211.

Gran, K.B., D.R. Montgomery, and J.C. Halbur. 2011. Long-term elevated post-eruption sedimentation at Mount Pinatubo. *Geology* 39: 367–370.

Hardison, J.H. III. 2000. *Post-lahar channel adjustment, Muddy River, Mount St. Helens, Washington*. M.S. thesis. Fort Collins: Colorado State University.

Hoblitt, R.P., C.D. Miller, and J.W. Vallance. 1981. Origin and stratigraphy of the deposit produced by the May 18 directed blast. In *The 1980 eruptions of Mount St. Helens, Washington*, Professional Paper 1250, ed. P.W. Lipman and D.R. Mullineaux, 401–420. Washington, DC: U.S. Geological Survey.

Janda, R.J., K.M. Scott, K.M. Nolan, and H.A. Martinson. 1981. Lahar movement, effects, and deposits. In *The 1980 eruptions of Mount St. Helens, Washington*, Professional Paper 1250, ed. P.W. Lipman and D.R. Mullineaux, 461–478. Washington, DC: U.S. Geological Survey.

Janda, R.J., D.F. Meyer, and D. Childers. 1984. Sedimentation and geomorphic changes during and following the 1980–1983 eruptions of Mount St. Helens, Washington. *Shin Sabo* 37 (2): 10–21 and 37(3): 5–19.

Johnson, M.G., and R.L. Beschta. 1980. Logging, infiltration capacity, and surface erodibility in western Oregon. *Journal of Forestry* 78: 334–337.

Jones, J.A. 2000. Hydrologic processes and peak discharge response to forest removal, regrowth, and roads in 10 small experimental basins, western Cascades, Oregon. *Water Resources Research* 36: 2621–2642.

Leavesley, G.H., G.C. Lusby, and R.W. Lichty. 1989. Infiltration and erosion characteristics of selected tephra deposits from the 1980 eruption of Mount St. Helens, Washington, USA. *Hydrological Sciences Journal* 34: 339–353.

Lipman, P.W., and D.R. Mullineaux, eds. 1981. *The 1980 eruptions of Mount St. Helens, Washington*, Professional Paper 1250. Washington, DC: U.S. Geological Survey.

Lisle, T.E. 1995. Effects of coarse woody debris and its removal on a channel affected by the 1980 eruption of Mount St. Helens, Washington. *Water Resources Research* 31: 1797–1808.

Lombard, R.E., M.B. Miles, L.M. Nelson, D.L. Kresch, and P.J. Carpenter. 1981. The impact of mudflows of May 18 on the lower Toutle and Cowlitz Rivers. In *The 1980 eruptions of Mount St. Helens, Washington*, Professional Paper 1250, ed. P.W. Lipman and D.R. Mullineaux, 693–699. Washington, DC: U.S. Geological Survey.

Magirl, C.S., R.C. Hilldale, C.A. Curran, J.J. Duda, T.D. Straub, M. Domanski, and J.R. Foreman. 2015. Large-scale dam removal on the Elwha River, Washington, USA: Fluvial sediment load. *Geomorphology* 246: 669–686.

Major, J.J. 2004. Posteruption suspended sediment transport at Mount St. Helens: Decadal-scale relationships with landscape adjustments and river discharges. *Journal of Geophysical Research* 109: F01002. https://doi.org/10.1029/2002JF000010.

Major, J.J., and L.E. Lara. 2013. Overview of Chaitén Volcano, Chile, and its 2008–2009 eruption. *Andean Geology* 40: 196–215.

Major, J.J., and L.E. Mark. 2006. Peak flow responses to landscape disturbances caused by the cataclysmic 1980 eruption of Mount St. Helens, Washington. *Geological Society of America Bulletin* 118: 938–958.

Major, J.J., and K.M. Scott. 1988. *Volcaniclastic sedimentation in the Lewis River valley, Mount St. Helens, Washington—processes, extent, and hazards*, Bulletin 1383-D. Washington, DC: U.S. Geological Survey.

Major, J.J., and T. Yamakoshi. 2005. Decadal-scale change of infiltration characteristics of a tephra-mantled hillslope at Mount St. Helens, Washington. *Hydrological Processes* 19: 3621–3630.

Major, J.J., T.C. Pierson, R.L. Dinehart, and J.E. Costa. 2000. Sediment yield following severe volcanic disturbance—a two-decade perspective from Mount St. Helens. *Geology* 28: 819–822.

Major, J.J., J.E. O'Connor, C.J. Podolak, M.K. Keith, G.E. Grant, K.R. Spicer, S. Pittman, H.M. Bragg, J.R. Wallick, D.Q. Tanner, A. Rhode, and P.R. Wilcock. 2012. *Geomorphic response of the Sandy River, Oregon, to removal of Marmot Dam*, Professional Paper 1792. Washington, DC: U.S. Geological Survey.

Manville, V. 2002. Sedimentary and geomorphic responses to ignimbrite emplacement: Readjustment of the Waikato River after the AD 181 Taupo eruption, New Zealand. *Journal of Geology* 110: 519–541.

Manville, V., K.A. Hodgson, and I.A. Nairn. 2007. A review of break-out floods from volcanogenic lakes in New Zealand. *New Zealand Journal of Geology and Geophysics* 50: 131–150.

Manville, V., B. Segschneider, E. Newton, J.D.L. White, B.F. Houghton, and C.J. Wilson. 2009. Environmental impact of the 1.8 ka Taupo eruption, New Zealand—Landscape responses to a large-scale explosive rhyolite eruption. *Sedimentary Geology* 220: 318–336.

Martinson, H.A., S.D. Finneran, and L.J. Topinka. 1984. *Changes in channel geomorphology of six eruption-affected tributaries of the Lewis River, 1980–82, Mount St. Helens, Washington*, Open-File Report 84-614. Washington, DC: U.S. Geological Survey.

Martinson, H.A., H.E. Hammond, W.W. Mast, and P.D. Mango. 1986. *Channel geometry and hydrologic data for six eruption-affected tributaries of the Lewis River, Mount St. Helens, Washington, water years 1983–84*, Open-File Report 85-631. Washington, DC: U.S. Geological Survey.

Maune, D.F., J. Binder Maitra, and E.J. McKay. 2001. Accuracy standards. In *Digital elevation model technologies and applications: The DEM users manual*, ed. D.F. Maune, 61–82. Bethesda: American Society for Photogrammetry and Remote Sensing.

McDonnell, J.J. 2003. Where does water go when it rains? Moving beyond the variable source area concept of rainfall-runoff response. *Hydrological Processes* 17: 1869–1875.

McGuire, K.J., J.J. McDonnell, M. Weiler, C. Kendall, B.L. McGlynn, J.M. Welker, and J. Siebert. 2005. The role of topography on catchment-scale water residence time. *Water Resources Research* 41: W05002. https://doi.org/10.1029/2004WR003657.

Meadows, T. 2014. Forecasting long-term sediment yield from the upper North Fork Toutle River, Mount St. Helens, USA. PhD. Thesis. Nottingham: University of Nottingham.

Meyer, D.F. 1995. *Stream-channel changes in response to volcanic detritus under natural and augmented discharge, South Coldwater Creek, Washington*, Open-File Report 94-519. Washington, DC: U.S. Geological Survey.

Meyer, D.F., and J.E. Dodge. 1988. *Post-eruption changes in channel geometry of streams in the Toutle River drainage basin, 1983–85, Mount St. Helens, Washington*, Open-File Report 87-549. Washington, DC: U.S. Geological Survey.

Meyer, D.F., and H.A. Martinson. 1989. Rates and processes of channel development and recovery following the 1980 eruption of Mount St. Helens, Washington. *Hydrological Sciences Journal* 34: 115–127.

Meyer, D.F., K.M. Nolan, and J.E. Dodge. 1986. *Post-eruption changes in channel geometry of streams in the Toutle River drainage basin, 1980–82, Mount St. Helens, Washington*, Open-File Report 85-412. Washington, DC: U.S. Geological Survey.

Miller, J.F., R.H. Frederick, and R.J. Tracey. 1973. *NOAA atlas 2, precipitation—frequency atlas of the western United States, Volume IX–Washington*. Silver Springs: U.S. Department of Commerce, National Oceanic and Atmospheric Administration, National Weather Service.

Moore, J.G., and T.W. Sisson. 1981. Deposits and effects of the May 18 pyroclastic surge. In *The 1980 eruptions of Mount St. Helens, Washington*, Professional Paper 1250, ed. P.W. Lipman and D.R. Mullineaux, 421–438. Washington, DC: U.S. Geological Survey.

Mosbrucker, A.R. 2014. *High-resolution digital elevation model of Mount St. Helens crater and upper North Fork Toutle basin, Washington, based on an airborne lidar survey of September 2009*, Data Series 904. Washington, DC: U.S. Geological Survey. https://doi.org/10.3133/ds904 .

Mosbrucker, A.R., K.R. Spicer, J.J. Major, D.R. Saunders, T.S. Christianson, and C.G. Kingsbury. 2015. *Digital database of channel cross-section surveys, Mount St. Helens, Washington*, Data Series 951. Washington, DC: U.S. Geological Survey. https://doi.org/10.3133/ds951.

Newhall, C.G., and R.S. Punongbayan, eds. 1996. *Fire and mud—eruptions and Lahars of Mount Pinatubo, Philippines*. Seattle: University of Washington Press.

O'Connor, J.E., P.F. McDowell, P. Lind, C.G. Rasmussen, and M.K. Keith. 2013. *Geomorphology and flood-plain vegetation of the Sprague and lower Sycan Rivers, Klamath basin, Oregon*, Scientific Investigations Report 2014-5223. Washington, DC: U.S. Geological Survey. http://doi.org/10.3133/sir20145223.

Paine, A.D.M., D.F. Meyer, and S.A. Schumm. 1987. Incised channel and terrace formation near Mount St. Helens, Washington. In *Erosion and Sedimentation in the Pacific Rim*, Publication 165, ed. R.L. Beschta, T. Blinn, G.E. Grant, F.J. Swanson, and G.G. Ice, 389–390. Christchurch: International Association of Hydrological Sciences.

Pierson, T.C., and J.J. Major. 2014. Hydrogeomorphic effects of explosive volcanic eruptions on drainage basins. *Annual Review of Earth and Planetary Sciences* 42: 469–507.

Pierson, T.C., P.T. Pringle, and K.A. Cameron. 2011. Magnitude and timing of downstream channel aggradation and degradation in response to a dome-building eruption at Mount Hood, Oregon. *Geological Society of America Bulletin* 123: 3–20.

Pitlick, J., J.J. Major, and K. Spicer. 2006. Morphodynamics of the North Fork Toutle River near Mount St. Helens. *EOS, Transactions of the American Geophysical Union*, 87: Abstract H51G-0564.

Pringle, P.T., and K.A. Cameron. 1999. Eruption-triggered lahar on May 14, 1984. In *Hydrological consequences of hot-rock/snowpack interactions at Mount St. Helens Volcano, Washington 1982–84*, Professional Paper 1586, ed. T.C. Pierson, 81–103. Washington, DC: U.S. Geological Survey.

Pringle, P., and K. Scott. 2001. Postglacial influence of volcanism on the landscape and environmental history of the Puget Lowland, Washington—A review of geologic literature and recent discoveries, with emphasis on the landscape disturbances associated with lahars, lahar runouts, and associated flooding. In *Proceedings of the 2001 Puget Sound research conference*. ed. T. Droscher. Olympia: Puget Sound Water Quality Action Team. http://archives.eopugetsound.org/conf/2001PS_ResearchConference/sessions/oral/4d_pring.pdf. Last accessed 10 Feb 2014.

Roering, J.J., J. Marshall, A.M. Booth, M. Mort, and Q. Jin. 2010. Evidence for biotic controls on topography and soil production. *Earth and Planetary Science Letters* 298: 183–190.

Rosenfeld, C.L., and G.L. Beach. 1983. *Evolution of a drainage network—remote sensing analysis of the North Fork Toutle River, Mount St. Helens, Washington*, Water Resources Research Institute Report WRRI-88. Corvallis: Oregon State University Press.

Rowley, P.D., M.A. Kuntz, and N.S. MacLeod. 1981. Pyroclastic-flow deposits. In *The 1980 eruptions of Mount St. Helens, Washington*, Professional Paper 1250, ed. P.W. Lipman and D.R. Mullineaux, 489–512. Washington, DC: U.S. Geological Survey.

Sarna-Wojcicki, A.M., S. Shipley, R.B. Waitt Jr., D. Dzurisin, and S.H. Wood. 1981. Areal distribution, thickness, mass, volume, and grain size of air-fall ash from the six major eruptions of 1980. In *The 1980 eruptions of Mount St. Helens, Washington*, Professional Paper 1250, ed. P.W. Lipman and D.R. Mullineaux, 577–600. Washington, DC: U.S. Geological Survey.

Schuster, R.L. 1981. Effects of the eruptions on civil works and operations in the Pacific Northwest. In *The 1980 eruptions of Mount St. Helens, Washington*, Professional Paper 1250, ed. P.W. Lipman and D.R. Mullineaux, 701–718. Washington, DC: U.S. Geological Survey.

Scott, K.M. 1988. *Origins, behavior, and sedimentology of lahars and lahar-runout flows in the Toutle-Cowlitz River system*, Professional Paper 1447-A. Washington, DC: U.S. Geological Survey.

Sigurdsson, H., S.N. Carey, and J.M. Epindola. 1984. The 1982 eruptions of El Chichón Volcano, Mexico: Stratigraphy of pyroclastic deposits. *Journal of Volcanology and Geothermal Research* 23: 11–37.

Sigurdsson, H., B. Houghton, S. McNutt, H. Rymer, and J. Stix, eds. 2015. The encyclopedia of volanoes, 2nd edition. Amsterdam: Academic Press.

Simon, A. 1992. Energy, time, and channel evolution in catastrophically disturbed fluvial systems. *Geomorphology* 5: 345–372.

———. 1999. *Channel and drainage-basin response of the Toutle River system in the aftermath of the 1980 eruption of Mount St.*

Helens, Washington, Open-File Report 96–633. Washington, DC: U.S. Geological Survey.

Simon, A., and C.R. Thorne. 1996. Channel adjustment of an unstable coarse-grained stream: Opposing trends of boundary and critical shear stress, and the applicability of extremal hypotheses. *Earth Surface Processes and Landforms* 21: 155–180.

Surono, P. Jousset, J. Pallister, M. Boichu, M.F. Buogiorno, A. Budisantoso, F. Costa, S. Andreasuti, F. Prata, D. Schneider, L. Clarisse, H. Humaida, S. Sumarti, C. Bignami, J. Griswold, S. Carn, C. Oppenheimer, and F. Lavigne. 2012. The 2010 explosive eruption of Java's Merapi volcano—A '100-year' event. *Journal of Volcanology and Geothermal Research* 241-242: 121–135.

Swanson, F.J., and J.J. Major. 2005. Physical events, environments, and geological-ecological interactions at Mount St. Helens—March 1980–2004. In *Ecological Responses to the 1980 Eruption of Mount St. Helens*, ed. V.H. Dale, F.J. Swanson, and C.M. Crisafulli, 27–44. New York: Springer.

Swanson, F.J., B.D. Collins, T. Dunne, and B.P. Wicherski. 1983. Erosion of tephra from hillslopes near Mount St. Helens and other volcanoes. In *Proceedings of symposium on erosion control in volcanic areas*, Public Works Research Institute Technical Memorandum 1908, 183–221. Tsukuba: Ministry of Construction.

Townsend, J.R. 2013. The development of a geomatics-based toolkit to assess the impact of engineered grade building structures on the North Fork Toutle River, Mt. St. Helens. MS thesis. Nottingham: University of Nottingham.

Voight, B. 1981. Time scale for the first moments of the May 18 eruption. In *The 1980 eruptions of Mount St. Helens, Washington*, Professional Paper 1250, ed. P.W. Lipman and D.R. Mullineaux, 69–86. Washington, DC: U.S. Geological Survey.

Voight, B., H. Glicken, R.J. Janda, and P.M. Douglass. 1981. Catastrophic rockslide avalanche of May 18. In *The 1980 eruptions of Mount St. Helens, Washington*, Professional Paper 1250, ed. P.W. Lipman and D.R. Mullineaux, 347–377. Washington, DC: U.S. Geological Survey.

Waitt, R.B. 1981. Devastating pyroclastic density flow and attendant air fall of May 18—stratigraphy and sedimentology of deposits. In *The 1980 eruptions of Mount St. Helens, Washington*, Professional Paper 1250, ed. P.W. Lipman and D.R. Mullineaux, 439–460. Washington, DC: U.S. Geological Survey.

———. 1989. Swift snowmelt and floods (lahars) caused by great pyroclastic surge at Mount St. Helens volcano, Washington, 18 May 1980. *Bulletin of Volcanology* 52: 138–157.

———. 2015. *In the path of destruction—eyewitness chronicles of Mount St. Helens*. Pullman: Washington State University Press.

Waitt, R.B., T.C. Pierson, N.S. MacLeod, R.J. Janda, B. Voight, and R.T. Holcomb. 1983. Eruption-triggered avalanche, flood, and lahar at Mount St. Helens—Effects of winter snowpack. *Science* 221: 1394–1397.

Warrick, J.A., J.A. Bountry, A.E. East, C.S. Magirl, T.J. Randle, G. Gelfenbaum, A.C. Ritchie, G.R. Pess, V. Leung, and J.J. Duda. 2015. Large-scale dam removal on the Elwha River, Washington, USA: Source-to-sink sediment budget and synthesis. *Geomorphology* 246: 729–750.

West Consultants, Inc. 2002. Mount St Helens engineering reanalysis: Hydrologic, hydraulic, and sedimentation analysis. Technical report prepared for U.S. Army Corps of Engineers Portland District. Bellevue, WA: West Consultants.

Western Regional Climate Center. 2015. Historical climate information, western U.S. climate data summaries. www.wrcc.dri.edu/climatedata/comparative/. Last accessed 14 May 2015.

Wheaton, J.M., J. Brasington, S.E. Darby, and D.A. Sear. 2010. Accounting for uncertainty in DEMs from repeat topographic surveys: Improved sediment budgets. *Earth Surface Processes and Landforms* 35: 136–156.

White, J.D.L., B.F. Houghton, K.A. Hodgson, and C.J.N. Wilson. 1997. Delayed sedimentary response to the AD 1886 eruption of Tarawera, New Zealand. *Geology* 25: 459–462.

Wickert, A.D., J.M. Martin, M. Tal, W. Kim, B. Sheets, and C. Paola. 2013. River channel lateral mobility: Metrics, time scales, and controls. *Journal of Geophysical Research – Earth Surface* 118: 396–412. https://doi.org/10.1029/2012JF002386.

Wilcox, A.C., J.E. O'Connor, and J.J. Major. 2014. Rapid reservoir erosion, hyperconcentrated flow, and downstream deposition triggered by breaching of 38 m tall Condit Dam, White Salmon River, Washington. *Journal of Geophysical Research – Earth Surface* 119: 1376–1394.

Willingham, W.F. 2005. The Army Corps of Engineers' short-term response to the eruption of Mount St. Helens. *Oregon Historical Quarterly* 106 (2): 174–203.

Winner, W.E., and T.J. Casadevall. 1981. Fir leaves as thermometers during the May 18 eruption. In *The 1980 eruptions of Mount St. Helens, Washington*, Professional Paper 1250, ed. P.W. Lipman and D.R. Mullineaux, 315–320. Washington, DC: U.S. Geological Survey.

Wohl, E. 2000. *Mountain rivers*, Water Resources Monograph 14. Washington, DC: American Geophysical Union.

Yamakoshi, T., Y. Doi, and N. Osanai. 2005. Post-eruption hydrology and sediment discharge at the Miyakejima volcano, Japan. *Zeitschrift für Geomorphologie Supplement Band* 140: 55–72.

Zehfuss, P.H., B.F. Atwater, J.W. Vallance, H. Brenniman, and T.A. Brown. 2003. Holocene lahars and their by-products along the historical path of the White River between Mount Rainier and Seattle. In *Western cordilleran and adjacent areas*, Field Guide 4, ed. T.W. Swanson, 209–223. Boulder: Geological Society of America.

Zheng, S., B. Wu, C. Thorne, and A. Simon. 2014. Morphological evolution of the North Fork Toutle River following the eruption of Mount St. Helens, Washington. *Geomorphology* 208: 102–116.

Geomorphic Response of the Muddy River Basin to the 1980 Eruptions of Mount St. Helens, 1980–2000

3

Thomas E. Lisle, Jon J. Major, and Jasper H. Hardison III

3.1 Introduction

Volcanic eruptions can impose overwhelming changes on landscapes, and responses to those changes can play out over decades as recently deposited volcanic sediment, mobilized older sediment, and large wood are routed through drainage basins. Disturbances to and subsequent responses of hillslopes, stream channels, and depositional landforms set the physical template for trajectories of terrestrial, riparian, and aquatic ecosystem evolution. The 18 May 1980 eruption of Mount St. Helens, which altered a broad swath of landscape, consisted of an ensemble of volcanic processes that included a large rockslide–***debris avalanche***, a devastating laterally directed pyroclastic density current (commonly called the lateral blast and herein called the ***blast PDC***), ***lahars*** (volcanic debris flows), localized pumiceous ***pyroclastic flows***, and widespread fall of volcanic ash (***tephrafall***) downwind of the volcano (Lipman and Mullineaux 1981; see also Pierson and Major 2014 for volcanic process terminology). Numerous studies document the intensities, distributions, and time trends of subsequent geomorphic responses, including hillslope erosion (Collins et al. 1983; Swanson et al. 1983; Collins and Dunne 1986, 1988; Leavesley et al. 1989), drainage development (Rosenfeld and Beach 1983; Janda et al. 1984; Collins and Dunne 1986; Simon 1999), evolution of channel morphology (Lisle et al. 1983; Janda et al. 1984; Meyer and Martinson 1989; Lisle 1995; Simon 1999; Hardison 2000; Frenzen et al. 2005), sediment yield (Lehre et al. 1983; Smith and Swanson 1987; Dinehart 1998; Major et al. 2000; Major 2004), and peak runoff (Major and Mark 2006).

This eruption provides an exceptional opportunity to observe and document responses to types of disturbances and their gradients of intensity that extend away from a volcano, and the propagation of those disturbances downstream by fluxes of sediment and wood. Disturbance gradients caused by the blast PDC, lahars, and tephrafall are well represented in the Muddy River drainage basin, which has headwaters that lie east–northeast of the volcano (Figs. 3.1 and 3.2). There, these processes deposited only tens of centimeters to a few meters of sediment on the landscape, compared with the tens to hundreds of meters of fill deposited in the adjacent Toutle River basin by the debris avalanche and subsequent pyroclastic flows and lahars (Voight et al. 1981; Janda et al. 1981; Rowley et al. 1981; Fig. 3.2). Thus, responses to disturbances caused by the eruption in the Muddy River basin may be more representative of the aftermath of eruptions elsewhere, in which volcanic deposits are relatively thin (≤5 m).

This chapter reports on geomorphic change in the Muddy River basin over the first 20 years following the 18 May 1980 eruption. In the Muddy River basin, transport and deposition of sediment and wood were great enough during this period to reveal the salient behaviors of the primary components of the sedimentary system, a system that consists of an array of sediment sources and temporary storage reservoirs connected by transport pathways having various lengths and degrees of connectivity. Fluxes of sediment and wood were contingent initially on an altered flow regime, and later on the timing and magnitude of large floods as well as on evolv-

Glossary terms appear in ***bold italic face***.

T.E. Lisle (✉)
U.S. Department of Agriculture, Forest Service, Pacific Southwest Research Station,
1700 Bayview Drive, Arcata, CA 95521-6013, USA
e-mail: thomas.lisle@gmail.com

J.J. Major
Cascades Volcano Observatory, U.S. Geological Survey,
1300 SE Cardinal Court, Bldg 10, Suite 100, Vancouver, WA 98683, USA
e-mail: jjmajor@usgs.gov

J.H. Hardison III
Realty and Natural Resources Division, Water Resources Branch, U.S. Fish and Wildlife Service,
1011 E. Tudor Rd., MS 211, Anchorage, AK 99503, USA
e-mail: Jasper_hardison@fws.gov

C.M. Crisafulli, V.H. Dale (eds.), *Ecological Responses at Mount St. Helens: Revisited 35 years after the 1980 Eruption*,
https://doi.org/10.1007/978-1-4939-7451-1_3

Fig. 3.1 Pre-1980 topographic map of the Mount St. Helens area showing the fill of volcaniclastic sediment that has subdued the rugged topography atop which the volcano has grown. Locations of study reaches in the Clearwater Creek catchment and sites of cross sections in the Smith Creek and Muddy River corridors are shown. *1* Clearwater #1. *2* Clearwater #2. *3* Clearwater #3. *4* Clearwater #4. *LC* lower Clearwater. Swift Reservoir is not shown because the map predates construction of Swift Dam.

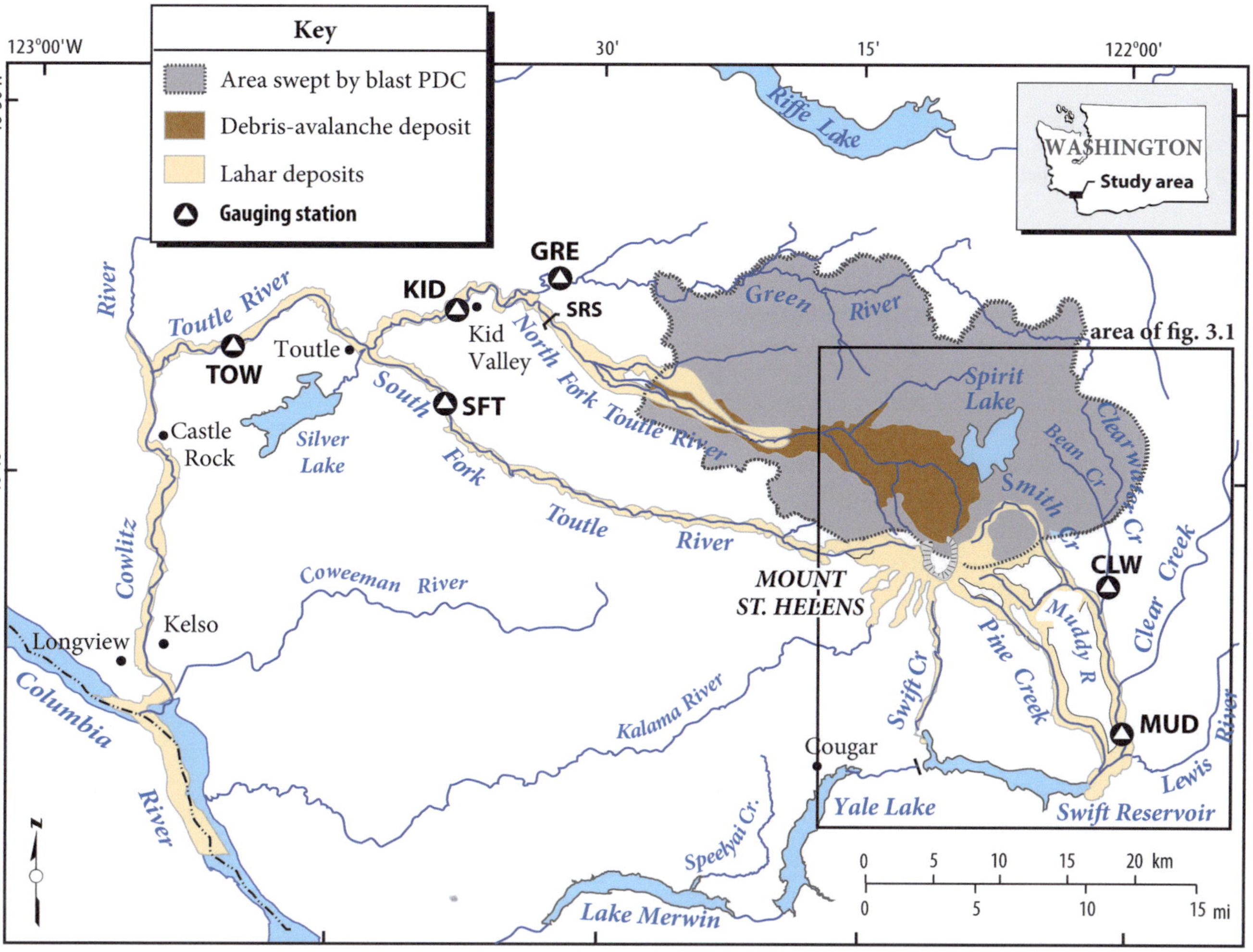

Fig. 3.2 Schematic depiction of distribution of volcanic deposits and disturbance zones of the 18 May 1980 eruption (adapted from plate 1 in Lipman and Mullineaux 1981). Box outlines the area shown in Fig. 3.1. Gauging stations (e.g., MUD) indicate sites where the USGS made sediment and water discharge measurements.

ing geomorphic and vegetative conditions after the initial disturbance. What makes this study particularly interesting is (1) the influence of gradients of eruptive disturbance on channel response and (2) the occurrence of an intense storm and large flood 16 years after the eruption. That storm and flood accelerated geomorphic processes in the sedimentary system, interrupted the recovery trajectories of channels and riparian ecosystems, and highlighted influences on sediment routing that were inherent in the physiography of the basin.

A broad overview of physical mechanisms causing landscape disturbance and of interactions between geomorphic and ecological responses in the entire area affected by the 18 May eruption from 1980 to 2004 is provided by Swanson and Major (2005). Here we focus on responses in the Muddy River basin, which includes five major tributary catchments (Muddy River, Smith Creek, Bean Creek, Clearwater Creek, and Clear Creek) that are arrayed normal to a southeast-to-northwest trending gradient of primary volcanic disturbance (Figs. 3.1 and 3.2). The types and degrees of disturbance in these catchments affected hydrologic and geomorphic processes that governed the subsequent flux and storage of sediment, which in turn affected valley-floor ecosystems. The eruption activated basin-scale responses in the sedimentary system which can be explained by considering basin physiography. At the reach scale, hydrogeomorphic responses to eruptive disturbances created contrasting conditions and trajectories of change for aquatic and riparian ecosystems.

In this chapter, we provide a broad description of the study area and the primary effects of the eruption on catchments within the Muddy River basin. We then provide sources and methods of data collection, and describe post-eruption changes in each of the catchments. Finally, we integrate results by considering landscape geography as a template that influences geomorphic disturbance and the trajectory of geomorphic response. We also highlight the influence of stochastic events and touch on the consequences of physical changes for aquatic and riparian ecosystems. We close with an evaluation of trajectories of geomorphic processes in the study area.

3.2 Study Area

Mount St. Helens is the most active volcano of the Cascade Range, and its edifice is a relatively youthful feature on the landscape (Mullineaux and Crandell 1981; Clynne et al. 2008). Over the last 4000 years, the volcano has entered significant eruptive periods about every two to seven centuries. The last eruption prior to 1980 occurred in the early half of the nineteenth century.

The volcano sits on a deeply eroded terrane of gently folded and altered volcanic and plutonic rocks (Fig. 3.1; Evarts et al. 1987; Evarts and Swanson 1994). Close to the volcano, fragmental deposits of eruption products that flow from the cone, such as debris avalanches, pyroclastic flows, and lahars, have accumulated thickly along drainages and buried much of the rugged topography. Deposits of other products of explosive eruptions, such as tephrafalls and highly mobile PDCs, are thinly draped over a wider area. The majority of the Muddy River basin, including the catchments of Smith Creek, Bean Creek, Clearwater Creek, and Clear Creek, is unburied and deeply carved into the old volcanic terrane (Fig. 3.1).

The Muddy River basin extends 20 km from Mount St. Helens. (Fig. 3.1). The rugged basin topography has been modified by volcanism and glaciation. Elevation ranges from 315 to 1500 m. Most drainages heading above 900 m were glaciated (Mundorff 1984), which created a stepped topography composed of shallow-gradient reaches hosting alluvial deposits, interspersed with steeper-gradient reaches. The topography is somewhat muted by accumulated tephra deposits, except for close to the volcano where valleys are thickly buried in volcanic sediment. Coniferous forest, including large mature stands and young conifers that were planted following clearcutting, dominated the landscape before the eruption; deciduous trees were prevalent in both riparian and disturbed areas (Swanson et al. 2005).

3.2.1 Basin Physiography

Owing to variations in basin topography, contrasting tendencies to transport or store sediment set the stage for different responses to eruption impacts. Volcanic processes transporting sediment down the flanks of the cone generally spread sediment widely over fans at the base of the edifice (Fig. 3.1). Subsequent reworking of that sediment is commonly confined to a small portion of the original depositional surface. In contrast, sediment-transport pathways in the older more dissected terrain, which includes our main study area, are concentrated in narrow valleys (Fig. 3.1) where a greater proportion of deposited sediment is subject to reworking. In this dissected terrain, channels and adjacent hillslopes are strongly coupled, allowing sediment to move directly from hillslopes to channels. In reaches where channels are incised into broad valley floors, sediment inputs from hillslopes as well as from channels upstream are buffered from entering the channel, because adjacent floodplains, terraces, and fans offer ample space for sediment storage.

3.2.1.1 Smith Creek–Muddy River Corridor

The Smith Creek–Muddy River corridor (Fig. 3.1) has been formed in deposits of lahars and pyroclastic flows (Crandell 1987). A fan of sediment deposited by Muddy River at the confluence of Smith Creek constricted the valley bottom and created a swampy, low-gradient reach upstream. The valley bottom between the Muddy River fan and Cedar Flats is 250–400 m wide. It is formed on a fill of pre-1980 pyroclastic-flow and lahar deposits tens of meters thick (Crandell 1987) and contains terraces, floodplains, channels with multiple threads, and alluvial fans. The channel at Cedar Flats had incised into the thick fill of pre-1980 deposits at the junction of Pine Creek and Muddy River and had locally scoured to bedrock, creating a narrow valley bottom 5–250 m wide (Hardison 2000). Valley gradients immediately above Cedar Flats and the Muddy River fan at Smith Creek are shallower than those farther upstream.

3.2.1.2 Bean Creek–Clearwater Creek Corridor

The Bean Creek and Clearwater Creek corridors (Fig. 3.1) show greater contrasts in sediment-routing characteristics than does the Smith Creek–Muddy River corridor. Bean Creek consists of a gently sloping upper catchment having high sediment-storage potential. The upper catchment empties into a bedrock gorge (2.2 km long) which leads into another depositional reach (2.2 km) between steep valley walls. This lower depositional reach connects to a longer gorge (2.5 km) that feeds into the junction with Clearwater Creek (Fig. 3.1). These gorges afford direct connectivity between sediment sources on the steep valley walls and the channel. Because the Bean Creek channel is generally steep and confined, it transmits most of its sediment directly to lower Clearwater Creek.

The broad catchment of middle Clearwater Creek valley upstream of Paradise Falls (Fig. 3.1) has a large storage capacity. An active floodplain commonly several channel-widths wide is bounded by terraces and alluvial fans. High-gradient tributaries drain the steep valley walls. However, extensive sediment-storage capacity of the channel bars, floodplain, and alluvial fans buffers the channel from direct sediment inputs from hillslopes and tributaries. Sediment tends to be spread among a variety of storage reservoirs and does not transit rapidly through the system. The channel in this reach is sinuous and has an alluvial bed and banks. Large individual pieces of wood and wood jams influence the formation and location of bars and pools. The lower Clearwater Creek valley downstream of the junction with Bean Creek is also broad and has a high potential for storage of wood and sediment.

3.2.2 Climate

The climate in the Pacific Northwest is maritime, with generally wet winters and dry, cool summers. Mean annual precipitation at Mount St. Helens is approximately 3000 mm, with 75% falling between October and April. Much of the precipitation above 1000 m in elevation falls as snow. Typical snowpack is 2–3 m thick, and the area above 1000 m is under snow cover for about 6 months each year. The largest floods are produced by intense atmospheric rivers of tropical water vapor directed at the region (Neiman et al. 2011) and by rain-on-snow events in winter; snowmelt produces moderate peaks and sustained recessional flows in late spring.

3.3 Primary Effects of the 1980 Eruptions

The 18 May 1980 eruption consisted of an ensemble of volcanic processes that caused a variety of disturbances among several watersheds (Fig. 3.2). Details of the eruption are provided in Lipman and Mullineaux (1981). Here, we briefly summarize the events. The eruption began with a massive failure of the volcano's north flank, which triggered an enormous rockslide–debris avalanche (Voight et al. 1981). That debris avalanche deposited 2.5 km^3 of rock, ice, soil, and organic debris in the upper North Fork Toutle River valley, Spirit Lake, and South Coldwater Creek valley (Fig. 3.2). Furthermore, it suddenly released pressure on the magma that had intruded high into the volcano. Unroofing of that magma and consequent release of volatiles in the magmatic system and flashing of the hydrothermal system triggered an energetic pyroclastic density current (blast PDC) that devastated ~570 km^2 of rugged, forested landscape in a roughly 180-degree arc north of the volcano (Fig. 3.2). That PDC stripped vegetation and soil from the landscape close to the volcano and toppled mature forest farther out. It also blanketed the landscape with as much as 1.5 m of graded gravel to silty sand, including organic debris; locally, some of the deposit slid off steep hillslopes and onto valley floors (Hoblitt et al. 1981; Waitt 1981; Brantley and Waitt 1988). Extensive lahars followed the blast PDC, swept all major channels draining the volcano (Fig. 3.2), and deposited tens to hundreds of centimeters of gravelly sand on valley floors and floodplains (Janda et al. 1981; Pierson 1985; Scott 1988). Tephrafall from a billowing eruption plume, which developed shortly after the onset of the eruption, blanketed proximal areas east–northeast of the volcano with a layer of ash (particles $\leq$2 mm) and gravelly pumice as thick as tens of centimeters (Waitt and Dzurisin 1981). Beyond about 15 km east of the volcano, accumulations of tephrafall constituted the primary landscape disturbance in many catchments.

Disturbances caused by the eruption east of the volcano varied by catchment (Fig. 3.2). The Muddy River and Smith Creek catchments were affected by the blast PDC, lahars, and tephrafall. The Bean Creek and Clearwater Creek catchments were affected only by the blast PDC and tephrafall. They were not affected by lahars, but were affected by sediment-laden water flows triggered by snowmelt. Of the 570 km^2 of landscape ravaged by the blast PDC, 20% lies within the Muddy River basin. The blast PDC extended over about 70% of the Smith Creek catchment, 100% of the Bean Creek catchment, and about 45% of the Clearwater Creek catchment above the Bean Creek confluence (Fig. 3.2). It uprooted and broke trees in a complex, topographically influenced pattern and introduced large volumes of wood to stream channels as mature trees toppled. Its deposits mantled hillslopes and are as thick as 60–70 cm in upper Smith Creek, more than 50 cm thick in the headwaters of Bean Creek, and less than 1 cm thick near their northeastern limit in Clearwater Creek catchment (Hoblitt et al. 1981; Lipman and Mullineaux 1981; Moore and Sisson 1981).

During and shortly after the eruption, fresh blast-PDC and tephrafall deposits blanketing the landscape were shed locally from steep (>70%) slopes by shallow-seated landslides (Hoblitt et al. 1981; Brantley and Waitt 1988). Deep landslides that mobilized more than the new surficial sediment and the immediately underlying soil formed in older tephrafall deposits were rare (Swanson et al. 1983; Smith and Swanson 1987). Most new landslides identified in June 1980 aerial photographs of the upper Green, Quartz, Smith, Bean, and Clearwater Creek catchments appear to have been triggered by groups of root-thrown trees blown over on steep slopes by the blast PDC (Swanson et al. 1983).

Except for the Smith Creek and Muddy River valleys, channels in catchments affected by the blast PDC were inundated by moderate, sediment-charged floods triggered by eruption-induced snowmelt or during rainfall events soon afterward. Sediment transported by these floods was deposited in channels, on floodplains, and in alluvial fans, commonly around trees toppled by the PDC. Moderately to poorly sorted sand deposits usually <2 m thick and having weak horizontal bedding suggest that the sediment-charged floods may have been hyperconcentrated flows (Pierson 2005).

Tephrafall from the billowing eruption column blanketed much of the Muddy River, Smith Creek, Bean Creek, and Clearwater Creek catchments with ash and pumice gravel.

This tephra deposit is distributed more uniformly across the watersheds than the deposit from the blast PDC, but it ranges in thickness from only about 2 to 40 cm (Waitt and Dzurisin 1981). The plume from another major eruption on 22 July 1980 followed approximately the same downwind trajectory across the Muddy River basin as did the 18 May plume (Waitt et al. 1981). Similar to the tephrafall deposit of 18 May, the 22 July deposit is composed of ash and pumice gravel, but is no more than 10 cm thick.

3.3.1 Smith Creek–Muddy River Lahars

High-velocity lahars were triggered in the Muddy River catchment during the first few minutes of the eruption when the blast PDC scoured, entrained, and melted snow and ice on the volcano's eastern flank (Janda et al. 1981; Pierson 1985; Brantley and Waitt 1988). A notably large lahar (approximately 10^7 m^3 in volume) swept through the Muddy River corridor and reached Swift Reservoir (Pierson 1985); a smaller lahar swept down Smith Creek (Brantley and Waitt 1988) (Fig. 3.2).

In Smith Creek valley, the eruption left a complex suite of PDC, lahar, and secondary pyroclastic-flow deposits (Brantley and Waitt 1988). The blast PDC was followed, within minutes, by snowmelt-triggered lahars that swept down steep tributaries and coalesced into a single flow that moved down valley. The lahar deposit on the valley floor overlies the blast-PDC deposit but is mixed with and overlain by deposits of secondary pyroclastic flows shed from valley walls when unstable blast-PDC deposits on steep slopes were rapidly remobilized (Brantley and Waitt 1988). On the valley floor of Smith Creek, this complex, intricate suite of deposits, as thick as 15 m, shows that concurrent processes affected the landscape.

Lahars triggered by the blast PDC entered the Muddy River corridor along four different tributary channels, coalesced, and surged down the Muddy River channel as the Muddy River lahar (Janda et al. 1981; Pierson 1985). Peak flow along the channel ranged from 10 to 20 m deep, and it moved at peak velocities of 3–23 m s^{-1} (Pierson 1985). Once coalesced, peak discharge, velocity, and depth of the lahar attenuated downstream: maximum discharge decreased from 22,000 m^3 s^{-1} to 5000 m^3 s^{-1} in the 7-km reach from the confluence with Clearwater Creek to the confluence with Clear Creek (Fig. 3.2; Pierson 1985). The deposit of this lahar is typically less than 0.5 m thick along the flow margin and 1–2.5 m thick along the channel axis (Janda et al. 1981; Pierson 1985). The thickest channel deposits are found behind valley bedrock constrictions (Hardison 2000). A tongue of the Muddy River lahar traveled 400 m up Clearwater Creek, deposited 1–2.5 m of coarse sandy sediment behind a wood-laden front (Janda et al. 1981), and raised the base level of the alluvial reach of lower Clearwater Creek, which became a study reach for this investigation. The lahar scoured vegetation and some upper soil layers from the broad fan separating Pine Creek and Muddy River, 2–4 km from the crater rim (Figs. 3.1 and 3.2), where its peak velocity was perhaps 30 m s^{-1} or more (Pierson 1985). Along the margins of inundated areas, riparian vegetation was toppled and some was abraded; in contrast, trees closer to the main axis of flow were sheared off and transported along the advancing flow front as a massive log jam, much of which was deposited at Cedar Flats downstream of Clear Creek (Figs. 3.1 and 3.2; Janda et al. 1981). Much wood was deposited on surfaces that were isolated from later channel processes. The lahar scoured the channel to bedrock in places in the vicinity of Cedar Flats. Deposits of smaller lahars triggered later in the day locally overlie the principal lahar deposit.

3.3.2 Sediment-Laden Flows in the Bean Creek–Clearwater Creek Catchments

No lahars occurred in the Bean Creek or Clearwater Creek catchments, but large volumes of ash, blast-PDC sediment, and pumice gravel entered Clearwater Creek directly from volcanic processes and by subsequent fluvial transport. The head of middle Clearwater Creek valley received the largest volume of fluvial sediment, which was contributed by steeper, more confined tributary channels upstream. Despite the occurrence of large inputs of wood and sediment (mostly sand, along with fine lithic and pumiceous gravel), and flood levels that were elevated by these inputs during and shortly after the eruption, streambanks were not eroded immediately nor were channel patterns affected substantially.

3.4 Data Collection

Our studies of post-eruption geomorphic changes were informed by repeated measurements of water and sediment discharges, channel and floodplain morphology, bed-material size, and wood delivered to channels. Measurements of water and sediment discharges were obtained from US Geological Survey (USGS) records at gauges operated on the Muddy River upstream of Lewis River (#14216500, drainage area 350 km^2) since 1982 and on Clearwater Creek below Bean Creek (#14216300, drainage area 85 km^2) from water years[1] 1982 to 1989. Other data reflect a mix of our own measurements and those of the US Forest Service and USGS.

[1]A water year runs from October 1 to September 30.

3.4.1 Smith Creek–Muddy River

We focus our investigation of channel changes in the Smith Creek–Muddy River corridor on the main valley above and below the confluence of Muddy River with Smith Creek. Multiple cross sections were established in September and October 1980 to monitor geomorphic change and channel adjustments along Smith Creek and Muddy River (Martinson et al. 1984). Cross sections were established in clusters of two to six in order to be able to link changes in riparian conditions, channel slope, channel pattern, and changing sediment caliber with changes in channel geometry (Martinson et al. 1984). Several cross sections, one per cluster, were designated as "indicator" cross sections, which were surveyed after significant storms (Martinson et al. 1986). The network of cross sections was resurveyed annually through 1985. Indicator cross sections were resurveyed in 1994, and a more thorough resurvey of many cross sections was done in 1995 (Hardison 2000). Several cross sections were resurveyed in 1996 following the meteorological flood of record on the Muddy River, and limited resurveys were done in 2002.

3.4.2 Clearwater Creek

The US Forest Service established three study reaches (middle Clearwater #1, middle Clearwater #4, and lower Clearwater) along Clearwater Creek in summer 1980 (Fig. 3.1). Reaches were chosen to observe changes in channel morphology and sediment storage along various sediment-storage reservoirs. Study reaches were established in the area affected by the blast PDC (middle Clearwater #1 and #4) and downstream where sediment eroded from disturbed areas of Clearwater and Bean Creek catchments was expected to accumulate (lower Clearwater). Clustered cross sections and longitudinal channel profiles were monumented and surveyed. Surveys were repeated annually with few exceptions until 1990, and again in 2000 or 2001.

Two additional monitoring reaches (middle Clearwater #2 and #3; Fig. 3.1) were established in 1982 to monitor effects of experimental removal of wood on channel morphology (Lisle 1995). At that time, middle Clearwater reaches #1 and #4 were also extended downstream. Each of the reaches was adapted for experimentation to include three sub-reaches: (1) an upstream control sub-reach where all wood was retained, (2) a sub-reach where all wood was removed, and (3) a sub-reach where only some wood was removed, according to guidelines for commercial salvage. Those guidelines recommended removing wood more than 3 m long and 0.3 m in diameter but not embedded in the channel. Wood removal began in 1982, after an annual survey, and continued episodically through 1985. These reaches were surveyed on the same schedule as that of the earlier established reaches.

Other data including wood volume and location, particle-size distributions of bed material, the area of channel bed covered with sand, and bedload transport rate were collected occasionally. Photos were taken repeatedly at cross sections, although vegetation growth limited comparisons at many locations.

Repeated channel surveys of Bean Creek are limited owing to its remote location. However, the USGS monitored cross sections in two reaches of Bean Creek from November 1981 to June 1984 (Martinson et al. 1984, 1986). Broader observations of geomorphic change, sediment input, and changes in storage within the catchment relied on air photo interpretation.

3.4.3 Clear Creek

Clear Creek represents the outer limit of catchments in this study affected directly by the eruption. It lies outside the area affected by the blast PDC, but it received limited tephrafall consisting of ash and pumice gravel. Five channel cross sections were monumented and surveyed by the US Forest Service in summer 1980, and resurveyed in summer 1981. In 1980, much of the pre-eruption gravel bed was covered by a thin layer of mobile sand and pumiceous gravel. In 1981, more of the pre-eruption gravel bed was exposed, and transport of remobilized sand and pumiceous gravel had diminished. Because the cross-section surveys showed little topographic change at that time, subsequent surveys were discontinued.

3.4.4 Landslide History from Aerial Photographs

Landslide history in Bean Creek and upper Clearwater Creek was constructed from aerial photographs. These catchments were chosen for landslide inventories to investigate downstream effects of large sediment inputs in channels not affected by lahars. Photographs, commonly at a scale of 1:9600, were obtained beginning about 1 month after the eruption and were acquired annually through 1986. Additional photography was flown in 1995, 1996, and 1998. Aside from the first suite, photographs were commonly obtained during lowest flow in September or October; some were obtained following winter high flows in the first few years after the eruption. Areas of individual landslides greater than 100 m^2 in these catchments were measured, and the total area of landslides obtained from photographs taken in 1981, 1985, 1995, and 1998 was summed for each year.

3.5 Post-eruption Geomorphic Response: 1980–2000

3.5.1 Changes in Runoff, Flow Routing, and Erosion

In the aftermath of the 1980 eruptions, the landscape in the upper Muddy River basin was largely deforested and blanketed by unconsolidated deposits of blast PDC and tephrafalls. As a result, rates of runoff, erosion, and suspended-sediment yield increased abruptly and then diminished over the following several years.

Runoff increased as a result of loss of foliar interception of rainfall and snow, reduced evapotranspiration, and the low infiltration capacity of tephra-deposit surfaces (Swanson et al. 1983; Leavesley et al. 1989; Major and Mark 2006; Pierson and Major 2014). The potential increase in precipitation reaching the ground surface owing to loss of foliar interception can be roughly estimated from interception rates of other forests, such as an advanced second-growth redwood forest in north coastal California (Reid and Lewis 2009). There, interception rates averaged 15–20% during prolonged rainstorms and did not vary significantly with rainfall intensity or storm duration. Equivalent increases in stream discharge after clearcut logging have been detected over a broad range of flow magnitudes. Those observations suggest that destruction of a broad swath of forest by the 18 May eruption could have significantly increased the amount of precipitation reaching the ground surface in catchments affected by the blast PDC. Forest destruction also greatly reduced evapotranspiration, which allowed subsurface sediment to remain wetter than usual. Infiltration capacities of new tephra deposits measured at various locations near Mount St. Helens (<10 mm h^{-1}) were an order of magnitude lower than capacities typical of forest soils (Swanson et al. 1983; Leavesley et al. 1989) and less than the intensities of frequent, low-magnitude rainfalls. As a result, atypical overland flow occurred even under common, low-intensity rainfalls.

Atypical overland flow provided abundant direct runoff from hillslopes to channels. Exceptional runoff, combined with changes in the hydraulic efficiency of channels and finer-grained bed material, promoted larger, more erosive floods moving through the affected catchments (Major and Mark 2006; Pierson and Major 2014). Increased peak discharges are attributed to increased flow volume reaching channels and to higher flow velocities that concentrated flood waves. Flow velocities were increased by a loss of hydraulic friction caused by: (1) straightening and smoothing of channels by the 1980 lahars; (2) deposition of sand from lahars, the blast PDC, and tephrafalls; and (3) persistent high concentrations of fine sediment, which reduced turbulent resistance within flows (Janda et al. 1984; Coleman 1986; Cloutier et al. 2006; Major and Mark 2006; Gran et al. 2006).

Changes in runoff and flood routing triggered significant sediment erosion in basins affected by the eruptions. Extensive rilling and gullying developed on hillslopes affected by the blast PDC and tephrafall during the first rainy season after the eruption (Swanson et al. 1983; Collins and Dunne 1988; Swanson and Major 2005). In addition, large amounts of sediment were eroded from channels significantly affected by the eruptions (e.g., Janda et al. 1984; Meyer and Martinson 1989; Simon 1999; Major et al., Chap. 2, this volume). Exceptional amounts of both hillslope and channel erosion triggered sediment yields from disturbed basins that exceeded probable pre-eruption yields by several orders of magnitude (Dinehart 1998; Major et al. 2000).

In general, changes to the hydrological regimes of disturbed basins were short lived compared with the duration of the sediment-erosion response. Within a few years, hillslope infiltration increased as fine tephra at the ground surface was winnowed and the surface was churned by salvage logging, burrowing animals, frost action, vegetation growth, and trampling by large animals (Lehre et al. 1983; Swanson et al. 1983; Collins and Dunne 1988; Major and Yamakoshi 2005). As a result, the landscapes in the disturbed catchments were able to absorb the intensities of most rainfalls, and infiltration-excess overland flow greatly diminished. Consequently, rills and gullies stabilized and began to fill within a few tens of months, even before areas became significantly vegetated, and sediment delivery from hillslopes diminished sharply (Collins and Dunne 1986, 1988; Major et al. 2000). Recovery rates of evapotranspiration and foliar interception have proceeded more slowly as vegetation has become reestablished, and these slower rates of recovery have influenced occurrences of landslides in several catchments. Prolonged sediment runoff owing to altered channel hydraulics and channel erosion, however, has persisted much longer and declined more gradually over the decades since the cataclysmic eruption (e.g., Major et al. 2000; Major 2004). Similarly rapid peaks and declines of hillslope and initial channel erosion followed by more prolonged channel adjustments have been documented following other large volcanic eruptions (e.g., Gran et al. 2011; Pierson and Major 2014).

Although infiltration rates increased rapidly after the eruptions and erosion from rills and gullies swiftly declined, landslides episodically delivered hillslope sediment for several years. Soil stability was reduced by the loss of root strength, first by root-throw of trees toppled by the blast PDC and later by root decay. At four study sites at the heads of landslides, Swanson et al. (1983) found that root-thrown trees accounted for 57–76% of the total diameter-at-breast-height of the stand, evidence of a large loss of soil root strength. With the death of forests in the area affected by the blast PDC, root decay further reduced soil strength even as roots from successional vegetation gradually developed and expanded (Swanson and Major 2005). Measurements from similar forest types following logging suggest that root strength may have decreased to a minimum 7–10 years after the eruption, and would have increased gradually thereafter, depending on vegetation type and rate of growth (Ziemer 1981; Roering et al. 2003). Simultaneously, recovering infiltration capacity and depressed rates of evapotranspiration from new, non-mature vegetation allowed for increased soil moisture. The loss of root strength and gradual increase of soil moisture thus provided a window of increased vulnerability to mass failure of the soil mantle and underlying, pre-1980 tephra in the Muddy River basin in the late 1980s–1990s.

An analysis of mean annual flow shows that a period of relatively dry climatic conditions from 1980 to the mid-1990s preceded a period of wetter-than-average conditions in the mid-to-late 1990s (Major 2004). During the wetter period, shallow landsliding increased in the Muddy River basin (as described in a later section) and suspended-sediment yields in all gauged rivers draining the areas affected by the 1980 eruptions increased. Annual peak flows since 1975 have varied widely with no clear trends (Fig. 3.3), but in early February 1996, record floods were recorded at gauging stations on Muddy River below Clear Creek (USGS #14216500, 1984–2013), East Fork Lewis River (USGS #14222500, 1930–2013), and Cispus River (USGS #14231900, 1930–2013).

3.5.2 Geomorphic Change in the Smith Creek–Muddy River Catchment

3.5.2.1 Post-eruption Adjustments Along Smith Creek

Post-eruption geomorphic change in the Smith Creek–Muddy River corridor was dominated by changes in channel morphology and particularly in bed elevation in Smith Creek (Hardison 2000). As in other catchments, the greatest geomorphic changes occurred swiftly after the eruption and then slowed considerably. Along the Smith Creek valley floor, channel changes began immediately after the eruption during summer low flows. In upper Smith Creek above the confluence with Ape Canyon Creek (Fig. 3.1), short, steep channel reaches (knickpoints) associated with rapid incision formed and migrated upstream during low flow. Incision owing to knickpoint migration locally accounted for up to half of the incision measured in the first 6 years after the eruption (Hardison 2000). Once autumn rains began in earnest in November 1980, very rapid channel incision (up to several meters) and widening (many tens of meters) ensued (e.g., SM030, Figs. 3.1 and 3.4). Beginning in water year 1982, channel changes slowed dramatically (Martinson et al. 1984; Hardison 2000), but persisted for several years, with rejuvenated incision recurring 3–6 years after the eruption. Along lower Smith Creek (below the Ape Canyon Creek confluence), summer low flow immediately after the eruption did not significantly modify the valley floor. The first high flows that began in November 1980 reworked the valley floor in that reach, and until April 1981, the channel along lower Smith Creek mostly aggraded and widened (e.g., SM080, SM100, Figs. 3.1 and 3.4). The valley floor immediately upstream of the Ape Canyon Creek confluence responded in a similar manner (SM060, Figs. 3.1 and 3.4). By September 1981, parts of lower Smith Creek had achieved maximum aggradation—up to 8 m—whereas other parts of the valley (e.g., SM080, Figs. 3.1 and 3.4) did not achieve maximum aggradation until several years later (Martinson et al. 1984; Hardison 2000). Approximately 2 years after the eruptions, much of the channel along lower Smith Creek began to incise, and a pattern of alternating incision and aggradation accompanied by continued widening persisted for several years. By the mid-1980s, the bed surface had changed from one dominated by sand to one armored with gravel and structured with frequent steps formed of cobbles and boulders (Hardison 2000).

3.5.2.2 Post-eruption Adjustments Along Muddy River

In contrast to the geomorphic response observed along Smith Creek, where knickpoint migration caused progressive upstream erosion and consequent downstream aggradation, channel adjustments along Muddy River were initiated simultaneously along the entire valley length (Hardison 2000). Those adjustments began immediately after the eruptions. After passage of the large lahar, the Muddy River below Smith Creek was a braided channel. During subsequent summer low flow, however, valley-wide incision of the Muddy River channel ensued. For the first few years following the 1980 eruptions, the channel bed locally aggraded or incised but largely maintained elevation. During water year 1981, about half of the cross sections monitored showed as much as 1–3 m of incision; the others generally maintained bed elevation or slightly aggraded (Fig. 3.5). During the same period, the channel widened and cross-sectional areas increased by a few tens to a few hundreds of square meters. These variations in incision and aggradation were due in part to subsidiary channels in braided reaches capturing flow and enlarging, leaving other channels less active. By the mid-1990s, moderate incision accompanied by as much as tens of meters of widening persisted locally (Fig. 3.5). Similar to adjustments along Smith Creek, those along Muddy River occurred swiftly and slowed rapidly after the eruptions.

Despite the delivery of the large pulse of sediment from Smith Creek, no discrete sediment wave was observed to spread or translate down the Muddy River corridor. Channel

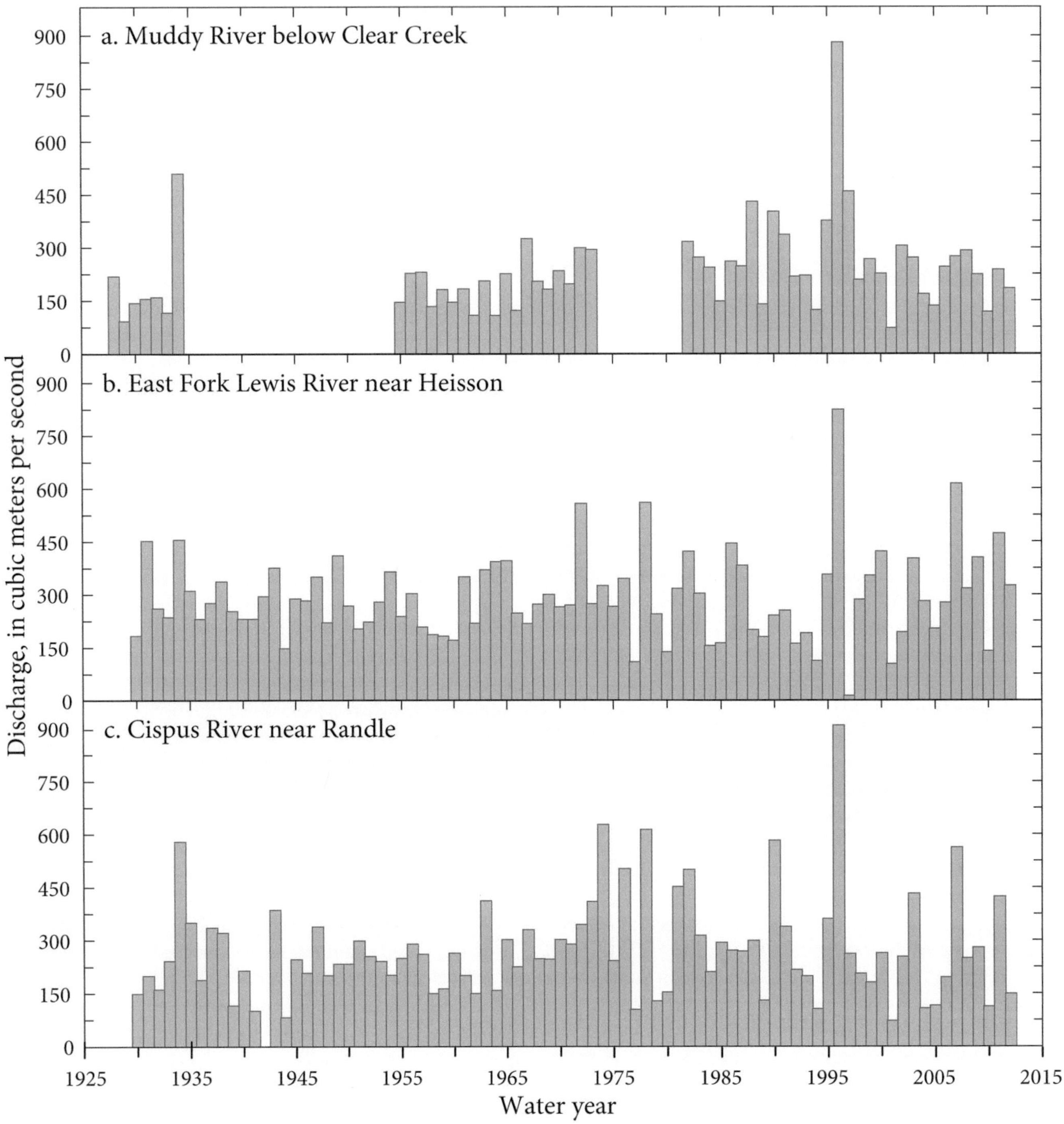

Fig. 3.3 Annual peak discharge measured at various regional USGS gauging stations. Muddy River below Clear Creek (#14216500), East Fork Lewis River near Heisson (#14222500), and Cispus River near Randle (#14231900) in the Cowlitz River basin north of the study area.

bed aggradation followed by degradation did not migrate progressively downstream past cross sections along the Muddy River valley. Instead, channel changes along Muddy River were influenced more by tributary sediment inputs and spatial relationships with slope and valley constrictions than by advancement of a pulse of sediment from Smith Creek (Hardison 2000).

The 1996 flood, despite being the flood of record (1950–2000 for the Muddy River gauge, excluding a data gap from 1974 to 1981), caused only a modest degree of change along the Smith Creek–Muddy River corridor (Figs. 3.4 and 3.5). Between 1986 and 1996, the Smith Creek channel widened by only a few meters to a few tens of meters (e.g., SM030, Fig. 3.4) and mostly exhibited meter-scale incision, although

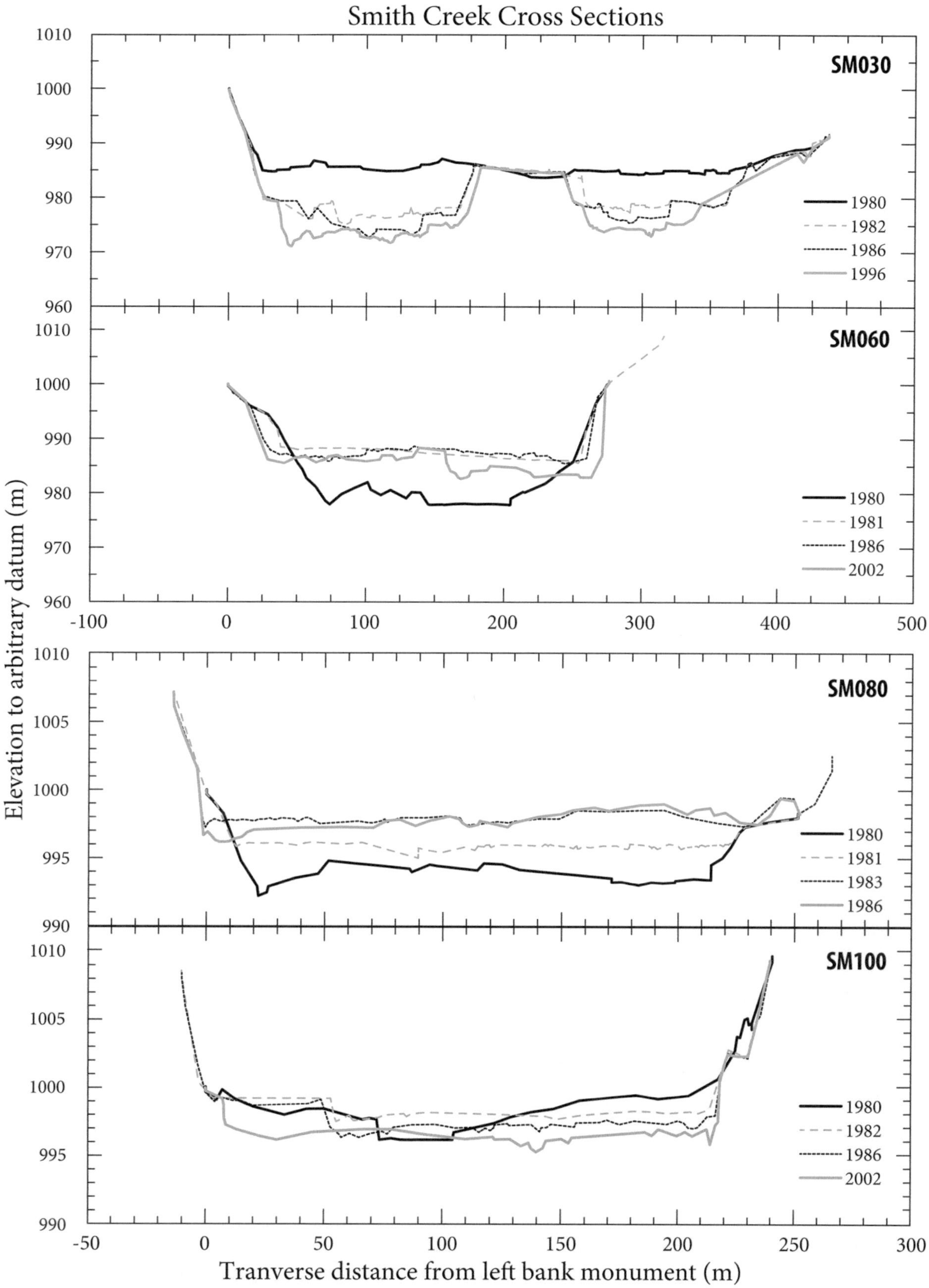

Fig. 3.4 Example cross sections measured along Smith Creek corridor. See Fig. 3.1 for section locations.

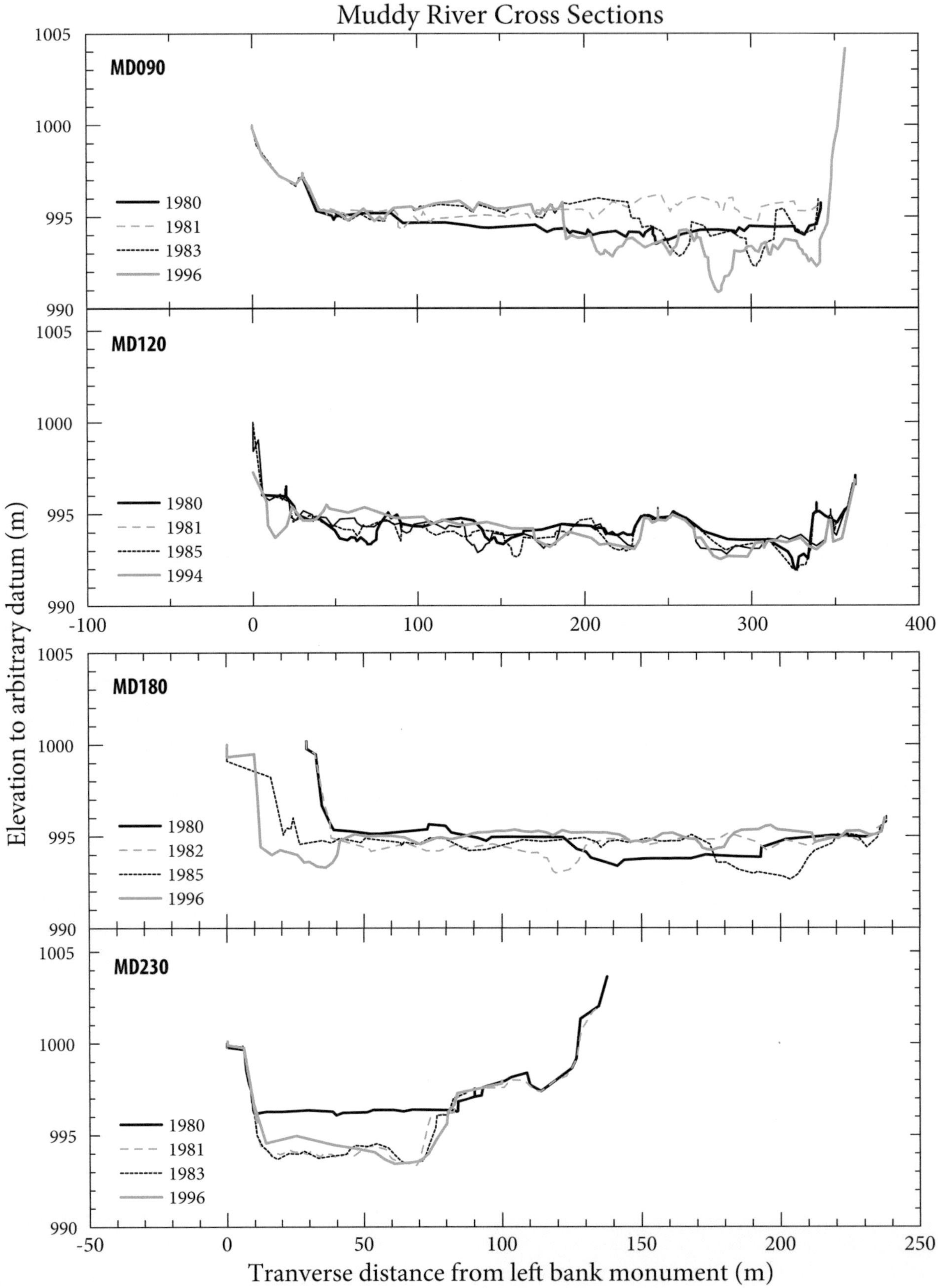

Fig. 3.5 Example cross sections measured along Muddy River corridor. See Fig. 3.1 for section locations.

some cross sections exhibited meter-scale aggradation. Locally, the flood triggered significant bank erosion and reworking of the floodplain surface along the lower Muddy River valley (e.g., MD180, Fig. 3.5; Frenzen et al. 2005), but overall geomorphic changes caused by the flood were relatively minor compared with initial changes caused by much lower-magnitude flows in the first few years following the eruptions.

3.5.3 Geomorphic Changes in the Bean Creek and Clearwater Creek Catchments

3.5.3.1 Landslides

Landsliding in the Clearwater Creek catchment in the first year after the eruption was limited primarily to layers of 1980 tephra and to small slides associated with toppled trees. In Boomer Creek (Fig. 3.1), a tributary to the middle Clearwater Creek valley which lies within the area affected by the blast PDC, shallow landsliding removed layers of 1980 tephra averaging 0.34 m thick from many slopes steeper than 70% (Smith and Swanson 1987). Even though several slides deep enough to remove trees and stumps occurred elsewhere in the Muddy River basin, they commonly were less than 100 m^2 in area and were difficult to discern from aerial photographs. Swanson et al. (1983), however, measured slides smaller than 100 m^2 in the blast-PDC area east of the Toutle drainage, including in the Smith, Bean, and Clearwater Creek catchments, from aerial photographs taken in June–August of 1967, 1979, 1980, and 1981. These areas included clearcuts as well as unlogged forests. They found a moderate increase in slide frequency (0.19 slides per square km per year) in the period bracketing the eruption. Landslide frequency was greater in downed forests, where small slides were apparently triggered by trees blown off steep slopes with shallow soils, than in areas clearcut before the eruption (Swanson and Major 2005). Most sediment mobilized by landslides and by rill and sheet erosion in Boomer Creek remained in storage on hillslopes below breaks in slope and behind fallen wood (Smith and Swanson 1987). The location of this sediment since the study was conducted is unknown, but sediment retention has probably diminished as wood decays.

The frequency and size of shallow landslides (less than about 2 m in depth) generally increased after 1981 (Fig. 3.6) but varied geographically. Landslides were nearly absent in the glaciated topography of the upper Clearwater and upper Bean Creek catchments, despite those areas being affected by the blast PDC. Landslide areas increased for each photo period in the middle Clearwater Creek valley, and particularly so in middle Bean Creek valley, but were negligible in lower Clearwater Creek valley, which was outside of the blast-PDC area. By 1998, approximately 3% of the

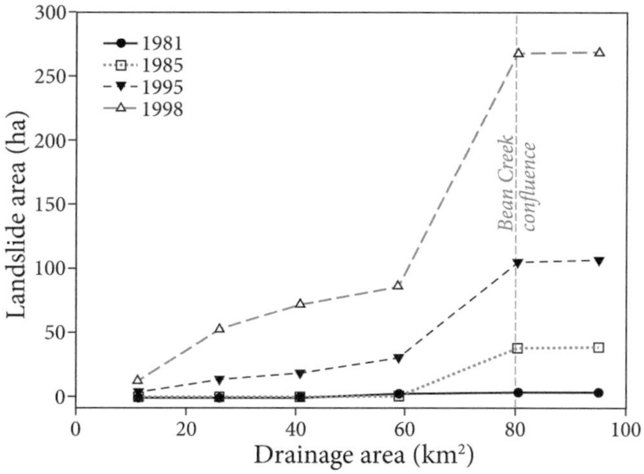

Fig. 3.6 Time series of landslide area as function of drainage area in the Clearwater and Bean Creek catchments as measured from aerial photographs. Landslide area increases sharply where the drainage area of Clearwater Creek includes the Bean Creek catchment.

Clearwater catchment upstream of the junction with Bean Creek was affected by landslides.

Particularly intense precipitation in the headwaters of Clearwater Creek and nearby catchments in February 1996 triggered several landslides. Large, planar landslides typically a few meters thick were released from the valley walls of middle Clearwater Creek valley (Fig. 3.7) and entrained abundant ash and pumice gravel deposited by the 1980 and earlier eruptions. Most landslide material was distributed across the valley floor without reaching the channel. Differences in landslide susceptibility after the 1980 and earlier eruptions may explain the occurrence of the large planar failures after 1980 (Fred Swanson, pers. comm., 2014): Tephrafalls from earlier eruptions were deposited in standing forests, whereas sediment from the 1980 blast PDC and tephrafalls was deposited among and atop trees toppled or killed during the 18 May eruption. If we assume that forests largely survived tephrafalls from earlier eruptions, then living roots would have helped bind shallow soil layers and reduce the propensity for landsliding. In contrast, most root strength was lost after the 18 May eruption owing to forest mortality, thereby increasing the propensity for landslides that incorporated 1980 and older tephra deposits. Unlike the Clearwater Creek catchment, where little landslide debris reached the channel, widespread landsliding and ***debris flows*** in the lower Bean Creek catchment contributed large volumes of sediment directly to the channel (Fig. 3.8).

3.5.3.2 Middle Clearwater Creek

Tributaries from steep valley sides transported large volumes of sediment into middle Clearwater Creek valley in 1980. Despite this abundant sediment input, most of the newly deposited tephra was not mobilized, and much of what was

Fig. 3.7 Landslides in the middle Clearwater Creek valley. Photograph taken August 2001 (T. Lisle). Clearwater Creek channel is located at "C".

Fig. 3.8 Landslide and debris-flow scars in lower Bean Creek valley. Photograph taken August 2001 (T. Lisle). Examples of shallow landslides (LS) and debris-flow tracks (DF) are labeled.

mobilized did not reach middle Clearwater Creek. For example, of the 960×10^3 t of sediment deposited in the 2.4-km^2 catchment of Boomer Creek, 88% remained on hillslopes, 6% was stored in channels, and 6% was exported to Clearwater Creek after 1 year (Smith and Swanson 1987). Despite the small fraction of sediment reaching Clearwater Creek, the yield (24 000 t km^{-2}) was still an order of magnitude greater than the upper limit of ambient suspended-sediment yields (~1000 t km^{-2}) of Cascade Range basins (Major et al. 2000). Sediment inputs from Boomer Creek declined sharply by 1982, as hillslope erosion declined and the channel became armored and stabilized over log steps.

Abundant fine sediment (mostly sand with small fractions of fine lithic and pumiceous gravel) was flushed from the gravel-bed channel of middle Clearwater Creek as the channel stabilized around new wood (Lisle 1995). In summer

1980, fine sediment covered all but the riffles and was transported as bedload during base flow throughout the middle Clearwater Creek valley. Within a few years, much of this sediment had flushed from the main channel, but it remained in pools, backwaters, and bars. By 1986, most of the bed surface became paved with pebbles and cobbles. This pavement of coarse sediment reemerged in areas where fine sediment was winnowed from the preexisting substrate, and developed in areas where the channel scoured and filled around wood.

The middle Clearwater Creek channel remained largely stable up to the 1996 flood because most large wood that influenced flow and channel morphology remained in place until then (Lisle 1995). Channel banks remained strong and cohesive because new tephrafall was shed from steep streambanks, allowing remnant riparian shrubs and herbs to sprout and grow soon after the eruption. Therefore, root strength was retained. Red alder (*Alnus rubra*), green alder (*Alnus viridis*), and willow (*Salix* spp.) sprouted on fresh bar surfaces by 1982, and stands became dense by 1990 (Lisle 1995). This vegetation stabilized new riparian surfaces and increased hydraulic friction. As a result, bank erosion was limited.

Wood in the Clearwater Creek channel is abundant, large, and influential on channel morphology and sediment routing (Lisle 1995). Prior to the 1980 eruptions, the catchment was forested by old-growth conifers or regenerated species following clearcut logging. Old-growth forests along the main stem of Clearwater Creek were dominated by enormous Douglas-fir (*Pseudotsuga menziesii*) and western redcedar (*Thuja plicata*). The blast PDC toppled large trees into the channel (Fig. 3.9a) (Lisle 1995). Some pieces (many >60 m long) spanned the channel (having an average bankfull width of 15–25 m) and had maximum diameters (~1 m) roughly equal to the mean bankfull depth (1.0–1.2 m). Because of their size, large pieces of wood remained as they fell until the 1996 flood, which transported large wood in some reaches (described below). Wood loadings measured in 1982 in a 1.8-km-long reach of middle Clearwater Creek ranged from 280 to 650 m^3 ha^{-1}, which is in the upper range of loading in Douglas-fir–dominated forests in the Cascade Range (Lisle 1995).

Fig. 3.9 Wood loading in Clearwater Creek valley. **a** Middle Clearwater reach #2 in 1982. **b** Lower Clearwater Creek in 2000. **c** Channel incised in former floodplain of lower Clearwater Creek in 2000. (Photos by T. Lisle).

Commercially valuable wood was salvaged from the channel and floodplain of much of middle Clearwater Creek and upper Bean Creek from 1982 to 1984. Most salvaged wood consisted of trunks of large trees that were not embedded in the channel, and salvage operations generally caused little disturbance to the bed and banks (Lisle 1995). Experimental removal of wood revealed the strong influence wood had on channel morphology and sediment storage (Lisle 1995). Total wood removal decreased pool volume and frequency, and reduced the amount of fine sediment in the bed. Partial wood removal typical of commercial operations, however, had no detectable effect on channel morphology or sediment storage.

Topographic surveys of the Clearwater Creek channel from 1980 to 1990 reveal that shifting wood and occasional debris flows produced only moderate changes in channel morphology (Fig. 3.10; Lisle 1995). Debris flows entered the Clearwater #1 and Clearwater #3 reaches (Fig. 3.1) between 1984 and 1986 and created debris jams that trapped sediment and raised local base levels. Elsewhere, large wood forced

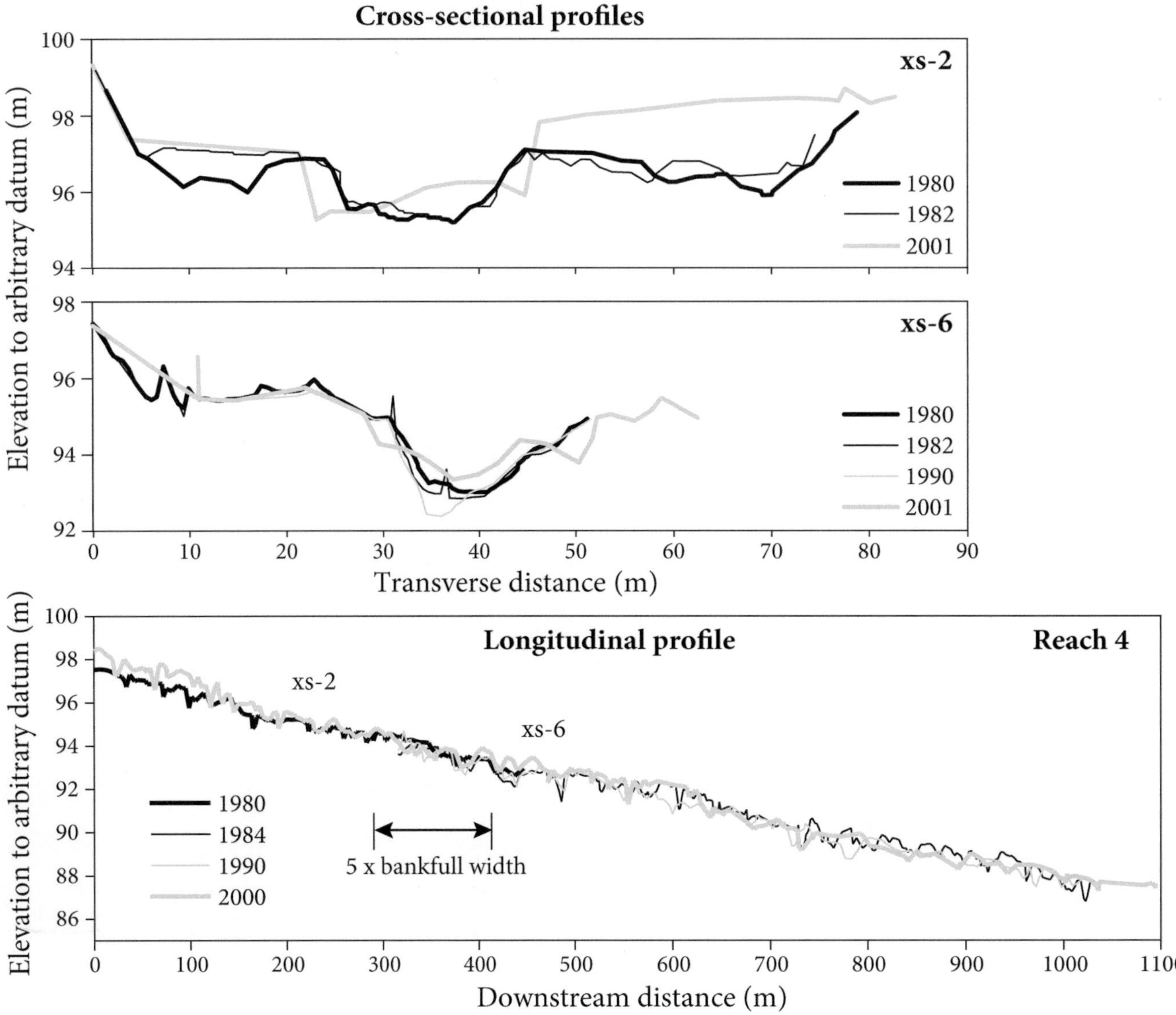

Fig. 3.10 Selected cross sections (above) and longitudinal profiles of middle Clearwater Creek reach #4. Locations of the cross sections are marked above the profiles. The undulations in the profiles are shorter than the common bar–pool spacing of 5 channel widths owing to local scour around wood.

profile undulations that are generally shorter than the common pool-riffle spacing of about five channel widths (Keller and Melhorn 1978), and shifting wood forced local changes in ***thalweg*** elevation. Most cross sections showed net fill in the first year after the eruption, followed by net scour by 1990. Most change occurred in the bed, as banks remained relatively stable.

The 1996 flood produced channel changes in middle Clearwater Creek valley that were unprecedented since the 18 May 1980 eruption (xs-2 Fig. 3.10). This is partly because the flood entrained some large wood that had not moved since that eruption. For example, sediment deposition upstream of a large log jam that formed in the Clearwater #4 reach in 1996 represented some of the greatest morphological change measured among cross sections since the eruption (xs-2, Fig. 3.10). Although most cross sections along this reach aggraded from 1990 to 2001, those not affected by wood movement generally showed little effect of the 1996 flood. Elsewhere in the middle Clearwater Creek valley, channel changes were punctuated by episodic sediment inputs. Deposits from a debris flow buried the channel (and survey monuments) behind a wood jam in the Clearwater #1 reach (Fig. 3.1) and rerouted the channel across the valley flat. A landslide buried the channel in the Clearwater #2 reach (Fig. 3.1) and caused a new channel to incise around its terminus. Local deposition was observed upstream of debris jams, landslide deposits, aggraded tributary junctions, and beaver dams.

3.5.3.3 Bean Creek

Bean Creek received large sediment inputs from steep valley walls during and after the 1980 eruptions and transmitted large volumes of sediment downstream to Clearwater Creek. Most fine sediment introduced to Bean Creek during the 18 May eruption was evacuated from in-channel storage within 2 years (Meyer and Martinson 1989). The channel subsequently received large volumes of landslide material, composed of a broad range of particle sizes, from steep valley walls. Much of that landslide debris was coarser than the sediment delivered by the 18 May eruption. In the first year after the eruption, sediment was contributed from shallow, planar landslides off the steepest slopes and from rills and gullies on gentler slopes. Cross-section surveys upstream of the Bean Creek gorges show thalweg incision of approximately 1 m or less from 1981 to 1984 (Martinson et al. 1984, 1986). Debris flows and deeper landslides occurred along the gorges downstream in the late 1980s, particularly during the 1996 event (Fig. 3.8). Aerial photographs and field inspections indicated that these landslides involved 1980 volcanic deposits, older tephras, and weathered bedrock, the latter containing substantial gravel-sized particles. Landslides and debris flows also incorporated large volumes of wood, which formed massive debris jams between the gorges. Large inputs of sediment and wood that were routed through the steep, storage-limited section of Bean Creek directly above lower Clearwater Creek (Fig. 3.1) provide evidence that Bean Creek has contributed much greater volumes of material to lower Clearwater Creek than has the much higher storage-capacity reach of middle Clearwater Creek.

3.5.3.4 Lower Clearwater Creek

Channel surveys show much greater changes in lower Clearwater Creek reach in the years following the eruptions than those in the middle Clearwater reaches, despite the fact that lower Clearwater was not directly affected by the blast PDC. In summer 1980, lower Clearwater Creek showed little overt sign of the 18 May eruption. Riparian vegetation, including red alder (*Alnus rubra*), big-leaf maple (*Acer macrophyllum*), and large conifers (e.g., western hemlock *Tsuga heterophylla*, *Pseudotsuga menziesii*, *Thuja plicata*), was dusted with volcanic ash but was otherwise unaffected. Bar–pool channel morphology was intact, but most of the cobble armor layer was covered with mobile, fine-grained volcanic sediment (80% of a 500-m section of the study reach).

During the first rainy season after the 18 May eruption, the lower Clearwater Creek valley floor was inundated with 2–4 m of sand and fine gravel (Fig. 3.11). At that time, sand and pumice gravel covered 68% of the 500-m study section of channel, which had become a shallow and braided reach with little cross-sectional or longitudinal relief. Between summer 1980 and summer 1981, channel gradient decreased 40%, from 0.59 m km^{-1} to 0.35 m km^{-1}. During a visit in August 1981, fine sand and small pumice pebbles (D_{50} ~ 8 mm) were actively transported by summer base flow. Riparian trees were dead but standing, and new wood had not been transported into the reach.

The wedge of aggradation of this fine bed material extended 1100 m upstream of the lahar deposit that plugged the downstream end of Clearwater Creek. Deposition of that sediment is attributed in part to the rise in base level caused by the plug of laharic sediment. Elevated base level persisted for at least two cycles of aggradation, one beginning in 1980 and another in 1996. The plug of debris was partly removed by wood salvage in 1984, but the Muddy River channel near the confluence with Clearwater Creek, which strongly influenced base level in lower Clearwater Creek, showed little change in mean bed elevation from about 1981 until the mid-1990s (Martinson et al. 1986; Hardison 2000) and thus continued to influence sediment accumulation.

The lower Clearwater Creek channel degraded and coarsened between 1981 and 1989 (Fig. 3.11). Bar–pool topography was partly restored and floodplain surfaces emerged as the active channel incised. By 1986, the channel had armored with large pebbles (D_{50} ~ 32 mm), and bedload transport during base flow ceased. However, the area of the bed covered by fine sediment in the 500-m study section increased from 18% in 1983 to 25% in 1989 as coarse and fine components of the bed material became more segregated. Local scour and deposition were enhanced by inputs of wood from dead riparian trees. Willows and alders colonized relatively stable mesic habitats in flood deposits.

A survey of lower Clearwater Creek in 2000 showed that sediment transported by the 1996 flood had accumulated and exceeded the level of aggradation achieved in 1981 by as much as 1–3 m (Fig. 3.11). Coarse gravel and large wood from Bean Creek were deposited as a wedge of sediment that extended into the lower Clearwater Creek study reach and increased its gradient to 0.79 m km^{-1}, which is steeper than the post-eruption gradient measured in 1980 (0.59 m km^{-1}). The steeper gradient following the 1996 flood suggests that this wedge of coarse sediment made a stronger contribution to sedimentation in the study reach than did the remnant wedge of fine material that extended upstream from the plug of laharic sediment near the lower end of the channel.

Abundant wood in a variety of configurations, including large jams (Fig. 3.9b), forced pronounced undulations in channel topography of lower Clearwater Creek. Wood jams forced the channel to wander from one side of the valley to the other. Channels that cut across the former floodplain became deep and narrow where they scoured around groves of large snags (Fig. 3.9c). As a result, the thalweg profile shows greater irregularity in 2000 than in any other survey, including that of 1980 (Fig. 3.11). Valley-spanning jams, including one at the downstream end of the study reach, appear to be responsible for some of the increased storage of sediment. Alluvial deposits

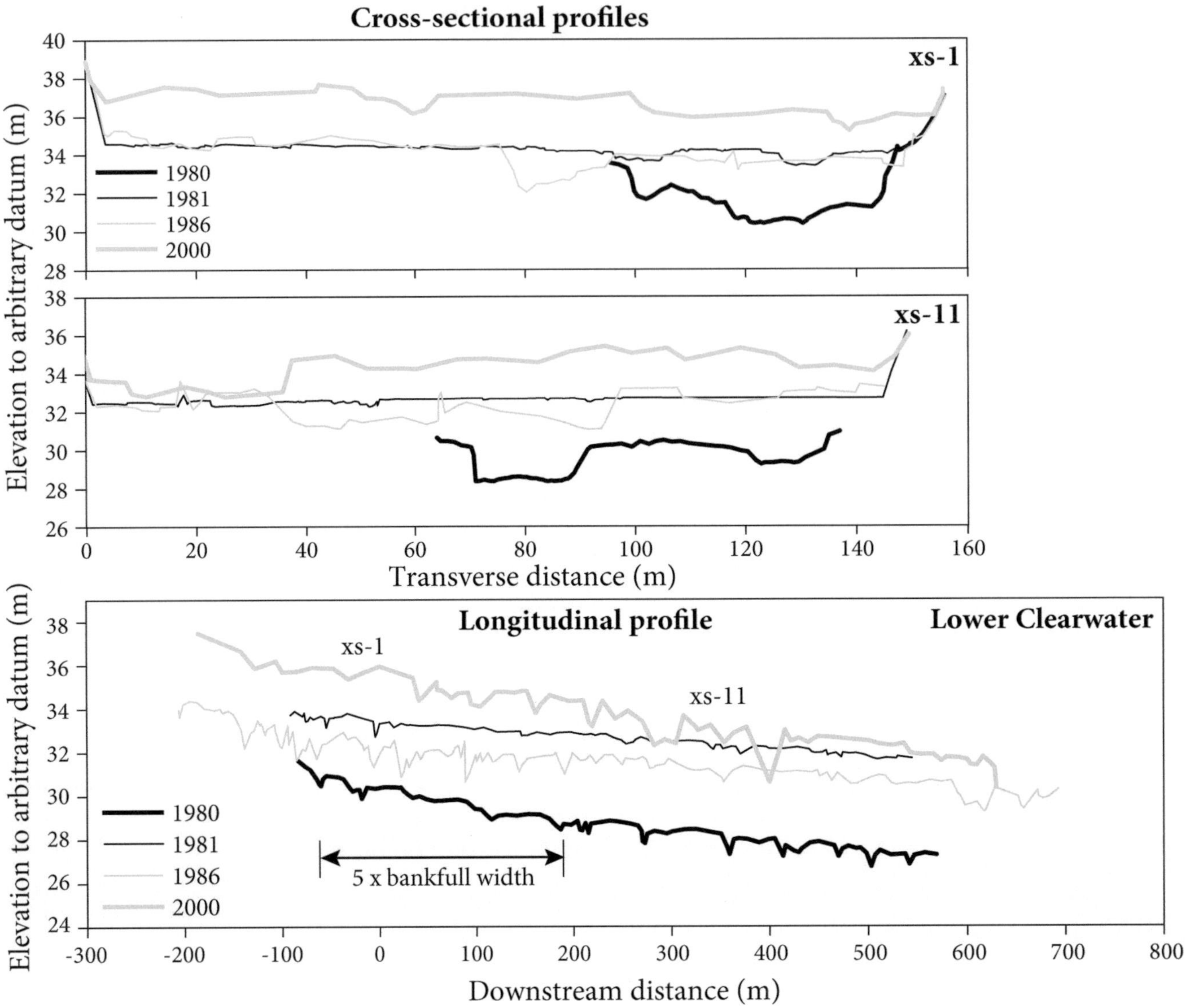

Fig. 3.11 Selected cross sections (above) and longitudinal profiles of lower Clearwater Creek. Locations of the cross sections are marked above the profiles. The undulations in the profiles are shorter than the common bar–pool spacing of 5 channel widths owing to local scour around wood.

ranged from those dominated by lithic gravel to those dominated by sand, pumice gravel, and wood.

3.6 Discussion

3.6.1 The Landscape as a Template for Geomorphic Outcomes

The 1980 eruptions of Mount St. Helens disturbed the mobility and routing of water, sediment, and wood over large portions of the Muddy River basin. Geomorphic responses to this disturbance were strongly influenced by hydrogeomorphic changes that affected the routing of sediment and wood through the basin and by linkages among sediment sources, pathways, and sediment reservoirs (Fig. 3.12). Imprinted on this template of linkages between landforms and the routing of sediment and wood were stochastic events (storm magnitudes and frequencies) that occurred across a landscape undergoing evolving geomorphic and vegetative conditions.

Effects of geomorphic responses to the 1980 eruptions on sediment reservoirs (e.g., channel bars, floodplains, and alluvial fans) were mediated by the relationships between volumes of sediment input to the systems relative to reservoir storage capacity and by the distances over which sediment inputs were dispersed. For example, large landslides in Bean Creek in the late 1980s and 1990s contributed large volumes of coarse sediment to a channel having little storage capacity. Hence, that sediment was transported downstream to lower Clearwater Creek, where it inundated a wider valley bottom.

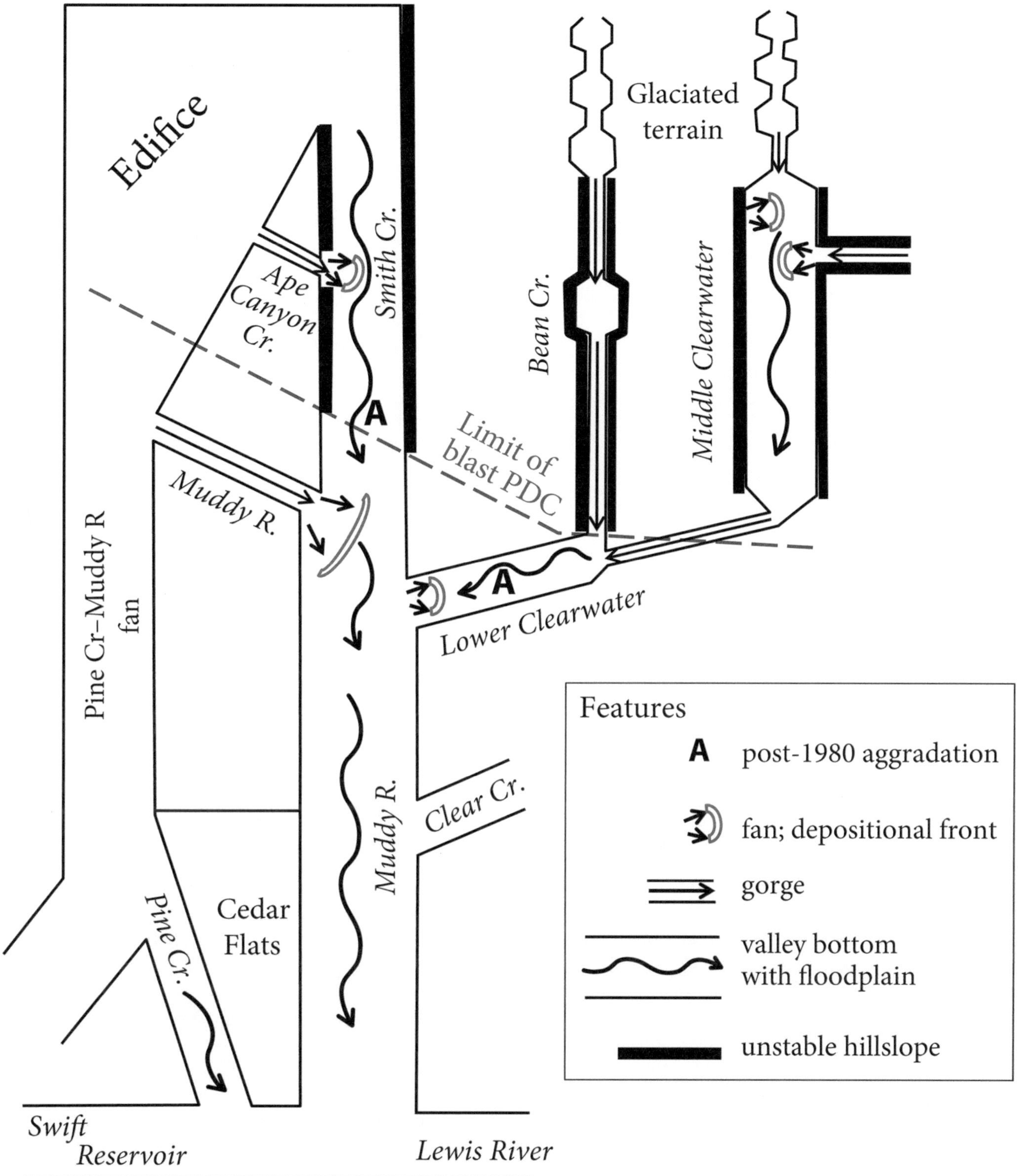

Fig. 3.12 Conceptual depiction of linkages among sediment sources, transport pathways, and sediment reservoirs in the Muddy River basin. These linkages influence interactions between evolving geomorphic and ecological trajectories following disturbances by the 1980 eruption of Mount St. Helens.

In contrast, sediment input to the middle Clearwater Creek valley encountered sediment reservoirs so vast that the lower portions of the valley were effectively buffered from hillslope sediment input (landslides and debris flows). Here, the channel is dynamic primarily because of shifting wood, but that dynamism is chiefly lateral; channel-bed elevation has remained nearly static since fine sediment deposited by the eruptions was mostly winnowed away in the early 1980s. A low-gradient reach upstream of the Muddy River–Smith Creek junction became a major depositional zone for lahars and remobilized eruption deposits (e.g., SM060, SM080 in Fig. 3.4). Farther downstream in the Muddy River valley, the channel was apparently so overwhelmed by the 1980 lahar that subsequent hillslope–channel interactions and associated variations in valley erosion or deposition were relatively minor compared with the initial effects of the lahar.

3.6.2 Downstream Propagation of Disturbances

Primary eruption effects were strongly nonuniform and exhibited both radial (away from the volcano) and tangential (upstream to downstream from sources distributed radially) gradients (Fig. 3.2). Radial gradients in the Muddy River basin were expressed as diminishing types and intensities of primary effects away from the volcano normal to a generally southwest to northeast direction. For example, the headwaters of Smith Creek and Muddy River catchments were affected by the blast PDC and tephrafall, and the Smith Creek–Muddy River valley floor was inundated by a major lahar and remobilized deposits from the blast PDC. The lahar overwhelmed the influence of all but the largest-scale morphologic features of the valley (namely, valley constrictions) and completely changed the valley-floor morphology (Janda et al. 1981). Subsequent floods reworked the channel that had developed across the lahar deposit (Martinson et al. 1984; Hardison 2000). In contrast, the Bean Creek and Clearwater Creek catchments were draped by thinner deposits of the blast PDC and tephrafall, but were not affected by lahars. Clear Creek basin received only minor tephrafall. Tangential gradients (generally trending north to south) were created by the distribution of the blast PDC. That event transected the upstream reaches of the catchments in the Muddy River basin and created upstream sources of sediment and wood that were subsequently routed downstream beyond the limit of the area of impact (Fig. 3.2).

Propagation of sediment pulses in all catchments led to the spreading of excess sediment along channels and across valley floors without causing downstream migration of discrete sediment waves. This behavior is attributed to four factors:

1. The overwhelming sediment load of the 1980 lahar in the Smith Creek–Muddy River corridor affected the entire length of the river system. Reaches upstream of constrictions (e.g., Muddy River fan at the Smith Creek confluence, Cedar Flats) aggraded preferentially, but initial streamwise contrasts in aggradation and degradation were not strong.
2. High Froude numbers (>0.4) that characterize steep, gravel-bed channels during sediment-transporting flows suppress wave translation and promote dispersion (Cui and Parker 2005; Lisle 2008). Froude numbers <0.4 at discharges capable of transporting gravel generally occur only in channels with gradients much less than 1 m km^{-1} (Grant 1997), a gradient that is approached or exceeded in several reaches of Muddy River (Hardison 2000) and Clearwater Creek (Lisle 1995).
3. Valley slope along lower Smith Creek increased from 0.7 m km^{-1} just above the confluence with Muddy River to 2.2 m km^{-1} just below the confluence (Hardison 2000). This change in gradient increased stream power, which increased the transport capacity of the river and likely kept sediment persistently mobile along the Muddy River corridor below Smith Creek.
4. Sediment storage, particularly deposition on bars and floodplains, tends to disperse sediment pulses (Nakamura and Kikuchi 1996). Such storage reservoirs are prevalent along middle Clearwater Creek and lower Muddy River.

Despite the general tendency for sediment dispersion in the Muddy River basin, enhanced deposition occurred in some catchments far downstream of sediment sources, where reductions in transport capacity were forced by valley-scale features. For example, lower Clearwater Creek aggraded more than any other study reach in the Clearwater catchment. This aggradation was triggered initially by a plug of sediment and wood emplaced in the downstream end of the valley by the Muddy River lahar. That plug of debris forced deposition of a wedge of sandy sediment, thinning upstream along Clearwater Creek. Subsequently, large fluxes of coarse sediment from Bean Creek accumulated in a wedge extending downstream from the Bean Creek confluence, as sediment encountered the decreased gradient and confinement of lower Clearwater Creek.

3.6.3 Stochastic Events: The Flood and Landslides of 1996

Climatic events acted on diverse landforms, gradients of eruption disturbance, and evolving geomorphic conditions, which affected runoff rates and sediment mobility on hillslopes and in channels. Both timing and magnitude of climatic events influenced their effects. During early stages of response to the 1980 eruptions, rain storms were moderate in magnitude (Swanson and Major 2005). Nevertheless, enhanced runoff eroded hillslopes and routed large volumes of fine sediment in channels. Surface runoff and hillslope erosion rates declined sharply within a few years after the 1980 eruptions (Collins and Dunne 1986; Major and Mark 2006) as tephra surfaces coarsened and infiltration capacity began to recover. However, sediment delivery remained higher than that typical of forested watersheds in the Cascade Range for several years after the eruptions (Major et al. 2000). Soil strength in areas of heavy tree mortality declined and probably reached a minimum approximately a decade or more after the eruptions as roots of destroyed forest decayed. Depressed interception rates and soil strength in the 1990s set the stage for hillslope response to a major rainfall event in 1996 in some sectors of the area affected by the blast PDC.

The atmospheric-river event of 1996 delivered intense rainfall to a geomorphically and ecologically evolving landscape. That event produced another major pulse of sediment and wood in the catchments of the Muddy River basin. The intensity and duration of rainfall during that event locally exceeded thresholds for hillslope stability. The greatest number of landslides and the highest river stages in the Muddy River basin since the eruptions occurred at that time. Significantly elevated suspended-sediment yield at the Muddy River gauging station during the consequent flood (~50% of the annual load transported in a single day) revealed that the landscape was still highly erosive (Major et al. 2000; Major 2004). However, changes in channel morphology and bed elevation along the Muddy River were minor relative to changes caused by the 1980 lahar and by earlier smaller floods in the aftermath of eruptive disturbance. Only locally did the 1996 flood cause significant geomorphic and ecologic change (Frenzen et al. 2005). In contrast, in the Clearwater Creek and Bean Creek catchments, the 1996 flood exceeded the 1980 events in flow magnitude, and it caused the greatest change in channel morphology in many reaches since the eruptions (Fig. 3.10). Large wood was moved for the first time since its introduction to channels by the blast PDC, and its organization into channel-spanning jams disturbed floodplains and altered channel courses (Fig. 3.9). Likewise, the proportion of catchment hillslopes destabilized by landslides peaked between 1995 and 1998 (with many landslides observed following the 1996 event) (Fig. 3.6). Those landslides incorporated soil and tephra layers below the 1980 tephra (Figs. 3.7 and 3.8), demonstrating that the response to the 1980 eruptions involved mobilization and routing of both new material and older sediment that had been stable for millennia.

3.6.4 Geomorphic Influences on Aquatic and Riparian Ecosystems

The 18 May 1980 eruption profoundly affected terrestrial, riparian, and aquatic ecosystems in the Muddy River basin and initiated new trajectories of geomorphic and ecologic processes that will evolve for years to come. Terrestrial ecosystems were affected most by the blast PDC and to a lesser degree by tephrafall (Antos and Zobel 2005; Swanson and Major 2005). Aquatic and riparian ecosystems downstream of the landscape impacted by the blast PDC have been affected by the routing of sediment and wood from that disturbed area. After 30 years, sediment yield from the Muddy River basin is similar to yields from some sediment-rich rivers draining Mount Rainier, Mount Baker, and the North Cascades of Washington, but it exceeds those of most other basins in the Cascade Range (Major et al. 2000; Czuba et al. 2011; Major et al., Chap. 3, this volume). High sediment concentrations probably influence the composition if not the productivity of benthic communities (Waters 1995).

Despite substantial dynamism, channels in the Muddy River basin appear to be maintaining generally single-thread meandering or wandering planforms while scouring, filling, and shifting their courses. This is due in large measure to the great reduction of sediment loading and to recovery of riparian vegetation. Locally, however, channel widening and deflection around log jams causes channel braiding. Influences of the competition between geomorphic reworking and vegetative stability on channel form have been observed along many fluvial channels having variable sediment flux, including those responding to the 1991 eruption of Mount Pinatubo, Philippines (e.g., Corenbilt et al. 2009; Gran et al. 2015).

There are several indications of increased productivity associated with disturbed channels and new alluvial deposits. With the toppling of late-seral coniferous overstory, insolation has increased in the blast-PDC area as well as in valley reaches downstream that had experienced disturbance of valley-floor forests. Although regrowth of successional species is restoring some shade, these changes in vegetative structure can be expected to increase primary productivity and summer water temperatures in streams. Most of the successional species that are replacing coniferous forest include deciduous trees and shrubs (e.g., alders and willows). These species contribute litterfall that yields nutrients and organic matter that are more readily available to aquatic and riparian food webs than that derived from litterfall from conifer forests (Petersen and Cummins 1974; Cummins et al. 1989). Prevalent orange-stained seeps along new bar margins highlight emerging and oxidizing constituents that are mobilized by anaerobic conditions in the substrate. These biotic conditions are widespread in the channel systems of the Muddy River basin and will remain so as long as channels remain highly dynamic, and non-coniferous, early-seral plant communities dominate riparian areas.

Large wood that was introduced to channels by the 18 May eruption, subsequent landsliding, and channel migration enhances channel complexity and the diversity of aquatic habitats. Wood jams, herbivory, and post-eruption perturbations by floods have maintained a ribbon of diversity in riparian corridors through conifer plantations in the middle Clearwater Creek valley. Dense streamside stands of alder, willow, and cottonwood (*Populus* spp.) that established by 1990 were thinned by beavers by 2000; browsing by elk has further influenced plant community structure. Punctuated floodplain disturbance caused by shifting wood jams has created new surfaces for riparian succession. Residual pool depths in middle Clearwater Creek decreased to a minimum by 1985 as a result of wood salvaging but recovered by 1990 (Lisle 1995). In lower Clearwater Creek, channel relief was greatest in 2000 when the channel was at its highest level of aggradation and wood loading was at its peak.

3.6.5 Trajectories of Change

Assessments of trajectories of geomorphic and ecological change rely partly on trends in geomorphic stability thresholds. Vegetation growth continues to gradually raise thresholds of slope instability by replacing lost root strength in soils and by increasing foliar interception. However, the most severely disturbed catchments where natural forest growth (as opposed to growth of commercial plantations) is slow remain in a highly active condition that may persist for decades.

Geomorphic stability of channels is influenced by topographic linkages. Reaches of channel unbuffered from ongoing landslides and erosion of large gravel stores, such as Bean Creek and lower Clearwater Creek, will remain highly mobile. In buffered reaches, such as middle Clearwater Creek, gravel movement has decreased to levels that maintain channel stability. Depositional landforms downstream of large sediment stores and recovering hillslopes are likely to undergo episodic disturbance. Reworking of volcanic deposits in Smith Creek and Muddy River continues to furnish large supplies of a wide range of sediment sizes. These differences in the frequency, extent, and intensity of disturbance can be expected to strongly influence aquatic and riparian ecosystems.

Trajectories of geomorphic processes are also influenced by the magnitude and timing of hydrologic (and volcanic) events that exceed or reset geomorphic thresholds. As new forests grow, the potential impact of another storm of magnitude similar to that of 1996 should diminish, although stability thresholds may still be widely exceeded. Some projections of climate change in this region, which include decreased snow cover, increased mean annual precipitation, increased frequency of heavy precipitation, increased runoff, and increased frequency of rain-on-snow events, indicate that the frequency of large floods may increase (Milly et al. 2002; Hamlet and Lettenmaier 2007).

Even without large sediment inputs, wood-rich reaches of Clearwater Creek will remain dynamic as peak flows act on large pieces of decaying wood that date from 1980. Breakup, entrainment, and aggregation of logs into jams can be expected to diversify the distribution of wood by flushing it from some reaches and concentrating it in others, as has been observed elsewhere after large inputs of wood (e.g., Bryant 1980; Ulloa et al. 2015). Over the coming decades, the sizes and volumes of wood in channels will decrease as the large 1980 wood decays and new wood from younger and smaller riparian and terrestrial sources replaces it. However, landslides from forested slopes still have the potential to locally augment wood loading.

3.7 Conclusions

The 18 May 1980 eruption of Mount St. Helens produced a mosaic of primary landscape disturbances that decreased in intensity away from the volcano. East of the volcano, that mosaic of disturbances initiated contrasting responses in the catchments of Muddy River, Smith Creek, Bean Creek, and Clearwater Creek. A devastating pyroclastic density current—here called the blast PDC—swept across headwaters of these catchments, toppled mature forest, and mantled hillslopes with as much as 70 cm of graded gravel- to silt-size sediment. Rapidly melted glaciers and snowfields on the edifice of the volcano produced lahars (volcanic debris flows) that coursed down Smith Creek and Muddy River, routed large volumes of wood and sediment down the entire lengths of the systems, and draped the channels and floodplains with up to 3 m of gravelly sand. No lahars occurred in Bean Creek or Clearwater Creek. Instead, floods bearing large loads of sand deposited sediment in channels and on floodplains. Tephrafall from the billowing eruption column that followed the blast PDC subsequently blanketed the catchments with 2–40 cm of ash and pumice gravel.

Subsequent geomorphic responses to these landscape disturbances were influenced by evolving hillslope and channel conditions that affected fluxes of water, sediment, and wood, as well as by stochastic occurrences of exceptional storms. These fluxes moved from the area affected by the blast PDC to downstream reaches where, except in lahar-affected areas, primary eruption disturbances were minor or insignificant. The initial highly erosive conditions in the area of the blast PDC, caused by loss of foliar interception and deposition of silty sediment that significantly reduced surface infiltration rates, were ameliorated somewhat in the first years after the eruption as the fresh tephra surface coarsened through a variety of processes and infiltration increased. During the following decades, increased infiltration into hillslopes and loss of soil strength owing to root decay lowered the threshold for landsliding, and landslide area increased sharply during an exceptional rainstorm in 1996. Owing partly to this enhanced susceptibility to landsliding, and to persistent channel reworking, sediment yield has remained high in the Muddy River basin for more than two decades after the eruption. These results demonstrate that under evolving conditions, the timing as well as the magnitude of hydrologic events influences fluxes of sediment and wood from disturbed terrains.

The progression of change in river corridors in the decades after the eruption has differed strongly between the Smith Creek–Muddy River and Bean Creek–Clearwater Creek catchments owing to differences in topography, primary disturbances, and magnitudes of subsequent hydrogeomorphic response processes. The following contrasts in initial impacts and subsequent channel responses were observed:

- Lahars in the Smith Creek–Muddy River catchments inundated a large fan on the southeast flank of the volcano (Pine–Muddy fan) and valley bottoms downstream with coarse gravelly sand. An emergent large lahar resulting from coalescence of smaller lahars moving down multiple tributaries stripped riparian corridors and locally obliterated the preexisting channel. In contrast, preexisting channels in the Bean Creek–Clearwater Creek catchments remained essentially intact after the eruption because sand and pumice gravel from the blast PDC and subsequent tephrafall only temporarily buried gravel beds and floodplains with much shallower thicknesses of sediment than did the lahar along the Smith Creek–Muddy River corridor.
- Most of the large wood introduced or entrained in the Muddy River corridor was swept by the lahar from the active channel and floodplain into marginal deposits and a large jam that temporarily spanned the channel at Cedar Flats. In contrast, most large trees toppled by the blast PDC into Clearwater Creek lay as they fell throughout the channel system, owing to the absence of lahars and the large size of fallen trees relative to channel dimensions.
- Only a fraction of the large volume of sediment introduced to the Muddy River was reworked and redeposited or carried out of the system within a few years after the eruption. Post-eruption changes had less geomorphic impact on the Muddy River corridor than did the eruption-triggered lahar. Nevertheless, the volume of sediment eroded from the Smith Creek–Muddy River corridors after the eruption exceeded that eroded from the Bean Creek–Clearwater Creek corridors. Although reworking of the valley fill in the Smith Creek–Muddy River corridor led to incision and widespread lateral erosion of new channels in the first few years after the eruption, the overall post-eruption channel patterns were a product of changes caused by deposition of a thick fill of blast-PDC sediment and the lahar. In contrast, the impacts of the eruption on Clearwater Creek were more limited and temporary. Most of the sediment that initially covered the channel bed was winnowed rapidly, exposing the pre-eruption gravel bed. Channel banks remained intact and little lateral erosion occurred.
- Wood had little effect on post-eruption channel changes along the Muddy River corridor because most had been swept away by the lahar or deposited on surfaces isolated from later channel processes. In contrast, new wood introduced to Clearwater Creek frequently modified local channel morphology in the area affected by the blast-PDC by forcing new zones of scour and deposition. That scour and deposition of sediment occurred despite the fact that in most reaches, channel dimensions and patterns have remained essentially constant. Wood continues to be a major driver of channel change in the Bean Creek and Clearwater Creek catchments as it gets concentrated in log jams.

A large storm and flood in 1996 tested the resilience of recovering catchments across the mosaic and gradients of geomorphic disturbance in the Muddy River basin. The flood reworked some of the blast PDC- and lahar-affected areas of Smith Creek and Muddy River, but did not accomplish as much geomorphic work as did much smaller floods during the first few years after the eruption. In the Bean Creek and Clearwater Creek catchments, the 1996 event triggered numerous shallow landslides and mobilized large wood that had not moved since the 18 May eruption. Although no significant reach-scale bed-elevation change occurred, transport and deposition of wood locally modified channel morphology more than did the disturbance by the 18 May eruption. Widespread landsliding in Bean Creek ultimately resulted in deposition of coarse sediment and wood in lower Clearwater Creek. Aggradation in lower Clearwater Creek in 1996 exceeded the levels observed 1 year after the eruption, when fine bed material that had accumulated behind a blockage caused by the Muddy River lahar buried the lower 2 km of the channel and floodplain. Steering of the channel by wood jams altered the channel form across the entire valley bottom. Although much of the change wrought by the 1996 flood can be attributed to conditions set up by the 1980 eruptions, the changes observed in the Clearwater Creek catchment after the 1996 flood were comparable to or locally greater than those observed in 1980.

Geomorphic change and biotic activity in the Muddy River basin continue to interact in terrestrial, riparian, and aquatic ecosystems, and in many cases diversify ecosystem conditions. New tephra remains in place over most of the area affected by the 1980 eruptions, but older soil and tephra layers have been exposed by surface erosion, landsliding, and bioturbation. This exposure has promoted the spread and growth of vegetation, especially of deciduous species that support a more diverse food web than the conifer species that dominated forests before the eruption. Fluvial activity creates variations in age and characteristics of channel and floodplain surfaces that diversify riparian plant communities. Abundant fine sediment will continue to affect benthic communities in streams (e.g., Waters 1995), but channel wood will steer hydraulic forces that diversify streambed conditions. Wood will also have a persistent effect on aquatic habitats, especially in the Clearwater Creek catchment. There, wood contributed by the 18 May eruption and a cascade of post-eruption processes continues to be routed in dynamic and complex arrangements.

Acknowledgment We acknowledge the special contributions of our late colleague Dick Janda, who fostered many of the initial studies of geomorphic response to the eruption and promoted communications among scientific communities.

Glossary

Blast pyroclastic density current A form of pyroclastic density current initiated by rapid decompression of lava domes or cryptodomes (magma bodies cooled high within a volcanic edifice) owing to sudden collapse. Rapid decompression results in a directed explosion that initially impels the current laterally before it becomes a gravity-driven flow. In the case of the Mount St. Helens 1980 eruption, failure of the volcano's north flank unroofed pressurized magma and superheated water. Rapid exsolution of magmatic gases and conversion of superheated water to steam produced a laterally directed blast, which formed a density current that flowed across rugged topography. The current contained fragmented rock debris as well as shattered forest material.

Debris avalanche A rapid granular flow of an unsaturated or partly saturated mixture of volcanic rock particles (± ice) and water, initiated by the gravitational collapse and disintegration of part of a volcanic edifice. Debris avalanches differ from debris flows in that they are not water-saturated. Although debris avalanches commonly occur in association with eruptions, they can also occur during periods when a volcano is dormant.

Lahar An Indonesian term for a rapid granular flow of a fully saturated mixture of volcanic rock particles (± ice), water, and commonly woody debris. A lahar that has ≥50% solids by volume is termed a *debris flow*; one that has roughly 10–50% solids by volume is termed a *hyperconcentrated flow*. Flow type can evolve with time and distance along a flow path as sediment is entrained or deposited.

Pyroclastic density current (PDC) Rapid flow of a dry mixture of hot (commonly >700 °C) solid particles, gases, and air, which can range in character from a dense, ground-hugging flow (*pyroclastic flow*) to a turbulent, low-density cloud of mostly fine ash and superheated air (*pyroclastic surge*). A single PDC commonly involves both flow types as a result of gravitational segregation. Flows are generally gravity driven but may be accelerated initially by impulsive lateral forces of directed volcanic explosions. Flows typically move at high velocity (up to several hundred km/hr).

Pyroclastic flow See pyroclastic density current (PDC).

Tephrafall A rain of volcanic particles to the ground following ejection into the atmosphere by an explosive eruption. Tephra is a collective term for particles of any size, shape, or composition ejected in an explosive eruption.

Thalweg The trajectory of the connection of lowest points of the channel bed along the length of a stream channel. The thalweg marks the natural direction of a watercourse.

References

Antos, D.B., and J.A. Zobel. 2005. Plant responses in forests of the tephra-fall zone. In *Ecological Responses to the 1980 Eruption of Mount St. Helens*, ed. V.H. Dale, F.J. Swanson, and C.M. Crisafulli, 47–58. New York: Springer.

Brantley, S.R., and R.B. Waitt. 1988. Interrelations among pyroclastic surge, pyroclastic flow, and lahars in Smith Creek valley during first minutes of 18 May 1980 eruption of Mount St. Helens, USA. *Bulletin of Volcanology* 50: 304–326.

Bryant, M.D. 1980. *Evolution of large organic debris after timber harvest: Maybeso Creek, 1949–1978.* General Technical Report PNW-101. Portland: U.S. Department of Agriculture, Forest Service, Pacific Northwest Forest and Range Experiment Station.

Cloutier, D., M.N. LeCouturier, C.L. Amos, and P.R. Hill. 2006. The effects of suspended sediment concentration on turbulence in an annular flume. *Aquatic Ecology* 40: 555–565.

Clynne, M.A., A.T. Calvert, E.W. Wolfe, R.C. Evarts, R.J. Fleck, and M.A. Lanphere. 2008. The Pleistocene eruptive history of Mount St. Helens, Washington, from 300,000 to 12,800 years before present. In *A volcano rekindled: The renewed eruption of Mount St. Helens, 2004–2006*, Professional Paper 1750, ed. D.R. Sherrod, W.E. Scott, and P.H. Stauffer. 593–627. Washington, DC: U.S. Geological Survey.

Coleman, N. 1986. Effects of suspended sediment on open-channel velocity distribution. *Water Resources Research* 22 (10): 1377–1384.

Collins, B.D., and T. Dunne. 1986. Erosion of tephra from the 1980 eruption of Mount St. Helens. *Geological Society of America Bulletin* 97: 896–905.

———. 1988. Effects of forest land management on erosion and revegetation after the eruption of Mount St. Helens. *Earth Surface Processes and Landforms* 13: 193–205.

Collins, B.D., T. Dunne, and A.K. Lehre. 1983. Erosion of tephra-covered hillslopes north of Mount St. Helens: May 1980–May 1981. *Zeitschrift für Geomorphologie* 46: 103–121.

Corenblit, D., J. Steiger, A.M. Gurnell, E. Tabacchi, and L. Roques. 2009. Control of sediment dynamics by vegetation as a key function driving biogeomorphic succession within fluvial corridors. *Earth Surface Processes and Landforms* 34: 1790–1810.

Crandell, D.R. 1987. *Deposits of pre-1980 pyroclastic flows and lahars from Mount St. Helens volcano, Washington.* Professional Paper 1444. Washington, DC: U.S. Geological Survey.

Cui, Y., and G. Parker. 2005. Numerical model of sediment pulses and sediment supply disturbances in mountain rivers. *Journal of Hydraulic Engineering* 131 (8): 646–656.

Cummins, K.W., M.A. Wilzbach, D.M. Gates, J.B. Perry, and W.B. Taliaferro. 1989. Shredders and riparian vegetation. *BioScience* 39: 24–30.

Czuba, J.A., and C.S. Magirl, C.R. Czuba, E.E. Grossman, C.A. Curran, A.S. Gendaszek, R.S. Dinacola. 2011. *Sediment load from major rivers into puget sound and its adjacent waters*. Fact Sheet 2011–3083. Tacoma: U.S. Geological Survey. http://pubs.usgs.gov/fs/2011/3083/pdf/fs20113083.pdf. Accessed 10 Dec 2015.

Dinehart, R.L. 1998. *Sediment transport at gaging stations near Mount St. Helens, Washington, 1980–1990: Data collection and analysis*. Professional Paper 1573. Washington, DC: U.S. Geological Survey.

Evarts, R.C., and D.A. Swanson. 1994. Geologic transect across the tertiary Cascade Range, southern Washington. In *Geologic field trips in the Pacific Northwest*. Geological Society of American Annual Meeting, vol. 2, Chapter H, 1–31, ed. D.A. Swanson and R.A. Haugeraud. Seattle: University of Washington Department of Geological Sciences.

Evarts, R.C., R.P. Ashley, and J.G. Smith. 1987. Geology of the Mount St. Helens area: Record of discontinuous volcanic and plutonic activity in the Cascade Arc of southern Washington. *Journal of Geophysical Research* 92 (B10): 10155–10169.

Frenzen, P.M., K.S. Hadley, J.J. Major, M.H. Weber, J.F. Franklin, J.H.I. Hardison, and S.M. Stanton. 2005. Geomorphic change and vegetation development on the Muddy River mudflow deposit. In *Ecological responses to the 1980 eruption of Mount St. Helens*, ed. V.H. Dale, F.J. Swanson, and C.M. Crisafulli, 75–92. New York: Springer.

Gran, K.B., D.R. Montgomery, and D.G. Sutherland. 2006. Channel bed evolution and sediment transport under declining sediment inputs. *Water Resources Research* 42: 1–14. W10407. https://doi.org/10.1029/2005/WR004306.

Gran, K.B., D.R. Montgomery, and J.C. Halbur. 2011. Long-term elevated post-eruption sedimentation at Mount Pinatubo, Philippines. *Geology* 39: 367–370.

Gran, K.B., M. Tal, and E.D. Wartman. 2015. Co-evolution of riparian vegetation and channel dynamics in an aggrading braided river system, Mount Pinatubo, Philippines. *Earth Surface Processes and Landforms*. https://doi.org/10.1002/esp.3699.

Grant, G.E. 1997. Critical flow constraints flow hydraulics in mobile-bed streams: A new hypothesis. *Water Resources Research* 33 (2): 349–358.

Hamlet, A.F., and D.P. Lettenmaier. 2007. Effects of 20th century warming and climate variability on flood risk in the western US. *Water Resources Research* 43 (6): W06427. https://doi.org/10.1029/2006WR005099.

Hardison, J.H. III. 2000. *Post-lahar channel adjustment, Muddy River, Mount St. Helens, Washington*. Master of Science thesis. Fort Collins: Colorado State University.

Hoblitt, R.P., C.D. Miller, and J.W. Vallance. 1981. Origin and stratigraphy of the deposit produced by the May 18 directed blast. In *The 1980 Eruptions of Mount St. Helens, Washington*. Professional Paper 1250, ed. P.W. Lipman and D.R. Mullineaux, 401–420. Washington, DC: U.S. Geological Survey.

Janda, R. J., K. M. Scott, K. M. Nolan, and H. A. Martinson. 1981. Lahar movement, effects, and deposits. In *The 1980 Eruptions of Mount St. Helens, Washington*. Professional Paper 1250, ed. P.W. Lipman and D.R. Mullineaux, 601–616. Washington: U.S. Geological Survey.

Janda, R.J., D.F. Meyer, and D. Childers. 1984. Sedimentation and geomorphic changes during and following the 1980–1983 eruptions of Mount St. Helens, Washington. *Shin-Sabo* 37 (3): 5r17.

Keller, E.A., and W.N. Melhorn. 1978. Rhythmic spacing and origin of pools and riffles. *Geological Society of America Bulletin* 89: 723–730.

Leavesley, G.H., G.C. Lusby, and R.W. Lichty. 1989. Infiltration and erosion characteristics of selected tephra deposits from the 1980 eruption of Mount St. Helens, Washington, U.S.A. *Hydrological Sciences* 34 (3): 339–353.

Lehre, A.K., B.D. Collins, and T. Dunne. 1983. Post-eruption sediment budget for the North Fork Toutle River Drainage, June 1980–June 1981. *Zietschrift fur Geomorphologie* 46: 143–163.

Lipman, P.W., and D.R. Mullineaux, 1981. *The 1980 eruptions of Mount St. Helens, Washington*. Professional Paper 1250. Washington, DC: U.S. Geological Survey.

Lisle, T.E. 1995. Effects of coarse woody debris and its removal on a channel affected by the 1980 eruption of Mount St. Helens, Washington. *Water Resources Research* 31 (7): 1797–1808.

———. 2008. The evolution of sediment waves influenced by varying transport capacity in heterogeneous rivers. In *Gravel bed rivers VI: From process understanding to river restoration*, ed. H. Habersack, H. Piegay, and M. Rinaldi, 443–472. Amsterdam: Elsevier.

Lisle, T.E., A.K. Lehre, H.A. Martinson, D.F. Meyer, K.M. Nolan, and R.D. Smith. 1983. Stream channel adjustments after the 1980 Mount St. Helens eruptions. In *Proceedings of the symposium on erosion control in volcanic areas, technical memorandum of the public works research institute 1908*, 33–72. Tsukuba: Ministry of Construction.

Major, J.J. 2004. Post-eruption suspended sediment transport at Mount St. Helens: Decadal-scale relationships with landscape adjustments and river discharges. *Journal of Geophysical Research* 109:1–22:https://doi.org/10.1029/2002JF000010.

Major, J.J., and L.E. Mark. 2006. Peak flow responses to landscape disturbances caused by the cataclysmic 1980 eruption of Mount St. Helens, Washington. *Geological Society of America Bulletin* 118: 938–958.

Major, J.J., and T. Yamakoshi. 2005. Decadal-scale change of infiltration characteristics of a tephra-mantled hillslope at Mount St. Helens, Washington. *Hydrological Processes* 19: 3621–3630.

Major, J.J., T.C. Pierson, R.L. Dinehart, and J.E. Costa. 2000. Sediment yield following severe volcanic disturbance—A two decade perspective from Mount St. Helens. *Geology* 28: 812–822.

Martinson, H.A., S.D. Finneran, and L.J. Topinka. 1984. *Changes in channel geometry of six eruption-affected tributaries of the Lewis River, 1980–82, Mount St. Helens, Washington*. Open-File Report 84–614. Portland: U.S. Geological Survey.

Martinson, H.A., H.E. Hammond, W.W. Mast, and P.D. Mango. 1986. *Channel geometry and hydrologic data for six eruption-affected tributaries of the Lewis River, Mount St. Helens, Washington, Water Years 1983–84*. Open-File Report 85-631. Portland: U.S. Geological Survey.

Meyer, D.F., and H.A. Martinson. 1989. Rates and processes of channel development and recovery following the 1980 eruption of Mount St. Helens, Washington. *Hydrological Sciences Journal* 34: 115–127.

Milly, P.C.D., R.T. Wetherald, K.A. Dunne, and T.L. Delworth. 2002. Increasing risk of great floods in a changing climate. *Nature* 415 (6871): 514–517.

Moore, J.G., and T.W. Sisson. 1981. Deposits and effects of the May 18 pyroclastic surge. In *The 1980 Eruptions of Mount St. Helens, Washington,* Professional Paper 1250, ed. P.W. Lipman and D.R. Mullineaux, 421–438. Washington, DC: U.S. Geological Survey.

Mullineaux, D.R., and D.R. Crandell. 1981. The eruptive history of Mount St. Helens. In *The 1980 Eruptions of Mount St. Helens, Washington.* Professional Paper 1250, ed. P.W. Lipman and D.R. Mullineaux, 3–15. Washington, DC: U.S. Geological Survey.

Mundorff, M.J. 1984. Glaciation in the lower Lewis River basin, southwest Cascade Range, Washington. *Northwest Science* 58: 269–281.

Nakamura, F., and S. Kikuchi. 1996. Some methodological developments in the analysis of sediment transport processes using age distribution of floodplain deposits. *Geomorphology* 16: 139–145.

Neiman, P.J., L.J. Schick, F.M. Ralph, M. Hughes, and G.A. Wick. 2011. Flooding in western Washington: The connection to atmospheric rivers. *Journal of Hydrometeorology* 12 (6): 1337–1358. https://doi.org/10.1175/2011JHM1358.1.

Petersen, R.C., and K.W. Cummins. 1974. Leaf processing in a woodland stream ecosystem. *Freshwater Biology* 4: 343–368.

Pierson, T.C. 1985. Initiation and flow behavior of the 1980 Pine Creek and Muddy River lahars, Mount St. Helens, Washington. *Geological Society of America Bulletin* 96: 1056–1069.

———. 2005. Hyperconcentrated flow—Transitional process between water flow and debris flow. In *Debris-flow hazards and related phenomena*, ed. M. Jakob and O. Hungr, 159–202. Chichester: Springer-Praxis.

Pierson, T.C., and J.J. Major. 2014. Effects of explosive volcanic eruptions and related processes on surrounding landscapes. *Annual Review of Earth and Planetary Sciences* 42: 469–507.

Reid, L.M., and J. Lewis. 2009. Rates, timing, and mechanisms of rainfall interception loss in a coastal redwood forest. *Journal of Hydrology* 375: 459–470.

Roering, J.J., K.M. Schmidt, J.D. Stock, W.E. Dietrich, and D.R. Montgomery. 2003. Shallow landsliding, root reinforcement, and the spatial distribution of trees in the Oregon Coast Range. *Canadian Geotechnical Journal* 40: 237–253.

Rosenfeld, C.R., and G.L. Beach. 1983. *Evolution of a drainage network: Remote sensing analysis of the North Fork Toutle River, Mount St. Helens, Washington.* Water Resources Research Institute Report WRRI-88. Corvallis: Oregon State University.

Rowley, P.D., M.A. Kuntz, and N.S. Macleod. 1981. *Pyroclasitic-flow deposits*. Professional Paper 1250, ed. P.W. Lipman D.R. Mullineaux, 489–512. Washington, DC: U.S. Geological Survey.

Scott, K.M. 1988. *Origins, behavior, and sedimentology of lahars and lahar-runout flows in the Toutle-Cowlitz River system.* Professional Paper 1447-A. Washington, DC: U.S. Geological Survey.

Simon, A. 1999. *Channel and drainage-basin response of the toutle river system in the aftermath of the 1980 eruption of Mount St. Helens, Washington.* Open-File Report 96–633. Vancouver: U.S. Geological Survey.

Smith, R.D., and F.J. Swanson. 1987. Sediment routing in a small drainage basin in the blast zone at Mount St. Helens, Washington, USA. *Geomorphology* 1: 1–14.

Swanson, F.J., and J.J. Major. 2005. Physical events, environments, and geological-ecological interactions at Mount St. Helens—March 1980–2004. In *Ecological responses to the 1980 eruption of Mount St. Helens*, ed. V.H. Dale, F.J. Swanson, and C.M. Crisafulli, 27–44. New York: Springer.

Swanson, F.J., Collins, B., Dunne, T., and B. P. Wicherski. 1983. Erosion of tephra from hillslopes near Mount St. Helens and other volcanoes. In *Proceedings of the symposium on erosion control in volcanic areas, technical memorandum of the public works research institute 1908*, 183–221. Tsukuba: Ministry of Construction.

Swanson, F.J., C.M. Crisafulli, and D.K. Yamaguchi. 2005. Geological and ecological settings of Mount St. Helens before May 18, 1980. In *Ecological responses to the 1980 eruption of Mount St. Helens*, ed. V.H. Dale, F.J. Swanson, and C.M. Crisafulli, 13–26. New York: Springer.

Ulloa, H., A. Iroumé, L. Mao, A. Andreoli, S. Diez, and L.E. Lara. 2015. Use of remote imagery to analyse changes in morphology and longitudinal large wood distribution in the Blanco River after the 2008 Chaitén volcanic eruption, southern Chile. *Geografiska Annaler. Series A, Physical Geography* 97 (3): 523–541.

Voight, B., H. Glicken, R.J. Janda, and P.M. Douglass. 1981. Catastrophic rockslide avalanche of May 18. In *The 1980 Eruptions of Mount St. Helens, Washington.* Professional Paper 1250, ed. P.W. Lipman and D.R. Mullineaux, 347–378. Washington, DC: U.S. Geological Survey.

Waitt, R.B., Jr. 1981. Devastating pyroclastic density flow and attendant air fall of May 18—Stratigraphy and sedimentology of deposits. In *The 1980 Eruptions of Mount St. Helens, Washington*. Professional Paper 1250, ed. P.W. Lipman and D.R. Mullineaux, 439–458, Washington, DC: U.S. Geological Survey.

Waitt, R.B., Jr., and D. Dzurisin 1981. Proximal air-fall deposits from the May 18 eruption—stratigraphy and field sedimentology. In *The 1980 Eruptions of Mount St. Helens, Washington.* Professional Paper 1250, ed. P.W. Lipman and D.R. Mullineaux, 601–616. Washington, DC: U.S. Geological Survey.

Waitt, R.B., Jr., V.L. Hansen, A.M. Sarna-Wojcicki, and S.H. Wood. 1981. Proximal air-fall deposits of eruptions between May 24 and August 7, 1980—stratigraphy and field sedimentology. In *The 1980 Eruptions of Mount St. Helens, Washington.* Professional Paper 1250, ed. P.W. Lipman and D.R. Mullineaux, 617–630. Washington, DC: U.S. Geological Survey.

Waters, T.F. 1995. *Sediment in streams: sources, biological effects, and control.* American Fisheries Society Monograph 7. Bethesda: American Fisheries Society.

Ziemer, R.R. 1981. Roots and the stability of forested slopes. In *Erosion and sediment transport in pacific rim steeplands*, ed. T.R.H. Davies and A.J. Pearce, 343–361. Christchurch: International Association of Hydrologic Science.

4 The New Spirit Lake: Changes to Hydrology, Nutrient Cycling, and Biological Productivity

James E. Gawel, Charles M. Crisafulli, and Rich Miller

4.1 Introduction

Biologically, chemically, and physically, Spirit Lake was radically transformed by the 1980 eruption of Mount St. Helens and the concurrent landslide that inundated and buried the former lake and created a new one in its place. A powerful pyroclastic density current (blast PDC) physically removed or leveled the forest surrounding Spirit Lake, and the debris avalanche generated a tsunami that deposited many of these felled trees into the lake (Dahm et al. 2005), where they have formed a floating log mat that covers close to one-fifth of the lake surface even 35+ years after the blast (Fig. 4.1). With the natural outflow of the lake blocked by the debris-avalanche deposit, an outlet tunnel (2600 m long, 3.3 m diameter) connecting the lake to the North Fork of the Toutle River was completed by the Army Corps of Engineers in 1985 to provide a new outlet for Spirit Lake and to prevent catastrophic breaching of the debris dam.

The massive changes to the lake ecosystem immediately after the eruption have been well documented (Larson and Geiger 1982; Wissmar et al. 1982a, b; Larson and Glass 1987), and subsequent, less intensive studies through 2005 provide additional insights into post-eruption lake responses (Larson 1993; Dahm et al. 2005; Larson et al. 2006). However, recent regrowth of terrestrial vegetation in the watershed and internal changes in the lake itself, such as development of extensive littoral vegetation and the continued existence and colonization of the log mat, continue to alter the biogeochemistry of Spirit Lake but have not been well documented.

Our study's objectives were as follows: (1) to add to the extant body of knowledge about the lake's responses to the 1980 eruption by extending the record of measurements through 2014, (2) to establish a better understanding of the processes that govern lake productivity by developing hydrologic and nutrient mass balance models for the lake, and (3) to compare recent measurements of lake productivity and chemistry to pre- and early post-eruption data. This first attempt to quantify nutrient reservoirs and fluxes in the lake will serve as a baseline for future tracking of changes to the ecosystem.

4.2 Methods

4.2.1 Water-Column Methods

Beginning in 2008, researchers from the University of Washington Tacoma (UWT) began intensive summer through autumn (June through October) sampling of Spirit Lake and its catchment, including physical, chemical, and biological parameters. They made single trips to Spirit Lake in August 2005, September 2007, and October 2013, and they undertook multiple sampling trips each year from 2008 through 2011, and in 2014. Water-column measurements and samples were collected from both the east and west arms of the lake (Fig. 4.2) when possible, allowing for the position of the floating log mat.

Glossary terms appear in ***bold italic face***.

J.E. Gawel (✉)
Environmental Sciences and Studies, School of Interdisciplinary Arts and Sciences, University of Washington Tacoma, 1900 Commerce St, Campus Box 358436, Tacoma, WA 98402, USA
e-mail: jimgawel@uw.edu

C.M. Crisafulli
U.S. Department of Agriculture, Forest Service, Pacific Northwest Research Station, Mount St. Helens National Volcanic Monument, 42218 NE Yale Bridge Rd, Amboy, WA 98601, USA
e-mail: ccrisafulli@fs.fed.us

R. Miller
Department of Environmental Science and Management, Portland State University, PO Box 751, Portland, OR 97207, USA
e-mail: richm@pdx.edu

C.M. Crisafulli, V.H. Dale (eds.), *Ecological Responses at Mount St. Helens: Revisited 35 years after the 1980 Eruption*, https://doi.org/10.1007/978-1-4939-7451-1_4

Fig. 4.1 Aerial photo of Spirit Lake showing wind-packed log mat. This image was used to estimate the surface area of the log mat for biofilm calculations.

Temperature, dissolved oxygen, pH, and specific conductivity were measured in situ using a multiparameter water-quality probe (Hydrolab Quanta) calibrated each day at altitude before use. Water transparency was measured using a Secchi disk. Long-term water-temperature measurements were also collected in the lake using a temperature data logger suspended from an imbedded tree 1 m below the lake surface at the time of deployment (Fig. 4.2). Changing water levels in the lake, however, resulted in variable measurement depths throughout each year, generally ranging between 1 and 3 m in depth.

Samples for the analysis of nutrients, alkalinity, chlorophyll, and phytoplankton were collected at 5-, 15-, and 25-m depths using a Niskin bottle. Nutrient samples (nitrate/nitrite, ammonia, total N, orthophosphate, and total P) were collected in acid-washed Nalgene bottles, either filtered on site (dissolved fraction, 0.4-μm syringe filter) or unfiltered (total). Samples for alkalinity and chlorophyll were collected in Nalgene bottles (125 mL and 250 mL, respectively). Phytoplankton samples were collected in 250-mL brown Nalgene bottles and fixed with 1% Lugol's solution. Zooplankton samples were collected by vertical tows of 5, 15, and 25 m using an 80-μm mesh plankton net (30.5-cm diameter). The net was rinsed thoroughly to transfer organisms into the sample cup, and the samples were then preserved in 70% ethanol in 250-mL Nalgene bottles. All samples were placed on ice and transported to UWT.

Water samples for the analysis of total N and P were digested by alkaline persulfate digestion (Standard Methods 4500-P J). Filtered and digested unfiltered nutrient concentrations in water samples (total N and P, soluble reactive phosphorus, nitrate/nitrite, and ammonia) were determined by standard wet chemical methods using either a flow-through autoanalyzer (Technicon AA-II, 2008–2009, University of Washington Marine Chemistry Laboratory; Valderrama 1981) or a discrete autoanalyzer (SmartChem

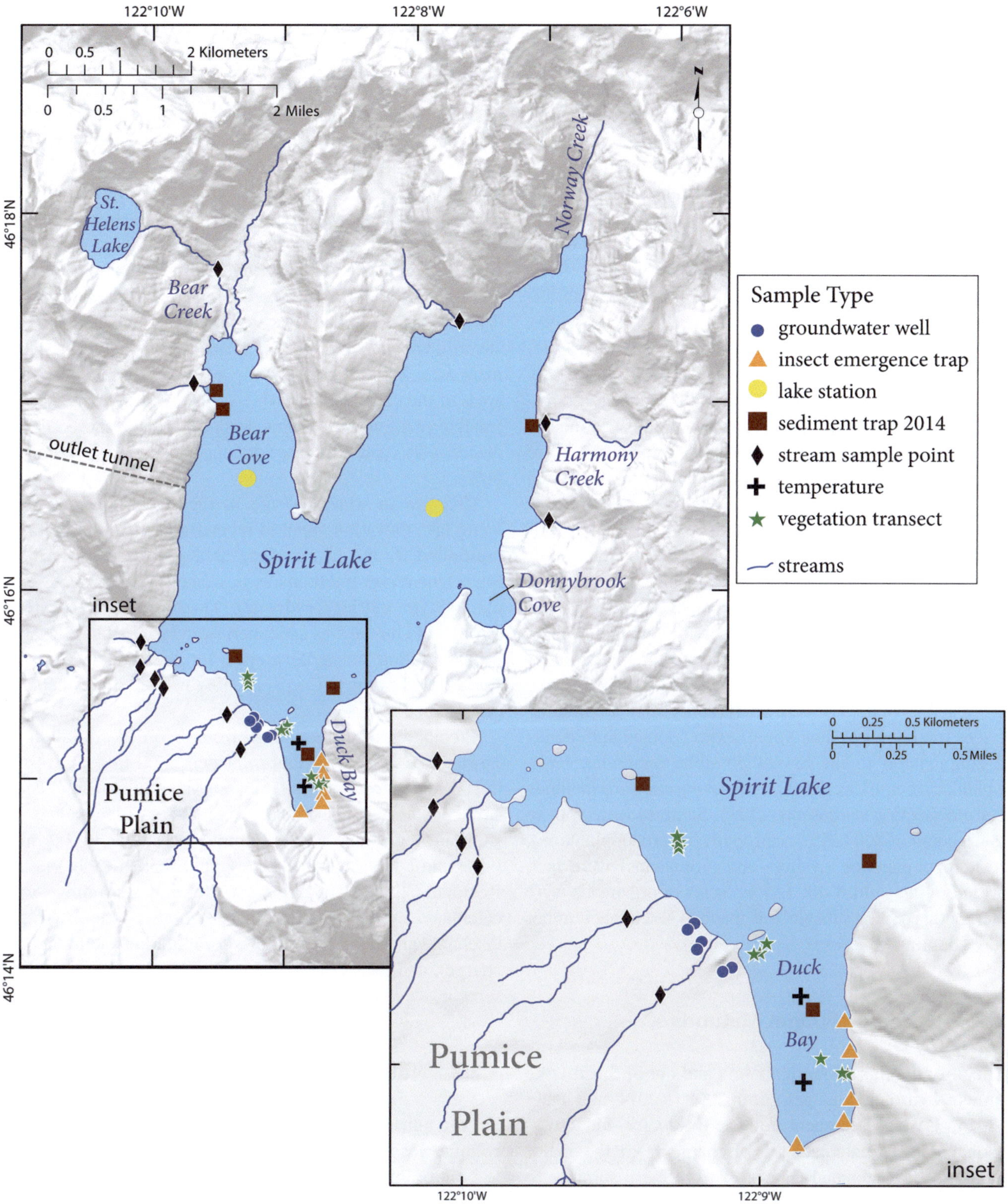

Fig. 4.2 Spirit Lake and surrounding watershed, showing sampling locations for water-column measurements, long-term temperature sensors, insect emergence traps, wells, streams, aquatic macrophytes, and sediment traps. (Cartography by Kathryn Ronnenberg, USDA Forest Service, Pacific Northwest Research Station).

200, 2010–2011 and 2014 UWT, Standard Methods 4500-P E and 4500-NH3 G and EPA 353.2).

Total alkalinity was determined by Gran titration. Chlorophyll *a* concentrations were determined using the method of Strickland and Parsons (1972). Generally, samples were filtered onto glass-fiber filters (Whatman GF/F), extracted in 90% acetone, and sonicated to disrupt the cells. Samples were then centrifuged, and the fluorescence of the supernatant was measured before and after addition of acid.

Phytoplankton samples were counted and identified by Aquatic Analysts Inc. (taxonomist Jim Sweet). Microscope slides were prepared from each sample by filtering an aliquot of the sample through a 0.45-μm membrane filter (APHA 1992, 10200.D.2; McNabb 1960), after which a section of the filter was cut out and placed on a glass slide. Immersion oil was added to make the filter transparent, followed by placing a cover slip on top. Algal units (defined as discrete particles—cells, colonies, or filaments) were counted along a measured transect of the microscope slide with a Zeiss standard microscope (1000×, phase contrast). Only those algae that were believed to be alive at the time of collection (intact chloroplast) were included, with a minimum of 100 algal units counted (APHA 1992, 10200.F.2.c.). Average biovolume for each taxon was estimated from measurements of each species. The number of cells per colony, or the length of a filament, was recorded to calculate biovolume per unit alga.

Zooplankton samples were analyzed by ZP's Taxonomic Services (taxonomist Allan Vogel), who counted and identified zooplankton to the lowest possible taxonomic level. Zooplankton counts were conducted using standard methods under a dissecting microscope (32×). Samples were split to obtain a count of 400 (±5%) arthropods (macrozooplankton). All microzooplankton (rotifers and protozoans) were then analyzed in this split if 400 (±5%) or fewer organisms were present; otherwise, a subsplit was made to reach that number before counting.

4.2.2 Hydrology Budget Methods

To quantify the flux of nutrients into and out of Spirit Lake requires a hydrologic budget. From 2008 to 2011, and again in 2014, UWT researchers collected velocity and cross-sectional data to calculate the discharge from the major streams in the watershed (Fig. 4.2) from early summer to early autumn. Local rainfall and regional evaporation rates, lake levels, and outflows through the lake's engineered outlet tunnel were determined from data available online (U.S. Geological Survey 2016). Easily accessible streams flowing north from the Pumice Plain into Spirit Lake and the stream entering the northeast arm of the lake near Harmony Falls were usually measured at least once a month from June to October. Other streams at the north end of both the west and east arms of Spirit Lake were measured only when the position of the log mat allowed.

Cross-sectional area for each stream was measured along relatively straight reaches within approximately 100 m of the lake shore. Stream depths and velocities were measured at 3–10 evenly spaced positions across the stream channel (depending on stream width) using either a SonTek FlowTracker acoustic flowmeter (2008) or a Swoffer flowmeter (2009–2011 and 2014). A calibration curve relating stream catchment area to discharge was established for the sampled streams, which accounted for approximately 44% of the catchment area. This relationship was used to estimate flow from areas of the catchment for which direct measurement was not feasible (e.g., where streams dropped off lake shore cliffs in inaccessible parts of the drainage basin—approximately 56%).

Changes in water storage in the lake were calculated using lake elevation changes recorded at the outflow tunnel multiplied by the lake surface area measured by Portland State University (methods as described in Sect. 4.2.3.7 for macrophyte nutrient estimate). Discharge from the lake through the tunnel was estimated from lake surface elevation using a rating curve developed for the tunnel (Christine Budai, U.S. Army Corps of Engineers, 2010, personal communication).

Precipitation was estimated from recorded rainfall at the Spirit Lake SNOTEL station located ~500 m southwest of the lake (USDA-NRCS 2016), and evaporation from the lake surface was estimated using regional average values for Bumping Lake (the nearest monitoring station, ~100 miles [~160 km] NE of Spirit Lake) published by the Western Regional Climate Center (2016). Groundwater inflows were calculated by the difference between other inflows (stream discharge and rainfall), outflows (evaporation and tunnel discharge), and lake storage changes. No attempt was made to separate shallow (interflow) from deep groundwater inputs.

4.2.3 Nutrient Budget Methods

For the nutrient mass balance model, the Spirit Lake water column was taken as the control volume, and all fluxes into and out of this reservoir were estimated by various methods. In addition, standing stocks of several nutrient reservoirs within the lake were also estimated to examine their relative importance in nutrient storage within the control volume. The basic conceptual nutrient mass balance model is shown in Fig. 4.3.

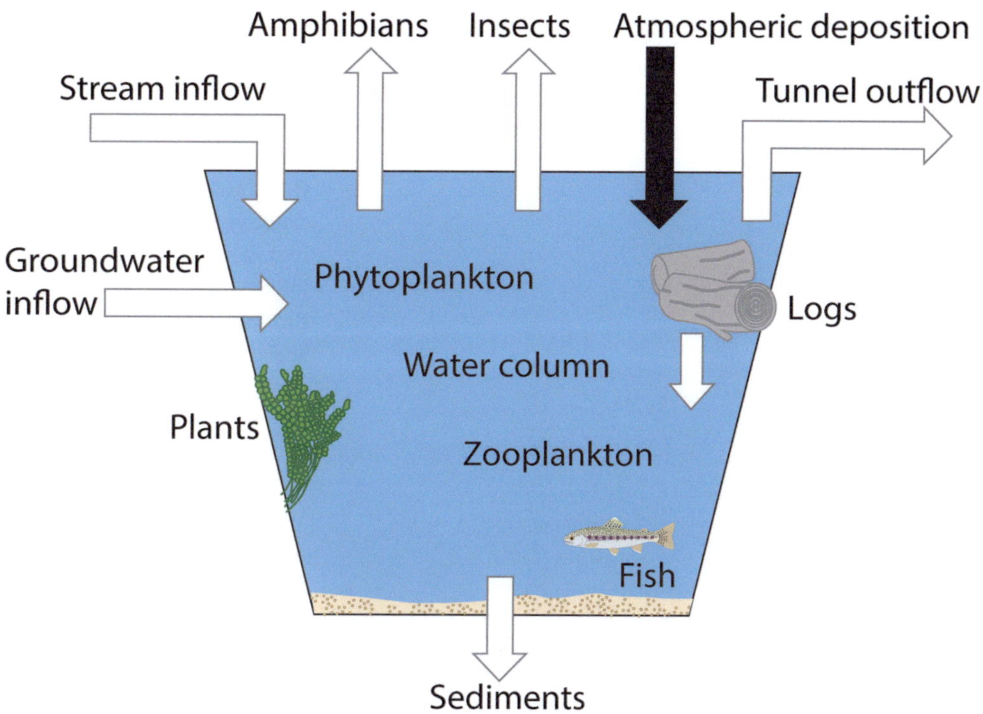

Fig. 4.3 Conceptual nutrient mass balance model for Spirit Lake. *Arrows* denote fluxes; *black arrow* is estimated atmospheric deposition from literature.

4.2.3.1 Advection

Nutrient inflows from surface- and groundwater into Spirit Lake were estimated by multiplying average total N and P concentrations, determined by measurement, by total flow rates on a monthly basis for the June–October season (loading rate = concentration x discharge). Samples for nutrient analysis were collected from streams monthly, concurrent with discharge measurements. Groundwater samples were collected monthly from six hand-driven (30.5 cm screen length, 2.1–2.4 m belowground), stainless-steel piezometers near the southern shore of the lake (Fig. 4.2). The piezometers were first pumped at low flow rates to evacuate stagnant water. Samples for total N and P were stored in acid-washed Nalgene bottles on ice for transport to UWT and then frozen prior to analysis.

Nutrient outflow through the tunnel was estimated using average ***epilimnetic*** total N and P concentrations and tunnel discharge. Total atmospheric deposition (wet + dry) for nitrogen was taken from an average of estimates calculated by the West-Wide Jumpstart Air Quality Modeling Study (Western Regional Air Partnership 2013) for White Pass and Mount Rainier National Park (70 km NE and 70 km NNE of Spirit Lake, respectively). Total atmospheric phosphorus deposition was calculated using North American average measured loading rates (Tipping et al. 2014). All nutrient inputs to the lake calculated on an areal basis were multiplied by a constant lake surface area measured by the authors in 2009–2010 as described in Sect. 4.2.3.7.

4.2.3.2 Insects

Nutrient loss as a result of insect emergence was estimated using floating emergence traps built by UWT undergraduate Cameron Marshall or modified from purchased traps (BugDorm amphibious emergence trap). Traps consisted of a white mesh tent with a collection bottle containing suspended insecticide vapor strips. Trap tents were secured to water-tight 1-m^2 floating, black polyvinyl chloride (PVC) frames tethered in place in the lake using an anchor, float, and nylon line (Fig. 4.4a). Floating emergence traps are preferable to benthic traps in lakes with dense aquatic macrophyte growth (Cushman 1983), such as in the Spirit Lake nearshore area, and efficiency of capture is comparable to or better than submerged traps (Davies 1984). Traps were located along the south shore of Spirit Lake at a water depth of 1–4 m and deployed for 24–72 h. Insect samples were sorted into broad (typically ordinal) taxonomic groups, dried, and analyzed for total N and P using standard methods (CHN analysis, Method 29-2, Weaver 1994, and ICP-OES, EPA 365.4, University of Washington Analytical Service Center). Total insect nutrient flux from the lake was determined by multiplying the average trap flux over the sampling season (g N or P m^{-2} day^{-1}) by the area of the lake less than 10 m in depth (i.e., the depth of the average ***thermocline*** in Spirit Lake from 2008 through 2011) and the length of the sampling season. The use of this depth as a cutoff was based on the work of Winnell and White (1985), who found that insect emergence in Lake Michigan was very low below the thermocline.

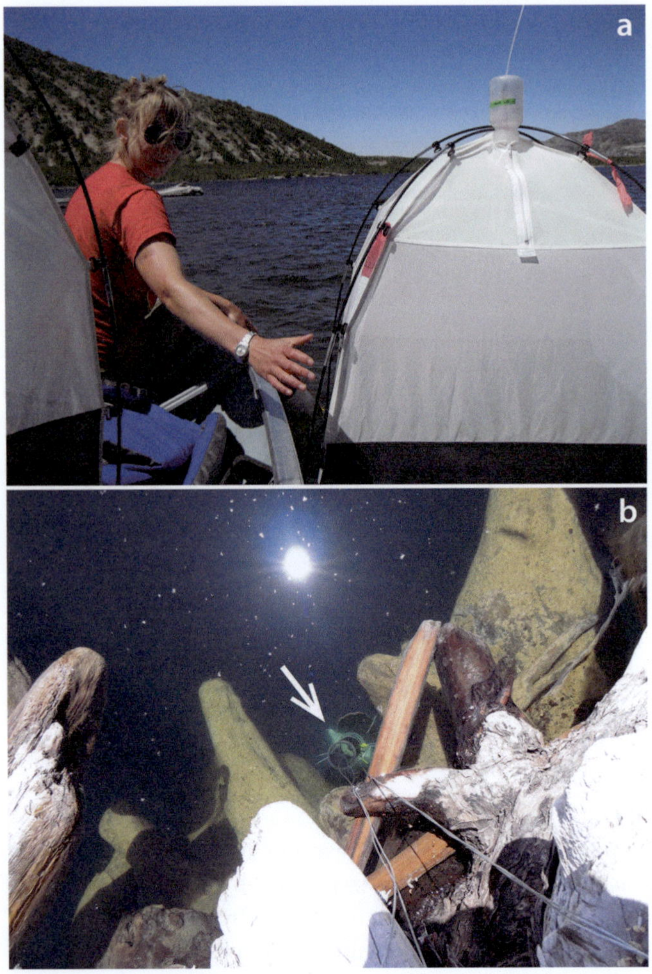

Fig. 4.4 Field photos of (**a**) floating insect emergence trap (BugDorm) in 2010 on Spirit Lake and (**b**) sediment trap (*white arrow*) suspended from a floating log in the log mat on Spirit Lake in 2014.

4.2.3.3 Sedimentation

Nutrient loss to the sediments was estimated from sediment traps. Sediment traps were constructed from 61-cm-long (24 in), 10-cm (4 in) diameter black acrylonitrile butadiene styrene (ABS) pipe with a conical funnel glued to the bottom, fitted with a removable 125-mL acid-washed Nalgene bottle. Sediment traps were suspended ~2 m above the lake bottom attached to nylon line anchored to the bottom and buoyed by near-surface floats (to avoid damage by the floating log mat). In 2014, sediment traps were located near embedded logs and tethered to the logs for increased likelihood of recovery. Sediment samples collected were filtered (0.4 μm), dried, and analyzed for total N and P as described for insects in Sect. 4.2.3.2.

4.2.3.4 Log Mat/Biofilm

In 2014, we made the first estimates of the nutrient flux from the floating log mat to the Spirit Lake water column. To do this, we suspended sediment traps (constructed as described in Sect. 4.2.3.3) approximately 1.5–3.0 m below three random large logs floating in the lake (Fig. 4.4b). The logs were marked with 1.5-m-tall flags for visibility, and the traps traveled within and under the log mat for 1–2 months before the accumulated material was collected for analysis. We assumed that the nutrients collected were primarily from biofilm production subsequently shed by the grinding action of the log mat. The total flux of nutrients from the log mat was calculated using the average flux of nutrients into the traps, and the total area of the log mat was calculated from a recent aerial photo (Fig. 4.1). This two-dimensional estimate for surface area of the log mat underestimates the three-dimensional surface of the logs below the water line.

4.2.3.5 Amphibians

We also estimated the loss of nutrients by amphibian emigration. To derive biomass estimates for amphibians leaving Spirit Lake, we sampled recently metamorphosed population segments in replicated plots during late summer 2010. Nine 15 × 50-m plots were installed in the riparian zone immediately adjacent to the shoreline. The long axis of each plot (50 m) was positioned parallel to the shoreline and extended from the current waterline 15 m inland. The rationale for using a 15-m band from the water edge was based on predispersal habitat use by recently metamorphosed amphibians at Spirit Lake (Crisafulli et al. 2005). The corners of each 750-m^2 plot were marked with rebar, and a measuring tape was stretched taut around the corner bars to delineate the plot boundaries. Plots were sampled on three occasions between 19 August and 10 September. We selected the sampling date that yielded the greatest cumulative amphibian biomass in nine plots, based on the schedules of species metamorphosis derived from sampling at Spirit Lake and other lakes at Mount St. Helens from 1981 through 2010 (Crisafulli et al. 2005). This method results in a minimum estimate of the actual population observed at the sites and provides the best estimate of amphibian biomass leaving the lake. Within each plot, researchers carefully searched the ground surface, vegetation, and beneath cover objects (tree boles and fragments) for amphibians, which were captured, identified to species, weighed to the nearest 0.01 g, measured (snout-vent and total length, mm), and placed in a bucket until plot sampling was complete. All animals were released back to their exact point of capture. Sampling generally occurred from 0900 to 1600 h. Mean daily air temperature on the 3 days surveyed was 7.7 °C (±0.95 standard deviation), and conditions leading up to and on the first two survey days were dry. On the day preceding the last survey day, 2.5 mm of rain fell, but the day of survey was dry. During the last sampling effort, a subsample of individuals of each species was retained, euthanized by either pithing or freezing, weighed for wet weight, and oven-dried at 45 °C for 48 h. To extrapolate sample estimates to the entire lake, a GIS (geographic information system) was used to measure the length of shoreline that contained suitable amphibian habitat (e.g., shallow littoral zone) and avoided steep topography such as cliffs and where water depth increased precipitously. There were 4083 m of shoreline with suitable amphibian habitat, and we then multiplied this value by 15 to correspond to the width of our study plots, yielding a total estimated suitable habitat area of 61 245 m^2. Representative samples of amphibian species were ground and analyzed for total N and P as described for insects in Sect. 4.2.3.2.

4.2.3.6 Water Column, Benthic Sediment, and Plankton Reservoirs

Standing-stock reservoirs of nutrients estimated in Spirit Lake include the water column, benthic sediments, phytoplankton, zooplankton, aquatic macrophytes, fish, and biofilm growing on the floating log mat. We estimated the reservoir of nutrients in the water column by using the average total N and P concentrations measured over the sampling season and multiplying by the lake volume as calculated by Portland State University (methods as described in Sect. 4.2.3.7 for macrophytes).

We collected benthic sediments in 2010–2011 from 24 random locations spaced throughout the lake using a hand-hauled petit ponar dredge (Wildco). Samples were homogenized on the boat and a subsample collected and stored on ice prior to processing. Sediments were dried for 3 days at 80 °C, homogenized in a Wiley mill, and analyzed for total N and P by the University of Washington Analytical Service Center (see details in Sect. 4.2.3.2). We estimated the total reservoir of nutrients in the sampled layer (approximately 10 cm) of benthic sediments using the average sediment concentration and the area of the lake.

As described in Sect. 4.2.1, we estimated nutrient mass in phytoplankton from biovolume measurements of samples collected. We multiplied average phytoplankton biovolume per liter by the volume of the lake photic zone (estimated by thermocline depth) and converted to total N and P using published values for the relationship between biovolume and dry weight (0.1 g dry wt cm^{-3} biovolume; Wilson 2003) and between the N and P content by dry weight for phytoplankton (5.5% N and 1.14% P by dry weight; Duarte 1992).

On 14 September 2010, we determined zooplankton nutrient mass per organism by collecting zooplankton samples from triplicate 25-m tows. One replicate was stored in ethanol and counted as described in Sect. 4.2.1. We filtered the other two replicates onto 0.45-μm polycarbonate filters, dried and analyzed them for total N and P as outlined for insects in Sect. 4.2.3.2. We established an average mass of N and P per organism and calculated the total mass of N and P in zooplankton in the lake by multiplying this conversion factor by the number of organisms per tow area and the total lake area.

4.2.3.7 Macrophyte Reservoir

We estimated aquatic macrophyte biomass by diver surveys and acoustic vegetation mapping. In both 2009 and 2010, divers collected macrophyte samples along three transects from the southern shore of Spirit Lake (Fig. 4.2). Macrophytes were sparse in less than 2 m of water in Spirit Lake owing to the scouring action of the log mat, and only patchy periphyton were found below 8-m depth as a result of light limitation. We collected triplicate samples from randomly placed 900-cm^2 quadrats near the center line of each transect at approximately 3-, 5-, and 7-m depths. We estimated total sample biovolume in situ from average plant height and quadrat area for each sample (to be used with acoustic mapping), then removed aboveground biomass, dried samples at 60 °C for 3 days, and analyzed them for total N and P content. We applied average areal biomass measurements from the areas surveyed in 2009 and 2010 to the entire lake area from 2 to 8 m in depth.

In 2010, we also estimated total biovolume of macrophytes in Spirit Lake by acoustic mapping. We carried out bathymetric and aquatic-plant coverage surveys on 24–27 August 2009 and 17–18 July 2010, using a Biosonics DE Scientific Echosounder equipped with a 420-kHz, narrow-beam (8° beam width) transducer. Persistent logs along the western shore of the northeastern arm of the lake and the eastern shore of the northwestern arm prevented collection of hydroacoustic data in those areas. We collected additional data in shallow areas along the southern shore with dense aquatic vegetation. We integrated and saved location data and hydroacoustic return signal data stronger than −147 dB using Biosonics Visual Acquisition software.

Based on characteristics of each acoustic ping, we used Biosonics Visual Analyzer software to determine the vertical location of the top of the macrophyte canopy and the sediment/water interface. In areas with very dense aquatic plants and abundant submerged logs, we adjusted the bottom track in Visual Analyzer to correct for strong return signals off objects above the sediment surface. We calculated sediment elevations from depth recordings, taking into account transducer face depth and the water-surface elevation at the lake outlet (USGS site number 14240304), obtained for each sample date.

We combined the resulting sediment elevation points with LIDAR elevation points (EarthData International of Maryland 2003) within 10 m of the lake shoreline and interpolated a bathymetric surface and aquatic-plant canopy height from the points using the ArcGIS Geostatistical Analyst kriging algorithm (ESRI 2001). Based on diver transect/quadrat vegetation data (described in the first paragraph of this section) as well as inspection of the acoustic signal, data suggested that plant canopy points at depths greater than 8 m were signals from submerged logs rather than aquatic plants and thus we did not include them in interpolations. We calculated surface area and volume estimates at 0.1-m vertical intervals from the interpolated surfaces using the ArcGIS "Surface Volume" tool. Using diver-collected samples from 2010 (described above, this section), we converted plant biovolume to N and P mass.

4.2.3.8 Log-Mat Biofilm Reservoir

We estimated the nutrient mass in the biofilm growing on the floating log mat in 2010, 2011, and 2014. Because logs of any appreciable size in the mat are waterlogged and stable, with little to no change in the side submerged, biofilm has established on the submerged portion of each log. We sampled random logs during August each year by flipping the log and cutting out 5 × 5 × 0.5-cm wood samples with a vibrating saw. Samples were placed in plastic bags and stored on ice for transport. In the lab, samples were transferred into glass jars and dried for 3 days at 80 °C, homogenized in a Wiley mill, and analyzed for total N and P. We calculated the total mass of nutrients in the biofilm on the log mat by multiplying the average mass per sample area by the total area of the log mat determined from a recent satellite image.

4.2.3.9 Fish Reservoir

During the summer months from 2008 through 2011, we sampled rainbow trout (*Oncorhynchus mykiss*), the only fish species known to inhabit post-eruption Spirit Lake, using hook and line. It is believed that all fish in Spirit Lake perished during the 1980 eruption and that rainbow trout were illegally introduced to the lake by an unknown party in the early 1990s (Bisson et al. 2005). Sampling was focused in Duck Bay in the southeast portion of the lake (Fig. 4.2), where we captured 904 fish, processed them for snout-fork length, mass, sex, and reproductive status, and then fitted them with a Floy tag and released them. To estimate the rainbow trout population size, we used a closed population, multiple capture–recapture analysis (Schnabel 1938). We summarized recapture data by total captures (n), recaptures (m), marked fish extant in population (M), and the number of marked fish returned to the population after each survey (R). We used the mrClosed vignette in the FSA package built under R version 0.4.44 (Ogle 2015) to calculate population estimates and 95% confidence intervals. Tagged fish were never recaptured more than 3 years after the original marking date; therefore, we used only individuals marked between 2008 and 2011 to calculate the number of marked individuals in the population. For nutrient assays on 10 September 2010, we retained and euthanized a male trout that was kept on ice until transferred to a freezer within 24 h. This individual had a snout-fork length of 438 mm and wet weight of 950 g. The fish was later thawed at room temperature, cut into small pieces (~1 cm^3) and placed in a drying oven for 48 h at 45 °C. Dried tissue was analyzed for total N and P as described for insects in Sect. 4.2.3.2.

4.3 Lake Conditions Prior to the 1980 Eruption

Spirit Lake was not well studied prior to the 1980 eruption, although some scant and basic chemistry, biology, and bathymetry data were gathered in 1974 (Dion et al. 1976) and again on 4 April 1980, just prior to the eruption (Wissmar et al. 1982a). Spirit Lake's relatively small drainage basin (roughly seven times the lake surface area) was densely forested with mixed conifer species, suggesting that few nutrients were available for mobilization into the lake (Vitousek et al. 1979). Pre-eruption Spirit Lake was steep-sided and deep (Dion et al. 1976), with mean and maximum depths of 37 m and 58 m, respectively (Table 4.1 and Fig. 4.5).

Nutrient concentrations in Spirit Lake were low before the eruption: total nitrogen was <170 μg L^{-1} and total phosphorus was <12 μg L^{-1} in near-surface waters (Table 4.2; Dion et al. 1976). In general, water-quality parameters in Spirit Lake were similar to other forested montane lakes in Washington's Mount Rainier National Park, where average values ranged from 13 to 181 μg L^{-1} for total N, 5–47 μg L^{-1} for total P, 4.3–57.9 μS cm^{-1} for conductivity, and 8–409 μeq L^{-1} for alkalinity (Larson et al. 1994).

Water clarity was high in pre-eruption Spirit Lake but variable throughout the summer, with ***Secchi depths*** ranging from 7 to 14 m (Fig. 4.6; Dion et al. 1976). Chlorophyll *a* concentrations were less than 1.3 μg L^{-1} (Table 4.2; Dion et al. 1976). Based on Secchi depth and chlorophyll *a*, the lake's trophic state index (TSI) was less than 30 (Table 4.3) and would have been categorized as ***oligotrophic*** (Carlson 1977).

Because of the lake's steep-sided and deep bathymetry and because it was largely shaded along the shoreline by both cliffs and forest cover, emergent and submergent plants covered less than 1% of the lake area (Dion et al. 1976). In June 1974, the main rooted aquatic plant observed was *Sphagnum* moss, and the phytoplankton community was described simply as "diatoms, greens, and yellow-browns" (Dion et al. 1976).

Other interesting pre-eruption data include the dissolved oxygen and temperature profiles collected in 1974 (Dion et al. 1976). Maximum surface temperature in August

Table 4.1 Spirit Lake morphometry and watershed area before and after the 1980 eruption of Mount St. Helens.

	Prior to 1980 eruption	After 1980 eruption	Percent change
Surface area (km^2)	5.3	11.0	+108
Volume (m^3)	1.97×10^8	2.56×10^8	+30
Mean depth (m)	37	23	−38
Max depth (m)	58	39	−33
Drainage area (km^2)	38.6	26.4	−32

Data from Dion et al. 1976, or this study collected in 2009–2010.

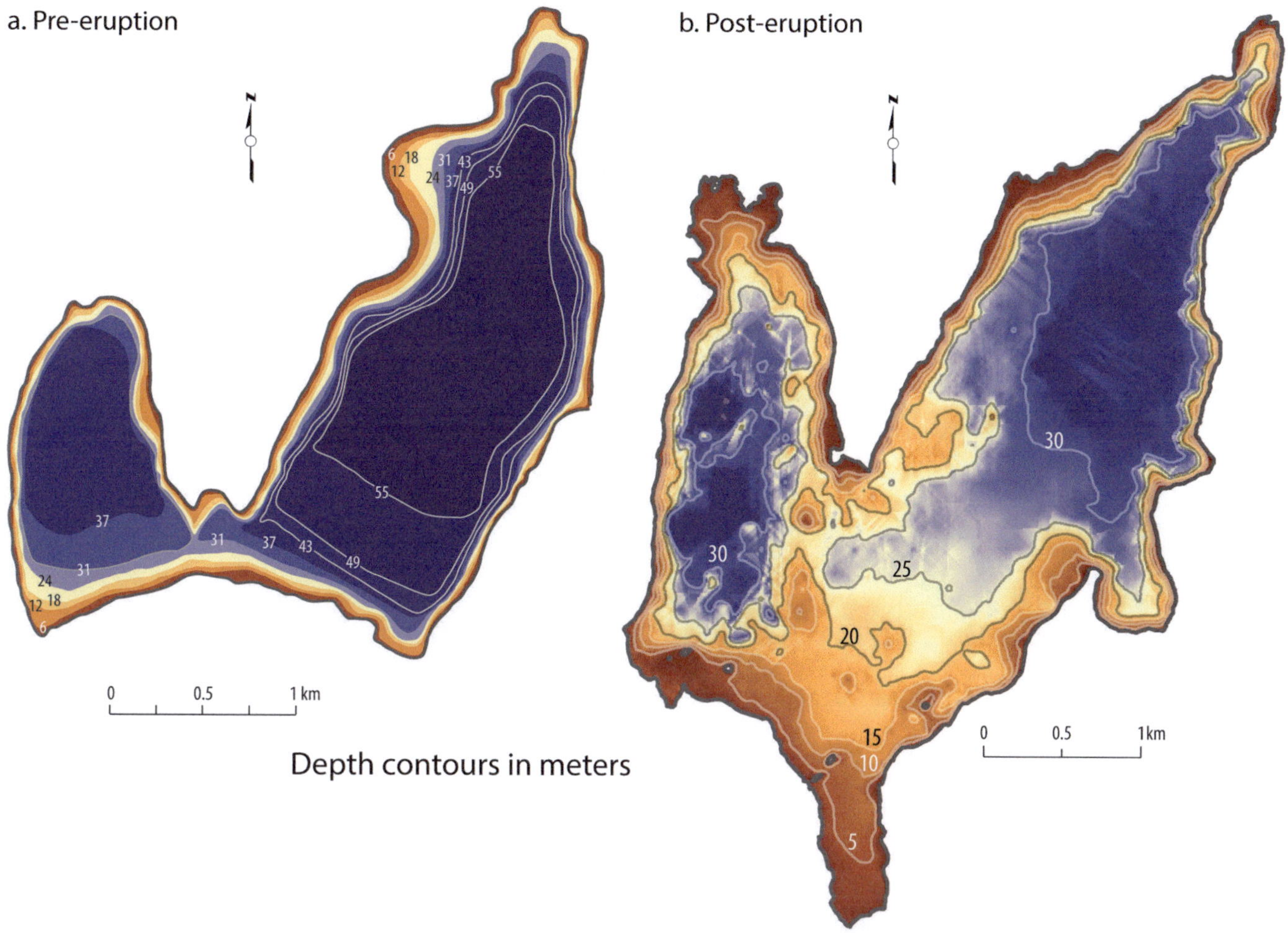

Fig. 4.5 Bathymetric maps of Spirit Lake (**a**) prior to (Dion et al. 1976) and (**b**) after the 1980 eruption of Mount St. Helens.

Table 4.2 Historic Spirit Lake surface water-quality data compared to current study. Where multiple months were sampled and results were variable, the range of values (low–high) is given.

Dates measured	Dissolved N: $NO_3^- + NO_2^- + NH_3$ (μg N L^{-1})	Particulate organic N (μg N L^{-1})	Dissolved PO_4^{3-}–P (μg P L^{-1})	P_{TOT} (μg P L^{-1})	ALK (μeq L^{-1})	SpC (μS cm^{-1})	Chlorophyll a (μg L^{-1})
June–October 1974[a]	10–40	30–130	0–3	2–12	260	19–27	0.3–1.3
4 April 1980[b,c]	2.38	47	0.93	7.12	139	NM	NM
30 June 1980[b,c]	18.63	70	230	236	3010	NM	NM
August 1980[d]	>2000	NM	250	260	3400	793	NM
August 1985[e]	3	NM	5	24	1702	690	NM
September 2004[f]	NM	NM	5	13	768	250	0.26–0.34
31 August 2005f	2	NM	<1	16	726	236	2.1–17.2
4 August 2005	NM	NM	NM	NM	NM	221–232	NM
11 September 2007	NM	NM	NM	NM	NM	200–202	NM
June–October 2008	0.9–9.1	NM	0.5–1.2	NM	NM	150–165	0.175–14.7
June–October 2009	3–1613	NM	<2	11	305–611	159–181	0.144–3.21
June–October 2010	NM	NM	<9	11	438–586	150–171	0.178–5.26
June–October 2011	27–1545	NM	<2.8	14	375–627	126–157	0.174–1.60
12 October 2013	NM	NM	NM	NM	NM	111–117	NM
June–October 2014	3–1003	NM	0–32	11	375–501	100–108	NM

NM not measured.

Sources: [a]Dion et al. (1976), [b]Wissmar et al. (1982a), [c]Larson (1994), [d]Larson and Geiger (1982), [e]Larson and Glass (1987), and [f]Larson et al. (2006). Discrepancies were noted in Larson (1994) reporting of Wissmar et al. (1982a) data.

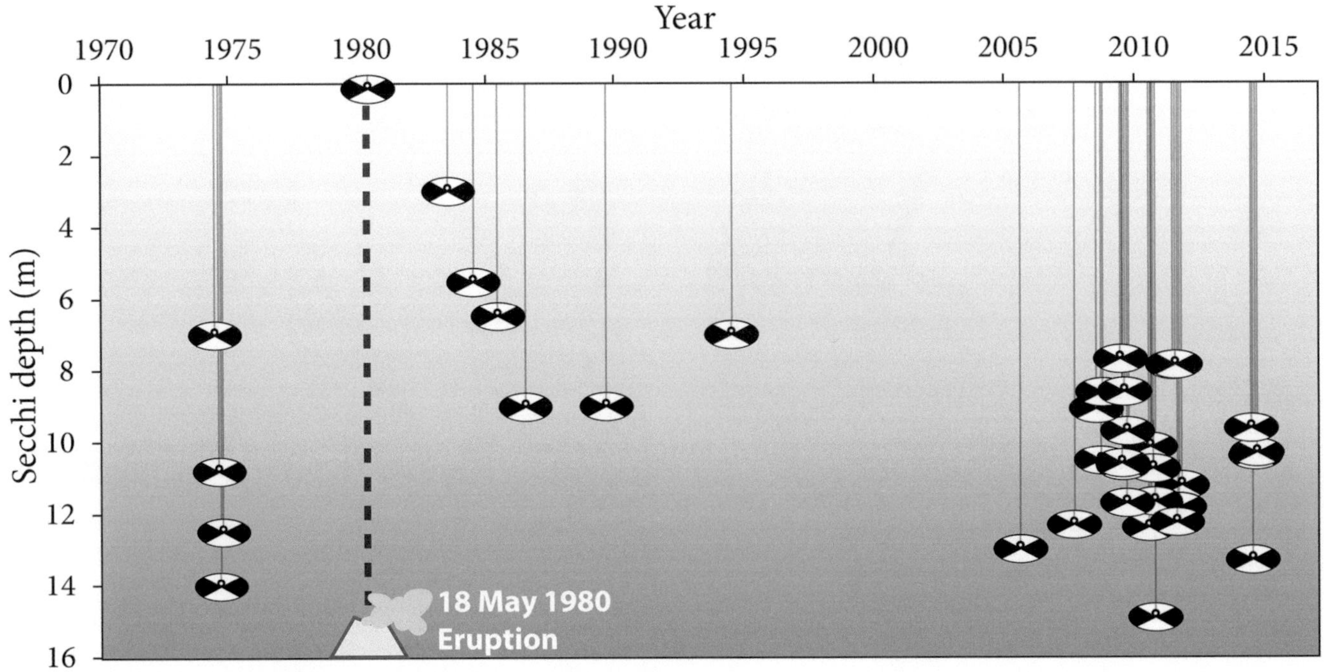

Fig. 4.6 Secchi depth recorded from 1974 through 2014.

Table 4.3 Trophic state index (TSI) calculated for Spirit Lake prior to and after the 1980 eruption of Mount St. Helens using summertime average Secchi depth and chlorophyll *a* concentrations as productivity indicators.

	TSI based on indicator	
Year	Secchi depth (m)	Chlorophyll *a*
1974[a] pre-eruption	25	27
2005[b]	23	43
2008	28	37
2009	27	33
2010	24	34
2011	27	31

Sources: [a]Dion et al. (1976) and [b]Larson et al. (2006).

reached 16 °C, whereas dissolved oxygen concentrations at the bottom of the lake dropped below 2 mg L^{-1}. The low oxygen concentrations suggest that there was significant sediment oxygen demand before the eruption owing to organic-rich sediments. Moreover, it is likely that the sediments became anoxic during the summer in pre-eruption Spirit Lake, as evidenced by higher dissolved PO_4-P, Fe, and Mn, as well as elevated specific conductivity in the bottom waters (Dion et al. 1976).

4.4 Changes in Lake Properties After Eruption, 1980–2005

4.4.1 Water Clarity

Immediately following the 1980 eruption, Secchi depth was greatly reduced to "a few centimeters" by avalanche deposits and charred forest debris that entered the lake and anaerobic microbial production of metal–sulfide particulates (Larson and Glass 1987). Water clarity improved steadily thereafter, reaching 9 m by 1987 (Fig. 4.6). Secchi depth in July 1994 was similar to 1987 (Larson 1994), but measurements made in 2004 and 2005 (Larson et al. 2006) varied seasonally between 10 and 15 m, which were similar to those reported prior to the 1980 eruption (Fig. 4.6).

4.4.2 Nutrient Concentrations

Shortly after the eruption in June 1980, dissolved and total P concentrations skyrocketed (Table 4.2) as a result of the combined effects of P-enriched volcanic material and charred forest debris (Dahm et al. 1983). Dissolved PO_4-P concentrations increased approximately 80 times above pre-eruption levels, and total P concentrations were 20 times greater (Dion et al. 1976; Larson and Geiger 1982). By August 1980, dissolved inorganic N concentrations had also increased dramatically, reaching levels 50 times higher than pre-eruption levels. This increase was short-lived as N and P concentrations fell to near pre-eruption levels by 1985, although total P and orthophosphate concentrations remained slightly elevated (approximately twofold higher than pre-eruption levels) until 2004 (Larson et al. 2006).

4.4.3 Dissolved Oxygen

The large influx of organic matter and reduced chemical compounds immediately after the eruption fueled intense microbial activity. A combination of abiotic redox reactions and accelerated aerobic respiration (owing to warmer water temperatures and an abundance of organic matter) depleted oxygen in all but the top 1 or 2 m of the water column, where dissolved oxygen concentrations were 2–3 mg L^{-1} during summer 1980 (Dahm et al. 2005). By 1983, however, inflows of fresh water, the infusion of oxygen by wind mixing, and a decrease in organic matter (as a result of oxidation by microbes and the settling of suspended solids) resulted in an increase in dissolved oxygen concentrations in the lake (Dahm et al. 2005). Water-column profiles in 1986 showed continued oxygen demand below the thermocline, with dissolved oxygen concentrations ranging between 1 and 2 mg L^{-1} from below the thermocline to near the lake bottom at 30 m (Dahm et al. 2005).

4.4.4 Alkalinity

Total alkalinity increased by a factor of 10 or more immediately after the eruption and has decreased more slowly than many other water-quality parameters since the eruption (Table 4.2). Alkalinity in 2005 was still almost three times greater than pre-eruption levels.

4.4.5 Temperature

Surface temperatures in post-eruption Spirit Lake spiked immediately after the eruption in May 1980 to ~34 °C (Wissmar et al. 1982b), before decreasing to around 17 °C by August 1983 (Dahm et al. 2005).

4.4.6 Chlorophyll

Post-eruption chlorophyll *a* concentrations, not measured until September 2004 (Larson et al. 2006), were similar to pre-eruption values (Table 4.2). Concentrations measured in 2005 were more than ten times pre-eruption levels. Pre- and post-eruption comparisons are questionable since pre-eruption data are sparse and chlorophyll concentrations can vary greatly from year to year (as may have occurred between 2004 and 2005) and even within a season (Knowlton and Jones 2006).

4.4.7 Trophic Status

Since summer 1980, Spirit Lake has evolved from a ***hyper-eutrophic*** system to one that can be regarded as oligotrophic (Table 4.3). This has resulted partly from the steady decline in nutrient concentrations, especially nitrogen and phosphorus. Further evidence of the lake's recovery are current Secchi depths, which can reach pre-eruption levels. In 2005, however, chlorophyll *a* concentrations were quite a bit higher, resulting in TSI values ranging from 23 to 43 (Table 4.3), suggesting the lake may now be more productive, at least periodically, compared with its oligotrophic pre-eruption state.

4.5 The New Spirit Lake: 2005–2014

4.5.1 General Characteristics of the Water Column

Spirit Lake was stratified by the time we arrived in June of each year and remained stratified into October (2009 shown in Fig. 4.7a as an example). Judging from the generally weak stratification on 22 June 2011 (data not shown) and the deepening and weakening of the thermocline at the end of September and the beginning of October 2008, 2010, and 2011 (data not shown), we assume that the lake began to stratify in late May–early June and that autumn turnover occurred in late October–early November. The long-term temperature data collected in the lake (Fig. 4.8) show that surface water temperatures below 4 °C (the temperature of maximum water density) were reached during winter, resulting in inverse temperature stratification with water <4 °C at approximately 1–5-m depth. Therefore, although we do not have direct observations that the lake was ice-covered in winter (access to Spirit Lake in winter is very difficult, requiring either a full-day ski through avalanche-prone terrain or a helicopter flight), the lake can be classified as ***dimictic*** (having two separate mixing periods—autumn and spring). The thermocline was situated around 7–9-m depth each year, with epilimnetic (warmer surface layer of the lake) temperatures reaching 21 °C in late July 2009 and ***hypolimnetic*** (cold bottom layer of the lake) temperatures averaging between 4 and 7 °C.

Dissolved oxygen profiles indicate a regularly occurring summertime oxygen maximum in the ***metalimnion***, most

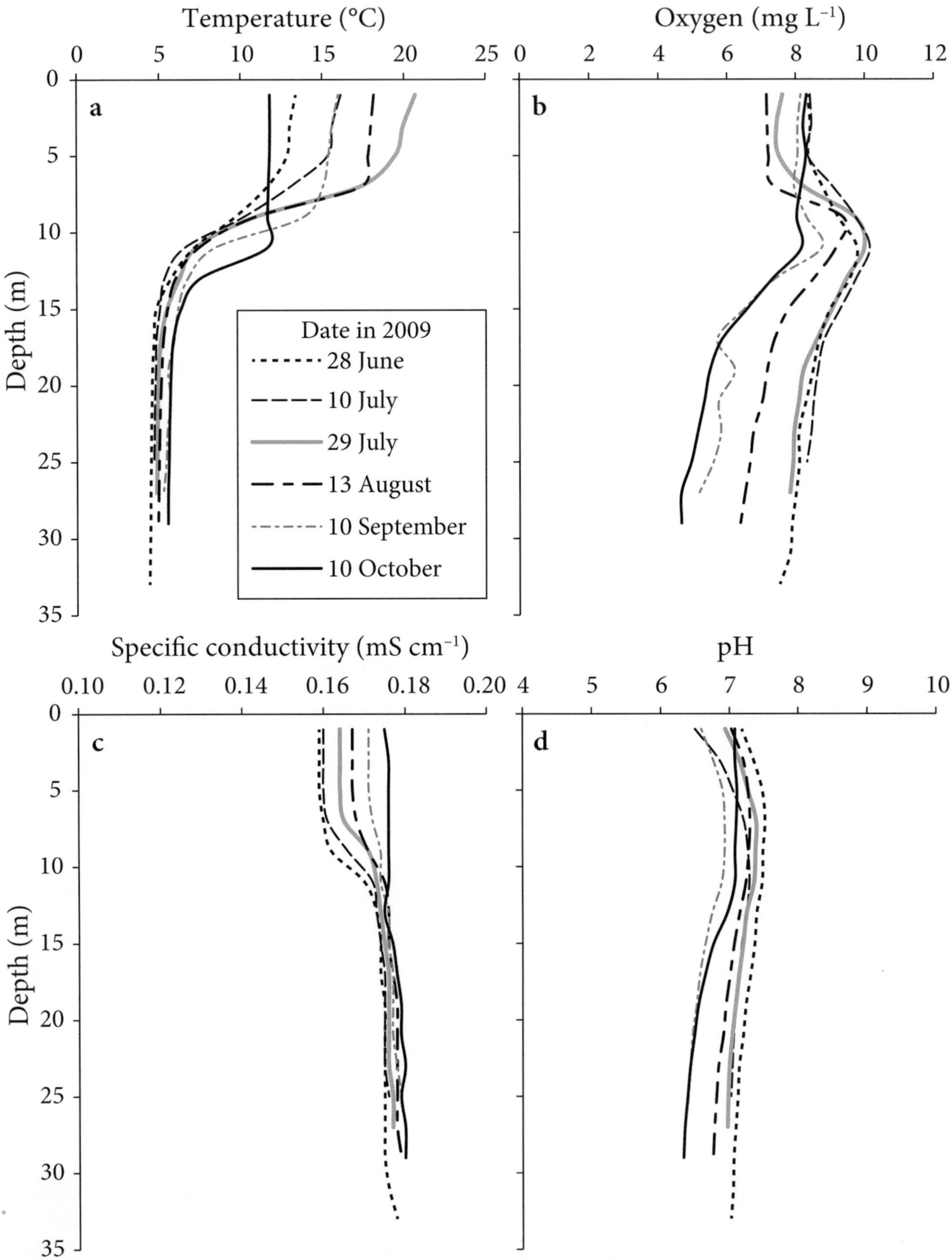

Fig. 4.7 Water-column profiles for (**a**) temperature, (**b**) dissolved oxygen, (**c**) specific conductivity, and (**d**) pH in the west arm of Spirit Lake in 2009.

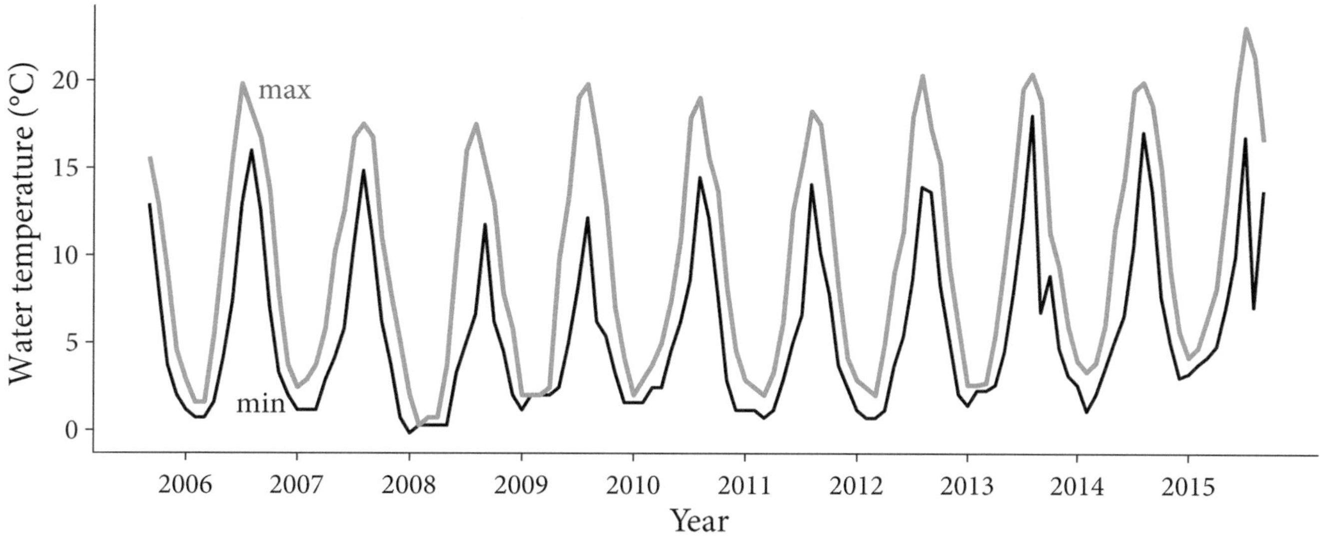

Fig. 4.8 Monthly minimum and maximum temperature recorded in Duck Bay surface water from 2006 through 2015.

likely resulting from a combination of higher numbers of phytoplankton, lower temperatures, and a more stable water column (2009 data shown in Fig. 4.7b as an example). The metalimnion is the middle layer of the lake where the temperature changes rapidly from warmer to cooler. The lowest dissolved oxygen concentration we recorded was 2.0 mg L^{-1} in October 2009 near the lake bottom, although the minimum concentration stayed above 3.3 mg L^{-1} from 2010 through 2014. Hypolimnetic oxygen concentrations showed oxygen depletion during the summer, although oxygen demand was less than that recorded in the 1980s. In 2010 and 2011, the areal hypolimnetic oxygen deficit (AHOD)—the rate of oxygen consumption by biotic and abiotic processes in the lower layer of the stratified lake—for Spirit Lake was 0.88 and 0.76 g O_2 m^{-2} day^{-1}, respectively, a 12–24% reduction from 1986 (Larson and Glass 1987). This AHOD is in the same range as a set of 11 ***eutrophic*** lakes measured in a Swiss study, where the authors found an average range of 0.90 ± 0.30 g O_2 m^{-2} day^{-1} (Müller et al. 2012). However, it has been shown that a thicker hypolimnion may artificially indicate a more productive lake than is justified by other measures, suggesting a problem with the application of AHOD as a comparative measure of trophic status across lakes with different morphologies (Cornett and Rigler 1980).

Specific conductivity—a measure of dissolved ions in the water—in Spirit Lake has continued to decline since the eruption (2009 shown in Fig. 4.7c as an example). Measurements throughout the water column declined steadily from a maximum value of 202 μS cm^{-1} in 2007 to 108 μS cm^{-1} in 2014, much lower than values recorded in 1986 (range 375–550 μS cm^{-1}) and lower even than 2005, when values were recorded in the 220–240 μS cm^{-1} range by UWT and Dahm et al. (2005) (Table 4.2). As mentioned by Larson and Glass (1987), conductivities continue to be higher in the hypolimnion than in the epilimnion by 10–20 μS cm^{-1}. The conductivity profiles we measured in the lake (Fig. 4.7c) suggest a hypolimnetic source of dissolved ions via mineral-laden groundwater entering the lake below the thermocline, rather than the diffusion of dissolved ions from anoxic sediments.

From 2008 through 2014, water-column profiles of pH ranged from 6.1 to 8.0 (2009 shown in Fig. 4.7d as an example). The highest pH values were generally found in the metalimnion, coinciding with higher dissolved oxygen, suggesting photosynthetic depression of CO_2 as the source of higher pH. Our more recent pH measurements exhibited a greater range and reached more acidic levels than in 1986 (pH 7.0–8.0) but were similar to 1983 and within the range of values recorded in the pre-eruption lake in 1974 (Larson and Glass 1987).

From 2008 through 2011, maximum chlorophyll concentrations decreased from 14.7 to 1.6 μg L^{-1} (Table 4.2). Concentrations varied by depth and date, with lower values generally found in the epilimnion (5 m) and the highest values found just below the metalimnion at 15 m or occasionally as deep as 25 m (Fig. 4.9). The highest recorded chlorophyll concentration in our study was 14.7 μg L^{-1} measured in 2008; this falls in the range measured in 2005 by Larson et al. (2006). Although we measured higher chlorophyll concentrations from 2008 through 2010 than were found prior to the 1980 eruption, concentrations in 2011 were lower than the 3 previous years and again in the concentration range recorded in 1974 (Table 4.2). Using average summer chlorophyll values from our study, Spirit Lake's trophic state index (TSI) fluctuated year to year between 31

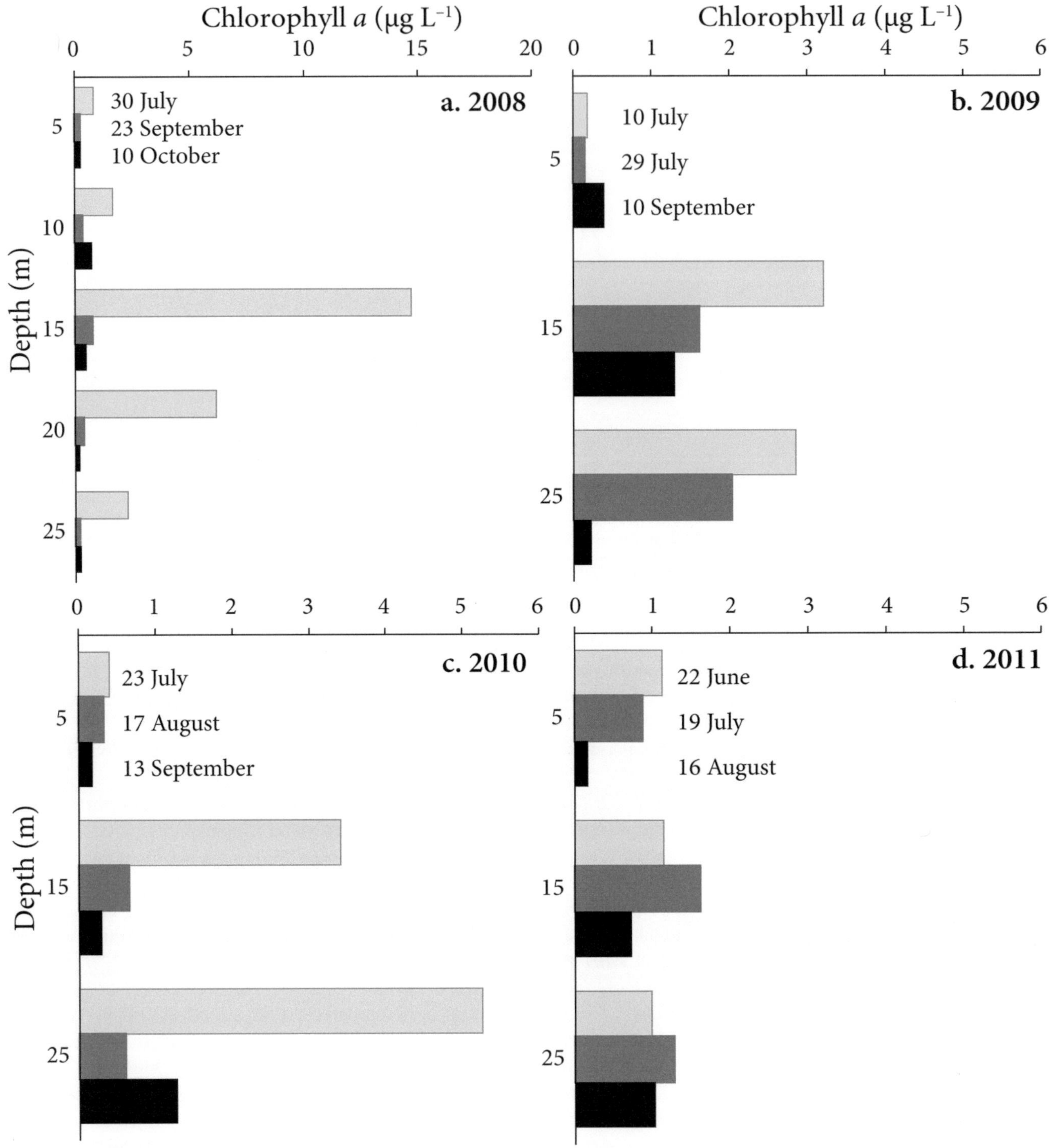

Fig. 4.9 Chlorophyll *a* (μg L^{-1}) in the west arm of Spirit Lake, 2008–2011. Bars represent single measurements at each depth for each sampling date.

and 37, placing the lake in the lower ***mesotrophic*** to upper oligotrophic range (Table 4.3).

Secchi-depth values continued to fluctuate throughout the summer season from 2007 to 2014, ranging from a low of 7.7 m to a high of 14.9 m (Fig. 4.6). This range is almost identical to the range recorded prior to the eruption in 1974 (Dion et al. 1976). If Secchi depth can be used as a measure of phytoplankton densities in Spirit Lake, our measurements from 2008 through 2014 show that algal densities were often highest during June and July, usually lower in August and September, and then higher again in October during autumn turnover. Average Secchi-depth values decreased slightly from 2008 to 2011, with a corresponding decrease in calculated TSI (Table 4.3). TSI (Secchi) places the lake in the oligotrophic range across all years.

Alkalinity has continued to decrease after the eruption from over 3000 μeq L^{-1} in August 1980 to 747 μeq L^{-1} in 2004–2005 to an average of 461 μeq L^{-1} in 2009–2014 (Table 4.2). Alkalinity in 2014 (375–501 μeq L^{-1}) was within a factor of two of pre-eruption levels (260 μeq L^{-1} in 1974). This decrease in alkalinity suggests that dissolved carbonate concentrations, noted by Wissmar et al. (1990) to dominate alkalinity in Spirit Lake by 1981, are continuing to decline in hydrologic inputs to the lake. Alkalinity was almost always greater at depth, although the difference from top to bottom was generally less than 20% (data not shown). The higher alkalinity measured in the hypolimnion may again result from mineral-laden groundwater inflows entering the lake below the thermocline, as suggested by specific conductivity profiles (Fig. 4.7c). In comparison, alkalinity measured in St. Charles and Meta Lakes, two lakes in close proximity to Spirit Lake not affected by the debris avalanche, was 179 μeq L^{-1} for both lakes in early October 2014.

4.5.2 Biological Characteristics

Most of the efflux of insect biomass, and thus nutrients, was concentrated in July and August of all years sampled (e.g., 2009, Fig. 4.10a), with year-to-year estimates varying by about 30% from the averages of 144 kg P and 1682 kg N. Our insect data showed a predominance in numbers and

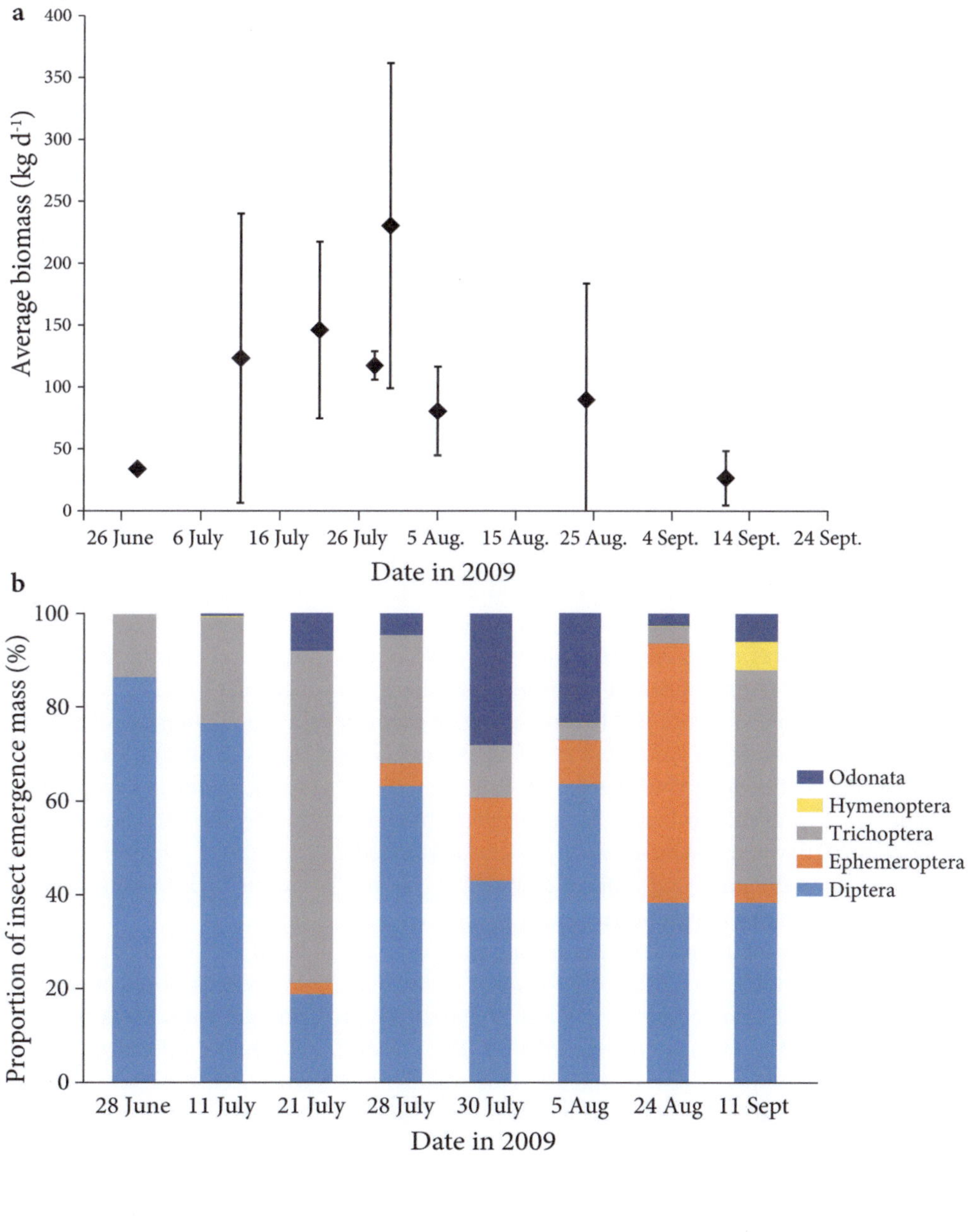

Fig. 4.10 Insect emergence biomass from 2009 shown as **(a)** average total biomass (error bars are ±1 standard deviation) and **(b)** % of total biomass by order.

biomass of Diptera species emerging from Spirit Lake in all years (2008–2011), followed in biomass by Trichoptera (caddisflies), then Ephemeroptera (mayflies), and Odonata (dragonflies and damselflies) (e.g., 2009, Fig. 4.10b). Other orders identified as minor contributors to insect biomass were Hemiptera, Lepidoptera, Hymenoptera, and Coleoptera. In 2009, Diptera dominated early emergence in June and early July and then contributed about half of the biomass throughout the rest of July and into September. Ephemeroptera emerged primarily in late July and August; Odonata emerged in July and early August; and Trichoptera emerged throughout the whole sampling period from June through September.

Phytoplankton trends were complex, as there were differences with depth, month, and year, and even occasionally from the west arm to the east arm of the lake on the same day. However, a few general noteworthy patterns are described here. Maximum biovolumes from 2008 through 2010 (Table 4.4) fell within the range previously recorded for Spirit Lake after the eruption, although falling short of the maximum biovolume recorded in 1983 (Larson et al. 2006). Plankton densities in 2008–2010 reached higher maximum numbers than seen in the 1990s and early 2000s, with values comparable to those recorded in 1983 and 1986.

Comparing 2008–2010 samples collected from 15-m depth (depth of highest phytoplankton density and biovolume) with samples collected from 2005 and earlier, *Rhodomonas minuta*, *Cryptomonas erosa*, and *Asterionella formosa* continued to be sizable components of the plankton community. These three species have been identified in Spirit Lake almost every year since 1983 (Larson et al. 2006). However, we also found some new species dominating (both in density and biovolume) the phytoplankton community in certain years, including *Ochromonas* sp., *Glenodinium* sp., and *Dinobryon sertularia* (e.g., 2010, Fig. 4.11). All of these species had been noted in the plankton community prior to our work in 2008 (Larson et al. 2006), but they had never been noted as dominant. *Ochromonas* sp. was found only one other year (1985), and *Glenodinium* sp. was only found a few times sporadically in the past. *Dinobryon sertularia* had become a regular member of the plankton community before our sampling began having been detected continuously since 1994.

Other notable events included a cyanobacteria bloom in September 2010, when *Aphanizomenon flos-aquae* reached a density of 220 cells mL^{-1} and a biovolume of 235 318 μm^3 mL^{-1} at 5 m in the east arm of Spirit Lake. This species dominated the sample at 5 m in 2010 after having been recorded only one other time, in 2005. *Gloeotrichia echinulata*, another species of cyanobacteria, was also found at a high biovolume in 2010, when it was recorded at 281 123 μm^3 mL^{-1} at 25 m in the west arm in August. This was the first time this species had been identified in Spirit Lake (Larson et al. 2006).

Overall in 2008–2010, Copepoda and Cladocera generally dominated the zooplankton community in the epilimnion (0–10 m) but also occurred throughout the water column. In contrast, Rotifera were the dominant taxa near the bottom of the lake (20–25 m) and also had the highest density of all zooplankton (data not shown). The maximum densities of zooplankton from 2008 through 2010 were higher than values recorded from 1986 through 2005 (Table 4.4), and in 2008 the maximum density approached the highest value recorded in 1983 (Larson et al. 2006).

Daphnia pulicaria by far dominated the cladoceran taxa all months and years, whereas *Leptodiaptomus tyrrelli*, diaptomid copepodites, and copepod ***nauplii*** dominated the copepod taxa throughout the 3 years sampled (Table 4.4). The rotifer community was more diverse than the cladocerans and copepods, with *Polyarthra doliochoptera*, *Gastropus stylifer*, *Keratella quadrata*, *Keratella cochlearis*, and *Conochilus unicornis* all showing dominance at different times and depths. *Gastropus stylifer*, a species not identified before 2008, was present in almost all samples and dominated the Rotifera more often than any other species.

4.5.3 Hydrologic Mass Balance Model

Although some year-to-year variation was seen from 2008 through 2014, in general precipitation and evaporation were almost equal during June–October (Table 4.5). Over that timeframe, outflows exceeded inflows, and therefore lake volume decreased by approximately $1.3–2.4 \times 10^7$ m^3. Surface inflows were generally lower than groundwater inflows by a factor of two to three times, except in especially wet years. Total precipitation from January to May in 2011 and 2014 was 140–142 cm (55–56 in), or 10–30% greater than in other years sampled (Table 4.6). This pattern was reflected in the greater surface inflows in 2011 and 2014 compared with 2008–2010 (Table 4.5). Note that precipitation levels during the June–October sampling season did not necessarily reflect winter–spring patterns.

The dominant role of groundwater in terms of hydrologic inputs to Spirit Lake was expected given the exceedingly high permeability and hydraulic conductivity of the material deposited by the 1980 eruption and the enormous quantity of water stored and released from the edifice and crater glacier. Also, our methods for estimating stream discharge may have overestimated summertime streamflow from areas of the drainage basin not directly measured, because many of these small subwatersheds would be expected to have little capacity for water storage, would lack a residual snowpack, and would not be connected to localized groundwater sources (e.g., springs and seeps). On the other hand, our monthly frequency of streamflow measurements may have underestimated total stream discharge, as storm flows may have been inadequately sampled,

Table 4.4 Phytoplankton and zooplankton summary data for Spirit Lake from 1983 through 2010.

	Total phytoplankton density (no. mL^{-1})	Total phytoplankton biovolume (μm^3 mL^{-1})	Dominant phytoplankton species (by density) at 15 m in 2008–2010	Zooplankton assemblage			Zooplankton density (total orgs m^{-3})
				Dominant Copepoda	Dominant Cladocera	Dominant Rotifera	
April–October 1983[a]	1–2794	381–1,104,410	*Asterionella formosa*, *Diatoma tenue elongatum*		*Ceriodaphnia* spp.	*Keratella* spp.	3720–128,480
July–November 1984[a]	9–670	1579–122,248	*Achnanthes minutissima*, *Nitzschia linearis*, *Rhodomonas minuta*, *Cryptomonas erosa*				
January–September 1985[a]	1–1786	236–316,538	*Rhodomonas minuta*, *Asterionella formosa*, *Achnanthes minutissima*, *Cryptomonas erosa*				
April–September 1986[a]	30–4131	3740–985,439	*Asterionella formosa*, *Rhodomonas minuta*	*Leptodiaptomus tyrrelli*	*Daphnia pulex*	*Keratella* spp.	2555–21,887
August–September 1989[a]	110–516	208,729–884,531	*Cyclotella comta*, *Rhodomonas minuta*, *Cryptomonas erosa*		*Daphnia pulex*		1006–1196
July 1994[a]	122–242	84,956–529,742	*Cyclotella comta*, *Rhodomonas minuta*	*Diaptomus tyrrelli*	*Daphnia schødleri*	*Filinia terminalis*	1272–2719
September 2004[a]	4–288	2069–18,014	*Rhodomonas minuta*	*Leptodiaptomus tyrrelli*	*Daphnia pulicaria*		11,816–24,545
July–August 2005[a]	17–789	7606–731,913	*Rhodomonas minuta*, *Epithemia sorex*, *Achnanthes minutissima*				2982–11,739
July–September 2008[b]	78–2322	18,655–247,348	*Ochromonas* spp., *Rhodomonas minuta*, *Dinobryon sertularia*	*Leptodiaptomus tyrrelli*, diaptomid copepodites, nauplii	*Daphnia pulicaria*	*Polyarthra doliochoptera*, *Gastropus stylifer*	8934–85,604
June–October 2009[b]	20–1866	8972–176,034	*Rhodomonas minuta*, *Dinobryon sertularia*, *Asterionella formosa*, *Cryptomonas erosa*	*Leptodiaptomus tyrrelli*, diaptomid copepodites, nauplii	*Daphnia pulicaria*	*Gastropus stylifer*, *Keratella quadrata*, *Keratella cochlearis*, *Conochilus unicornis*	221–25,136
July–October 2010[b]	23–1061	11,655–719,781	*Glenodinium* spp., *Rhodomonas minuta*, *Dinobryon sertularia*, *Asterionella formosa*, *Cryptomonas erosa*, *Ankistrodesmus falcatus*	*Leptodiaptomus tyrrelli*, diaptomid copepodites, nauplii	*Daphnia pulicaria*	*Gastropus stylifer*, *Keratella quadrata*, *Keratella cochlearis*	1545–46,400

Source: [a]Larson et al. (2006) and [b]this study.

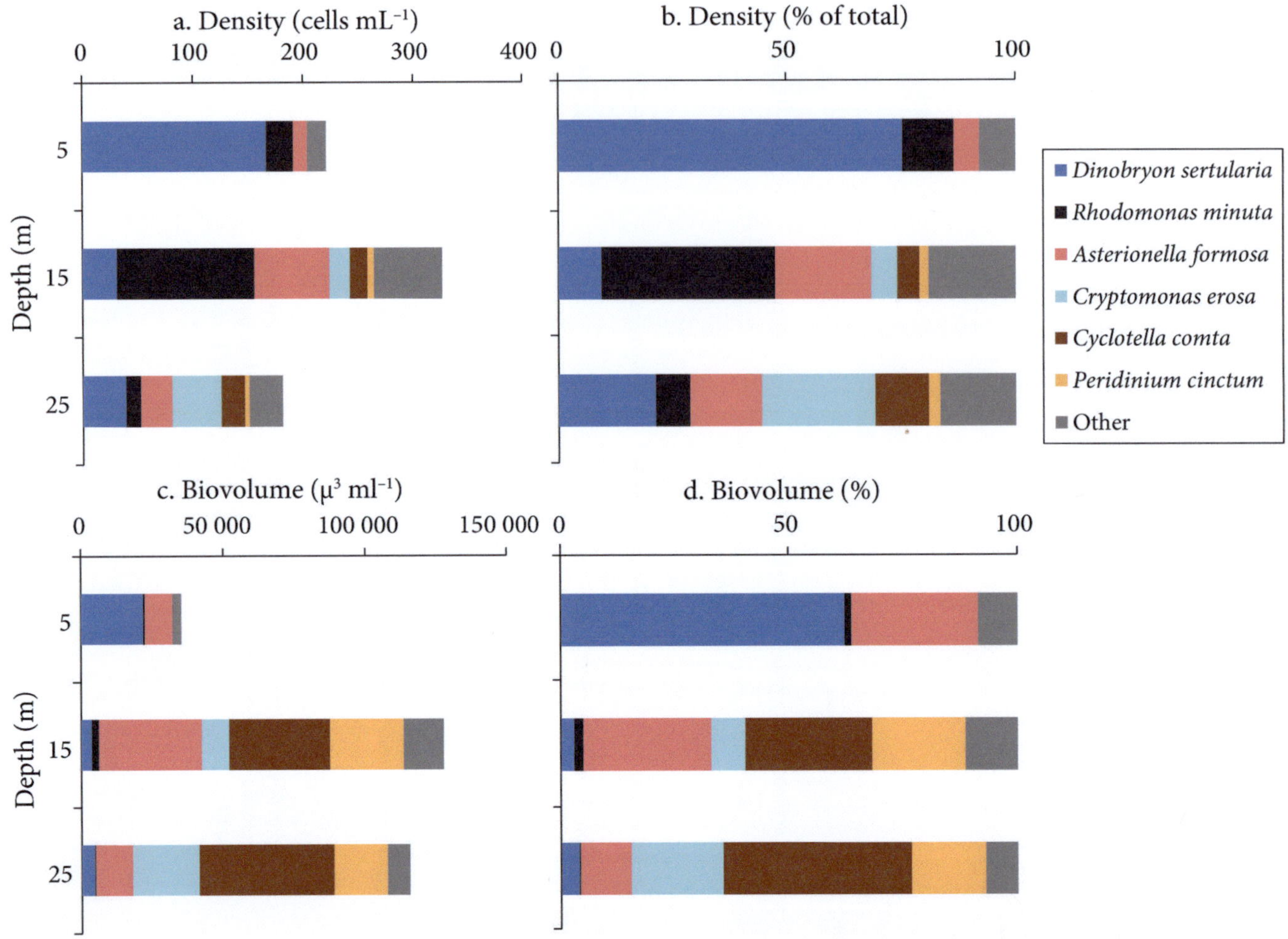

Fig. 4.11 Dominant phytoplankton species density in (**a**) cells mL^{-1} and (**b**) % of total, and biovolume in (**c**) $\mu m^3\ mL^{-1}$ and (**d**) % of total for samples collected 8 July 2010. The top six species by biovolume are graphed individually, with remaining species grouped together as "Other".

Table 4.5 Hydrologic budget calculated for Spirit Lake showing total flux (m^3) for June–October for each year measured.

	2008	2009	2010	2011	2014
Tunnel outflow	6.58E + 07	5.35E + 07	6.69E + 07	7.44E + 07	5.90E + 07
Surface inflow	1.05E + 07	1.20E + 07	1.72E + 07	3.65E + 07	3.34E + 07
Precipitation	4.50E + 06	5.06E + 06	6.34E + 06	4.16E + 06	6.26E + 06
Evaporation	5.55E + 06	5.55E + 06	5.55E + 06	5.55E + 06	5.55E + 06
Lake volume change	−2.35E + 07	−1.68E + 07	−1.34E + 07	−1.34E + 07	−1.34E + 07
Groundwater inflow	3.29E + 07	2.52E + 07	3.55E + 07	2.58E + 07	1.14E + 07

possibly adding significant stream discharge in discrete events. We do not know how these water fluxes change in the winter and early spring because we have not been able to access the lake when roads are impassable because of snow.

4.5.4 Nutrient Mass Balance Model

4.5.4.1 Inflows

For nutrient inputs to Spirit Lake (Table 4.7), surface water generally carried higher concentrations of phosphorus than groundwater (average 52 μg P L^{-1} in streams and 24 μg P L^{-1} in groundwater). On the other hand, groundwater carried higher concentrations of nitrogen than streams (average 254 μg N L^{-1} in groundwater and 134 μg N L^{-1} in streams). As the water flows were similar in magnitude for the two sources, the total influx of P into Spirit Lake was greater from surface waters (1147 kg P from groundwater and 635 kg P from streams), and the total influx of N was greater from groundwater (6664 kg N from streams and 2349 kg N from groundwater). The uncertainty in measuring total stream discharge discussed in Sect. 4.4.3 also created

Table 4.6 Precipitation data (including snow–water equivalent) from the SNOTEL site located on the pumice plain near Spirit Lake from November 2007 through October 2014 (http://www.wcc.nrcs.usda.gov/nwcc/site?sitenum=777&state=wa).

	Precipitation (cm)				
Month	**2008**	**2009**	**2010**	**2011**	**2014**
November of previous year	21.1	34.8	40.6	34.3	22.9
December of previous year	56.9	26.9	16.8	40.1	24.6
January	26.4	39.9	29.7	31.2	27.2
February	19.8	7.9	20.6	23.4	31.5
March	29.5	27.7	23.6	39.9	43.2
April	16.3	15.7	28.4	30.2	21.6
May	16.3	18.0	24.1	18.0	17.0
Total Jan–May	**108.3**	**109.2**	**126.4**	**142.7**	**140.5**
June	13.5	7.9	22.1	10.4	12.2
July	2.5	0.0	0.5	2.0	2.8
August	7.1	6.1	1.0	0.3	2.5
September	2.0	10.7	13.7	6.4	8.1
October	15.7	21.8	20.3	18.8	31.2
Total June–Oct	**40.9**	**46.5**	**57.7**	**37.8**	**56.9**

uncertainty in our estimates of total N and P fluxes in stream water and groundwater for the nutrient budget. In order to improve these estimates, it would be necessary to install an automated system for recording stream discharge and collecting water samples, a difficult task in a system with dynamic channel migration and multiple inflows into the lake.

Although we did not directly measure the flux of N and P into Spirit Lake by atmospheric deposition, the estimated magnitude of N loading via the atmosphere (1937 kg N from June through October) was similar in magnitude to other hydrologic inputs (e.g., 2349 kg N from streams). Atmospheric loading of P (133 kg P from June through October), on the other hand, was almost five times less than the smallest hydrologic source (635 kg P from groundwater) but of similar magnitude to some other fluxes, including insect losses (144 kg P). However, given that atmospheric deposition of nitrogen varied by at least a factor of two for nearby locations (Western Regional Air Partnership 2013; Tipping et al. 2014), we recommend that deposition samples should be collected directly at the lake to improve the model. The nutrient influx to the lake from dry deposition may be higher than wet deposition, considering the windblown dust and arthropods transported from the Pumice Plain and deposited on the lake, as we witnessed regularly in the summer. For example, Edwards and Sugg (1993) estimated that from 1983 through 1985 the mean summer season deposition of wind-deposited N and P was 81.8 mg m^{-2} and 5.5 mg m^{-2} based on terrestrial arthropod collectors located on the adjacent Pumice Plain. Based on Edwards and Sugg's measurements from the early 1980s, the arthropod contribution alone could be 900 kg N and 61 kg P. Such a contribution would increase the estimate of the aerial deposition of nutrients to the lake by approximately 46% for both N and P and account for some of the discrepancy between measured sediment-loading rate and the load calculated by difference (Table 4.7). In addition, a sizable contribution from plants (foliage, seeds, spores) and inorganic materials (soil and dust) is likely added to the lake each summer. Moreover, wind-borne loading to the lake by arthropods and plant materials is likely much higher than that calculated by Edwards and Sugg (1993) in the 1980s because of the increased vegetative cover in the contributing airshed 30+ years later (Dale et al. 2005). Therefore, direct measurements of aerial inputs to Spirit Lake in the future could substantially reduce the considerable uncertainty in our current estimates, which likely underestimate the influx of nutrients from aerial deposition.

The log-mat biofilm input to the lake was estimated at 184 kg P and 1596 kg N over the 5-month period. This input was of the same order of magnitude as the atmospheric input estimates. Moreover, these quantities may be an underestimate of nutrient inputs from the log mat. In summer, the log mat regularly drifted over one stationary sediment trap anchored near the steep, rocky shore in the east arm of Spirit Lake; that trap received 2.5–42 times the total mass input per day compared with the floating traps attached to the underside of logs drifting away from shore (data not shown). This observation suggests that abrasion of the log mat against the rocky substrates present along a large portion of the lake's shoreline may result in even larger inputs of nutrients from the log mat to the lake.

What is not known at this time is whether the log mat is a source of new nutrients to the lake through release of nutrients stored in the logs and/or N_2-fixation in the biofilm, or whether inputs from the log mat are just a recycling of nutrients taken up by the biofilm from nutrients already present in the epilimnion. If inputs from the log mat are a significant source of new nutrients into the lake, this result may have important implications for management of large woody debris in Spirit Lake and other nutrient-poor lakes common in the Cascade Range and elsewhere. If left in place, floating woody debris could increase N inputs to oligotrophic lakes and therefore increase primary productivity and fish biomass. We plan to conduct a more intensive study of the log mat in the future, including an examination of $\delta^{15}N$ values to determine the source of nitrogen incorporated into the biofilm.

4.5.4.2 Outflows

Nutrient outflows from the water column of Spirit Lake included losses resulting from insect emergence, amphibian emigration, sediment losses, and lake outflows through the tunnel (Table 4.7). Sampling for amphibians yielded 233 individuals representing five lentic-breeding species: northwestern salamander (*Ambystoma gracile*), rough-skinned

Table 4.7 Nutrient mass balance model for Spirit Lake using data collected in 2008–2014. Aqueous concentrations (*white cells at top*) and nutrient budget fluxes and reservoirs (*gray cells*) are based on total P or N. Additional data (*white cells at bottom*) based on average values above.

	Phosphorus Budget						Nitrogen Budget					
	2008	2009	2010	2011	2014	*P average*	2008	2009	2010	2011	2014	*N average*
Epilimnetic avg (μg L^{-1})	NM	11	11	14	11	*12*	NM	347	104	0	73	*131*
Streams avg (μg L^{-1})	NM	57	74	30	46	*52*	NM	310	121	20	86	*134*
Groundwater avg (μg L^{-1})	NM	23	42	9	21	*24*	NM	470	292	107	145	*254*
Water column avg (μg L^{-1})	NM	7	14	11	8	*10*	NM	326	140	8	95	*142*
Atmospheric deposition (kg)	133	133	133	133	133	*133*	1937	1937	1937	1937	1937	*1937*
Tunnel outflow (kg)	NM	588	736	1041	649	*753*	NM	18 554	6956	0	4305	*7454*
Surface inflow (kg)	NM	683	1269	1096	1537	*1147*	NM	3717	2076	731	2873	*2349*
Groundwater inflow (kg)	NM	580	1486	232	240	*635*	NM	11 849	10 388	2760	1659	*6664*
Insect flux (kg)	210	96	110	158	NM	*144*	2519	1113	1272	1823	NM	*1682*
Log mat flux (kg)	NM	NM	NM	NM	184	*184*	NM	NM	NM	NM	1596	*1596*
Total sediment flux (kg)	NM	NM	6033	3574	1607	*3738*	NM	NM	32 439	10 594	16 414	*19 816*
Amphibian flux (kg)	NM	NM	NM	NM	0.012[a]	*0.012*	NM	NM	NM	NM	0.1[a]	*0.1*
Fish reservoir (kg)	NM	NM	56	NM	NM	*56*	NM	NM	370	NM	NM	*370*
[b]Water column reservoir (kg)	NM	1795	3590	2821	2051	*2564*	NM	83 592	35 898	2051	24 360	*36 475*
Phytoplankton reservoir (kg)	26	20	45	NM	NM	*30*	127	98	219	NM	NM	*148*
Zooplankton reservoir (kg)	258	71	65	NM	NM	*131*	1947	535	488	NM	NM	*990*
Aquatic macrophyte reservoir (kg)	NM	136	218/364[c]	NM	NM	*177*	NM	1233	1424/2446[c]	NM	NM	*1329*
Biofilm reservoir (kg)	NM	NM	29	124	43	*65*	NM	NM	1451	1459	779	*1230*
Benthic sediment reservoir (kg)	NM	NM	736 853/633 821[d]		NM	*736 853*	NM	NM	763 000/595 000[d]		NM	*763 000*

	Phosphorus Budget	Nitrogen Budget
Total nutrient influx into lake basin	2098 kg	12 546 kg
Total nutrient efflux from lake basin	897 kg	9136 kg
Calculated flux to sediments (by difference)	1201 kg	3411 kg calculated
Measured flux to sediments	3738 kg	19 816 kg measured
Ratio measured/calculated	3.1 times	5.8 times
Reservoir total	2863 kg	39 404 kg
Turnover rate (reservoir total/(total outflows+sediment flux))	1.6 times per sampling period	0.7 times per sampling period
Biotic reservoir (% total reservoir)	16 %	10 %
N:P molar ratio in biota	**19.6**	
N:P molar ratio in inflows	**13.2**	

[a]Measured in August 2013.
[b]Includes plankton.
[c]Second estimate using sonar data.
[d]Second estimate uses kriging to spatially average N and P.

newt (*Taricha granulosa*), Pacific treefrog (*Pseudacris regilla*), western toad (*Anaxyrus boreas*), and Cascades frog (*Rana cascadae*). The maximum biomass of amphibians was 0.159 g wet wt m^{-2}, which resulted in an efflux of 9.73 kg wet weight when adjusted for the available amphibian habitat in Spirit Lake. Based on the average nutrient content of 35.7% C, 8.3% N, and 1.0% P (dry mass) for amphibians, these values represented an efflux of 100 ± 45 g N and 12 ± 5 g P. Although we likely have underestimated the amphibian biomass produced and exported from the lake, we believe that the relative contribution of amphibians to the total efflux of nutrients from the lake is minor given the limited suitable habitat, largely confined to small segments of the south shore and the very distal ends of the two northern arms, and the presence of predatory fish.

We estimated that insect emergence from Spirit Lake removed on average 144 kg P and 1682 kg N (Table 4.7). This outflow value may be an underestimate for several reasons. First, it has been shown in the literature that floating emergence traps are preferentially avoided by some insect species during emergence, including some Ephemeroptera and Diptera species (Malison et al. 2010). Second, we collected samples only from open water away from the floating log mat (Figs. 4.1 and 4.2). We suspect that the log mat increases emergence densities and possibly expands the lake area where significant insect emergence occurs. Anecdotally, our team encountered large numbers of remnant insect exoskeletons and live dragonfly larvae on floating logs while collecting biofilm samples. Our team will attempt to target insect emergence from the log mat in the future.

Overall, the average total P influx to the lake over our 5-month sampling period was approximately 2100 kg P (groundwater, surface water, precipitation, and log mat), whereas approximately 900 kg P was lost via the tunnel and insect emergence, leaving 1200 kg P (calculated by difference) to be removed to the sediments (Table 4.7). This efflux of P to the sediments is approximately three times less than the estimated flux of 3740 kg P based on empirical data. For N, the average total influx was approximately 12 500 kg N; the average lost was 9100 kg N, with 3400 kg N estimated to be removed to the sediment. This flux of N to the sediments is approximately sixfold lower than the measured flux of 19 800 kg N.

We suspect that the source of this discrepancy in balancing the nutrient budget resulted from underestimating nutrient inputs rather than outflows. This is because we have greater confidence in the accuracy of the output estimate, for the tunnel outflow was continuously monitored, the nutrient concentration in the epilimnion was nearly constant, and losses associated with insect and amphibian emigration were small. Although transport of nutrients associated with the removal of aquatic biota (i.e., fish, amphibians, and aquatic plants) by semiaquatic and terrestrial herbivores (e.g., waterfowl and beaver [*Castor canadensis*]) and predators (e.g., mink [*Neovison vison*], otters [*Lontra canadensis*], waterfowl, and raptors) was not quantified, we estimate that this flux would have been small, given the small biomass of fish, amphibians, and aquatic plants (Table 4.7).

Conversely, estimates for inflows rely on the assumption that stream discharges and concentrations measured in a few locations accurately represent the entire catchment. It is possible that stream discharge and nutrient concentrations are heterogeneous across the post-eruption landscape. This would introduce a large bias into our inflow estimates. The other sizable influx with suspected greater uncertainty is atmospheric deposition, including wet and dry deposition (i.e., arthropods, plant matter, and inorganic dust). Reducing this uncertainty will require the direct collection of wet and dry deposition samples from the lake.

4.5.4.3 Reservoirs

Nutrient reservoirs quantified in Spirit Lake included water, plankton, aquatic macrophytes, log-mat biofilm, sediments, and fish. In general, the size of the total reservoirs of N and P was within a factor of two of the measured outflows. We calculated a nutrient turnover rate of 0.7 and 1.6 times per 5-month sampling period for N and P, respectively (Table 4.7), suggesting a high turnover rate for all reservoirs of nutrients in the system.

By far the biggest reservoir of nutrients was in the lake water, but the total N and P concentrations were very dilute. The total nutrients in the water column (approximately 2400 kg P and 35 300 N) were judged to be mostly in non-planktonic (organic and inorganic) particulate form, as total N and P water-column estimates far exceeded the reservoir estimated in the phytoplankton (30 kg P and 148 kg N) and zooplankton (131 kg P and 990 kg N), and dissolved nitrate, ammonia, and orthophosphate were mostly undetectable. Our estimates showed greater nutrient biomass in zooplankton than in phytoplankton, which may be counter to ecological expectations if phytoplankton are the primary source of food for the zooplankton. It is possible that non-planktonic particulate matter or bacteria may be another source of food for zooplankton. Alternatively, higher zooplankton biomass also may be indicative of high grazing rates commensurate with high primary production rates (e.g., Sommer et al. 2012). Estimates of primary production rates would be useful to assess these hypotheses in the future.

The nutrient reservoir in the aquatic macrophytes (177 kg P and 1329 kg N) was slightly larger than in the combined plankton (Table 4.7). The nutrient reservoir in the log-mat biofilm was similar to the macrophytes in magnitude for nitrogen (1230 kg N), but smaller for phosphorus (65 kg P). Comparing these three reservoirs of nutrients in primary producers in the lake, we find that macrophytes and log-mat biofilm increase available nutrients for grazers in the lake almost 200% over plankton alone (Table 4.7). Therefore, it is possible that Secchi-depth comparisons, which reflect plankton abundance alone rather than overall lake productivity, may not accurately estimate changes in lake trophic status after the eruption.

The rainbow trout population was estimated at 13,005 individuals (95% CI (confidence intervals), 8719–20,228), and the mean fish mass in Spirit Lake during 2011 was estimated at 13,954 kg (95% CI, 9887–21 705 kg). Based on our lab assays, the fish contained 45.2% C, 12.9% N, and 1.9% P, which yielded a nutrient reservoir of 370 kg N and 56 kg P. This reservoir is two to three times less than the mass of nutrients in the zooplankton reservoir.

We do not have measurements for the reservoirs of nutrients in fully aquatic arthropods (e.g., amphipods), amphibians, or benthic macroinvertebrates in the lake, which may be a sizable pool of importance to trophic transfer in the Spirit Lake ecosystem. We hope to fill this data gap in future work.

Together, the biotic reservoirs of nutrients contain approximately 10% and 16% of the N and P, respectively, in the lake (Table 4.7). The N:P molar ratio in the biota was 19.6, but the ratio in the incoming waters was 13.2, suggesting a slight N deficiency in Spirit Lake. This measure is imperfect, as the bioavailable fraction in the incoming waters may vary significantly from the total, but it does make some intuitive sense, allowing for the higher P content of the volcanic deposits and the low concentration of N in the surrounding watershed (e.g., soils and vegetation). This result also suggests that additional N inputs from N_2-fixation by the log-mat biofilm, if significant, could be important in the overall trophic status of this disturbed ecosystem.

Our physical and chemical data suggested that the sediment reservoir was a net sink, not a source, of nutrients. As bottom waters were sufficiently oxygenated (Fig. 4.7b), it was unlikely that redox conditions favored the release of P from the sediments through reductive dissolution. In addition, specific-conductivity measurements (Fig. 4.7c) did not suggest the diffusion of dissolved ions from the benthic sediments was significant. However, it is likely that benthic macroinvertebrates and aquatic macrophytes in the nearshore region of the lake recycle some of the nutrients in this reservoir back into other reservoirs in the lake. An interesting spatial pattern emerged from our data, showing higher N and P concentrations in the north ends of the east and west arms of the lake and in Duck Bay compared to the south end of the lake (Fig. 4.12). We know that the log mat is forced to the north end of Spirit Lake by prevailing winds during most of the summer. An analysis of the images taken in 2014 with a time-lapse camera, set to capture a view of most of the lake, showed that logs were present in the northeast arm of the lake 60–95% of the time and in the northwest arm 35–75% of the time from July through October (Fig. 4.13). The log mat spent little time along the southwest shore. This result suggests that the log mat may be responsible for the greater N and P in the north end of Spirit Lake. In contrast, Duck Bay is a long, narrow extension of the lake to the SE in which the log mat almost never concentrates; rather, this area has an extensive, shallow littoral zone with a dense macrophyte community. We hypothesize that the macrophytes contribute to the comparatively high nutrient concentrations in the sediments in this portion of the lake.

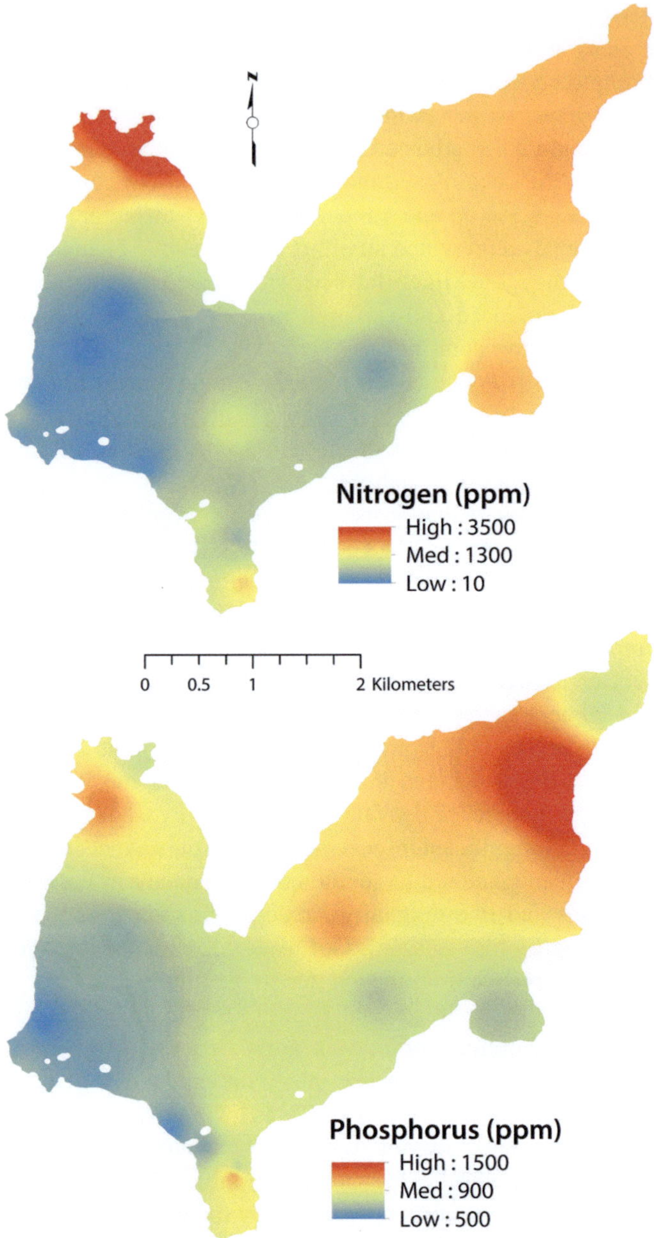

Fig. 4.12 Nutrient concentrations (ppm) in benthic sediments in Spirit Lake collected in 2010 and 2011.

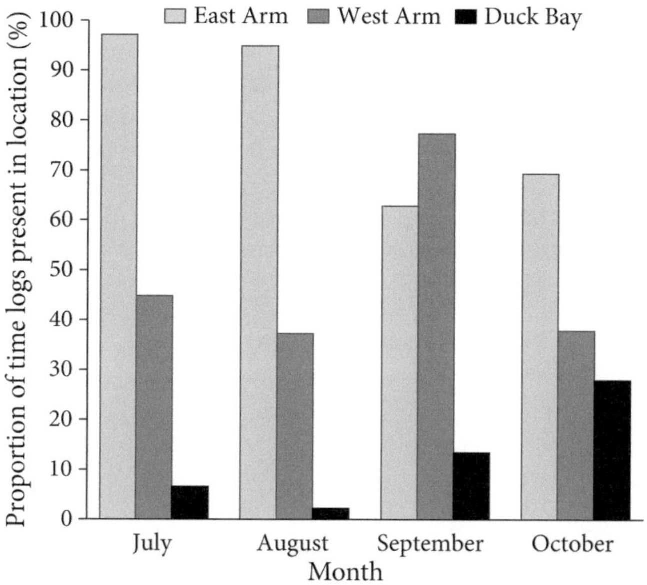

Fig. 4.13 Analysis of time-lapse photos of log-mat position in Spirit Lake, June–September 2014.

4.6 Conclusions

Comparing current conditions in Spirit Lake to what we know about the pre-eruption lake, we find that many parameters have returned to levels seen prior to 18 May 1980. Unfortunately, we only have significant summer data from 1974 (Dion et al. 1976) to compare to, and it is impossible to know whether that year represents normal or anomalous conditions for the pre-eruption lake. The maximum recorded temperature in Spirit Lake was 16 °C in August 1974, which is 5 °C lower than the maximum in 2009. More recent continuous temperature monitoring in the lake shows an increasing temperature trend over the last 10 years (Fig. 4.8). This trend is supported by other lake research in the Pacific Northwest. Annual average water temperature in Lake Washington increased ~1.6 °F (0.9 °C) from 1964 to 1998 (Arhonditsis et al. 2004). Higher temperatures can have significant effects on aquatic primary production and species composition. Therefore, the effects of climate change should be investigated in future Spirit Lake research.

Evidence from dissolved oxygen, phosphate, manganese, and iron measurements and specific conductivity values in the hypolimnion of Spirit Lake in 1974 (Dion et al. 1976) suggests the sediments in the pre-eruption lake may have gone anoxic in the late summer. Although we did record dissolved oxygen concentrations as low as 3.3 mg L^{-1} at 29 m depth in the water column in 2014, we did not see evidence of sediment anoxia either in dissolved phosphate concentrations or specific conductivity values in the bottom waters. With the exception of Duck Bay, sediments in Spirit Lake at present are low in organic matter and therefore may not currently provide fuel for high levels of biological oxygen demand. This situation may change in the future as the surrounding watershed continues to develop, potentially increasing allochthonous carbon inputs.

Both alkalinity and specific conductivity have continued to decrease in Spirit Lake 30+ years after the eruption (Table 4.2). These changes may both be related to decreasing weathering rates from geological parent material in the lake basin and the watershed. Values for alkalinity and specific conductivity are still 44% and 270% higher in 2014 than in 1974 (Dion et al. 1976), but they are continuing to decline and may reach pre-eruption levels, barring significant additional disturbance in the near future.

Indicators of Spirit Lake productivity, Secchi depth (Fig. 4.6), total P and N, and chlorophyll *a* (Table 4.2) are now at levels comparable to those measured in Spirit Lake in 1974. Thus, the TSI values calculated for the lake in 2011 are comparable to 1974 (Table 4.3). This suggests that planktonic productivity has returned to pre-eruption levels. However, greater primary production by aquatic macrophytes and log-mat biofilm may mean greater lake productivity in the new Spirit Lake than in the pre-eruption lake, although we have no data for making such a conclusion for the pre-eruption lake.

This study serves as a baseline for understanding future changes in the hydrology, chemistry, and ecology of Spirit Lake. As vegetation in the watershed continues to expand and cover more of the landscape, surface and stream-channel erosion should decrease and organic matter should develop on the ground, increasing the retention of nutrients on land and potentially decreasing external nutrient inputs to the lake. In addition, as the vegetation matures and succession continues, the system should shift from wind-driven to gravity-driven, and there should be an associated decrease in aerially transported inputs. However, terrestrial grazing rates and the relative proportion of N_2 fixed and exported by terrestrial plants will also affect the rate of this change. As terrestrial succession proceeds, we predict a decrease in N_2 fixed by plants, as the species responsible for these processes (*Lupinus* spp., *Alnus* spp.) require sunlight and do not occur in great abundance in the understory of forests. Moreover, effects of climate change, including temperature and precipitation shifts, have the potential to alter the hydrology and ecology of the watershed, with unknown consequences for Spirit Lake ecosystem structure and functioning. Our future work aims (1) to continue long-term measurements of the physical, chemical, and biological properties of the lake, to establish a long-term data series, (2) to better understand the role of the log mat in the chemistry and ecology of the lake, and (3) to work with others to develop a coupled terrestrial/aquatic nutrient model for the lake. Through this work we hope to chronicle the long-term response of the Spirit Lake ecosystem to cataclysmic volcanism and to elucidate mechanisms applicable to other disturbed systems in order to increase our understanding of basic science principles and to inform management decisions for the future. Our work shows the potential for floating logs to provide additional nutrients to oligotrophic lakes and therefore to increase fish biomass, a desired outcome for many lake managers.

Acknowledgments The authors would like to thank all those students and staff members from University of Washington Tacoma and Bellarmine High School who put in so much effort from 2005 to 2014 to make this work possible: from UWT, Joe Chynoweth, Cameron Marshall, Ashley Datema, Bridget Mason, Heather Jennings, Nicole Butcher, Tiffany Wax, Laura Alskog, Danielle Dahlquist, Erika Klein, Katie Royer, Chelsie Strowbridge, Terri Hurlbut, Amy Leslie, Shantelle Reamer, Julie Palumbo, Jessica West, Brenda LeMay, Jessica Asplund, and Mindy Roberts; from Bellarmine High School, Ron Nielsen, Dave DeGroot, Meghan Blanchet, AJ Joseph, Sara Scott, Conor Flannigan, Jane Manley, Maya Cates-Carney, and Heidi Hirsh. We were also aided by many field crew members working for the US Forest Service at Mount St. Helens. A special thanks to Chuck Neudorf for his help with GIS. This work was supported by funding from the UWT Founders Endowment, the UWT School of Interdisciplinary Arts and Sciences Research and Teaching Fund, and the Mountaineers Foundation.

Glossary

Dimictic (adj) A lake having a pattern of stratification in which it undergoes two periods of mixing, one in the spring and one in the autumn.

Epilimnion (adj epilimnetic) The upper layer of a lake during thermal stratification, characterized by well-mixed, well-illuminated warmer water.

Eutrophic (adj) The condition of water being rich in plant nutrients and the results of this condition.

Hypereutrophic (adj) The condition of water being extremely productive, with abundant nutrients and very high primary producer biomass.

Hypolimnion (adj hypolimnetic) The poorly illuminated lower region of a stratified lake, characterized by denser, colder water protected from wind action; lies below the metalimnion.

Mesotrophic (adj) The condition of water being only moderately rich in plant nutrients and the results of this condition.

Metalimnion (n) The central layer of a stratified lake between the epilimnion and the hypolimnion, characterized by rapid temperature change with depth; the region where the thermocline is found.

Nauplii (n pl), nauplius (sing) The earliest free-living stage in the development of most crustaceans, including Copepoda.

Oligotrophic (adj) The condition of water being poor in plant nutrients and the results of this condition.

Secchi depth (n) A measure of water clarity equal to the mean depth of the point where a weighted black and white checkered disk (20 cm in diameter) disappears when viewed from the shady side of a vessel and that point where it reappears upon raising it after it has been lowered beyond visibility.

Thermocline (n) The plane of maximum rate of decrease of temperature with respect to depth.

References

American Public Health Association (APHA). 1992. *Standard methods for the examination of water and wastewater*. 18th ed. Washington, DC: American Public Health Association, American Water Works Association, Water Environment Federation.

Arhonditsis, G.B., M.T. Brett, C.L. DeGasperi, and D.E. Schindler. 2004. Effects of climatic variability on the thermal properties of Lake Washington. *Limnology and Oceanography* 49: 256–270.

Bisson, P.A., C.M. Crisafulli, B.R. Fransen, R.E. Lucas, and C.P. Hawkins. 2005. Responses of fish to the 1980 eruption of Mount St. Helens. In *Ecological responses to the 1980 eruption of Mount St. Helens*, ed. V.H. Dale, F.J. Swanson, and C.M. Crisafulli, 163–181. New York: Springer.

Carlson, R.E. 1977. A trophic state index for lakes. *Limnology and Oceanography* 22: 361–369.

Cornett, R.J., and F.H. Rigler. 1980. The areal hypolimnetic oxygen deficit: An empirical test of the model. *Limnology and Oceanography* 25: 672–679.

Crisafulli, C.M., L.S. Trippe, C.P. Hawkins, and J.A. MacMahon. 2005. Amphibian responses to the 1980 eruption of Mount St. Helens. In *Ecological responses to the 1980 eruption of Mount St. Helens*, ed. V.H. Dale, F.J. Swanson, and C.M. Crisafulli, 183–198. New York: Springer.

Cushman, R. 1983. An inexpensive, floating, insect-emergence trap. *Bulletin of Environmental Contamination and Toxicology* 3: 547–550.

Dahm, C.N., J.A. Baross, A.K. Ward, M.D. Lilley, and J.R. Sedell. 1983. Initial effects of the Mount St. Helens eruption on nitrogen cycle and related chemical processes in Ryan Lake. *Applied and Environmental Microbiology* 45: 1633–1645.

Dahm, C.N., D.W. Larson, R.R. Petersen, and R.C. Wissmar. 2005. Response and recovery of lakes. In *Ecological responses to the 1980 eruption of Mount St. Helens*, ed. V.H. Dale, F.J. Swanson, and C.M. Crisafulli, 255–274. New York: Springer.

Dale, V.H., D.R. Campbell, W.M. Adams, C.M. Crisafulli, V.I. Dains, P.M. Frenzen, and R.F. Holland. 2005. Response and recovery of lakes. In *Ecological responses to the 1980 eruption of Mount St. Helens*, ed. V.H. Dale, F.J. Swanson, and C.M. Crisafulli, 59–73. New York: Springer.

Davies, I.J. 1984. Sampling aquatic insect emergence. In *A manual on methods for the assessment of secondary productivity in fresh waters*, ed. J.A. Downing and F.H. Rigler, 161–227. Oxford: Blackwell Scientific.

Dion, N.P., G.C. Bortleson, J.B. McConnell, and J.K. Innes. 1976. *Data on selected lakes in Washington, part 5*. Water-Supply Bulletin 42, Part 5. U.S. Geological Survey, prepared for the Washington State Dept. of Ecology. https://fortress.wa.gov/ecy/publications/documents/wsb42e.pdf.

Duarte, C.M. 1992. Nutrient concentration of aquatic plants: Patterns across species. *Limnology and Oceanography* 37: 882–889.

EarthData International of Maryland. 2003. LIDAR mapping of the Toutle River, Washington, river basin. Data collected between 31

October and 2 November 2002. Available from https://wagda.lib.washington.edu/data/type/elevation/lidar/st_helens/toutle03.html.

Edwards, J.S., and P. Sugg. 1993. Arthropod fallout as a resource in the colonization of Mount St. Helens. *Ecology* 74: 954–958.

Environmental Systems Research Institute [ESRI]. 2001. *ArcGIS geostatistical analyst: Statistical tools for data exploration, modeling, and advanced surface generation, an ESRI white paper*. Available at http://www.esri.com/library/whitepapers/pdfs/geostat.pdf.

Knowlton, M.F., and J.R. Jones. 2006. Natural variability in lakes and reservoirs should be recognized in setting nutrient criteria. *Lake and Reservoir Management* 22: 161–166.

Larson, D.W. 1993. The recovery of Spirit Lake: A natural calamity provides scientists with a rare opportunity to study the rejuvenation of a once-pristine lake. *American Scientist* 81: 166–177.

———. 1994. A case of natural restoration of an aquatic ecosystem. *Lakeline* December: 26–31.

Larson, D.W., and N.S. Geiger. 1982. Existence of phytoplankton in Spirit Lake near active volcano Mt. St. Helens, Washington, U.S.A.: Post-eruption findings. *Archiv für Hydrobiologie* 93: 375–380.

Larson, D.W., and M.W. Glass. 1987. *Spirit Lake, Mount St. Helens, Washington: Limnological and bacteriological investigations*. Vol. 1. Portland: U.S. Army Corps of Engineers, Portland District.

Larson, G.L., A. Wones, C.D. McIntire, and B. Samora. 1994. Integrating limnological characteristics of high mountain lakes in the landscape of a natural area. *Environmental Management* 18: 871–888.

Larson, D.W., J. Sweet, R.R. Petersen, and C.M. Crisafulli. 2006. Posteruption response of phytoplankton and zooplankton communities in Spirit Lake, Mount St. Helens, Washington. *Lake and Reservoir Management* 22: 273–292.

Malison, R.L., J.R. Benjamin, and C.V. Baxter. 2010. Measuring adult insect emergence from streams: The influence of trap placement and a comparison with benthic sampling. *Journal of the North American Benthological Society* 29: 647–656.

McNabb, C.D. 1960. Enumeration of freshwater phytoplankton concentrated on the membrane filter. *Limnology and Oceanography* 5: 57–61.

Müller, B., L.D. Bryant, A. Matzinger, and A. Wüest. 2012. Hypolimnetic oxygen depletion in eutrophic lakes. *Environmental Science and Technology* 46: 9964–9971.

Ogle, D.H. 2015. *FSA fisheries stock analysis*. R package version 0.4.44. Accessed 2015. http://derekogle.com/fishR/packages.

Schnabel, Z.E. 1938. The estimation of total fish population of a lake. *The American Mathematical Monthly* 45: 348–352.

Sommer, U., R. Adrian, L. De Senerpont Domis, J.J. Elser, U. Gaedke, B. Ibelings, E. Jeppesen, M. Lürling, J.C. Molinero, W.M. Mooij, E. van Donk, and M. Winder. 2012. Beyond the Plankton Ecology Group (PEG) model: Mechanisms driving plankton succession. *Annual Review of Ecology, Evolution, and Systematics* 43: 429–448.

Strickland, J.D.H., and T.R. Parsons. 1972. *A practical handbook of seawater analysis*. Ottawa: Fisheries Research Board of Canada.

Tipping, E., S. Benham, J.F. Boyle, P. Crow, J. Davies, U. Fischer, H. Guyatt, R. Helliwell, L. Jackson-Blake, A.J. Lawlor, D.T. Monteith, E.C. Roweg, and H. Tobermanac. 2014. Atmospheric deposition of phosphorus to land and freshwater. *Environmental Science: Processes & Impacts* 16: 1608–1617.

U.S. Department of Agriculture, Natural Resources Conservation Service [USDA-NRCS]. 2016. *National water and climate center*. Accessed 2016 Feb 4. http://www.wcc.nrcs.usda.gov/nwcc/site?sitenum=777&state=wa.

U.S. Geological Survey. 2016. *National water information system data available on the World Wide Web* (Water Data for the Nation). Accessed 4 Feb 2016. http://waterdata.usgs.gov/usa/nwis/uv?site_no=14240304.

Valderrama, J.C. 1981. The simultaneous analysis of total nitrogen and total phosphorus on natural waters. *Marine Chemistry* 10: 109–122.

Vitousek, P.M., J.R. Gosz, C.C. Grier, J.M. Melillo, W.A. Reiners, and R.L. Todd. 1979. Nitrate losses from disturbed ecosystems. *Science* 204: 469–474.

Weaver, R.W. 1994. *Methods of soil analysis, part 2: Microbiological and biochemical properties*. Madison: Soil Science Society of America.

Western Regional Air Partnership. 2013. West-wide jumpstart air quality modeling study—Final report, appendix N. Fort Collins, CO. Accessed online: http://wrapair2.org/pdf/Appendix_N_N&S_Deposition_IMPROVE_SrcCat.xlsx.

Western Regional Climate Center. 2016. Average pan evaporation by State (Washington). Accessed 2016 Feb 2016. http://www.wrcc.dri.edu/htmlfiles/westevap.final.html#WASHINGTON.

Wilson, A.E. 2003. Effects of zebra mussels on phytoplankton and ciliates: A field mesocosm experiment. *Journal of Plankton Research* 25: 905–915.

Winnell, M.H., and D.S. White. 1985. Trophic status of southeastern Lake Michigan based on the Chironomidae (Diptera). *Journal of Great Lakes Research* 11: 540–548.

Wissmar, R.C., A.H. Devol, A.E. Nevissi, and J.R. Sedell. 1982a. Chemical changes of lakes within the Mount St. Helens blast zone. *Science* 216: 175–178.

Wissmar, R.C., A.H. Devol, J.T. Staley, and J.R. Sedell. 1982b. Biological responses of lakes in the Mount St. Helens blast zone. *Science* 216: 178–181.

Wissmar, R.C., D.M. McKnight, and C.N. Dahm. 1990. Contribution of organic acids to alkalinity in lakes within the Mount St. Helens blast zone. *Limnology and Oceanography* 35: 535–542.

Soil Carbon and Nitrogen and Evidence for Formation of Glomalin, a Recalcitrant Pool of Soil Organic Matter, in Developing Mount St. Helens Pyroclastic Substrates

5

Jonathan J. Halvorson, Kristine A. Nichols, and Charles M. Crisafulli

5.1 Introduction

Plant and soil community assembly on Mount St. Helens volcanic substrates has been a subject of considerable interest for more than 30 years. Early colonists such as lupines (*Lupinus latifolius* and *L. lepidus*) were successful because local source populations possessed morphological and physiological adaptations to harsh conditions (Braatne and Bliss 1999), including the ability to fix atmospheric nitrogen via symbiosis with *Rhizobium* (Halvorson et al. 1992). Organic matter, nutrient concentrations, microbial biomass, and other indicators of soil development increased relatively rapidly under lupines. These factors, together with biotic interactions such as competition with other plants or insect herbivory, abiotic erosive processes, and stochastic environmental drivers like the period of unusually wet conditions recorded for the pyroclastic site from 1995 through 1999, helped shape emerging patterns of plant community structure and soil characteristics (Bishop et al. 2005; Halvorson et al. 2005; del Moral et al. 2012).

Glossary terms are shown in ***bold italic face***.

J.J. Halvorson (✉)
U.S. Department of Agriculture, Agricultural Research Service, Northern Great Plains Research Laboratory, Mandan, ND 58554-0459, USA
e-mail: Jonathan.Halvorson@ars.usda.gov

K.A. Nichols
Rodale Institute, 611 Siegfriedale Road, Kutztown, PA 19530, USA

U.S. Department of Agriculture, Agricultural Research Service, Northern Great Plains Research Laboratory, Mandan, ND 58554-0459, USA
e-mail: Kristine.Nichols@rodaleinstitute.org

C.M. Crisafulli
U.S. Department of Agriculture, Forest Service, Pacific Northwest Research Station, Mount St. Helens National Volcanic Monument, 42218 NE Yale Bridge Road, Amboy, WA, USA
e-mail: ccrisafulli@fs.fed.us

One unexpected observation of early studies was evidence for significant accumulations of glomalin in pyroclastic deposits, especially under lupines (Halvorson and Smith 2009). Glomalin, produced by ***arbuscular mycorrhizal fungi*** (AMF), is thought to be an important pool of decomposition-resistant organic matter (Wright and Upadhyaya 1996, 1998; Treseder and Turner 2007; Singh 2012; Walley et al. 2013). As a mixture of compounds (Gillespie et al. 2011), glomalin is a glycoproteinaceous material defined operationally by alkaline salt extraction at high temperature and pressure, which can be classified as Bradford-reactive soil protein (BRSP), glomalin-related soil protein (GRSP), total glomalin (TG), easily-extractable glomalin (EEG), immunoreactive TG (IRTG), immunoreactive EEG (IREEG), immunoreactive GRSP, or immunofluorescent material (e.g., Rillig et al. 2003; Lovelock et al. 2004b; Singh 2012; Koide and Peoples 2013; Nichols et al. 2013). Despite issues concerning the exact molecular composition of glomalin, potential co-extractants measured during the protein analysis (Halvorson and Gonzales 2006; Gillespie et al. 2011), and variations in molecular structure with different arbuscular mycorrhizal species or decomposition processes in the soil, glomalin concentration is frequently correlated with water-stable aggregation, mycorrhizal activity, and soil organic C and N concentrations (Nichols and Wright 2004, 2005; Treseder and Turner 2007; Koide and Peoples 2013).

Arbuscular mycorrhizal fungi (AMF) are often found in early primary-successional substrates (Kikvidze et al. 2010), including those deposited by the 1980 eruption of Mount St. Helens (Allen et al. 2005; Titus et al. 2007), but their role in soil ecosystem development and interactions with plants is likely to be complex and variable (Singh et al. 2011; Kardol et al. 2013) and is still incompletely understood (Dickie et al. 2013). Diversion of ***photosynthate*** from growth to support the AMF-plant relationship may increase the availability of nutrients to plants (Van Der Heijden and Horton 2009; Neumann and George 2010). Mycorrhizal-plant associations

C.M. Crisafulli, V.H. Dale (eds.), *Ecological Responses at Mount St. Helens: Revisited 35 years after the 1980 Eruption*,
https://doi.org/10.1007/978-1-4939-7451-1_5

are often associated with phosphorus acquisition or availability (Richardson and Simpson 2011; Turner et al. 2013), but their role in nitrogen acquisition and cycling is increasingly appreciated (Smith and Smith 2011; Veresoglou et al. 2012; Hodge and Storer 2015). The efficiency of the trade-off of plant-derived carbon for mycorrhizally enhanced nutrient access has been used to differentiate whether AMF is a parasite or a mutualist (Johnson et al. 1997), but this distinction is still debated (Smith and Smith 2011). In several instances, AMF associations have produced no growth benefits in early successional species (Titus and del Moral 1998; Sikes et al. 2012).

Less is understood about the nature and strength of feedbacks between AMF, plants, and soil conditions. Fungal community structure has been found to develop along with aboveground vegetation, increasing in diversity with age of the terrain (Cutler et al. 2014) and affecting plant community composition and thus aboveground productivity (Rillig et al. 2014). Other research suggests that fungal activity may respond more to soil conditions than to plant identity (Sikes et al. 2014).

Aside from a paucity of critical plant nutrients like nitrogen, the biggest barriers plants face in pyroclastic substrates are the lack of water and soil organic matter. Glomalin and glomalin-like biomolecules produced by AMF may be instrumental in surmounting these barriers (Wu et al. 2013; Miransari 2014). In addition to boosting nutrient retention by providing organic binding sites (González-Chávez et al. 2002), the roles of these molecules in building soil structure and creating micro- and macro-sites for better water-holding capacity by stimulating aggregate formation may be critical to soil formation in pyroclastic substrates (Rillig and Mummey 2006; Nichols et al. 2013). As a glycoprotein, glomalin is hydrophilic, particularly where the glycosyl groups attach, but amino acid analysis plus water-stable aggregation analysis demonstrates that it has hydrophobic character as well, owing to more hydrophobic amino acids (K. Nichols, personal communication). Glycoproteins commonly exhibit different folding patterns, exposing hydrophilic sides in aqueous solutions and hydrophobic sides for binding and other activities. Glomalin may thus increase water repellency in soils (Wright and Upadhyaya 1998; Rillig 2005; Young et al. 2012). Increased water repellency is often related to aggregate stability and may enhance water and carbon storage in pyroclastic soils and affect mineralization of soil organic matter (Atanassova et al. 2014). Glomalin production may be stimulated in response to chemical, physical, or even biological soil factors (Purin and Rillig 2007). For example, Rillig and Steinberg (2002) observed less total hyphal length but more glomalin produced when roots were grown in small glass beads (<106 μm) simulating less-aggregated soil, compared with those grown in larger glass beads (710–1180 μm) simulating an aggregated soil. Changes in community structure of AMF and production of glomalin were induced as a response to toxic heavy metals like Al, Cr, or Cu, or to salinity (Na) (González-Chávez et al. 2004; Hammer and Rillig 2011; Seguel et al. 2013; Gil-Cardeza et al. 2014; Krishnamoorthy et al. 2014).

The objectives of this study were to gather information about organic matter that has accumulated in Mount St. Helens pyroclastic substrates since the 1980 eruption. This information is consistent with our long-term studies of soil development, and whereas previous studies emphasized simple binary comparisons between lupine-influenced zones and barren substrates, this work better reflects contemporary successional patterns across the landscape by comparing soil characteristics along a gradient of plant community complexity. A further objective was to measure the quantity and distribution of BRSP, hereafter referred to as glomalin, both relative to depth and in association with different plant community types, and to estimate its contribution to total soil carbon and nitrogen pools. We suggest that glomalin forms relatively early during soil development, and that even small amounts of such microbially produced substances may have important effects on soil processes and thus vegetation development prior to the accumulation of humified soil organic matter.

5.2 Materials and Methods

5.2.1 Site Descriptions and Sampling

We collected soil on the Mount St. Helens pyroclastic-flow deposits (Swanson and Major 2005) (46.23°N, 122.16°W) in September 2011 near study sites used extensively as part of a long-term investigation of small-mammal community reassembly on the Pumice Plain (Crisafulli et al. 2005) (Fig. 5.1). We sampled from (1) herb-dominated upland areas that varied in their abundance of bryophytes (mosses), grasses, and forbs, with diversity characterized as low (LDH), intermediate (IDH), or high (HDH), (2) dense shrub-dominated thickets of upland green alder (*Alnus viridis*) (SDU), and (3) riparian areas supporting dense thickets of willow (*Salix sitchensis*) (SDR) (Fig. 5.2a–e). The LDH, IDH, and HDH communities occupy largely intact surfaces created during the eruption and are well represented in the majority of the landscape. The shrub-dominated communities are more limited in their coverage, occupying areas adjacent to stream courses (shrub-dominated riparian, SDR) or areas subjected to post-eruption erosional processes with lower surface elevations and high topographic moisture (shrub-dominated upland, SDU). Collectively, these two shrub community types occupy <10% of the Pumice Plain. Litter accumulation in the herb-dominated sites relates to overall plant cover and decreases from the HDH to the LDH community types.

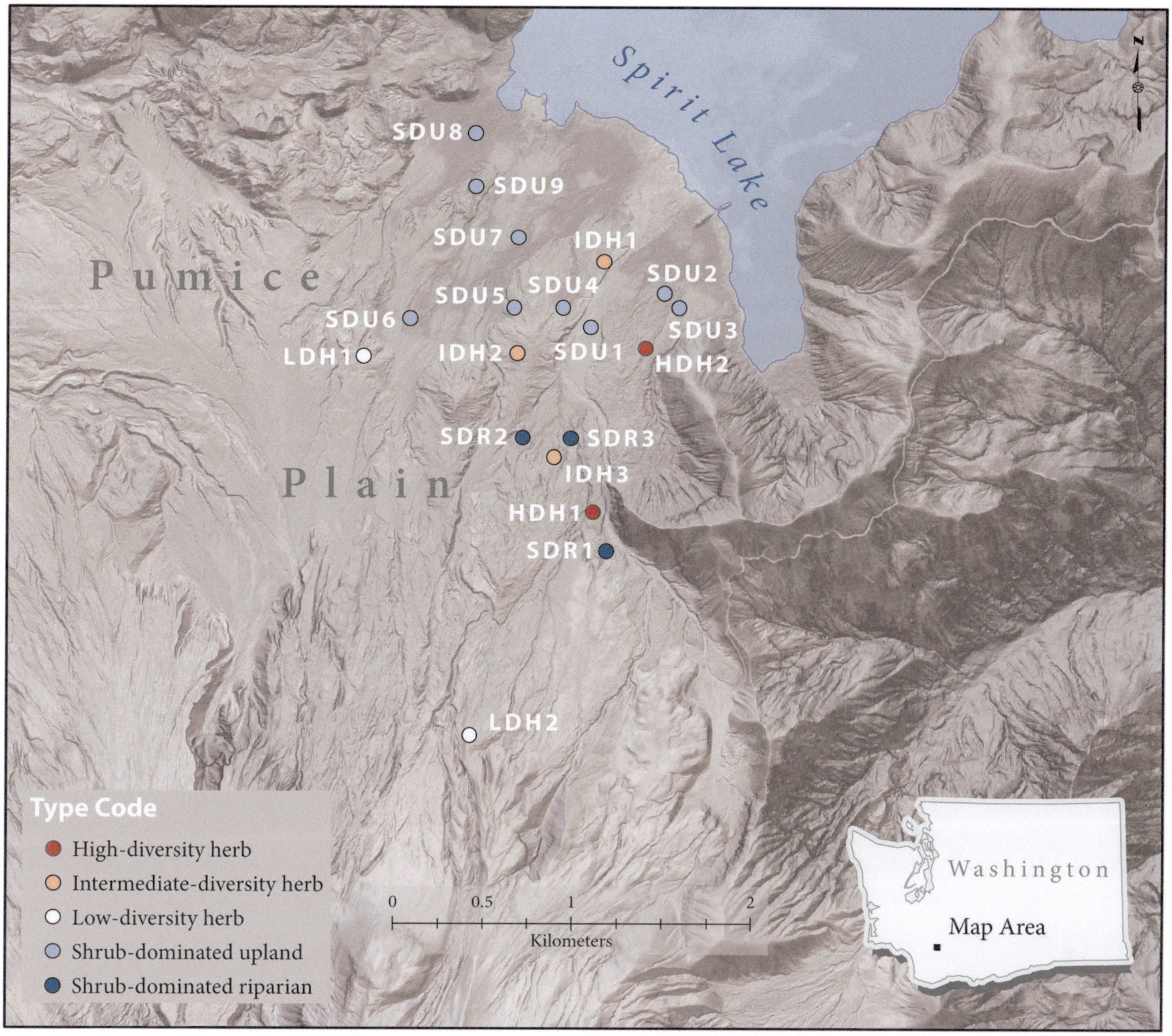

Fig. 5.1 Locations of study sites where soil samples were collected. (Cartography by Kelly Christiansen, USDA Forest Service, Pacific Northwest Research Station).

Distinct organic layers are present in both of the shrub community types.

We also sampled in two late-seral forest stands unaffected by the 1980 eruption and used these as reference sites to compare with the early primary-successional sites on the Pumice Plain. The reference sites are located in the Pacific silver fir zone (Franklin and Dyrness 1988) about 40 km northeast of Mount St. Helens, at 1120 m (ASL). Dominant overstory species in these stands are *Abies amabilis* and *Tsuga heterophylla*, with co-dominant *Pseudotsuga menziesii* and *Thuja plicata*. Mean stem density and basal area per hectare are 625 and 145.7 m^2, respectively. Understory community dominants are ericaceous shrubs (*Vaccinium* spp.) with high cover values (58%), *Abies amabilis* saplings, and a variety of herbs (e.g., *Linnaea borealis*, *Cornus canadensis*, *Rubus lasiococcus*).

These reference forest sites were selected because they were of similar age and vegetative structure to forest growing in the Spirit Lake basin (now the Pumice Plain) prior to the 1980 eruption (USDA Forest Service, 1979 unpublished data) and had similar physical site characteristics, such as slope gradient, elevation, and climate. Based on these similarities, we assume that these reference sites supported soils representative of our study area and more broadly similar to late-seral forest of the southern Washington Cascade Range, including the Mount St. Helens vicinity. Forest soils in this region are andisols of volcanic origin (Dahlgren et al. 2004), and a common feature of these soils is their development on

Fig. 5.2 Examples of (**a**) a low-diversity upland herb community (LDH) characterized by low species richness and cover values (5–10%) of forbs and grasses and an occasional small-stature shrub; (**b**) an intermediate-diversity upland herb community (IDH) characterized by moderate species richness and cover values (20–30%) of forbs and grasses and low shrub cover; (**c**) a high-diversity upland herb community (HDH) characterized by high species richness and cover values (~50%) of forbs and grasses with numerous scattered *Salix sitchensis* and *Alnus viridis* shrubs; (**d**) a shrub-dominated upland thicket (SDU) characterized by mesic upland sites that support dense monotypic stands of *A. viridis*; and (**e**) a shrub-dominated riparian thicket (SDR) characterized by wet/mesic seep and streamside sites that support dense growth of *S. sitchensis* and *A. viridis*. For both SDU and SDR, canopy height of 2–5 m and cover >70% create heavily shaded understories with low-diversity herb communities and contribute to high litter accumulation. (Photos by C.M. Crisafulli).

top of fresh volcanic surfaces. Indeed, soil pits at our reference forest sites reveal this typical pattern of tephra deposits interleaved with organic soils that developed between volcanic deposition events (see Fig. 11.5 in Allen et al., Chap. 11, this volume). Thus, our reference forest sites provide a relevant comparison to evaluate differences in key soil metrics (i.e., nutrients, organic matter, and glomalin) between undisturbed late-seral forest soils typical of the region and those developing on the Pumice Plain. Finally, the potential natural vegetation of the Pumice Plain is coniferous forest, and we assume that in the absence of additional volcanic activity, both vegetation and soils on our site will eventually converge toward something roughly similar to pre-eruption conditions, although this process could take centuries.

For each plant community, we collected mineral soil with a trowel at three depths (surface to 5, 5–10, and 10–20 cm) from three randomly located pits at two or three study sites. For SDU, we collected samples in nine randomly selected alder thickets. We also sampled the distinct organic layer present on top of the soil for SDU and SDR, but did not consider these samples in our analyses. In the lab, we sieved samples (2 mm sieve), dried them to a constant mass at 55 °C, and stored them at room temperature pending further analysis. Dried samples averaged less than 0.3% H_20 by weight. We estimated particle bulk density as the average mass of both "freely settled" (poured) and "tapped" soil needed to fill a container of known volume (25 cm^3).

5.2.2 Vegetation Structure

We distinguished vegetation structure at each study site using a modified "foliage height diversity" (FHD) method (MacArthur and MacArthur 1961). This method provides information on plant physiognomy by measuring the vertical distribution of vegetation by life form (see below). At each of our study sites, we arrayed nine transects at 15-m intervals. We fixed transect length at 30 m for LDH, IDH, and HDH sites, but varied the length at the SDR and SDU sites based on the width of those habitats (range 7.9–30 m). We placed FHD measurement stations (n = 66–144) every 2 m along each transect, and at each station we vertically positioned a 3-m graduated PVC pole and assessed the presence (contact with pole) of bryophytes, graminoids, forbs, shrubs, and trees in each of the following height classes: 0–10 cm, 11–25 cm, 26–50 cm, 51–100 cm, 101–150 cm, 151–200 cm, 200–300 cm, and >300 cm. Thus, across sites there were between 582 and 1152 points in space where we assessed vegetation presence. At each site, we calculated the product of the number of FHD stations on each transect and the eight height classes to derive the total number of points sampled for vegetation. Then we summed the number of intercepts with vegetation present (positive hits). We divided the number of positive hits along each transect by the total number of possible points along that transect to calculate the proportion of vegetation present.

5.2.3 Total and Soluble Soil Carbon (C) and Nitrogen (N)

We determined total soil C and N after dry oxidation (Nelson and Sommers 1996) using a FlashEA 1112 NC Analyzer (CE Elantech, Lakewood, New Jersey, USA) with CE Elantech NC reference soil (3.5% C and 0.37% N) and K-factor calibration with aspartic acid as a standard. We extracted water-soluble organic carbon (WSC) and nitrogen (WSN) using a sequential cool- (23 °C) and hot-water (80 °C) extraction procedure and analyzed the solutions with a Shimadzu TOC-VCPN analyzer equipped with a TNM-1 module (Shimadzu Scientific Instruments, Columbia, Maryland, USA) (Halvorson et al. 2013).

5.2.4 Glomalin (Bradford-Reactive Soil Protein)

We extracted subsamples (2 g) from each soil for glomalin with 100 mM tetrasodium pyrophosphate, pH 9.0, for 1-h cycles at 121 °C in an autoclave (Wright et al. 2006). We repeated the 1-h extraction cycle 2–15 (mean = 4) times until the extraction solution was colorless. After each 1-h extraction cycle, we centrifuged samples to pellet the soil and remove extract solution. We combined all extract solutions per subsample and measured the total volumes. We used the Bradford total protein assay as a measure of glomalin, with bovine serum albumin (BSA) as the standard (Bradford 1976; Nichols and Wright 2004). We added an aliquot of 100 mM tetrasodium pyrophosphate to the BSA standard to remove background cross-reactivity with the dye solution. We calculated total protein values as mg Bradford-reactive soil protein (BRSP) g^{-1} soil.

5.3 Data Analysis

5.3.1 Vegetation Structure

We examined the effect of site and plant community type on the mean proportion of vegetation using a two-way ANOVA. There was no site effect on the proportion of vegetation (P = 0.28), and there was strong evidence that the mean proportion of vegetation differed between at least two of the plant communities ($P < 0.001$ from a one-way ANOVA

F-test). This finding was our justification for pooling all sites for a given community type. We also summarized the mean proportion of vegetation for each life form. We calculated the mean using the proportion of each life form intersecting a transect at eight heights.

5.3.2 Soil

We conducted data analysis using SAS/STAT software, version 9.2 of the SAS System for Windows (SAS Institute 2002–2008). We used a linear mixed model with repeated measures (SAS PROC MIXED, SAS Institute, Cary, North Carolina, USA) to analyze patterns of total and soluble carbon and nitrogen and glomalin, and selected a model that contained both fixed (plant community) and random (sample location) effects, with depth as a repeated measure. We used the KR (Kenward-Roger) option to calculate degrees of freedom and selected covariance structures to minimize Akaike's information criterion. We performed multiple pairwise comparisons of means using Fisher's protected least significant difference test using a value of 5% ($P < 0.05$) as the minimum criterion for significance unless otherwise noted. We evaluated assumptions of normality and identified appropriate data transformations with SAS/ASSIST.

Using SAS PROC CORR, we characterized bivariate relationships between glomalin and the other soil properties by Spearman correlations. Unlike Pearson correlations, Spearman correlations do not assume bivariate normality and are less biased by outliers. Values given in text and graphs are the arithmetic mean ± the standard error of the mean expressed on an oven-dry soil basis.

5.4 Results and Discussion

Early work (through about 2005) was based on a simple binary sampling design comparing lupine-influenced zones to barren substrates. Beginning in 1990, living and dead lupines were delineated from each other and sample depths were standardized (0–5, 5–10 cm). By 2010, it was clear that there was a need to compare the initial herbaceous colonists (lupines) to other nitrogen-fixing species (upland alder) because woody species were becoming increasingly dominant in the landscape. This initial shift in woody-plant abundance has several important potential implications for plant succession (e.g., species turnover), microclimate (e.g., altered light regime, decreased wind, increased water retention), and soil genesis (e.g., heavy litter input, *Frankia* nitrogen influence). The current sampling approach, adopted in 2011, moved away from simple binary comparisons to multiple comparisons among several sites representative of a gradient of plant community complexity.

5.4.1 Vegetation Structure

At the coarsest level, vegetation analyses revealed an unsurprising dichotomy between structurally simple herb-dominated communities and comparatively complex shrub-dominated communities (Fig. 5.3). A significant difference was apparent between LDH and all other sites, and IDH and HDH were significantly different from both shrub community types, but not from one another. Shrub sites were similar to each other. However, when looking at the mean proportion of different life forms present at the sites, a more complex pattern emerged (Fig. 5.4). Bryophyte values were highest at IDH and HDH sites, lowest at shrub sites, and intermediate at LDH sites (Fig. 5.4a). Graminoids followed a more complex pattern, with the highest value in SDU followed in decreasing proportion from HDH > LDH > SDR > IDH (Fig. 5.4b). Forb values generally increased from LDH to SDU (Fig. 5.4c). All herb-dominated sites had low shrub values, whereas SDR and SDU had a similar mean proportion of shrubs (Fig. 5.4d).

The plant communities reported in this chapter are more complex in terms of species richness and physical structure as compared with the lupine communities studied in the first decades after the eruption (Halvorson et al. 1991, 2005; Halvorson and Smith 2009). In many locations on the Pumice Plain, the sequence of plant community development in upland habitats involved *L. lepidus* establishment, followed by rapid growth and spread of plants in expanding rings around the mother plant and maintenance of a near monoculture for a few to several years. In this regard, lupine commu-

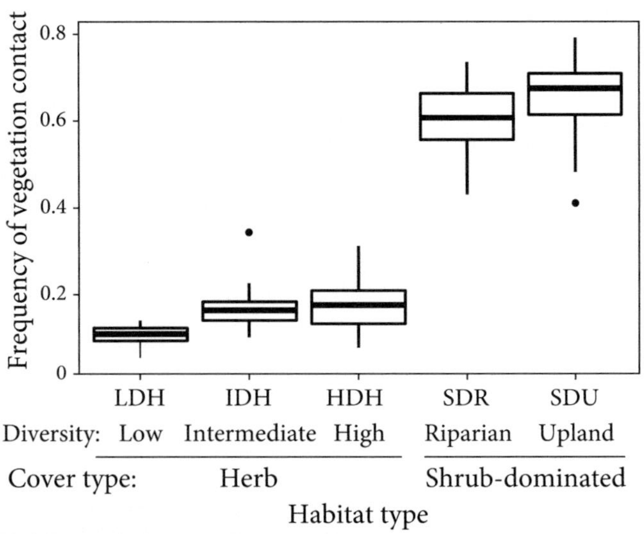

Fig. 5.3 Vegetation structural complexity across five plant community types on the Pumice Plain, Mount St. Helens, USA. Low-diversity herb (LDH), intermediate-diversity herb (IDH), high-diversity herb (HDH), shrub-dominated riparian (SDR), and shrub-dominated upland (SDU). Data points are mean values with error bars representing one standard error.

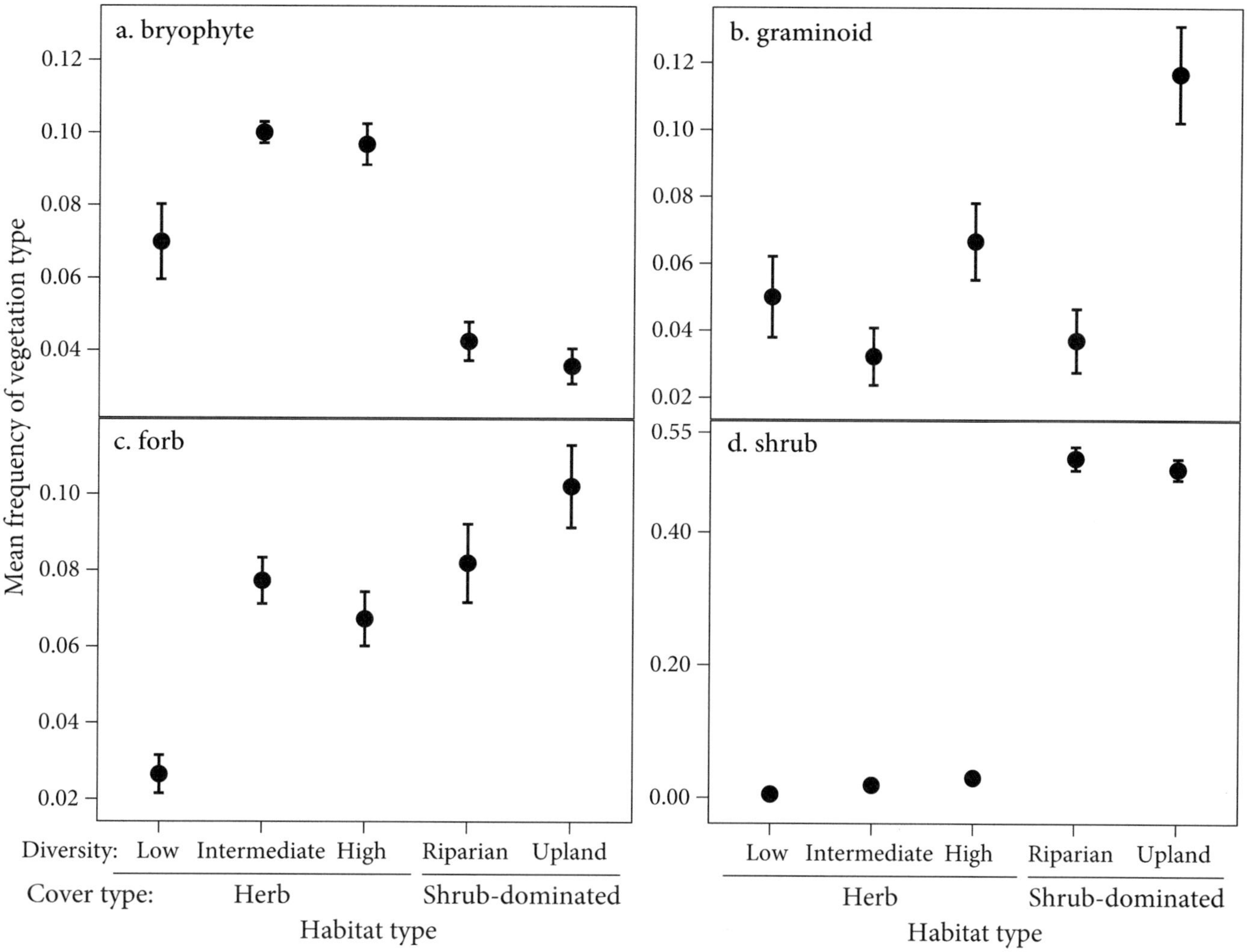

Fig. 5.4 Mean proportion of vegetation by life form across five plant community types on the Pumice Plain, Mount St. Helens, USA. Data points are mean values with error bars representing one standard error.

nities spread in something analogous to an advancing army front. These young and vigorously growing lupine populations appear to inhibit the establishment of other plant taxa (Morris and Wood 1989) until natural senescence ensues or is hastened by herbivores (Bishop et al. 2005). Once a dieback event occurs, numerous wind-dispersed herbs and graminoids establish, and lupine rarely attains dominance, yet tends to persist as a community member for decades. The upland communities in this study represent different states along this general trajectory.

5.4.2 Soil

Soil bulk density, total and soluble carbon and nitrogen, and glomalin were strongly affected by plant community type, but this effect also depended on sample depth (Table 5.1). In general, differences among the community types were most apparent at the 0–5 cm depth, but there was little distinction among sites at the 10–20 cm depth. Most soil characteristics varied with depth, but these changes were often least evident in the LDH.

5.4.2.1 Bulk Density

Bulk density increased significantly with depth under most community types. The lower values of bulk density near the surface were likely associated with inputs of organic matter from vegetation and were particularly evident at SDR. Conversely, soil bulk density for LDH did not vary with depth and showed less influence of organic inputs at the 0–5 cm depth. Values of average bulk density at the 10–20 cm depth were similar for all locations, 1.42 ± 0.02 Mg m^{-3}.

Surface values of bulk density in the LDH, IDH, and SDU community types were similar to those previously observed for barren pyroclastic sites, whereas those for the HDH and SDR sites were consistent with those found under lupine colonists

Table 5.1 Mean (SEM) bulk density, total and extractable soil carbon and nitrogen, and glomalin in mineral soil under different plant community types growing on Mount St. Helens pyroclastic substrates in 2011 ($n = 6–9$). Numbers denote differences among soil depths within each community type. Letters denote differences among community types at each depth (protected LSD, $P \leq 0.05$).

Community type	Depth (cm)	Bulk density[a] (mg m^{-3})	Total soil C (g kg^{-1})	Total soil N (g kg^{-1})	Cool-water-extractable C (mg kg^{-1})	Cool-water-extractable N (mg kg^{-1})	Hot-water-extractable C (mg kg^{-1})	Hot-water-extractable N (mg kg^{-1})	Glomalin (g kg^{-1})
LDH ($n = 6$)	0–5	1.34 (0.06) 1,a	2.52 (0.65) 1,c	0.25 (0.06) 1,b	34.9 (5.5) 1,c	4.4 (1.1) 1,c	117 (30) 1,c	9.0 (2.7) 1,c	0.33 (0.05) 1,c
	5–10	1.41 (0.05) 1,x	1.07 (0.42) 2,y	0.11 (0.04) 2,x	25.7 (5.4) 2,y	2.5 (0.7) 2,y	54.3 (18.7) 2,z	3.9 (1.7) 2,z	0.20 (0.04) 2,y
	10–20	1.41 (0.05) 1,q	0.94 (0.26) 2,q	0.09 (0.02) 2,q	29.8 (8.0) 12,q	1.8 (0.4) 3,q	43.7 (11.5) 2,r	2.2 (0.7) 3,r	0.20 (0.03) 2,q
IDH ($n = 9$)	0–5	1.27 (0.04) 2,ab	5.87 (1.32) 1,b	0.46 (0.09) 1,b	61.6 (6.8) 1,bc	8.5 (1.2) 1,bc	269 (52) 1,b	20.8 (4.6) 1,b	1.15 (0.30) 1,b
	5–10	1.39 (0.05) 1,xy	1.44 (0.27) 2,xy	0.14 (0.02) 2,x	36.7 (4.3) 2,xy	3.3 (0.5) 2,xy	75.8 (13.8) 2,yz	4.8 (1.0) 2,yz	0.27 (0.03) 2,xy
	10–20	1.43 (0.04) 1,q	0.90 (0.12) 3,q	0.09 (0.01) 3,q	30.3 (3.3) 3,q	2.2 (0.2) 3,q	50.6 (6.3) 3,r	2.9 (0.4) 3,qr	0.22 (0.02) 2,q
HDH ($n = 6$)	0–5	1.14 (0.02) 2,ab	5.67 (1.07) 1,bc	0.46 (0.09) 1,b	65.4 (8.9) 1,bc	7.3 (1.7) 1,bc	245 (46) 1,b	18.9 (4.8) 1,b	1.01 (0.23) 1,b
	5–10	1.21 (0.05) 2,y	2.99 (1.13) 2,x	0.23 (0.08) 2,x	56.9 (14.0) 1,x	4.0 (1.2) 2,xy	130 (43) 2,xy	7.7 (2.6) 2,xy	0.62 (0.26) 2,x
	10–20	1.31 (0.06) 1,q	1.21 (0.34) 3,q	0.10 (0.03) 3,q	39.8 (7.0) 2,q	2.2 (0.4) 3,q	64.4 (13.8) 3,qr	3.1 (0.7) 3,qr	0.25 (0.04) 3,q
SDU ($n = 8$)	0–5	1.28 (0.04) 2,ab	6.15 (1.21) 1,bc	0.45 (0.09) 1,ab	123 (24) 1,ab	14.0 (4.3) 1,ab	391 (90) 1,ab	30.9 (10.0) 1,b	1.22 (0.24) 1,b
	5–10	1.40 (0.03) 1,xy	1.88 (0.31) 2,xy	0.16 (0.02) 2,x	50.0 (3.3) 2,xy	4.2 (0.4) 2,xy	133 (11) 2,x	7.0 (0.8) 2,xy	0.34 (0.06) 2,x
	10–20	1.42 (0.03) 1,q	2.78 (1.02) 2,q	0.20 (0.06) 2,q	40.6 (4.8) 3,q	3.0 (0.4) 3,q	103 (15) 3,q	4.6 (0.7) 3,q	0.25 (0.02) 3,q
SDR ($n = 8$)	0–5	1.12 (0.06) 2,b	20.6 (5.3) 1,a	1.16 (0.28) 1,a	201 (32) 1,a	24.6 (4.1) 1,a	707 (151) 1,a	57.2 (11.3) 1,a	3.34 (0.65) 1,a
	5–10	1.46 (0.02) 1,x	2.16 (0.31) 2,x	0.14 (0.02) 2,x	55.6 (7.1) 2,x	4.9 (0.7) 2,x	128 (19) 2,x	8.3 (1.4) 2,x	0.48 (0.12) 2,x
	10–20	1.48 (0.02) 1,q	1.34 (0.30) 3,q	0.09 (0.02) 3,q	42.2 (5.7) 3,q	3.3 (0.7) 3,q	82.3 (12.3) 3,q	4.7 (1.0) 3,q	0.32 (0.09) 3,q
C X D	*P*	0.0004	0.0002	<0.0001	<0.0001	0.0002	<0.01	<0.005	<0.005
Reference forest ($n = 6$)	0–5	0.68 (0.04)	77.5 (10.3)	2.05 (0.25)	729 (105)	26.9 (4.9)	3051 (412)	100 (18)	8.37 (0.89)
	5–10	0.83 (0.04)	51.3 (12.4)	1.19 (0.20)	378 (57)	14.9 (1.5)	1838 (371)	58.3 (9.9)	5.51 (0.80)
	10–20	0.93 (0.05)	33.0 (10.1)	0.83 (0.21)	238 (41)	9.3 (1.5)	1016 (227)	32.0 (7.2)	4.02 (0.94)

[a]Determined as the average of mass for packed and unpacked soil of a known volume (25 cc).

(Halvorson and Smith 2009). Values of bulk density at the 10–20 cm depth changed little from those reported for young pyroclastic-flow deposits (Wilson and Head 1981).

5.4.2.2 Total Soil C and N

Concentrations of total soil carbon and nitrogen decreased significantly with depth under all community types, especially the SDR. Differences among the community types were most evident at the 0–5 cm depth, with the highest concentrations of C and N in SDR soil and the lowest in the LDH soil. We also observed significant differences among community types for total soil C at the 5–10 cm depth, but not at the 10–20 cm depth (average 1.46 ± 0.26 g kg^{-1}). Total soil nitrogen was less variable with depth and indistinguishable among community types at 5–10 and 10–20 cm, 0.16 ± 0.02 and 0.12 ± 0.02 g kg^{-1}, respectively.

Soil C/N ratios varied with main effects of depth ($P < 0.0001$) and plant community ($P < 0.01$). Higher ratios (13.3 ± 0.5) occurred at the surface, compared with 11.6 ± 0.4 and 11.7 ± 0.4 at the 5–10 and 10–20 cm depths, respectively, and ratios were arrayed among the different plant community types in association with larger amounts of organic matter: SDR (15.1 ± 0.4) > SDU (12.5 ± 0.5) ≥ HDH (12.2 ± 0.4) ≥ IDH (10.7 ± 0.4) ≥ LDH (9.4 ± 0.4). In comparison, C/N ratios in the reference forest sites were much higher, 37.5 ± 1.5, 40.7 ± 3.3, and 36.7 ± 3.2 at the 0–5, 5–10, and 10–20 cm depths, respectively. Increasing C/N ratios along the gradient of plant communities from LDH to reference forest sites suggest the accumulation of recalcitrant soil organic carbon, a result of greater inputs from vegetation over time, but may also derive from changes in fungal effects on decomposition during succession. Such changes may be associated with the increasing prevalence of ectomycorrhizal and ericoid mycorrhizal fungal associations, which produce enzymes (Phillips et al. 2014) that may result in greater competition for nitrogen between free-living decomposers and plant hosts. Recently, Averill et al. (2014) demonstrated that soils dominated by plants associated with ***ectomycorrhizal*** and ***ericoid mycorrhizal fungi*** store 1.7 times more C per unit N than do soils dominated by plants associated with arbuscular mycorrhizal fungi.

The LDH and HDH community types correspond most closely to the barren and lupine (*L. lepidus*)-influenced sites contrasted in earlier studies and show that organic matter has continued to accrue in developing pyroclastic soils since the eruption in 1980 (Fig. 5.5). From initial values near 0 in barren pyroclastic substrates (Engle 1983), concentrations of total C and N in the 0–5 cm depth of LDH sites increased after 31 years by an average of 81.3 ± 20.9 and 8.1 ± 2.0 mg kg^{-1} $year^{-1}$, almost twice as fast as those reported previously for bare sites (Halvorson and Smith 2009). In comparison, by 2011, the concentrations of total C and N increased in HDH sites by an average 183 ± 35 and

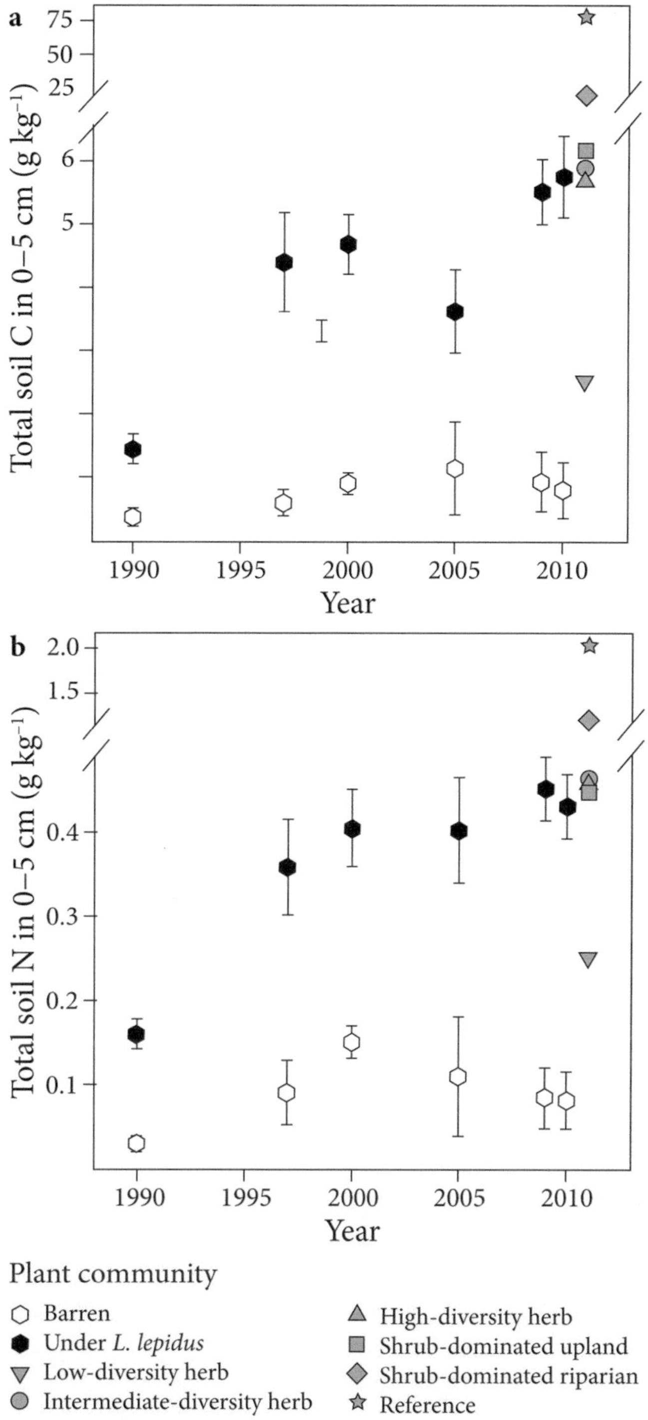

Fig. 5.5 Average values from the current study in relation to historic trends of (**a**) total soil carbon and (**b**) soil nitrogen in lupine-influenced (from under live or dead *Lupinus lepidus*) and barren pyroclastic soil. Pre-2011 values are averages with 95% confidence intervals determined by dry combustion. For barren samples n = 11, 18, 30, 10, 8, and 8 for 1990, 1997, 2000, 2005, 2009, and 2010, respectively. For lupine-influenced samples n = 22, 28, 29, 20, 8, and 8 for 1990, 1997, 2000, 2005, 2009, and 2010, respectively. Data for 1990–2005 are from Halvorson and Smith (2009). Data for 2009 and 2010, from Jonathan Halvorson (unpublished), are from soil samples collected on 1 September 2009 and 21 September 2010 in the same vicinity as the previous studies. Samples were collected, processed, and analyzed using methods and equipment identical to the current study.

14.8 ± 2.9 mg kg^{-1} year^{-1}, more comparable with earlier estimates under lupines (Halvorson et al. 2005; Halvorson and Smith 2009). The highest accumulations of total C and N in surface soil, equivalent to 665 ± 171 and 37.5 ± 9.1 mg kg^{-1} year^{-1}, were associated with the SDR community type.

Relatively high concentrations of carbon and nitrogen observed in surface samples from SDU and SDR sites undoubtedly relate to significant accumulations of organic matter from leaf litter at these shrub-dominated sites and perhaps to the difficulty in clearly delineating the surface of the mineral soil. Total carbon and nitrogen at both sites were about 380 and 19 g kg^{-1} in the coarse fractions of the organic layer (>2 mm) and about 200 and 11.5 g kg^{-1} in the finer fraction (<2 mm), respectively (Jonathan Halvorson, unpublished data).

5.4.2.3 Water-Extractable C and N

Cool-water-extractable C and N decreased with depth, especially in the SDU and SDR community types. The highest concentrations were extracted from the shrub-dominated community types, whereas lower concentrations were extracted from the herb-dominated communities, especially LDH. Differences among community types diminished with depth and were absent at 10–20 cm, averaging 36.5 ± 2.5 and 2.6 ± 0.2 mg kg^{-1} for extractable C and N, respectively. Hot water extracted two to three times more C and N in patterns similar to cool water; the highest concentrations and greatest variability among community types occurred at the 0–5 cm depth, and shrub-dominated community types had higher concentrations compared to LDH. However, unlike total or cool-water-extractable C and N, higher concentrations of hot-water-extractable C and N were found at the 5–10 and 10–20 cm depths under the shrub-dominated and HDH communities than under the LDH.

The C and N extracted from soil with cool water may correlate to soluble plant residues, whereas C and N recovered after hot-water incubation correlate with soil microbial biomass C and N, mineralizable N, and total carbohydrates (Ghani et al. 2003; Curtin et al. 2006). Both cool- and hot-water-extractable C have been found to be related to soil microbial activity in patterns affected by soil and plants (Uchida et al. 2012). Earlier studies of Mount St. Helens pyroclastic deposits by Halvorson and Smith (2009) reported higher water-extractable C and N values, decreasing with depth, under lupines, as compared with barren sites. More importantly, they found that the amount of hot-water-soluble C was indistinguishable from microbial biomass C measured by substrate-induced respiration. Estimates of soil microbial biomass C at 0–10 cm depth under herb-dominated community types, if still comparable to hot-water-extractable C, were slightly higher than those reported in 2005 and equivalent to about 5% of the total soil C.

5.4.2.4 Glomalin

Average concentrations of glomalin were greatest in the 0–5 cm depth but varied among community types with highest values for SDR and lowest for LDH. Concentrations of glomalin decreased with depth to a similar low value across all sites at the 10–20 cm depth (0.24 g kg^{-1} soil). Glomalin was strongly and positively correlated with total and soluble soil C and N at each depth (Table 5.2), but negatively correlated with soil bulk density. With the exception of LDH, estimates of glomalin in 0–5 cm depth were within the broad range of values reported for temperate and boreal forests (0.6–5.8 g kg^{-1}) (Singh 2012). However, these values were lower than at two old-growth reference forests located about 40 km to the NE of Mount St. Helens that ranged from 8.4 to 4.0 g kg^{-1} at 0–5 and 10–20 cm (Table 5.1).

In other studies (Rillig et al. 2003; Emran et al. 2012; Gispert et al. 2013; Vasconcellos et al. 2013), glomalin has been found to be positively correlated with soil C and N but negatively correlated with soil bulk density. Glomalin production is generally higher under conditions that favor AMF activity (Ryan and Graham 2002) or diversity (Helgason et al. 1998), such as minimal physical disruption of the soil (Wright et al. 1999; Borie et al. 2000; Wright and Anderson 2000), higher levels of plant diversity, or low to moderate soil fertility (Treseder and Allen 2002; Lovelock et al. 2004a; Treseder 2004). Changes in glomalin have also been related to the amount of physical disruption of soil or to shifting plant communities (often afforestation) (e.g., Treseder and Turner 2007; Fokom et al. 2012). For example, sites under arable land use showed an exponential decrease in glomalin levels by ~60% after 110 years (Spohn and Giani 2010, 2011). Conversely, glomalin in soil increased with the amount of time under conservation management, including reduced tillage, elimination of fallow periods, use of perennial crops, and greater annual crop diversity (Wright and Anderson 2000; Rillig et al. 2001; Harner et al. 2004; Staddon 2005).

Table 5.2 Spearman correlations between BRSP and selected soil properties for pooled data across vegetation community types.

Depth	*n*	Bulk density	Total soil C	Total soil N	Total water-extractable soil C	Total water-extractable soil N
0–5 cm	35	−0.62***	0.96***	0.94***	0.94***	0.92***
5–10 cm	36	−0.44**	0.89***	0.91***	0.90***	0.86***
10–20 cm	36	−0.51**	0.78***	0.78***	0.84***	0.84***

*$P < 0.05$; **$P < 0.01$; ***$P < 0.0001$.

5.4.2.5 Standing Pools of Total C, N, and Glomalin

Standing stocks of organic matter, including glomalin, are the net result of processes that affect its production, such as increased net primary productivity, and those that affect its decomposition, such as soil microbial community structure, moisture, temperature, or disturbance (Treseder and Turner 2007). Stocks of total C, N, and glomalin in the top 20 cm, calculated from data in Table 5.1, varied with plant community type (Fig. 5.6) and were similarly arrayed (SDR ≥ SDU ≥ HDH ≥ IDH ≥ LDH). Values ranged from 1406 ± 238, 84 ± 13, and 241 ± 36 g m^{-2} at SDR sites for total C, N, and glomalin, respectively, to 358 ± 89, 36 ± 7, and 62 ± 8 g m^{-2} at LDH sites. Vegetative complexity/diversity data showed no meaningful difference between SDU and SDR (Fig. 5.3), a pattern consistent with total C and N (Fig. 5.6a, b). However, higher glomalin values for SDR sites (Fig. 5.6c) may relate to the age of the plant communities. The SDR sites were among the first to establish (circa late 1980s), whereas the SDU sites initiated later (about 2005). Thus, SDR sites had been accumulating organic matter in comparatively large quantities for 20–25 years. In addition, SDU sites had formerly been colonized by forb and grass species that are slowly waning but still present, whereas in the SDR sites, the transition to shrubs has been underway for decades. The SDU community type was notable for the relatively large pool of total C and N at 10–20 cm depth, not observed in other community types. Increasing C and N with depth suggests the presence of substantial amounts of soil organic matter at depths greater than those examined in this study. Such patterns can result from deeper root distribution under upland shrubs, together with reduced rates of decomposition with depth related to changes in the activity and composition of microbial communities (Gill and Burke 2002; Fierer et al. 2003), and point to the need for a more complete sampling of the soil column in future studies. Similarly, the comparatively large pools of total C, N, and glomalin in the reference forest sites also increased with depth, to about 7455 ± 734, 189 ± 13, and 864 ± 71 g per m^{-2}, respectively.

5.4.2.6 Glomalin Carbon and Nitrogen

The carbon content of glomalin is known to vary with depth or concentration and to be affected by extraction method, ranging between 27% and 43%, although glomalin nitrogen is more narrowly defined, about 2–4% (Nichols and Wright 2005; Halvorson and Gonzalez 2006). Estimates based on an average composition of 35% C and 4% N showed little variation among community types but demonstrated that glomalin differentially accounted for about $6.2 \pm 0.3 < 7.3 \pm 0.3 \leq 8.0 \pm 0.5\%$ of the total soil C at 0–5, 5–10, and 10–20 cm depths or 0.52 ± 0.09, 0.13 ± 0.02, and 0.09 ± 0.01 g kg^{-1}, respectively. Glomalin accounted for about 9.6 ± 0.3% (0.028 ± 0.004 g kg^{-1}) of soil N, irrespective of depth. Though affected

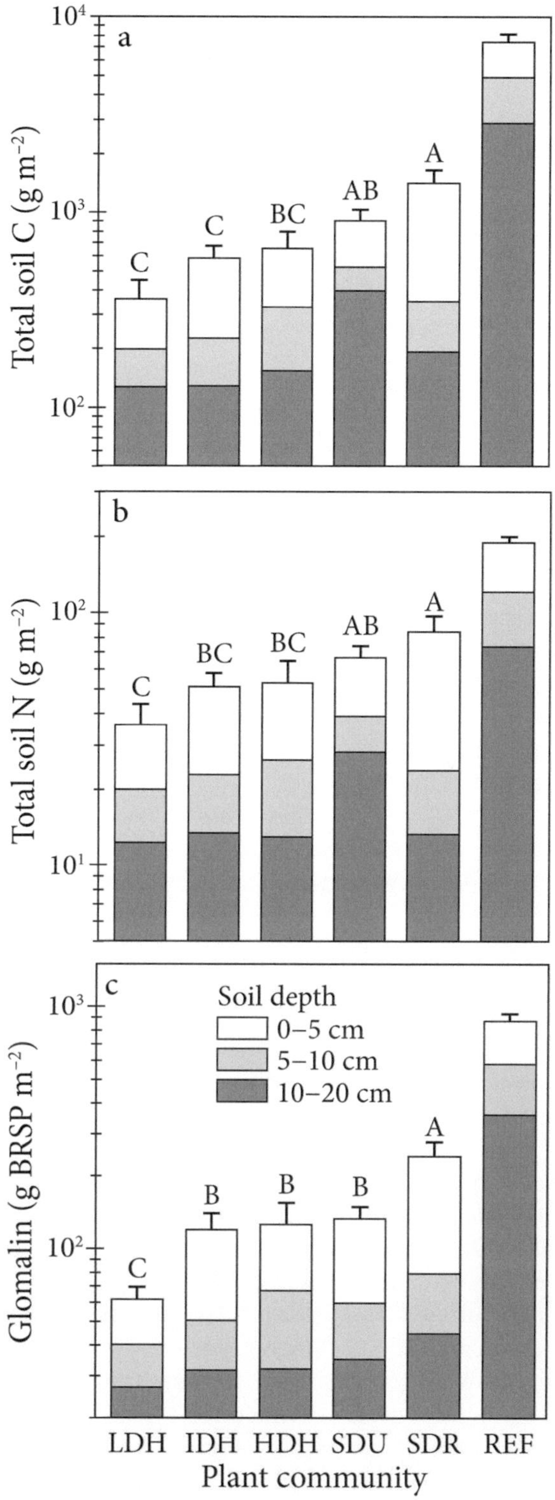

Fig. 5.6 Average (standard error) total pool sizes of (**a**) soil carbon, (**b**) soil nitrogen, and (**c**) glomalin (BRSP) in mineral soil under different plant community types growing on Mount St. Helens pyroclastic substrates in 2011. Values are depth-weighted and calculated from data in Table 5.1 (n = 6–9). Letters denote differences among community types at the 0–20 cm depth (log-10 transformed data, protected LSD, $P \leq 0.05$). Data for reference forest sites (REF) are shown for comparative purposes but not included in the statistical analyses. Note the use of log-10 scales on the y-axes.

by the method used to extract glomalin, these numbers are in reasonable agreement with earlier values for Mount St. Helens pyroclastic substrates (Halvorson and Smith 2009) and some other studies (Woignier et al. 2014), but higher than the 3–5% of the total C and N pools typical of temperate and tropical sites (Rillig et al. 2001, 2003; Nichols 2003; Lovelock et al. 2004a; Halvorson and Gonzalez 2006; Jorge-Araújo et al. 2015).

Enrichment of glomalin-C with depth can be attributed to older and more recalcitrant forms of organic matter that may be co-extracted with glomalin (Woignier et al. 2014) but can also be associated with soil disturbance or changes in plant communities. Studies by Spohn and Giani (2010, 2011) found that the ratio of glomalin-related soil protein to total organic carbon increased over time after conversion from forest soils to arable soils. A plot of the proportion of glomalin-C against total soil C (Fig. 5.7a) suggests production or retention of glomalin relative to other pools of soil organic matter may be especially significant when organic matter is low. The concentrations of glomalin observed in surface samples at pyroclastic sites (0–5 cm) accounted for a nearly constant proportion of the total pool of soil C.

Although different trends were observed among the sites, the proportion of glomalin-N also appeared to decrease as the concentration of total soil N increased to 0.2–0.3 g kg^{-1}, particularly at the 10–20 cm depth. However, there was also evidence for increasingly higher proportions of glomalin-N accompanying higher concentrations of soil N observed in surface (0–5 cm) samples from the pyroclastic sites and the reference forest sites (Fig. 5.7b). Relatively high enrichment of glomalin-N in pyroclastic substrates may reflect strong demands for a limited pool of soil N by plants that induce AMF production (Johnson 2010) and thus the amount of N stabilized in glomalin. Nitrogen incorporated into the structure of glomalin would be stabilized against microbial mineralization but likely still functionally active with respect to glomalin's effects on soil structure (Rillig et al. 2007). However, although buffered from losses, it is not clear whether the nitrogen in glomalin itself or occluded in glomalin-related aggregates would be accessible to plants. Thus, sequestration of N in the hyphae of symbiotic AMF might remove it from the pool of nutrients readily available to the host plant but could also form the basis of an exclusionary competitive mechanism beneficial to the host by limiting access to nutrients by competitors.

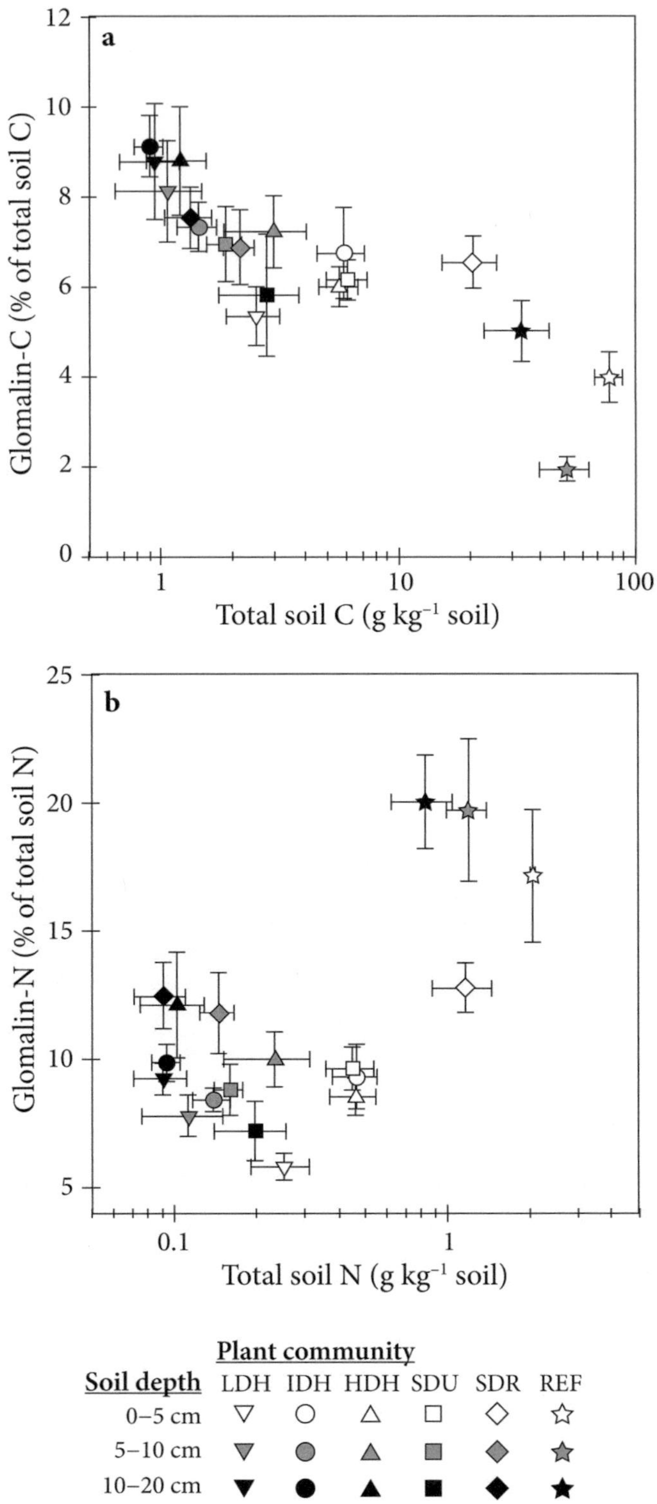

Fig. 5.7 Estimated average (standard error) values for (**a**) glomalin-C as a function of total soil C and (**b**) glomalin-N as a function of total soil N. Values were derived from measurements of BRSP assuming 35% C and 4% N. Note log-10 scales on the *x*-axes.

5.5 Conclusions

Through their effects on plant physiology, soil ecological interactions, and soil engineering, mycorrhizal fungi are important but incompletely understood drivers of pedogenic processes during primary succession (e.g., Rillig and Steinberg 2002; Rodriguez et al. 2009; Torres et al. 2012). Formation of stable soil organic matter is typically the result of a relatively slow series of decomposition processes that can be constrained, in early successional sites, by the availability of exogenous inputs or by environmental conditions that limit microbial activities. Alternatively, compounds such as glomalin, a glycoprotein produced by arbuscular mycorrhizal fungi, may form relatively early during soil development and improve aggregate stabilization, water infiltration, and carbon and nitrogen storage.

The presence of glomalin in Mount St. Helens pyroclastic substrates is further evidence of the positive feedbacks between plants, microorganisms, and soils thought to play a central role in early successional communities. Our data suggest that C, N, and glomalin concentrations increase together with vegetation complexity, and in this sense there appears to be a coupling between plant community development and soil genesis during early primary succession although the two may not synchronize. Although establishment of woody plants (shrubs) is an ecological threshold for plant community change and development, changes in soil genesis likely lag behind vegetative community structure. Even small amounts of microbially produced substances like glomalin may have disproportionate effects on soil processes and plant development prior to the accumulation of humified soil organic matter. However, the details of these effects, together with a more complete understanding of the evolving roles of fungi during primary succession, remain to be elucidated.

Acknowledgments The authors thank T. Blackman, E. Denton, J. Harrah, T. Robertson A. Smits, C. Thorson, and D. Wetch for sample collection and technical assistance. Photo credits: C.M. Crisafulli. Funding for CMC was provided by the USDA Forest Service, Pacific Northwest Research Station, and the National Science Foundation (LTREB Program DEB-0614538). This work is dedicated to Jeffrey L. Smith.

Glossary

Arbuscular mycorrhizal fungi A plant-fungal relationship, found in over 90% of vascular plants, in which the fungus penetrates the cortical cells of the roots and forms highly branched arbuscules and, in some species, vesicles. These fungi provide nutrients and water to plants as well as assisting in soil structure and disease prevention.

Ectomycorrhizal fungi A plant-fungal relationship, found in about 2% of plant species, particularly woody plants, in which the fungus, primarily basidiomycetes and some ascomycetes, forms a dense hyphal sheath, known as the mantle, around the root surface. The hyphal network may extend several centimeters into the surrounding soil to aid in water and nutrient uptake, often helping the host plant to survive adverse conditions, in exchange for carbohydrates.

Ericoid mycorrhizal fungi A plant-fungal relationship between members of the plant family Ericaceae, such as blueberries, cranberries, and *Rhododendron*, and several lineages of fungi in acidic and nutrient poor soils found in boreal forests, bogs, and heathlands. The fungus establishes loose hyphal networks around the outside of hair roots, from which they penetrate the walls of cortical cells to form intracellular coils that can densely pack individual plant cells. The coils function only for a period of a few weeks before the plant cell and fungal hyphae begin to degrade. The coils are where nutrients are exchanged for carbohydrates, and these fungi also have enzymatic capabilities to break down complex organic molecules and act as saprophytes, living on dead or decaying organic matter.

Mycorrhiza(e) A symbiotic association between a fungus and the roots of a host plant where the body of the fungi are thread-like hyphae that extend the root system. The mycorrhizal association is generally mutualistic, playing important roles in soil biology, chemistry, and physics, but in particular species or in particular circumstances, mycorrhizae may be variously pathogenic in host plants.

Photosynthate A chemical product of photosynthesis—carbohydrates that are synthesized from carbon dioxide and a source of hydrogen (usually water), using light as an energy source.

References

Allen, M.F., C.M. Crisafulli, S.J. Morris, L.M. Egerton-Warburton, J.A. MacMahon, and J.M. Trappe. 2005. Mycorrhizae and Mount St. Helens: Story of a symbiosis. In *Ecological responses to the 1980 eruption of Mount St. Helens*, ed. V.H. Dale, F.J. Swanson, and C.M. Crisafulli, 221–231. New York: Springer.

Atanassova, I.D., S.H. Doerr, and G.L. Mills. 2014. Hot-water-soluble organic compounds related to hydrophobicity in sandy soils. In *Soil carbon*, ed. A.E. Hartemink and K. McSweeney, 137–146. Switzerland: Springer International Publishing.

Averill, C., B.L. Turner, and A.C. Finzi. 2014. Mycorrhiza-mediated competition between plants and decomposers drives soil carbon storage. *Nature* 505: 543–545.

Bishop, J.G., W.F. Fagan, J.D. Schade, and C.M. Crisafulli. 2005. Causes and consequences of herbivory on prairie lupine (*Lupinus lepidus*) in early primary succession. In *Ecological responses to the*

1980 eruption of Mount St. Helens, ed. V.H. Dale, F.J. Swanson, and C.M. Crisafulli, 151–161. New York: Springer.

Borie, F.R., R. Rubio, A. Morales, and C. Castillo. 2000. Relationships between arbuscular mycorrhizal hyphal density and glomalin production with physical and chemical characteristics of soils under no-tillage. *Revista Chilena de Historia Natural* 73: 749–756.

Braatne, J.H., and L.C. Bliss. 1999. Comparative physiological ecology of lupines colonizing early successional habitats on Mount St. Helens. *Ecology* 80: 891–907.

Bradford, M.M. 1976. A rapid and sensitive method for the quantitation of microgram quantities of protein utilizing the principle of protein-dye binding. *Analytical Biochemistry* 72: 248–254.

Crisafulli, C.M., J.A. MacMahon, and R.R. Parmenter. 2005. Small-mammal survival and colonization on the Mount St. Helens volcano: 1980–2002. In *Ecological responses to the 1980 eruption of Mount St. Helens*, ed. V.H. Dale, F.J. Swanson, and C.M. Crisafulli, 199–218. New York: Springer.

Curtin, D., C.E. Wright, M.H. Beare, and F.M. McCallum. 2006. Hot water-extractable nitrogen as an indicator of soil nitrogen availability. *Soil Science Society of America Journal* 70: 1512–1521.

Cutler, N.A., D.L. Chaput, and C.J. van der Gast. 2014. Long-term changes in soil microbial communities during primary succession. *Soil Biology and Biochemistry* 69: 359–370.

Dahlgren, R.A., M. Saigusa, and F.C. Ugolini. 2004. The nature, properties and management of volcanic soils. In *Advances in Agronomy No. 82*, ed. D. Sparks, 113–182. Amsterdam: Academic Press, Elsevier.

del Moral, R., L.A. Thomason, A.C. Wenke, N. Lozanoff, and M.D. Abata. 2012. Primary succession trajectories on pumice at Mount St. Helens, Washington. *Journal of Vegetation Science* 23: 73–85.

Dickie, I.A., L.B. Martínez-García, N. Koele, G.A. Grelet, J.M. Tylianakis, D.A. Peltzer, and S.J. Richardson. 2013. Mycorrhizas and mycorrhizal fungal communities throughout ecosystem development. *Plant and Soil* 367: 11–39.

Emran, M., M. Gispert, and G. Pardini. 2012. Patterns of soil organic carbon, glomalin and structural stability in abandoned Mediterranean terraced lands. *European Journal of Soil Science* 63: 637–649.

Engle, M. 1983. *Carbon, nitrogen and microbial colonization of volcanic debris on Mount St. Helens. Environmental science and regional planning*. Pullman: Washington State University.

Fierer, N., J.P. Schimel, and P.A. Holden. 2003. Variations in microbial community composition through two soil depth profiles. *Soil Biology and Biochemistry* 35: 167–176.

Fokom, R., S. Adamou, M.C. Teugwa, A.D. Begoude Boyogueno, W.L. Nana, M.E.L. Ngonkeu, N.S. Tchameni, D. Nwaga, G. Tsala Ndzomo, and P.H. Amvam Zollo. 2012. Glomalin related soil protein, carbon, nitrogen and soil aggregate stability as affected by land use variation in the humid forest zone of South Cameroon. *Soil and Tillage Research* 120: 69–75.

Franklin, J.F., and C.T. Dyrness. 1988. *Natural vegetation of Oregon and Washington*. Corvallis: Oregon State University Press.

Ghani, A., M. Dexter, and K.W. Perrott. 2003. Hot-water extractable carbon in soils: A sensitive measurement for determining impacts of fertilisation, grazing and cultivation. *Soil Biology and Biochemistry* 35: 1231–1243.

Gil-Cardeza, M.L., A. Ferri, P. Cornejo, and E. Gomez. 2014. Distribution of chromium species in a Cr-polluted soil: Presence of Cr(III) in glomalin related protein fraction. *Science of the Total Environment* 493: 828–833.

Gill, R.A., and I.C. Burke. 2002. Influence of soil depth on the decomposition of *Bouteloua gracilis* roots in the shortgrass steppe. *Plant and Soil* 241: 233–242.

Gillespie, A.W., R.E. Farrell, F.L. Walley, A.R.S. Ross, P. Leinweber, K.-U. Eckhardt, T.Z. Regier, and R.I.R. Blyth. 2011. Glomalin-related soil protein contains non-mycorrhizal-related heat-stable proteins, lipids and humic materials. *Soil Biology and Biochemistry* 43: 766–777.

Gispert, M., M. Emran, G. Pardini, S. Doni, and B. Ceccanti. 2013. The impact of land management and abandonment on soil enzymatic activity, glomalin content and aggregate stability. *Geoderma* 202: 51–61.

González-Chávez, C., J. D'Haen, J. Vangronsveld, and J. Dodd. 2002. Copper sorption and accumulation by the extraradical mycelium of different *Glomus* spp. (arbuscular mycorrhizal fungi) isolated from the same polluted soil. *Plant and Soil* 240: 287–297.

González-Chávez, M.C., R. Carrillo-González, S.F. Wright, and K.A. Nichols. 2004. The role of glomalin, a protein produced by arbuscular mycorrhizal fungi, in sequestering potentially toxic elements. *Environmental Pollution* 130: 317–323.

Halvorson, J.J., and J.M. Gonzalez. 2006. Bradford reactive soil protein in Appalachian soils: Distribution and response to incubation, extraction reagent and tannins. *Plant and Soil* 286: 339–356.

Halvorson, J., and J. Smith. 2009. Carbon and nitrogen accumulation and microbial activity in Mount St. Helens pyroclastic substrates after 25 years. *Plant and Soil* 315: 211–228.

Halvorson, J.J., J.L. Smith, and E.H. Franz. 1991. Lupine influence on soil carbon, nitrogen and microbial activity in developing ecosystems at Mount St. Helens. *Oecologia* 87: 162–170.

Halvorson, J.J., E.H. Franz, J.L. Smith, and R.A. Black. 1992. Nitrogenase activity, nitrogen fixation and nitrogen inputs by lupines at Mount St. Helens. *Ecology* 73: 87–98.

Halvorson, J.J., J.L. Smith, and A.C. Kennedy. 2005. Lupine effects on soil development and function during early primary succession at Mount St. Helens. In *Ecological responses to the 1980 eruption of Mount St. Helens*, ed. V.H. Dale, F.J. Swanson, and C.M. Crisafulli, 243–254. New York: Springer.

Halvorson, J.J., J.M. Gonzalez, and A.E. Hagerman. 2013. Retention of tannin-C is associated with decreased soluble nitrogen and increased cation exchange capacity in a broad range of soils. *Soil Science Society of America Journal* 77: 1199–1213.

Hammer, E.C., and M.C. Rillig. 2011. The influence of different stresses on glomalin levels in an arbuscular mycorrhizal fungus—Salinity increases glomalin content. PLoS *One* 6: e28426.

Harner, M.J., P.W. Ramsey, and M.C. Rillig. 2004. Protein accumulation and distribution in floodplain soils and river foam. *Ecology Letters* 7: 829–836.

Helgason, T., T.J. Daniell, R. Husband, A.H. Fitter, and J.P.W. Young. 1998. Ploughing up the wood-wide web? *Nature* 394: 431.

Hodge, A., and K. Storer. 2015. Arbuscular mycorrhiza and nitrogen: Implications for individual plants through to ecosystems. *Plant and Soil* 386: 1–19.

Johnson, N.C. 2010. Resource stoichiometry elucidates the structure and function of arbuscular mycorrhizas across scales. *New Phytologist* 185: 631–647.

Johnson, N.C., J.H. Graham, and F.A. Smith. 1997. Functioning of mycorrhizal associations along the mutualism–parasitism continuum. *New Phytologist* 135: 575–585.

Jorge-Araújo, P., H. Quiquampoix, P.T. Matumoto-Pintro, and S. Staunton. 2015. Glomalin-related soil protein in French temperate forest soils: Interference in the Bradford assay caused by co-extracted humic substances. *European Journal of Soil Science* 66: 311–319.

Kardol, P., G.B. De Deyn, E. Laliberté, P. Mariotte, and C.V. Hawkes. 2013. Biotic plant–soil feedbacks across temporal scales. *Journal of Ecology* 101: 309–315.

Kikvidze, Z., C. Armas, K. Fukuda, L. Martínez-García, M. Miyata, A. Oda-Tanaka, F. Pugnaire, and B. Wu. 2010. The role of arbuscular mycorrhizae in primary succession: Differences and similarities across habitats. *Web Ecology* 10: 50–57.

Koide, R.T., and M.S. Peoples. 2013. Behavior of Bradford-reactive substances is consistent with predictions for glomalin. *Applied Soil Ecology* 63: 8–14.

Krishnamoorthy, R., K. Kim, C. Kim, and T. Sa. 2014. Changes of arbuscular mycorrhizal traits and community structure with respect to soil salinity in a coastal reclamation land. *Soil Biology and Biochemistry* 72: 1–10.

Lovelock, C.E., S.F. Wright, D.A. Clark, and R.W. Ruess. 2004a. Soil stocks of glomalin produced by arbuscular mycorrhizal fungi across a tropical rain forest landscape. *Journal of Ecology* 92: 278–287.

Lovelock, C.E., S.F. Wright, and K.A. Nichols. 2004b. Using glomalin as an indicator for arbuscular mycorrhizal hyphal growth: An example from a tropical rain forest soil. *Soil Biology & Biochemistry* 36: 1009–1012.

MacArthur, R.H., and J.W. MacArthur. 1961. On bird species diversity. *Ecology* 42: 594–598.

Miransari, M. 2014. Mycorrhizal fungi to alleviate salinity stress on plant growth. In *Use of microbes for the alleviation of soil stresses*, ed. M. Miransari, 77–86. New York: Springer.

Morris, W.F., and D.M. Wood. 1989. The role of lupine in succession on Mount St. Helens: Facilitation or inhibition? *Ecology* 70: 697–703.

Nelson, D.W., and L.E. Sommers. 1996. Total carbon, organic carbon and organic matter. In *Methods of soil analysis part 3: Chemical methods*. No 5. in the Soil Science Society of America Books Series, ed. D.L. Sparks, A.L. Page, P.A. Helmke, R.H. Loeppert, P.N. Soltanpour, M.A. Tabatabai, C.T. Johnson, and M.E. Sumner, 961–1010. Madison: Soil Science Society of America.

Neumann, E., and E. George. 2010. Nutrient uptake: The arbuscular mycorrhiza fungal symbiosis as a plant nutrient acquisition strategy. In *Arbuscular mycorrhizas: Physiology and function*, ed. H. Koltai and Y. Kapulnik, 137–167. Amsterdam: Springer.

Nichols, K.A. 2003. *Characterization of glomalin: A glycoprotein produced by arbuscular mycorrhizal fungi*. Ph.D. dissertation. College Park: University of Maryland.

Nichols, K.A., and S.F. Wright. 2004. Contributions of soil fungi to organic matter in agricultural soils. In *Functions and management of soil organic matter in agro-ecosystems*, ed. F. Magdoff and R. Weil, 179–198. Washington, DC: CRC.

Nichols, K., and S. Wright. 2005. Comparison of glomalin and humic acid in eight native US soils. *Soil Science* 170: 985–997.

Nichols, K.A., J.J. Halvorson, and T. Caesar. 2013. Roles of biology, chemistry, and physics in soil macroaggregate formation and stabilization. *Open Agriculture Journal* 7: 107–117.

Phillips, L.A., V. Ward, and M.D. Jones. 2014. Ectomycorrhizal fungi contribute to soil organic matter cycling in sub-boreal forests. *ISME Journal* 8: 699–713.

Purin, S., and M.C. Rillig. 2007. The arbuscular mycorrhizal fungal protein glomalin: Limitations, progress, and a new hypothesis for its function. *Pedobiologia* 51: 123–130.

Richardson, A.E., and R.J. Simpson. 2011. Soil microorganisms mediating phosphorus availability update on microbial phosphorus. *Plant Physiology* 156: 989–996.

Rillig, M.C. 2005. A connection between fungal hydrophobins and soil water repellency? *Pedobiologia* 49: 395–399.

Rillig, M.C., and D.L. Mummey. 2006. Mycorrhizas and soil structure. *New Phytologist* 171: 41–53.

Rillig, M.C., and P.D. Steinberg. 2002. Glomalin production by an arbuscular mycorrhizal fungus: A mechanism of habitat modification? *Soil Biology and Biochemistry* 34: 1371–1374.

Rillig, M.C., S.F. Wright, K.A. Nichols, W.F. Schmidt, and M.S. Torn. 2001. Large contribution of arbuscular mycorrhizal fungi to soil carbon pools in tropical forest soils. *Plant and Soil* 233: 167–177.

Rillig, M.C., P.W. Ramsey, S. Morris, and E.A. Paul. 2003. Glomalin, an arbuscular-mycorrhizal fungal soil protein, responds to land-use change. *Plant and Soil* 253: 293–299.

Rillig, M., B. Caldwell, H.B. Wösten, and P. Sollins. 2007. Role of proteins in soil carbon and nitrogen storage: Controls on persistence. *Biogeochemistry* 85: 25–44.

Rillig, M.C., S. Wendt, J. Antonovics, S. Hempel, J. Kohler, J. Wehner, and T. Caruso. 2014. Interactive effects of root endophytes and arbuscular mycorrhizal fungi on an experimental plant community. *Oecologia* 174: 263–270.

Rodriguez, R.J., J.F. White Jr., A.E. Arnold, and R.S. Redman. 2009. Fungal endophytes: Diversity and functional roles. *New Phytologist* 182: 314–330.

Ryan, M.H., and J.H. Graham. 2002. Is there a role for arbuscular mycorrhizal fungi in production agriculture? *Plant and Soil* 244: 263–271.

SAS Institute. 2002–2008. *SAS/STAT user's guide, version 9.2*. Cary: SAS Institute.

Seguel, A., J. Cumming, K. Klugh-Stewart, P. Cornejo, and F. Borie. 2013. The role of arbuscular mycorrhizas in decreasing aluminium phytotoxicity in acidic soils: A review. *Mycorrhiza* 23: 167–183.

Sikes, B.A., H. Maherali, and J.N. Klironomos. 2012. Arbuscular mycorrhizal fungal communities change among three stages of primary sand dune succession but do not alter plant growth. *Oikos* 121: 1791–1800.

Sikes, B., H. Maherali, and J. Klironomos. 2014. Mycorrhizal fungal growth responds to soil characteristics, but not host plant identity, during a primary lacustrine dune succession. *Mycorrhiza* 24: 219–226.

Singh, L.P., S.S. Gill, and N. Tuteja. 2011. Unraveling the role of fungal symbionts in plant abiotic stress tolerance. *Plant Signaling & Behavior* 6: 175–191.

Singh, P.K. 2012. Role of glomalin related soil protein produced by arbuscular mycorrhizal fungi: A review. *Agricultural Science Research Journal* 2: 119–125.

Smith, S.E., and F.A. Smith. 2011. Roles of arbuscular mycorrhizas in plant nutrition and growth: New paradigms from cellular to ecosystem scales. *Annual Review of Plant Biology* 62: 227–250.

Spohn, M., and L. Giani. 2010. Water-stable aggregates, glomalin-related soil protein, and carbohydrates in a chronosequence of sandy hydromorphic soils. *Soil Biology and Biochemistry* 42: 1505–1511.

———. 2011. Impacts of land use change on soil aggregation and aggregate stabilizing compounds as dependent on time. *Soil Biology and Biochemistry* 43: 1081–1088.

Staddon, P.L. 2005. Mycorrhizal fungi and environmental change: The need for a mycocentric approach. *New Phytologist* 167: 635–637.

Swanson, F.J., and J.J. Major. 2005. Physical events, environments, and geological-ecological interactions at Mount St. Helens: March 1980–2004. In *Ecological responses to the 1980 eruption of Mount St. Helens*, ed. V.H. Dale, F.J. Swanson, and C.M. Crisafulli, 27–43. New York: Springer.

Titus, J., and R. del Moral. 1998. The role of mycorrhizal fungi and microsites in primary succession on Mount St. Helens. *American Journal of Botany* 85: 370–375.

Titus, J.H., S. Whitcomb, and H.J. Pitoniak. 2007. Distribution of arbuscular mycorrhizae in relation to microsites on primary successional substrates on Mount St. Helens. *Canadian Journal of Botany* 85: 941–948.

Torres, M.S., J.F. White Jr., X. Zhang, D.M. Hinton, and C.W. Bacon. 2012. Endophyte-mediated adjustments in host morphology and physiology and effects on host fitness traits in grasses. *Fungal Ecology* 5: 322–330.

Treseder, K.K. 2004. A meta-analysis of mycorrhizal responses to nitrogen, phosphorus, and atmospheric CO_2 in field studies. *New Phytologist* 164: 347–355.

Treseder, K.K., and M.F. Allen. 2002. Direct nitrogen and phosphorus limitation of arbuscular mycorrhizal fungi: A model and field test. *New Phytologist* 155: 507–515.

Treseder, K.K., and K.M. Turner. 2007. Glomalin in ecosystems. *Soil Science Society of America Journal* 71: 1257–1266.

Turner, B., H. Lambers, L. Condron, M. Cramer, J. Leake, A. Richardson, and S. Smith. 2013. Soil microbial biomass and the fate of phosphorus during long-term ecosystem development. *Plant and Soil* 367: 225–234.

Uchida, Y., S. Nishimura, and H. Akiyama. 2012. The relationship of water-soluble carbon and hot-water-soluble carbon with soil respiration in agricultural fields. *Agriculture, Ecosystems & Environment* 156: 116–122.

Van Der Heijden, M.G.A., and T.R. Horton. 2009. Socialism in soil? The importance of mycorrhizal fungal networks for facilitation in natural ecosystems. *Journal of Ecology* 97: 1139–1150.

Vasconcellos, R.L., J.A. Bonfim, D. Baretta, and E.J. Cardoso. 2013. Arbuscular mycorrhizal fungi and glomalin-related protein as indicators of soil quality in a recuperation gradient of the Atlantic forest in Brazil. *Land Degradation & Development* 27: 325–334.

Veresoglou, S.D., B. Chen, and M.C. Rillig. 2012. Arbuscular mycorrhiza and soil nitrogen cycling. *Soil Biology and Biochemistry* 46: 53–62.

Walley, F.L., A.W. Gillespie, A.B. Adetona, J.J. Germida, and R.E. Farrell. 2013. Manipulation of rhizosphere organisms to enhance glomalin production and C sequestration: Pitfalls and promises. *Canadian Journal of Plant Science* 94: 1025–1032.

Wilson, L., and J.W. Head. 1981. Morphology and rheology of pyroclastic flows and their deposits, and guidelines for future observations. In *The 1980 eruptions of Mount St. Helens, Washington.* Professional Paper 1250, ed. P.W. Lipman and D.R. Mullineaux, 513–524. Washington, DC: U.S. Geological Survey.

Woignier, T., P. Etcheverria, F. Borie, H. Quiquampoix, and S. Staunton. 2014. Role of allophanes in the accumulation of glomalin-related soil protein in tropical soils (Martinique, French West Indies). *European Journal of Soil Science* 65: 531–538.

Wright, S.F., and R.L. Anderson. 2000. Aggregate stability and glomalin in alternative crop rotations for the central Great Plains. *Biology and Fertility of Soils* 31: 249–253.

Wright, S.F., and A. Upadhyaya. 1996. Extraction of an abundant and unusual protein from soil and comparison with hyphal protein of arbuscular mycorrhizal fungi. *Soil Science* 161: 575–586.

———. 1998. A survey of soils for aggregate stability and glomalin, a glycoprotein produced by hyphae of arbuscular mycorrhizal fungi. *Plant and Soil* 198: 97–107.

Wright, S.F., J.L. Starr, and I.C. Paltineanu. 1999. Changes in aggregate stability and concentration of glomalin during tillage management transition. *Soil Science Society of America Journal* 63: 1825–1829.

Wright, S.F., K.A. Nichols, and W.F. Schmidt. 2006. Comparison of efficacy of three extractants to solubilize glomalin on hyphae and in soil. *Chemosphere* 64: 1219–1224.

Wu, Q.-S., A.K. Srivastava, and Y.-N. Zou. 2013. AMF-induced tolerance to drought stress in citrus: A review. *Scientia Horticulturae* 164: 77–87.

Young, I.M., D.S. Feeney, A.G. O'Donnell, and K.W.T. Goulding. 2012. Fungi in century old managed soils could hold key to the development of soil water repellency. *Soil Biology and Biochemistry* 45: 125–127.

6 Forest Understory Buried by Volcanic Tephra: Inertia, Resilience, and the Pattern of Community Redevelopment

Donald B. Zobel and Joseph A. Antos

6.1 Introduction

6.1.1 Damage to Plants and Their Recovery After Disturbance

During a disturbance, some plants are killed or injured, reducing their importance (e.g., cover). With time after disturbance, their importance increases as vegetation redevelops. The amount and characteristics of vegetation development after disturbance depend on the degree of damage to the plants, the type and intensity of disturbance, the levels of other environmental factors, and the type of plant. Vegetation may "recover"—that is, develop properties similar to pre-disturbance communities; in contrast, changed conditions after the disturbance may trigger development of communities unlike those previously existing on the site.

In this chapter, we analyze long-term records of redevelopment of understory vegetation that was buried by tephra (aerially transported volcanic ejecta) from the eruption of Mount St. Helens, Washington, on 18 May 1980. In the tephrafall zone—the location of our study, outside the zones of more severe disturbance (Fig. 6.1)—the tree canopy remained intact and the eruption had less impact on the biota than in other disturbance zones where most trees were killed or removed (Swanson and Major 2005). However, damage to understory plants in the tephrafall zone was substantial (Antos and Zobel 1985a; Zobel and Antos 1997) (Fig. 6.2).

We consider three measures of importance (which we call *attributes*) of plants and plant communities: shoot cover, shoot density (number of shoots per 1-m^2 plot), and species diversity (expressed as species density = number of species per 1-m^2 plot). We based our analyses on the concepts of inertia and resilience (Westman 1978, 1986), which we also used to evaluate our 10-year results (Zobel and Antos 1997). *Inertia* (often termed *resistance*) is the proportion of pre-disturbance plant importance that survived the disturbance; *resilience* is the degree to which the plants regain the importance destroyed by the disturbance. Inertia and resilience have been used in relation to a variety of disturbance types (e.g., Halpern 1988; Rejmánková et al. 1999). We determined how vegetation redevelopment is modified by disturbance intensity (tephra depth), snow pack during tephra emplacement, plant growth form, species, and time.

6.1.2 Responses of Understory Plants to Tephra

We studied understories of old-growth subalpine conifer forest in the *Abies amabilis* vegetation zone (Franklin and Dyrness 1973) that were buried beneath tephra. These forests were dominated by large trees of *Abies amabilis*, *Pseudotsuga menziesii*, *Tsuga heterophylla*, *T. mertensiana*, and *Chamaecyparis nootkatensis*, with many individuals exceeding 600 years of age. Areas of less dense trees, characteristic of these old forests, supported dense populations of shrubs or herbs or both; in contrast, understory plants were sparse in areas with a dense tree canopy. Snowpack starts to accumulate in autumn at our sites, usually reaching a few meters deep, and melts in late spring to late summer, depending on the year and location. Sites with more herb species (Table 6.1, Fig. 6.1) had later snow melt, more concave topography, and higher concentrations of Ca, Mg, and total N in the pre-eruption soil than sites with few herb species (Zobel and Antos 1991a).

Tephra at our sites was 4.5–15 cm deep, mostly single-grained pumice of sand (0.0625–2 mm) to lapilli (>2–64 mm) size (Zobel and Antos 1991a). After wetting and drying, fine-textured tephra at the surface formed a crust,

D.B. Zobel (✉)
Department of Botany and Plant Pathology, Oregon State University, 2082 Cordley Hall, Corvallis, OR 97331-2902, USA
e-mail: zobeld@science.oregonstate.edu

J.A. Antos
Department of Biology, University of Victoria, Victoria, BC V8W 3N5, Canada

C.M. Crisafulli, V.H. Dale (eds.), *Ecological Responses at Mount St. Helens: Revisited 35 years after the 1980 Eruption*,
https://doi.org/10.1007/978-1-4939-7451-1_6

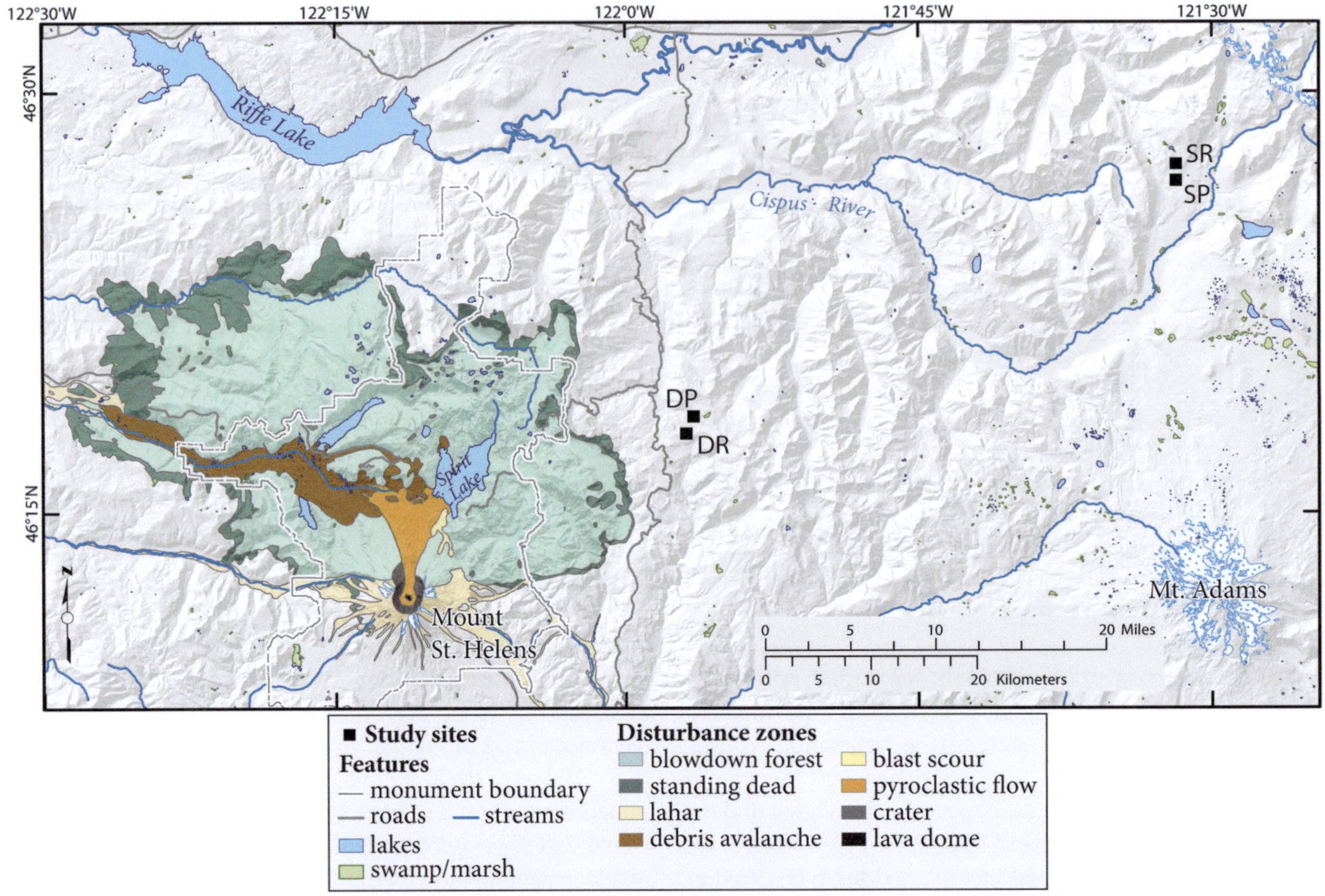

Fig. 6.1 Locations of tephrafall zone study sites in relation to areas with more severe damage from the 1980 eruption of Mount St. Helens. Cartography by Kathryn Ronnenberg, USDA Forest Service, Pacific Northwest Research Station.

Fig. 6.2 Thirty years of change in the understory of old-growth subalpine forest where 12–15 cm of tephra fell on snow and formed a substantial crust, obliterating most of the understory (Site DR, Fig. 6.1). *Left*, summer 1980; *right*, another part of site DR, 2010. Most small trees established on top of the tephra. (Photos by Joe Antos and Matt Blakely-Smith).

which impeded water infiltration and plant emergence, especially in deep tephra where this crust was thick and hard (Fig. 6.2). There were no overt signs of the tephra causing damage through heating, toxicity, abrasion, or branch loading. Tephra lost much of its sulfur and cations during the first few years after emplacement, whereas concentrations

Table 6.1 Study site characteristics. At each site we used 1-m² permanent plots, 100 on natural tephra and 50 with tephra removed in 1980. Sites DP and DR were 22 km NE of the crater, sites SP and SR, 58 km NE (Fig. 6.1). Site codes: *S* shallow tephra, *D* deep, *P* herb-poor, *R* herb-rich.

Site code	Tephra depth (cm)	Number of herb species	Elevation (m)	Snow cover 1980 (%)	% reduction in herb cover[a]	% reduction in shrub cover[a]
SP	4.5	9	1245	29	41	21
SR	4.5	26	1290	92	0	66
DP	15.0	12	1160	11	>99	30
DR	15.0	32	1240	88	>99	>99

[a]Based on 1981 cover in natural tephra plots compared to 1981 data from plots cleared of tephra in 1980.

of N and P increased slowly beneath the intact forest canopy (Zobel and Antos 1991a).

At our sites, most small plants, such as bryophytes and prostrate herbs, did not emerge after burial by tephra (Zobel and Antos 1997). In contrast, burial stimulated some small herbs that normally have limited belowground shoot growth to grow rapidly toward the surface (Antos and Zobel 1984, 1985b, c), allowing some small plants to emerge from burial. Many plants remained alive for years while buried, producing shoots where erosion removed tephra, and some shrubs emerged from un-eroded deep tephra after 7–10 years of burial (Zobel and Antos 1992).

Tephra depth varied within our sites, being thin or absent beneath logs on the forest floor and on steep microtopography. The crust broke around woody debris (Fig. 6.2). Such microsites enhanced plant emergence and served as centers from which survivors spread and post-eruption vegetation developed.

The 18 May eruption occurred when some sites remained largely buried by snowpack, although others were almost snow-free. Where tephra fell on snow, damage to small woody plants was much greater than where snow had already melted. With snowpack, plants much taller than the tephra depth became trapped under tephra, because they had been flattened beneath the snow (Antos and Zobel 1982). Therefore, we used the relative extent of snowpack on 18 May, along with tephra depth and herb richness, to define differences among our study sites (Table 6.1).

6.1.3 Plant Attributes and Measures of Damage and Recovery

At each site, we sampled 100 plots (permanent, 1 m²) on natural tephra and 50 plots from which tephra was removed in 1980; the cleared plots were sampled in 1982 and 1983 to estimate the pre-eruption vegetation on the site. All natural plots were sampled in 1981–1983, 2000, 2005, and 2010; in addition, only shallow tephra sites were sampled in 1989 and only deep tephra sites in 1984, 1987, and 1990. Plots were placed at 3-m intervals along several transects on relatively flat terrain at each site, with cleared plots along separate transects, interspersed among transects on natural tephra.

We considered two plant attributes, shoot cover and shoot density, along with species density (number of species per 1-m² plot), a measure of diversity. Cover was available for all growth forms: bryophytes, herbs, shrubs, and tree seedlings (small trees with foliage within 1.5 m of the soil surface). Shoot density and species density were available for herbs and shrubs for all sample years except 2005. We will use the term *importance* to represent all three attributes (cover, shoot density, and species density) collectively. Nomenclature follows Hitchcock and Cronquist (1973).

We analyzed four sets of data:

1. Cover for all growth forms at all sites
2. Shoot density and species density for herbs and shrubs at all sites
3. Cover and shoot density for three species that were common at all sites (*Rubus lasiococcus*, *Vaccinium membranaceum*, *V. ovalifolium*)
4. Cover and shoot density for four important herbs (*Erythronium montanum*, *R. lasiococcus*, *Tiarella unifoliata*, *Valeriana sitchensis*) at site DR (Fig. 6.1, Fig. 6.2)

We evaluated the effects of the disturbance and subsequent changes in an attribute (e.g., cover of herbs) by using four indices (Table 6.2) based on a simple model (Fig. 6.3), where *a* is the pre-disturbance value, *b* is the initial post-disturbance value, and *c* is the value at a given time following the disturbance. The proportion of importance (e.g., cover) remaining after disturbance can be called inertia (*b/a* in Fig. 6.3) (Westman 1978, 1986; Zobel and Antos 1997), and we use this term (Table 6.2).

The redevelopment of understory vegetation at our sites resulted from many processes and plant characteristics, some of them summarized through 2005 by Zobel and Antos

Table 6.2 Definitions of inertia, resilience, and related indices.

Index	Descriptor
b/a	Inertia
c/a	Relative plant importance
c/b	Expansion of survivors
$(c-b)/(a-b)$	Resilience

Each index is applied to three attributes: shoot cover, shoot density, and species density (diversity). In the text, the descriptor will be used without the formula. The variables *a*, *b*, *c* represent the key values (in Fig. 6.3) of the curve showing the changes in value of attributes: *a* is the *pre-eruption value*, measured in 1982 or 1983 from plots cleared in 1980; *b* is the *post-eruption value*, measured in 1981 in plots on natural tephra; and *c* is the *value in a later year*, measured in plots on natural tephra in the year under consideration.

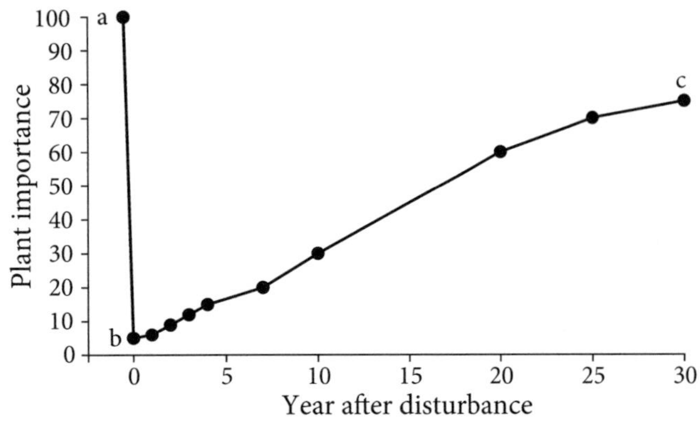

Fig. 6.3 Hypothetical pattern of recovery of plant importance after disturbance. The disturbance occurred at year 0. With time, importance increased. Values *a*, *b*, and *c*, noted on the graph, are used to calculate indices of inertia and resilience. *a* represents pre-disturbance plant importance; *b* represents post-disturbance plant importance; and *c* represents the importance at a specified later time.

(2007, 2009). In this paper, we view the process of redevelopment through 2010 in a more general manner, resulting in our calculation of three additional indices to describe it (Table 6.2). In the text, we use the terms defined in Table 6.2 to represent the formulas by which we calculated each index.

We ask two questions:

1. How did relative plant importance, expansion of survivors, and resilience change with time?
2. Which factors (among tephra depth, extent of snow, and growth form or species) were significantly related to the levels of the indices we calculated?

We used a combination of inertia and the maximal level of resilience to characterize the response of growth forms and species to tephra. The maximal resilience often occurred in 2010, but sometimes earlier, as far back as 1982.

Our calculations used means for each growth form or major species during each year at each study site. Values of importance for pre-disturbance vegetation (*a* in Fig. 6.3) came from cleared plots, from which tephra was removed in 1980 (Antos and Zobel 1985a). We used the lowest value in cleared plots in 1982 or 1983 as the pre-eruption value (Zobel and Antos 1997). Importance just after the disturbance (*b* in Fig. 6.3) was measured on the natural tephra plots in 1981 (Zobel and Antos 1997). In three cases at site SP (Fig. 6.1), where shallow tephra and minimal snowpack produced little damage, *b* exceeded *a*, so we set $a = b$; thus, inertia = 1.00, $(a-b) = 0$, and resilience was undefined.

If a variable could be transformed to normality, we used multifactor analysis of variance (ANOVA), with the following factors where appropriate: attribute (cover, shoot density, or species density), tephra depth (shallow or deep), extent of snowpack (little or much), year, and species or growth form. All ANOVAs included interaction terms for growth form (or species) × tephra depth and growth form (or species) × snow extent. When variables could not be transformed to normality, we used the Kruskal–Wallis test to determine the significance of each factor separately.

6.2 Inertia, Resilience, and Community Development

6.2.1 Changes in Relative Plant Importance Over Time

Larger growth forms had significantly higher relative plant cover (Fig. 6.4). Cover exceeded pre-eruption values only for tree seedlings after year 10. For all growth forms except herbs, relative cover increased significantly with time. Relative cover also differed among environments defined by tephra depth and extent of snow in 1980 (Fig. 6.5). The high values for deep tephra with much snow after year 10 are due to the high survival and expansion of tree seedlings in those conditions. In general, shallow tephra and little snow have higher relative cover than deep tephra and much snow. Relative cover increased significantly with year except where shallow tephra fell on snow.

Relative shoot density (Fig. 6.6) increased significantly with year for all situations except shallow tephra with little snow; it reached pre-disturbance values before year 10 for shallow but not for deep tephra. For each tephra depth, values were usually higher with little than with much snow.

Relative species density (Fig. 6.7) increased significantly with time for both tephra depths but was higher in shallow

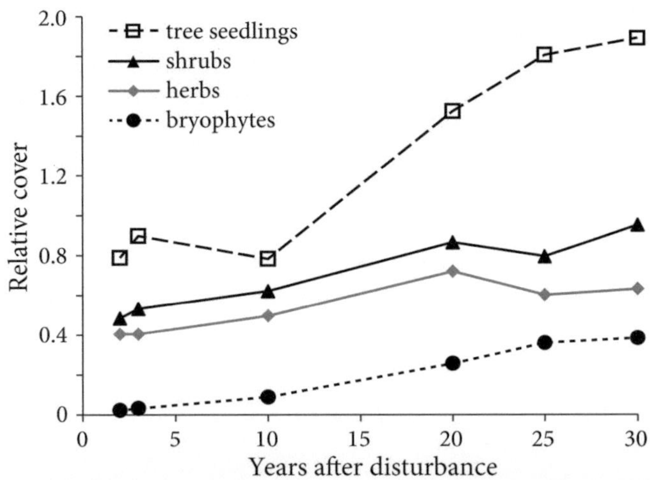

Fig. 6.4 Relative plant importance for cover for each growth form in each sampling year. Relative plant importance is defined in Table 6.2. Bryophytes and tree seedlings differed significantly from herbs and shrubs and from each other; herbs and shrubs did not differ significantly.

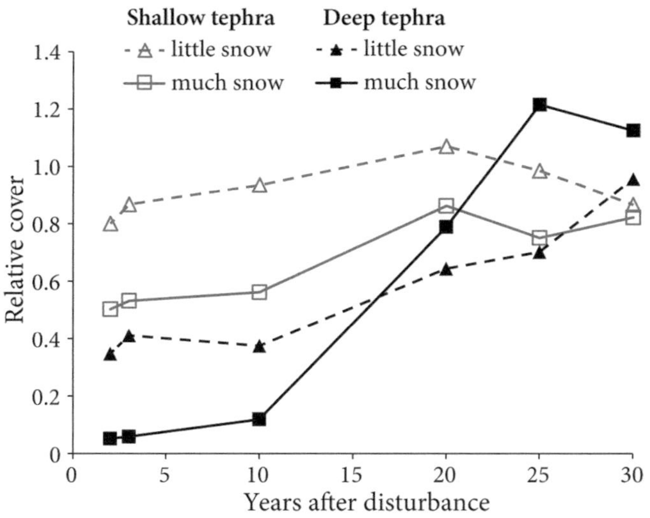

Fig. 6.5 Relative plant importance for cover for each combination of tephra depth and snow extent in each sampling year.

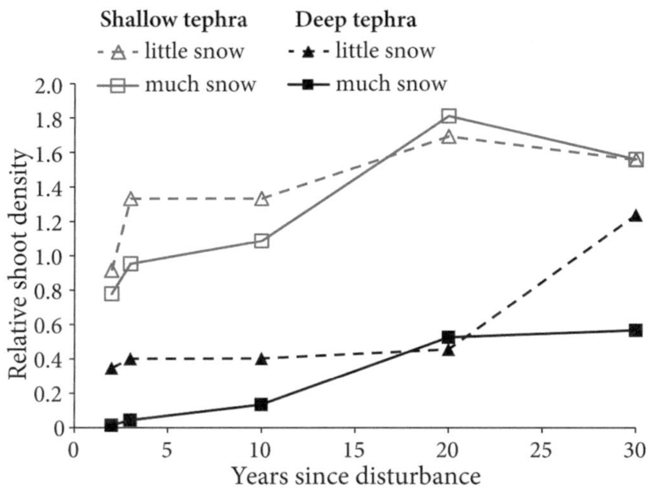

Fig. 6.6 Relative plant importance for shoot density for each combination of tephra depth and snow extent in each sampling year. Relative shoot density did not differ between growth forms.

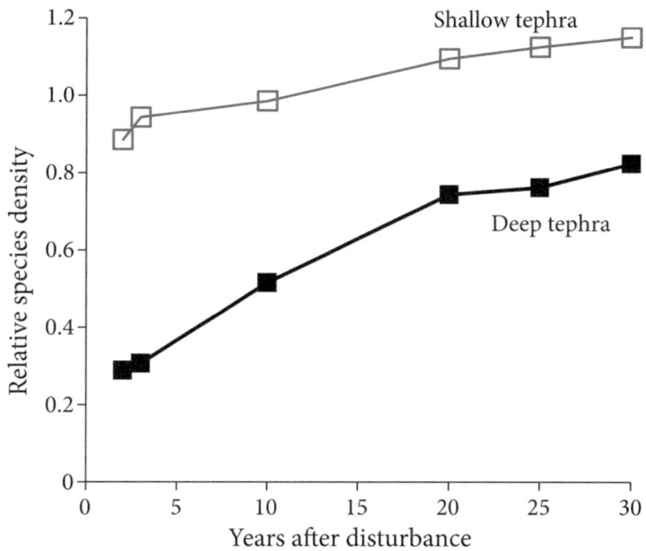

Fig. 6.7 Relative plant importance for species density for shallow and deep tephra.

tephra for all years, reaching pre-disturbance values after year 10. It did not differ significantly with growth form or extent of snow.

6.2.2 Variation in Inertia

Inertia was lower (damage greater) in deep tephra, with extensive snowpack, and for smaller plants (Table 6.3). Inertia for shoot density differed significantly for tephra depth, snow extent, and growth form, but for cover and species density the difference was not significant for any factor. Interactions among growth form and tephra depth were significant for inertia of cover ($p = 0.03$), but not for shoot density; interactions among growth form and snow extent were significant for both cover ($p = 0.04$) and shoot density ($p = 0.03$).

For the three species important at all sites, inertia for *Rubus lasiococcus* cover was lower in deep than in shallow tephra, whereas inertia for density of *Vaccinium* spp. was lower with much snow than with little (Table 6.4). There was no significant interaction of *Vaccinium* spp. with tephra depth or snow extent.

Among four important herbs at site DR, which had deep tephra and much snowpack, inertia was consistently very low (mean = 0.007) and did not differ significantly among attributes and species.

6.2.3 Variation Among Years in Resilience and Other Indices

Although inertia has only one value, resilience and other indices related to revegetation were calculated for each sampling year. Relative plant importance, expansion of survivors, and resilience usually increased significantly with time after disturbance. There were exceptions: most indices for *Rubus lasiococcus* and expansion of survivors for shoot density for growth forms. Values for separate years are not presented.

6.2.4 Variation in Expansion of Survivors

Expansion of survivors was significantly ($p = 0.02$) lower for species density (4.47) than for cover (14.55) and shoot density (12.69). Expansion of survivors differed with all factors

Table 6.3 Significance of differences and means of inertia for cover, shoot density, and species density, for the factors tephra depth, snow extent, and growth form. Inertia did not differ among attributes (based on the Kruskal–Wallis test). *NS* not significant.

		Cover		Shoot density		Species density	
Factor	**Level**	**Mean**	***p*-value**	**Mean**	***p*-value**	**Mean**	***p*-value**
Tephra depth	Shallow	0.472	NS	0.636	<0.05	0.762	NS
	Deep	0.185		0.169		0.243	
Snow extent	Little	0.443	NS	0.536	<0.05	0.606	NS
	Much	0.213		0.269		0.399	
Growth form	Bryophyte	0.014	NS		<0.05		NS
	Herb	0.306		0.285		0.414	
	Shrub	0.395		0.520		0.591	
	Tree seedlings	0.599					

Table 6.4 Means of inertia for *Rubus lasiococcus* and *Vaccinium* spp. for the factors tephra depth and snow extent. Means for different attributes and *Vaccinium* spp. did not differ significantly.

		R. lasiococcus		*Vaccinium* spp.	
Factor	**Level**	**Mean**	***p*-value**	**Mean**	***p*-value**
Tephra depth	Shallow	0.711	0.003	0.511	0.16
	Deep	0.011		0.335	
Snow extent	Little	0.461	0.13	0.625	0.007
	Much	0.261		0.221	

(tephra depth, snow, growth form, and year) for species density, with all factors except snow for cover, but only with tephra depth for shoot density. In general, expansion of survivors was high where damage was intense, giving a small post-eruption value: it was greater with deeper tephra and more snow, and for smaller growth forms (Table 6.5).

Interactions with growth form were significant for cover with both tephra depth and snow, and for species density with snow. Shoot density could not be transformed to normality; thus, interaction significance was not determined.

For *R. lasiococcus*, expansion of survivors was higher in deep tephra and with much snow, but for *Vaccinium* spp. it was higher only with more snow (Table 6.6). Both *R. lasiococcus* and *Vaccinium* spp. differed among years. Because values could not be transformed to normality, significance of interactions was not determined.

For four important herbs at site DR (Table 6.7), expansion of survivors was lowest for *E. montanum*, for which cover declined with time and shoot density was unchanged, and highest for *R. lasiococcus* for both cover and shoot density. Years differed for both attributes.

6.2.5 Variation in Resilience

Resilience describes the extent to which the importance lost during disturbance has been regained on each sampling date. Values >1 indicate that the losses for that variable have been more than replaced.

Resilience differed significantly among attributes, with cover often having the lowest values (Table 6.8). Differences among factor levels were significant for all factors for species density, for all factors except snow for shoot density, and only for growth form and year for cover. Resilience was higher for shallow than for deep tephra. For species density, resilience was higher where snowpack was extensive than where there was little snow. Resilience was higher for large than for small growth forms, except that species density was more resilient for herbs than for shrubs. Resilience was >1 for shoot density when disturbance effects were not severe and growth forms were large, for species density in some situations, and for cover of tree seedlings only (Table 6.8). Interactions of growth form with tephra depth and snow were significant, except for that between species density and snow.

For three species important at all sites, resilience was greater for shallow than for deep tephra (Table 6.9). Resilience varied significantly among years for *Vaccinium* spp., but not for *R. lasiococcus*, an unusual situation. Interaction of *Vaccinium* spp. was significant with tephra depth but not with snow extent.

Significant differences occurred among the four herbs at site DR for both cover and shoot density (Table 6.10). Resilience was essentially nil for *E. montanum* but approached 1 for *T. unifoliata*; values for the other two species were intermediate. The negative value for *E. montanum* reflects a loss of cover with time.

6.2.6 Maximal Resilience

The mean of maximal resilience was greater (p = 0.0006) in shallow tephra (1.911) than in deep tephra (0.788). The maximal resilience value did not differ significantly among attributes, snow extent, or growth form.

Table 6.5 Significance of differences and means of expansion of survivors for cover, shoot density, and species density, for the factors tephra depth, snow extent, growth form, and year. *NS* not significant.

		Cover		Shoot density		Species density	
Factor	**Level**	**Mean**	***p*-value**	**Mean**	***p*-value**	**Mean**	***p*-value**
Tephra depth	Shallow	6.18	<0.05	2.18	NS	1.35	<0.01
	Deep	19.76		24.9		7.59	
Snow extent	Little	10.63	NS	5.64	<0.05	3.00	<0.01
	Much	15.87		23.38		5.95	
Growth form	Bryophyte	18.16	<0.001		<0.05		<0.001
	Herb	23.12		18.82		4.97	
	Shrub	5.98		8.86		3.97	
	Tree seedlings	4.09					
Year			<0.001		<0.01		<0.01

Table 6.6 Means of expansion of survivors for *Rubus lasiococcus* and *Vaccinium* spp. for the factors tephra depth, snow extent, and year. Attributes and *Vaccinium* spp. did not differ significantly.

		R. lasiococcus		*Vaccinium* spp.	
Factor	**Level**	**Mean**	***p*-value**	**Mean**	***p*-value**
Tephra depth	Shallow	2.17	0.002	2.39	0.98
	Deep	44.21		24.17	
Snow extent	Little	3.65	0.03	1.62	<0.0001
	Much	44.64		24.94	
Year			0.045		0.004

Table 6.7 Significance of differences and means of expansion of survivors for three herb species at site DR for the factors species and year. Attributes did not differ significantly. *Tiarella unifoliata* had no survivors and is not included in the table.

		Cover		Shoot density	
Factor	**Level**	**Mean**	***p*-value**	**Mean**	***p*-value**
Species	*Erythronium montanum*	0.52	<0.001	1.05	<0.001
	Rubus lasiococcus	70.90		66.29	
	Valeriana sitchensis	20.56		5.00	
Year			<0.05		<0.01

6.3 Changing Influence of Initial Damage

One would expect early post-eruption values of attributes, like cover, to reflect closely the value of inertia, as there has been little time for recovery of lost importance. With time, this influence of inertia is likely to weaken as the recovery period becomes long enough for site and plant differences to exert an influence on patterns of community redevelopment. Over our 30-year study, the regression coefficients of inertia with the values of cover, shoot density, and species density all declined significantly ($p = 0.007$, 0.049 and 0.012, respectively), although change after 2000 was small (Fig. 6.8). After 30 years, inertia for cover (but not shoot density or species density) was still significantly correlated with 2010 cover values.

6.4 Correlations Among Indices of Recovery

Resilience was strongly and positively correlated with relative plant importance at a sampling date for all attributes (Table 6.11). Expansion of survivors was not significantly correlated with either resilience or relative plant importance.

6.5 Differentiating Growth Forms and Species

We used two indices of response to tephra to differentiate growth forms and species: inertia and the maximal value of resilience. These two criteria sometimes increased together (Figs. 6.9 and 6.10) (Spearman rank correlation (r) = 0.52, $p = 0.01$ for growth forms; $r = 0.53$, $p = 0.02$ for three widespread species). In contrast, maximal resilience declined, although not significantly, as inertia increased for the four herbs from site DR (Fig. 6.11) (correlation coefficient = −0.64, $p = 0.09$).

Growth forms differed somewhat in their combinations of inertia and maximal resilience (Fig. 6.9). Shrubs tended to have higher inertia, whereas herbs had the highest maximal resilience by a substantial amount (>>1; the value 1 indicates full compensation for the loss during disturbance). However, both growth forms were represented throughout most of the range of both variables. The nine values with low inertia included all six from site DR, in which all post-eruption attribute values were very low, and all three attributes for herbs at site DP.

Table 6.8 Significance of differences and means of resilience, which measures the proportion of importance lost that was regained by a particular sampling year, for cover, shoot density, and species density, for the factors tephra depth, snow extent, growth form, and year. *NS* not significant.

		Cover		Shoot density		Species density	
Factor	**Level**	**Mean**	***p*-value**	**Mean**	***p*-value**	**Mean**	***p*-value**
Tephra depth	Shallow	0.570	NS	3.111	<0.001	1.146	<0.001
	Deep	0.571		0.300		0.434	
Snow extent	Little	0.585	NS	2.417	NS	0.635	<0.01
	Much	0.559		0.883		0.944	
Growth form	Bryophyte	0.180	<0.001		<0.001		<0.001
	Herb	0.464		0.934		1.107	
	Shrub	0.615		2.368		0.473	
	Tree seedlings	1.176					
Year			<0.001		<0.001		<0.001

Table 6.9 Significance of variation and means of resilience for *Rubus lasiococcus* and *Vaccinium* spp. for the factors tephra depth and year. Resilience did not differ among attributes, with snow, or between *Vaccinium* spp.

		R. lasiococcus		*Vaccinium* spp.	
Factor	**Level**	**Mean**	***p*-value**	**Mean**	***p*-value**
Tephra depth	Shallow	6.342	0.0001	1.143	<0.0001
	Deep	0.379		0.569	
Year			0.23		<0.0001

Table 6.10 Significance of differences and means of resilience for four herb species at site DR for the factors species and year. Resilience did not differ significantly between cover and shoot density.

		Cover		Shoot density	
Factor	**Level**	**Mean**	***p*-value**	**Mean**	***p*-value**
Species	*Erythronium montanum*	−0.003	<0.001	0.0003	<0.001
	Rubus lasiococcus	0.597		0.554	
	Tiarella unifoliata	0.749		1.079	
	Valeriana sitchensis	0.188		0.080	
Year			<0.001		<0.001

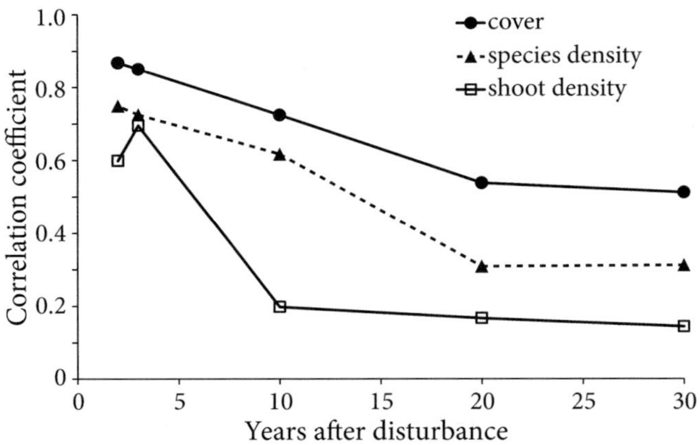

Fig. 6.8 Change with time of the correlation coefficient (r) between inertia and the value of the attribute (cover, shoot density, or species density). Cover and shoot density were not normally distributed and the Spearman rank correlation coefficient was used. Cover and species density have significant (p = 0.007, 0.012, respectively) negative simple regressions, and shoot density is significant (p = 0.021) in a double-reciprocal plot.

For cases with inertia <0.05 (great damage), maximal resilience did not reach 1.0; recovery of plant importance remained incomplete. Maximal resilience was high (>1.5) only when inertia exceeded about 0.5. On the other hand, cases with the highest inertia (shrub species density) had maximal resilience <1.0 (Fig. 6.9).

Among the three species important at all sites, inertia and maximal resilience were positively related (Fig. 6.10). Three species with inertia <0.5 did have maximal resilience >1.0. The three species did not differ in any obvious pattern.

Four major herbs at site DR differed greatly in their combinations of inertia and maximal resilience (Fig. 6.11). *Tiarella unifoliata* had zero inertia but the highest maximal resilience, with values >1.0. Thus, cover and shoot density had notably exceeded pre-disturbance values even though survivors were rare. *Rubus lasiococcus* had moderate values of both inertia and maximal resilience. *Erythronium montanum* had essentially zero maximal resilience and low inertia. Values of inertia were low, being >0.10 only for shoot density of *Valeriana sitchensis.* Cover and shoot density values were similar within a species, except for *V. sitchensis* (Fig. 6.11).

Table 6.11 Spearman rank correlation coefficients among index values for cover, shoot density, and species density.

Index	Correlated with	Cover	Shoot density	Species density
Relative plant importance	Resilience	0.927 ($p < 0.001$)	0.932 ($p < 0.001$)	0.899 ($p < 0.001$)
Expansion of survivors	Resilience	0.107	−0.076	0.242
Relative plant importance	Expansion of survivors	−0.191	−0.091	−0.122

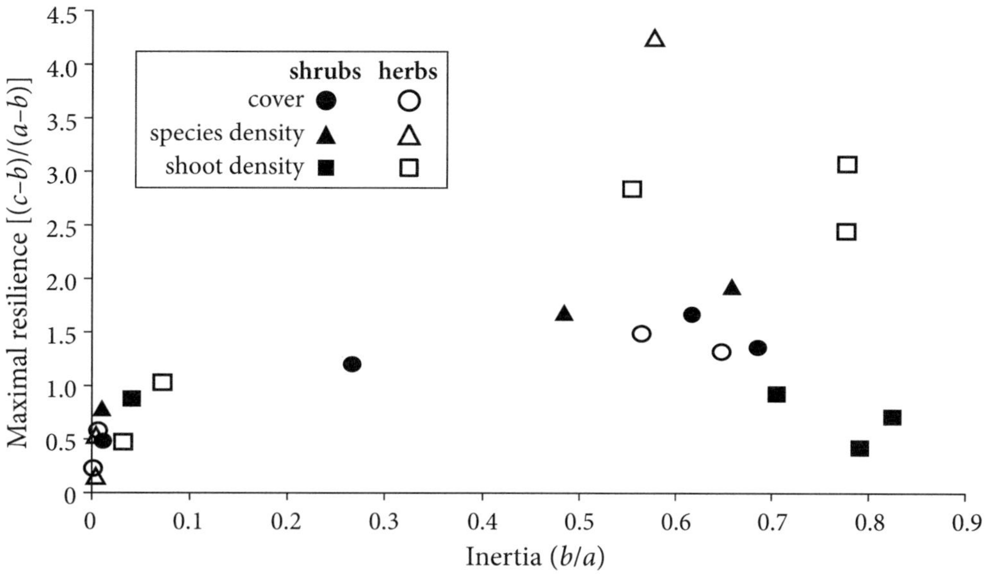

Fig. 6.9 Relationship between the maximal value of resilience among sample years and inertia for herbs and shrubs.

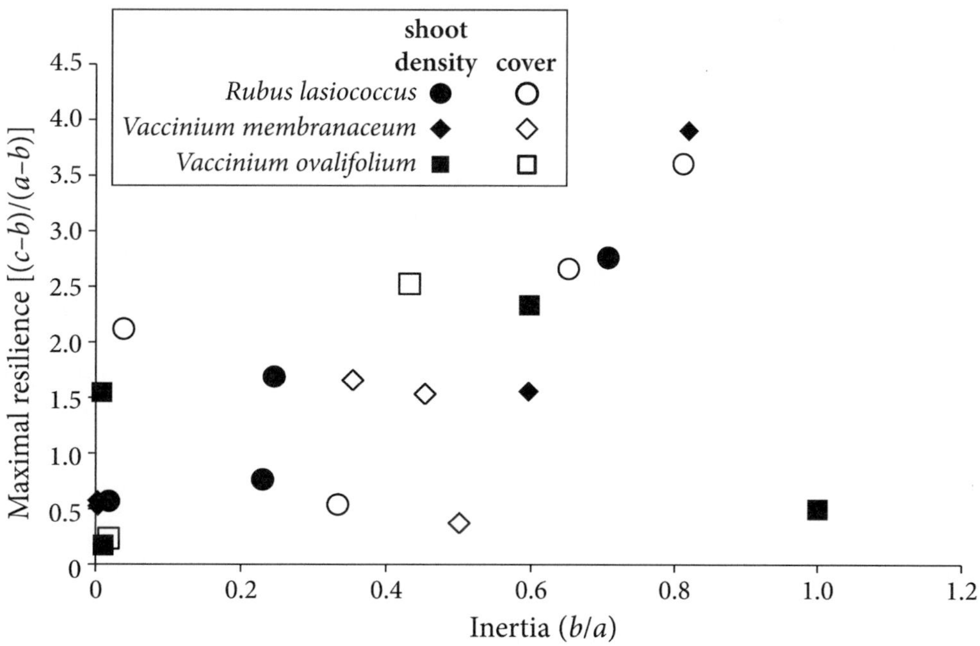

Fig. 6.10 Relationship of the maximal value of resilience to inertia for three widespread species.

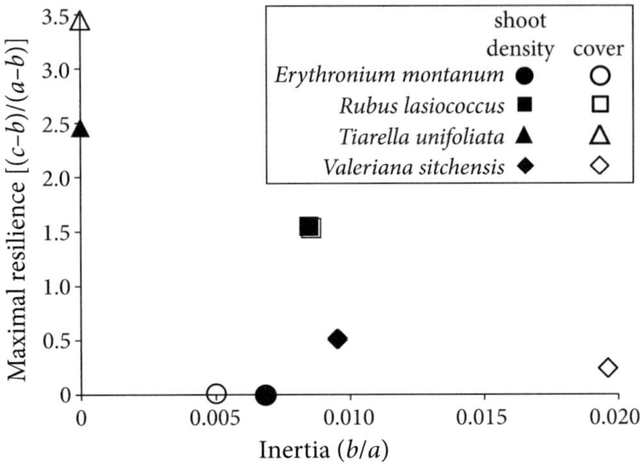

Fig. 6.11 Relationship of the maximal value of resilience to inertia for four herb species from site DR.

6.6 Discussion

6.6.1 Implications for Vegetation Change

Tephrafall was the least intense disturbance associated with the 1980 Mount St. Helens eruption (Swanson and Major 2005). Even so, understory cover 30 years later remained far below pre-disturbance levels in many situations (Figs. 6.4 and 6.5) and was still significantly correlated to the degree of damage done by the disturbance (i.e., inertia) (Fig. 6.8). A decade after the eruption, this relationship of cover to inertia caused us to emphasize the importance of beginning a study immediately after the disturbance (Zobel and Antos 1997). However, after 30 years, shoot density and diversity (species density) no longer showed a significant correlation with inertia. This different response of cover and shoot density may reflect the effects of new plant establishment, which immediately increased shoot density and diversity. In contrast, cover increased more slowly as the small, newly established plants grew. In addition, herbs that expanded their cover the most established later than those with less growth (Zobel and Antos 2009).

The responses of plants to disturbance are related to three major factors: disturbance intensity, other environmental properties, and plant characteristics. Here we demonstrate their clear, long-lasting effects:

1. Tephra depth. Deeper tephra produced more damage, thus lower inertia, and allowed less resilience.
2. Snow beneath tephra. Inertia of shrubs and small trees was less where snowpack occurred in 1980. Snow reduced resilience only for diversity (species density).
3. Plant size. Larger growth forms had higher inertia and resilience than smaller plants, in most cases. The exception was diversity (species density).

There were, however, notable exceptions and complications to these generalizations. Interactions can strongly modify the effects of disturbance, such as the often-significant ones between growth form, tephra depth, and extent of snow in 1980. Generally, tephra depth affected herb and moss importance more than that of woody plants; in contrast, snow extent primarily affected woody plants. Smaller plants would be more prone to burial by tephra if other factors were equal. At this elevation, however, woody understory plants are prostrate under snow during winter and spring, making them highly susceptible to burial. Herbaceous plants were affected little by the fall of the tephra on snow, as most survive only underground when snow is present, and evergreen herbs usually have low stature, reduced little by snow. This response of woody plants to the snow–tephra interaction was unexpected but dramatic (Antos and Zobel 1982, 1985a), and was still pronounced after 10 (Zobel and Antos 1997) and, here, after 30 years. It underscores the importance of observations just after disturbance—the cracking of tephra crust, which indicated where snow beneath it had melted, disappeared after the first winter.

Even after we accounted for conditions at the time of disturbance, plant size did not tell the whole story. Specific characteristics of growth form and life history induced differential responses among species. Forest herbs vary greatly in morphology and life history (Bierzychudek 1982; Antos and Zobel 1984; McCarthy 2003); thus, major differences in patterns of response to disturbance are to be expected (Antos and Zobel 1985a, b; Halpern 1989; Roberts and Gilliam 2003).

The four herb species studied at site DR differed in resilience (Table 6.10, Fig. 6.11), which may relate to several structural and reproductive differences (Antos and Zobel 1984, 1985b, c; Zobel and Antos 2007, 2009). *Rubus lasiococcus* spread rapidly via stolons, whereas the other three species had limited or no vegetative spread. *Tiarella unifoliata*, with the highest resilience, had the highest percentage of shoots flowering, moderate numbers of seedlings per flowering shoot, and effective dispersal. *Rubus lasiococcus* produced few flowers and seedlings but spread rapidly across the tephra via stolons within the first few years. Its importance did not change significantly with time, as it increased early and has waxed and waned during our study. In contrast, the low resilience of *V. sitchensis* was related to limited flowering and few seedlings.

Resilience was zero for *Erythronium montanum*, the pre-disturbance dominant, survivors of which lost cover, and it had no increase in shoot density. It failed to move its perennating organ from the buried soil into tephra; thus, we predicted *E. montanum* would have problems with continued survival (Antos and Zobel 1985c). Although *E. montanum* had a moderate percentage of shoots flowering, seedlings were sparse and it had limited dispersal. Our sites received a similar tephra layer in 1800 (Waitt and Dzurisin 1981), and it seems safe to assume that *E. montanum* was affected similarly then as in 1980. Thus, the species must have increased greatly during the ensuing 180 years; in the first 30 years after 1980, however, it has shown no recovery in our permanent plots. We have observed an increase in seedling density in spots outside our plots, suggesting that recent establishment of seedlings may allow its resilience to increase.

In contrast to herbs, we found no difference in inertia and resilience between the two shrub dominants, *Vaccinium membranaceum* and *V. ovalifolium*. These species are much more alike than the herbs we considered above: both are deciduous Ericaceae of similar stature and structure, traits shared with most shrubs in our sites.

6.6.2 Considerations for Long-Term Studies

Most studies of vegetation response to natural disturbance lack data about vegetation composition before the disturbance. We created good estimates of pre-disturbance values by clearing tephra from plots within 2–4 months after the eruption. Pre-disturbance values allowed us to calculate inertia, resilience, and other indices of redevelopment, and to determine how closely the trajectory of post-disturbance change is converging on the pre-disturbance condition. Several methods could theoretically be used to gather pre-disturbance data about vegetation composition. Ideally, one would have pre-existing, recently sampled plots that were disturbed at an intensity that induced variable responses among plants. In such a case, inertia and resilience could be calculated on a plot-by-plot basis, allowing small-scale study of the influence of environment. In our case, the cleared plots that provided pre-disturbance estimates were on transects interspersed among those with plots on natural tephra, allowing analysis on the scale of the site (ca. 1 ha), rather than the plot. A third approach is to compare disturbed vegetation with the nearest naturally unaffected areas of apparently similar vegetation. With tephra as deep as that on our plots, one cannot identify the understory vegetation beneath the tephra, preventing an appropriate choice of control stands. In addition, the tephra deposit was so large and understory vegetation varied so much and so locally that trying to make comparisons using control areas outside the tephra would be untenable. Another approach is to study tephra effects by adding tephra where none fell naturally or excluding tephra by covering plots during the eruption (e.g., Black and Mack 1986; Mack 1987). For small-scale studies of individual plant processes, this seemed to work well. But for a study at the community level, especially with deep tephra with complex stratification like that on our plots (Waitt and Dzurisin 1981), it would be virtually impossible to provide enough similar tephra to allow treatment of areas large enough to include the existing variation in vegetation and to escape edge effects.

In addition to having pre-disturbance information, our study was unusual for its long duration. Although many aspects of the current vegetation could be predicted from the first 10 years (Zobel and Antos 1997), temporal changes have been far from linear. The unexpected reduction, or even reversal, of the rate of increase in cover and shoot density after 1990 provided a glaring example (Figs. 6.4 and 6.6). We avoided confounding spatial with temporal change by using permanent plots and minimized sampling errors by having at least one author present throughout sampling.

Several issues increase the difficulty of interpreting our data. The sites differed substantially in pre-disturbance species composition (Zobel and Antos 1997); much of our interpretation, therefore, has been done at the level of growth forms. However, our studies of individual species show great disparity of post-disturbance behavior among herb species. In addition, in many cases, especially with tree seedlings and where inertia was high, current importance of a growth form (*c* in Fig. 6.3) exceeded that before the eruption (*a*) (Figs. 6.4, 6.5, 6.6, and 6.7, Tables 6.8, 6.9, and 6.10). In some cases the explanation is clear: at site DR, tree seedlings established and survived well on the surface of deep tephra and have grown into a layer that is much denser than before the eruption at any of our sites (Zobel and Antos 1991b, 1997; Antos and Zobel 2005). In other cases, the explanation is not clear, but several possibilities exist: secondary disturbance (e.g., tree death) stimulated understory growth; the soil environment has become more favorable than before the eruption; or increases in temperature with a longer growing season at these subalpine sites, associated with global warming, have increased productivity.

With appropriate attention to such changes in environment, we believe our study offers a well-documented, widely useful example of vegetation change after disturbance. It illustrates that effects of a single disturbance on different vegetation strata may be profoundly different. It shows the effect of differences in disturbance intensity, and their interaction with another factor (snow).

The tephra disturbance itself is important. Tephrafall is the most frequent and widespread volcanic impact, important in extensive regions of the earth, and has substantially affected vegetation in the Pacific Northwest (Ayris and Demelle 2012; Franklin and Dyrness 1973). There was a similar tephra deposit on our sites in 1800 and many deposits before that (Mullineaux 1986). Carefully designed and consistent long-term studies, initiated as soon after disturbance as possible, are critical to evaluating such large-scale disturbances, which profoundly affect vegetation patterns (Turner et al. 1998). One notable difference in vegetation near Mount St. Helens is the absence of the regionally dominant understory plant *Xerophyllum tenax* from our deep tephra plots and from most areas closer to the volcano, where deep tephra has repeatedly fallen (Swanson et al. 2005). This may reflect the inability of this species' thin, flexible new leaves to penetrate tephra that is deep enough to bury the plants completely.

6.6.3 Using Indices of Inertia and Resilience

Westman (1978, p. 705) defined resilience as "…the degree, manner, and pace of restoration of initial structure and function in an ecosystem after disturbance." Our index of resilience considers the "degree" to which measures of importance approached pre-disturbance values, and their change over time represents the "pace." We (Zobel and Antos 1997) also calculated elasticity (Westman 1978), another measure of "pace," for the first decade after the eruption, but the leveling off of increases in importance (Figs. 6.4 and 6.6) made this calculation ambiguous. Other analyses (Zobel and Antos 2007, 2009) considered the "manner" of herb restoration, including the source, permanence, and timing of establishment of new plants, their mechanism of expansion, and the influence of nearby conspecific plants, flowering frequency, and seedling production. In these analyses we emphasized comparison among species.

Resilience, inertia, and related indices contributed to understanding vegetation change, beyond graphing importance over time, by quantifying aspects of the pattern: how much of the importance survived the disturbance (inertia), how much of the original importance has been regained (relative plant importance), how much of the importance lost has been recovered (resilience), and how effectively survivors expanded. These indices could be statistically analyzed to determine the significance of patterns in vegetation change and the influence of plant and environmental factors on these aspects of vegetation change. Relative plant importance and resilience were highly correlated (Table 6.11), and as post-eruption importance approached zero, they became identical.

We previously computed the indices used in this paper, and one more, referring to four indices as varieties of resilience (Zobel and Antos 1997). Further consideration led us, for this paper, to select $(c-b)/(a-b)$ as our only measure of resilience, to interpret relative plant importance and expansion of survivors differently and to omit a fourth index defined as $(c-b)/a$ (Fig. 6.3). Resilience, as we define it here, $(c-b)/(a-b)$, is a clear statement of the degree of resilience (Westman 1978). Relative plant importance was useful for plotting the time-course of change in plant importance, allowing easy comparison among growth forms or species with different absolute values. We found, however, that $(c-b)/a$ was often strongly correlated with the other indices and did not add to our understanding of vegetation change; thus, we excluded it.

The concept of expansion of survivors provided information not correlated with resilience or relative plant importance. Expansion of survivors was often large, >10, emphasizing how much the importance of survivors increased. This index varied from <1 to >70, differing among environments, growth forms, species, and measures of importance. It was often greatest where there was low inertia (and thus limited plant occupancy); in such situations, a large expansion of importance may reflect high resource availability owing to lack of competition, or the capacity of non-woody plants to reproduce or grow rapidly. Expansion of survivors for shoot density and cover was higher than for species density.

Indices related to recovery also may be correlated with inertia. Among growth forms and widespread species, resilience was higher for those cases in which inertia was high (Figs. 6.9 and 6.10); in contrast, among four herbs at one site, the reverse seems to be true (Fig. 6.11). A general positive relationship between inertia and resilience could be expected, as plants that survived better (higher inertia) are likely to have more seed production or meristems for clonal spread. However, other factors may distinguish individual species' responses, as noted for herbs above.

One unforeseen problem was selecting which year's data to choose for the values of pre- and post-eruption importance (a and b in Fig. 6.3). Although 1980 data seemed an obvious choice for both a and b, these values proved inappropriate. Plant importance in cleared plots was low in 1980–1981 because some live plants with the capacity to survive and grow were not yet apparent: melt of snow beneath tephra was delayed for months, plots were cleared in late summer 1980, and plants may have been damaged by clearing. Because plant importance in some cleared plots declined from 1981 to 1982 to 1983, we selected the lesser of the 1982 and 1983 cleared plot values as pre-eruption importance (Zobel and Antos 1997). Consequently our estimate of pre-eruption importance is likely to be conservative. We used data from natural plots in 1981 to define post-eruption importance, owing to late snow melt beneath the tephra in 1980.

6.7 Conclusions

1. Inertia and resilience generally both declined as plant size decreased, as tephra depth increased, and where tephra fell on snow (i.e., the amount of damage to plants increased and the degree of recovery declined).
2. Resilience differed among measures of importance, with shoot density usually having larger values than cover. It varied most often with environment for species density, but did not vary significantly for cover.

3. Maximal values of resilience were greater in shallow than in deep tephra.
4. Resilience increased with time after disturbance for most attributes and plant types, although it leveled off after 20 years in many cases.
5. Responses of major herb species differed substantially, whereas shrub species did not.
6. Interactions of growth form with tephra depth and extent of snow in 1980 were often significant. Herbs were affected mostly by tephra depth, woody plants by extent of snow.
7. Expansion of survivors was higher where damage was greater—for small growth forms and sites with deep tephra and extensive snowpack in 1980.
8. Values of cover in 2010 remained significantly correlated with inertia.
9. Use of indices of inertia and resilience aided interpretation of vegetation change by allowing statistical analysis of features of the trajectories of community change that are otherwise difficult to analyze.

References

Antos, J.A., and D.B. Zobel. 1982. Snowpack modification of volcanic tephra effects on forest understory plants near Mount St. Helens. *Ecology* 63: 1969–1972.

———. 1984. Ecological implications of belowground morphology of nine coniferous forest herbs. *Botanical Gazette* 145: 508–517.

———. 1985a. Recovery of forest understories buried by tephra from Mount St. Helens. *Vegetatio* 64: 103–111.

———. 1985b. Plant form, developmental plasticity, and survival following burial by volcanic tephra. *Canadian Journal of Botany* 63: 2083–2090.

———. 1985c. Upward movement of underground plant parts into tephra deposits from Mount St. Helens. *Canadian Journal of Botany* 63: 2091–2096.

———. 2005. Plant responses in forests of the tephra-fall zone. In *Ecological responses to the 1980 eruption of Mount St. Helens*, ed. V.H. Dale, F.J. Swanson, and C.M. Crisafulli, 47–58. New York: Springer.

Ayris, P.M., and P. Delmelle. 2012. The immediate environmental effects of tephra emission. *Bulletin of Volcanology* 74: 1905–1936.

Bierzychudek, P. 1982. Life histories and demography of shade-tolerant temperate forest herbs: A review. *New Phytologist* 90: 757–776.

Black, R.A., and R.N. Mack. 1986. Mount St. Helens ash: Recreating its effects on the steppe environment and ecophysiology. *Ecology* 67: 1289–1302.

Franklin, J.F. and C.T. Dyrness. 1973. *Natural vegetation of Oregon and Washington*. General Technical Report PNW-8. Portland, OR: U.S. Department of Agriculture, Forest Service, Pacific Northwest Forest and Range Experiment Station.

Halpern, C.B. 1988. Early successional pathways and the resistance and resilience of forest communities. *Ecology* 69: 1703–1715.

———. 1989. Early successional patterns of forest species: Interactions of life history traits and disturbance. *Ecology* 70: 704–720.

Hitchcock, C.L., and A. Cronquist. 1973. *Flora of the Pacific Northwest*. Seattle: University of Washington Press.

Mack, R.N. 1987. Effects of Mount St. Helens ashfall in steppe communities of eastern Washington: One year later. Chapter 11. In *Mount St. Helens 1980: Botanical consequences of the explosive eruptions*, ed. D.E. Bilderback, 262–281. Berkeley: University of California Press.

McCarthy, B.C. 2003. The herbaceous layer of eastern old-growth deciduous forests. In *The herbaceous layer in forests of eastern North America*, ed. F.S. Gilliam and M.R. Roberts, 163–176. New York: Oxford University Press.

Mullineaux, D.R. 1986. Summary of pre-1980 tephra-fall deposits erupted from Mount St. Helens, Washington State, USA. *Bulletin of Volcanology* 48: 17–26.

Rejmánková, E., M. Rejmánek, T. Djohan, and C.R. Goldman. 1999. Resistance and resilience of subalpine wetlands with respect to prolonged drought. *Folia Geobotanica* 34: 175–188.

Roberts, M.R., and F.S. Gilliam. 2003. Response of the herbaceous layer to disturbance in eastern forests. In *The herbaceous layer in forests of eastern North America*, ed. F.S. Gilliam and M.R. Roberts, 302–320. New York: Oxford University Press.

Swanson, F.J., and J.J. Major. 2005. Physical events, environments, and geological-ecological interactions at Mount St. Helens: March 1980–2004. In *Ecological responses to the 1980 eruption of Mount St. Helens*, ed. V.H. Dale, F.J. Swanson, and C.M. Crisafulli, 27–44. New York: Springer.

Swanson, F.J., C.M. Crisafulli, and D.K. Yamaguchi. 2005. Geological and ecological settings of Mount St. Helens before May 18, 1980. In *Ecological responses to the 1980 eruption of Mount St. Helens*, ed. V.H. Dale, F.J. Swanson, and C.M. Crisafulli, 13–26. New York: Springer.

Turner, M.G., W.L. Baker, C.J. Peterson, and R.K. Peet. 1998. Factors influencing succession: Lessons from large, infrequent natural disturbances. *Ecosystems* 1: 511–523.

Waitt, R.B., Jr., and D. Dzurisin. 1981. Proximal air-fall deposits from the May 18 eruption-stratigraphy and field sedimentology. In *The 1980 eruptions of Mount St. Helens, Washington*. Professional Paper 1250, ed. P.W. Lipman and D.R. Mullineaux, 601–616. Washington, DC: United States Geological Survey.

Westman, W.E. 1978. Measuring the inertia and resilience of ecosystems. *BioScience* 28: 705–710.

———. 1986. Resilience: Concepts and measures. In *Resilience in mediterranean-type ecosystems*, ed. B. Dell, A.J.M. Hopkins, and B.B. Lamont, 5–19. Dordrecht: Junk.

Zobel, D.B., and J.A. Antos. 1991a. 1980 tephra from Mount St. Helens: Spatial and temporal variation beneath forest canopies. *Biology and Fertility of Soils* 12: 60–66.

———. 1991b. Growth and development of natural seedlings of *Abies* and *Tsuga* in old-growth forest. *Journal of Ecology* 79: 985–998.

———. 1992. Survival of plants buried for eight growing seasons by volcanic tephra. *Ecology* 73: 698–701.

———. 1997. A decade of recovery of understory vegetation buried by volcanic tephra from Mount St. Helens. *Ecological Monographs* 67: 317–344.

———. 2007. Flowering and seedling production of understory herbs in old-growth forests affected by 1980 tephra from Mount St. Helens. *Canadian Journal of Botany* 85: 607–620.

———. 2009. Species properties and recovery from disturbance: Forest herbs buried by volcanic tephra. *Journal of Vegetation Science* 20: 650–662.

7 Primary Succession on Mount St. Helens: Rates, Determinism, and Alternative States

Roger del Moral and Jonathan H. Titus

7.1 Introduction

The 1980 eruption of Mount St. Helens created a denuded landscape that was ideally suited to study ecosystem recovery. Post-eruption conditions ranged from the total destruction of areas disturbed by the debris avalanche and ***pyroclastic flows*** to moderate damage caused by tephrafall. In this chapter, we focus on how chance affected primary plant succession under a range of disturbance severity. We documented vegetation change in mid-elevation (980–1550 m) sites on Mount St. Helens between 1980 and 2010. We explore what controls successional rates and species patterns and whether different plant communities might exist under similar habitat conditions.

There are many questions about what controls, constrains, or promotes vegetation development (Walker and del Moral 2003). Our studies exploring the importance of ***microsites*** (Wood and del Moral 1987), ***mycorrhizae*** (Titus and del Moral 1998; Titus et al. 2007), biotic interactions (del Moral and Rozzell 2005), and ***dispersal*** (Fuller and del Moral 2003; del Moral and Eckert 2005) generated broader questions:

- What factors determine ***successional rates*** and how may they be assessed?
- Is there a predictable relationship between species composition and the environment?
- Do similar habitats support similar plant communities?

Answering these questions is critical in a world where dwindling resources and increasingly uncertain climate make ecosystem restoration ever more challenging (Walker et al. 2014).

7.2 Mount St. Helens Study Sites and Plots

The pyroclastic density current (referred hereafter as the ***blast PDC***) of 18 May 1980 created asymmetric impacts (Swanson and Major 2005) and a diverse, largely barren landscape near the volcano. Before the eruption, its slopes were a ***mosaic*** of meadows and coniferous woodlands (del Moral and Wood 1988). Today they express responses to volcanic disturbances of varying severity (Fig. 7.1).

Vegetation change at several sites was documented over decades, and the Muddy River lahar deposit and the Pumice Plain were surveyed extensively during selected years (see Table 7.1 for details of the number of plots and the duration of each monitoring period).

7.2.1 Lahar Sites

Lahars form when eruptions melt snow and glaciers to unleash gravity-driven slurries of rock and mud. Two physically similar lahars formed from old volcanic material on the south slope of the mountain. Lahar 1 terminated near the meadow above Butte Camp; its eastern edge is bordered by woodlands (Fig. 7.2a; del Moral 1998). Subalpine fir (*Abies lasiocarpa*) and lodgepole pine (*Pinus contorta*) quickly recolonized to dominate the vegetation, with an understory of partridgefoot (*Luetkea pectinata*), Davis's

Key terms in ***bold italic face*** are defined in the glossary.

R. del Moral (✉)
Department of Biology, University of Washington, Seattle, WA, USA
e-mail: moral@uw.edu

J.H. Titus
Department of Biology, State University of New York at Fredonia, Fredonia, NY, USA

C.M. Crisafulli, V.H. Dale (eds.), *Ecological Responses at Mount St. Helens: Revisited 35 years after the 1980 Eruption*, https://doi.org/10.1007/978-1-4939-7451-1_7

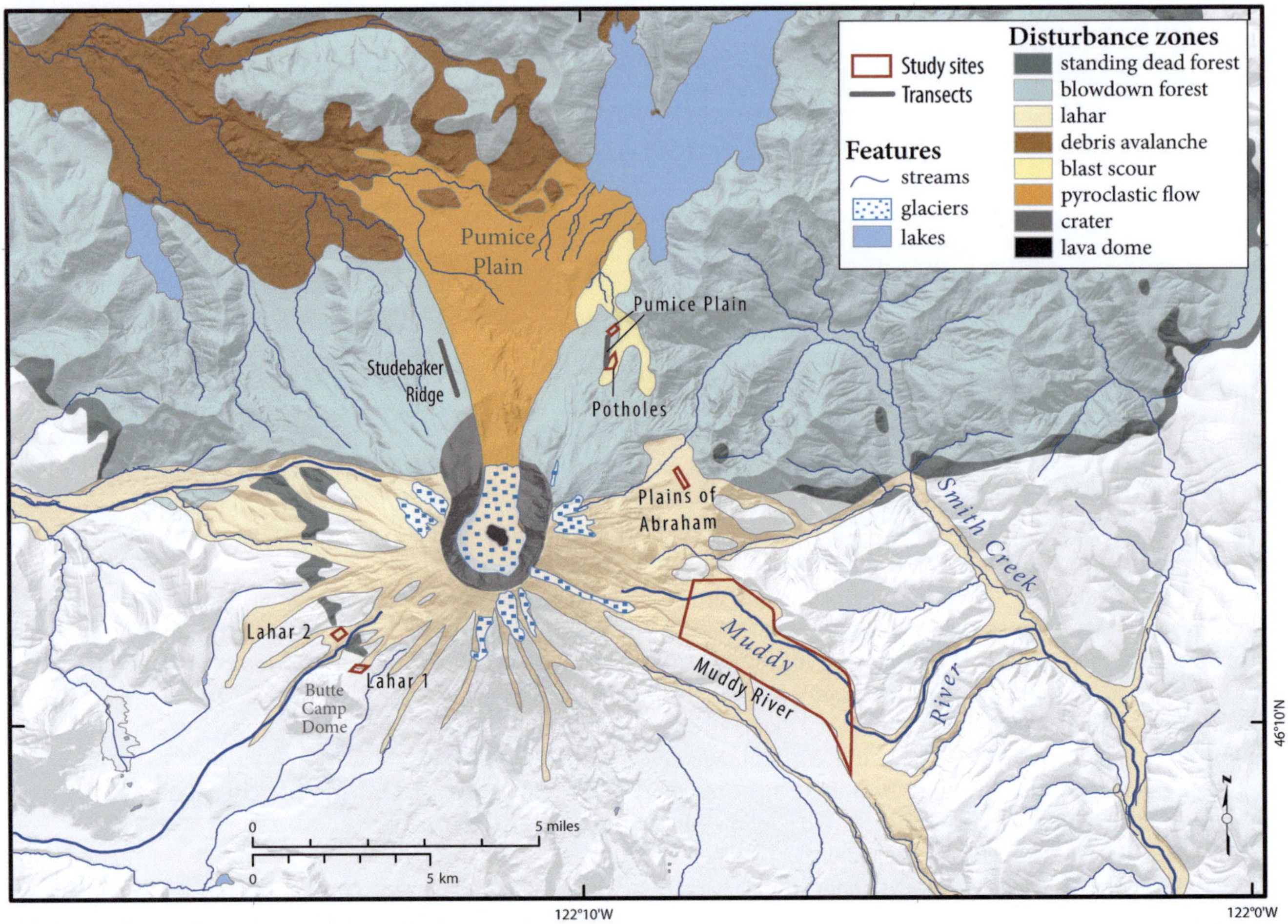

Fig. 7.1 Study area locations at Mount St. Helens. (Cartography by Kathryn Ronnenberg, USDA Forest Service, Pacific Northwest Research Station).

fleeceflower (*Polygonum davisiae*), and Shasta buckwheat (*Eriogonum pyrolifolium*). Lahar 2 (Fig. 7.2b), 1 km to the west, is >500 m from woodlands. Such isolation ordinarily constrains ***colonization*** and alters species composition (Granzow-de la Cerda et al. 2016). This site's vegetation is currently dominated by Pacific lupine (*Lupinus lepidus*), Cardwell's beardtongue (*Penstemon cardwellii*), *Eriogonum*, and *Polygonum*. Small, sparsely distributed conifers also occur.

The Muddy River lahar deposits were generally several meters thick in the areas studied, although thinner along the margins (Fig. 7.2c). Halpern and Harmon (1983) documented early colonization of this lahar. A vegetation survey conducted by del Moral et al. (2009) found that lower sites were dominated by *Lupinus* spp. and rock moss (*Racomitrium canescens*) and upper sites supported a complex mosaic of low shrubs. Conifer size and density decreased with elevation and with distance from lahar-deposit margins (del Moral and Ellis 2004).

7.2.2 Blast-PDC and Pyroclastic-Flow Sites

The Plains of Abraham, a nearly flat expanse, forms the eastern side of the volcano. The blast PDC removed vegetation, a lahar scoured its surface, and ***pumice*** rained down (***tephrafall***) to generate a persistent, desolate habitat (Fig. 7.3a, b; del Moral et al. 2010). The vegetation remains sparse and homogeneous, dominated by forbs (*Cistanthe*), shrubs (e.g., *Penstemon*, *Luetkea*), and graminoids (*Agrostis* spp., *Juncus* spp., *Carex* spp.).

The north flank of the cone supports the rolling Pumice Plain (Fig. 7.4a) from which the debris avalanche, blast PDC, and pyroclastic flows removed all vegetation except in a few ***relict sites***, and coarse pumiceous tephra covered most of the new deposits. The eastern Pumice Plain remained barren well into the 1990s, but is now dominated by *Lupinus*, *Penstemon*, *Luetkea*, *Salix*, and graminoids, often with mosses (Fig. 7.5a; del Moral and Jones 2002). Much of the Pumice Plain contains wetlands and is dissected by stream drainages.

Table 7.1 Characteristics of study sites. Latitude/longitude coordinates are for the center of the study area. Impact = principal volcanic events (pumice refers to coarse tephra deposits composed of this rock type). x = site was involved in the study. *Rate* studies of succession rate, *Det.* studies of deterministic control, *Alter. State* studies of alternative states. Locations are labeled in Fig. 7.1.

							Year		Studies		
Name	Design	N plots	Latitude	Longitude	Impact	Elevation (m)	Start	End	Rate	Det.	Alter. State
Lahar-1	Transect	2	46.17630	−122.22349	Lahar	1418–1460	1982	2005	x		
Lahar-2	Transect	5	46.18289	−122.22626	Lahar	1430–1470	1982	2004	x		
Plains of Abraham	Transect	10	46.21199	−122.13993	Blast PDC, lahar, tephrafall	1365–1367	1990	2010	x		
Pumice Plain	Transect	12	46.23413	−122.15966	Blast PDC, tephrafall	1248–1306	1989	2010	x		
Studebaker Ridge-1	Transect	10	46.23291	−122.19482	Blast PDC	1218–1341	1984	2010	x		
Studebaker Ridge-2	Transect	10	46.22807	−122.19417	Blast PDC	1354–1467	1989	2010	x		
Muddy Lahar	Survey	151	46.22990	−122.11881	Lahar	908–1320	N.A.	2007		x	x
Pumice Plain	Survey	271	46.24345	−122.17124	Pyroclastic flow, tephrafall	1056–1387	1992	2004		x	x
Plains of Abraham	Grid	400	46.21167	−122.14102	Blast PDC, lahar, tephrafall	1362–1367	1988	2010		x	x
Pumice Plain	Grid	200	46.23449	−122.15929	Blast PDC, tephrafall	1215–1245	1989	2010		x	x

At the Pumice Plain and Plains of Abraham sites, we conducted long-term studies using grids, ***permanent plots***, and surveys of depressions on the Pumice Plain known as "potholes" (Fig. 7.4b). Most potholes are dominated by *Agrostis* spp., *Lupinus lepidus*, *Racomitrium*, *Penstemon*, *Luetkea*, and other graminoids.

7.2.3 Blast-PDC Site

Studebaker Ridge forms the western boundary of the Pumice Plain and leads up to the edge of the crater. The ridge endured searing forces from the blast PDC that removed soil from the upper portion, but some remained on lower, more protected sites (Fig. 7.5b), where *Lupinus* with mosses and graminoids became common (del Moral 2007).

7.2.4 Study Plots

The initial goals of each study dictated sampling design (Table 7.1). Circular plots were designed to efficiently monitor vegetation development whereas square plots were arranged in grids to assess local spatial pattern and its development. Surveys described patterns on a landscape scale and were used to relate variation to environmental factors. Potholes were sampled repeatedly to investigate the development of consistent relationships between composition and environment.

Permanent plot data, methods, and syntheses were detailed by del Moral (2010). Plots were circular, 250 m^2 in area, and arrayed along transects in several habitats. Annually, near midseason, we determined species cover to the nearest 0.1% in 24 0.25-m^2 quadrats in each plot. Species present on the plot but absent from any quadrat were assigned a value of 0.1%.

We established and repeatedly surveyed grids of 10 × 10-m plots on the Plains of Abraham (10 by 40 plots; del Moral et al. 2010) and on the eastern Pumice Plain (10 by 20 plots; del Moral et al. 2012). We estimated vegetation cover with a logarithmic index described by del Moral and Wood (2012).

We surveyed the Pumice Plain in 1992 with 141 100-m^2 circular plots (del Moral et al. 1995); then in 2004, del Moral and Lacher (2005) led a survey using 271 250-m^2 circular plots spread over 7 km^2. In 2007, we surveyed the Muddy River lahar deposit (del Moral et al. 2009) with 151 200-m^2 plots arrayed over 5 km^2, ranging in elevation from 900 to 1320 m.

Several hundred potholes occur on the northeast slope. They are likely to have formed by differential melting of

Fig. 7.2 Lahar deposits (**a**) Lahar 1 (2002), (**b**) Lahar 2 (2003), (**c**) Muddy River lahar deposits (2007). (Photos by R. del Moral).

snow trapped beneath pumice that fell during the eruption (del Moral 1999a). Initially empty, they came to support vegetation that was denser than that of their surroundings. In 1993, we measured cover of all species in 111 potholes; most were resurveyed in 1997. From 2001 through 2009, plant cover of 80 intact potholes was determined. They averaged 27.3 m^2 in area and averaged 30 cm in depth. In 2008, we also sampled these 80 potholes with four 1-m^2 quadrats and collected environmental and geographic coordinate data.

7.3 Methods

7.3.1 Successional Rates

Ecologists measure the rate of species composition change with variants of two methods: time needed to reach equilibrium or a benchmark (Sparrius et al. 2013), or measures of floristic turnover (Chapman and McEwan 2013). The former is useful in chronosequence studies (Lindig-Cisneros et al. 2006) and for short-term experiments (Scott and Morgan 2012). Species turnover offers ways to test the effects of factors that regulate ***successional rates*** (Rebele 2013).

We asked a phenomenological question—what factors influence successional rates?—and a methodological one—what criteria best measure successional rates? We used several ways to assess community dynamics between 1980 and 2010 and to explore rate-regulating factors. These criteria include vegetation structure and development, trajectory complexity, measures of community similarity, and measures of floristic turnover on permanent plots established in these initially barren habitats: lahars, the Plains of Abraham, the Pumice Plain, and Studebaker Ridge.

7.3.1.1 Community Patterns

We classified permanent plots into ad hoc units termed community types (CTs) to assess how plot species composition changed over time (see Ujházy et al. 2011). Classifications used species percent cover with an agglomerative clustering strategy (Sørensen's index and flexible sorting, $\beta = -0.25$; McCune and Mefford 2006). Plots were combined until <50% of the information remained, yielding CTs with similar internal variation in each of the study area. Although dominants of two CTs might be similar, their cover differences alone could differentiate them (e.g., L-D and L-E, Table 7.2). A plot undergoing succession normally moves through multiple CTs (its ***trajectory***). ***Trajectory complexity*** is calculated from the number of CTs and their temporal persistence using the information theory statistic (H′).

Within each disturbance type, communities were assigned a code to represent the site and letters to specify the sequence of CT occurrence. For example, L-A identifies the first CT to establish on lahar deposits. Many CTs were transient, quickly progressing to more advanced ones. Persistent CTs were those that remained extant at the end of the study.

7.3.1.2 Vegetation Structure

We used plant cover in several assessments. Community types were described by their initial and maximum cover, net cover increment, rate of cover increase, and time needed to reach 90% of the maximum cover observed in the study

Fig. 7.3 Blast sites, east of crater. (**a**) The Plains of Abraham, which also received a lahar and pumice (1980). (**b**) Vegetation of the Plains of Abraham (2009). (Photos by R. del Moral).

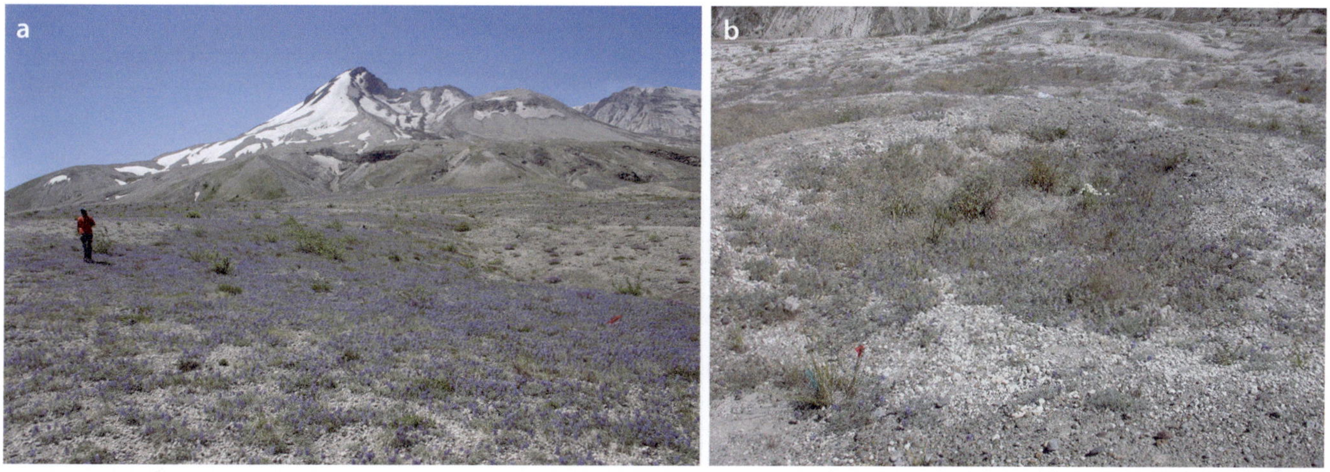

Fig. 7.4 Blast sites, north of crater: (**a**) The Pumice Plain (2010), (**b**) Potholes (2002). (Photos by R. del Moral).

Fig. 7.5 (**a**) Pyroclastic-flow site (2010). (**b**) Lower Studebaker Ridge (2009). (Photos by R. del Moral).

Table 7.2 Persistent community types on primary surfaces. *First cover* initial sampled vegetation cover, *max cover* maximum cover value, *year to 90%* time to reach 90% of maximum cover, *rate* cover percent increase per year to maximum, *mean no. CTs* number of transitional community types leading to the last community type, *H'* the complexity of the trajectory (see text).

CT[a]	1st cover (%)	Max cover (%)	Year to 90%	Rate ($year^{-1}$)	mean no. CTs	H′	Leading dominants[b,c]
L-D	0.2	61.3	25	2.04	4.0	1.119	*Pincon, Abilas, Luepec, Eripyr, Poldav*
L-E	0.2	54.0	25	1.79	4.0	1.207	*Abilas, Pincon, Luepec, Poldav, Eripyr*
L-G	0.1	13.5	17	0.5	4.0	1.142	*Luplep, Eripyr, Pencar, Poldav, Junpar*
PA-C	0.0	7.2	23	0.31	3.0	0.823	*Agrpal, Cisumb, Junpar, Luepec*
PA-D	0.1	11.7	23	0.51	3.25	0.833	*Pencar, Cisumb, Junpar, Agrpal*
PA-E	0.0	11.5	27	0.42	3.0	0.778	*Junpar, Luepec, Pencar, Agrpal*
PA-F	0.1	10.4	27	0.38	3.0	0.667	*Pencar, Raccan, Luepec, Junpar*
PP-F	0.5	14.8	22	0.65	5.5	1.483	*Pencar, Luplep, Agrpal, Carmer, Luzpar*
PP-G	0.6	21.1	22	0.93	5.0	1.424	*Agrpal, Pencar, Raccan, Luepec, Luplep*
PP-Ia	0.5	58.9	22	2.65	6.3	1.779	*Luplep, Agrpal, Pencar, Salsit, Raccan*
PP-Ib	0.5	93.5	22	4.23	6.7	1.867	*Luplep, Agrpal, Pencar, Salsit, Raccan*
SR-F	0.25	4.5	27	0.16	4.0	1.052	*Agrpal, Luepec, Junpar, Luplep, Agrsca*
SR-G	0.1	33.2	21	1.57	7.0	1.870	*Agrpal, Luplep, Raccan, Cisumb*
SR-Ha	0.63	15.1	21	0.69	6.5	1.563	*Agrpal, Raccan, Luplep Anamar, Luepec*
SR-Hb	0.35	5.4	27	0.18	4.3	1.271	*Agrpal, Raccan, Luplep, Agrsca*
SR-I	0.45	20.5	21	0.95	6.0	1.365	*Raccan, Agrpal, Luplep, Carmer, Junpar*
SR-J	0.1	55.8	19	2.93	5.0	1.361	*Raccan, Agrpal, Luplep, Carmem*
SR-K	0.5	10.9	16	0.65	6.0	1.502	*Cisumb, Agrpal, Luepec, Junpar, Luplep*
SR-La	0.15	69.1	16	4.33	4.3	1.231	*Raccan, Agrpal, Luplep, Pencar, Achmil*
SR-Lb	0.15	31.3	21	1.18	6.0	1.669	*Raccan, Agrpal, Luplep, Pencar, Salsit*

[a]L–D and L-E are on Lahar 1; L-G is on Lahar 2; PA communities are on the Plains of Abraham; PP communities are on the Pumice Plain; SR communities are on Studebaker Ridge.

[b]Species listed in order of dominance in last 3 years of plots sampled.

[c]Abbreviations followed by scientific name (common name): *Abilas* = *Abies lasiocarpa* (subalpine fir); *Achmil* = *Achillea millefolium* (yarrow); *Agrpal* = *Agrostis pallens* (dune bentgrass); *Agrsca* = *Agrostis scabra* (rough bentgrass); *Carmer* = *Carex mertensii* (Merten's sedge); *Cisumb* = *Cistanthe umbellata* (Mt. Hood pussypaws); *Eripyr* = *Eriogonum pyrolifolium* (Shasta buckwheat); *Junpar* = *Juncus parryi* (Parry's rush); *Luepec* = *Luetkea pectinata* (partridgefoot); *Luplep* = *Lupinus lepidus* (prairie lupine); *Luzpar* = *Luzula parviflora* (small-flowered woodrush); *Pencar* = *Penstemon cardwellii* (Cardwell's beardtongue); *Pincon* = *Pinus contorta* (lodgepole pine); *Poldav* = *Polygonum davisiae* (Davis' fleeceflower); *Raccan* = *Racomitrium canescens* (roadside rock moss); *Salsit* = *Salix sitchensis* (Sitka willow).

(Alvarez-Molina et al. 2012). Cover development is a useful way to summarize successional rates under some conditions (Prach et al. 2007), but becomes less effective as it approaches its maximum (Baasch et al. 2012).

7.3.1.3 Floristic Similarity

Floristic similarity relates the species composition of two samples. For example, when comparing two plots, the plot with lower similarity between initial and final samples has developed more quickly. We assessed change in a matrix of species ***percent similarity (PS)*** between each successive year (Kovach 1999) by calculating mean similarity of each plot matrix, averaged for all plots of each CT. Plots with lower mean similarity had progressed further than did those with high similarities. We also determined how long it took for plots initially within a CT to become floristically distinct, that is, to reach 10% similarity among them.

7.3.1.4 Floristic Turnover

Species turnover is a progressive change in species composition. We used species percent cover to provide a detailed analysis (cf. Chaideftou et al. 2012). We used ***detrended correspondence analysis*** (DCA) because changes are scaled so that 100 units approximate a 50% change in species composition (Freeman and Kobziar 2011). Therefore, the scores of the same plot over time indicate the degree of floristic change and the time interval indicates the rate of floristic change. We analyzed all permanent plots simultaneously to allow direct rate comparisons.

As sample composition changes, its position traverses DCA space, producing ***succession trajectories*** (Magnússon et al. 2009); therefore, the Euclidean distance (*d*) between DCA scores summarizes floristic change (Baniya et al. 2009). To calculate *d*, the DCA axes were weighted by the amount of overall variation contained in each (see del Moral and Magnusson 2014) and the change between successive years was determined. ***Principal components analysis*** (***PCA***) was used to smooth these vectors and to estimate rates of change. The slope of the regression of PCA with time also measures succession rate (del Moral and Chang 2015).

7.3.2 Determinism

Our studies of developing determinism compared vegetation patterns to predictive variables seeking statistically valid relationships. They were conducted at several scales, disturbance types, and locations: 2007 Muddy River lahar deposit survey, 1992 and 2004 Pumice Plain surveys, sequential studies of the Plains of Abraham and Pumice Plain grids, and the Pumice Plain potholes. Factors included both spatial and habitat variables and were analyzed by ***redundancy analysis*** (see Sect. 7.3.2.2).

7.3.2.1 Explanatory Variables

Vegetation responds to factors such as moisture, temperature, length of growing season, and fertility (Baeten et al. 2010), but it is rare that such explanatory variables can be measured directly or comprehensively. Often more readily measured surrogates that reflect potential explanatory variables are used to explore causal patterns between species composition and environmental and landscape factors.

Each determinism study had different available variables (Table 7.3). These included position, elevation, and distance to dense vegetation. We determined solar insolation from aspect and slope (McCune and Keon 2002) and measured sunlight near solar noon using a light meter. Soils collected between 5- and 15-cm depths were analyzed to determine moisture, pH, and texture classes. Fertility was estimated using *Lupinus* cover in preceding years, as verified in pilot studies. We noted where topography (rills and gullies) had exposed old soil. Exposure to desiccation was estimated from aspect and prevailing wind. Method details are in papers cited in Sect. 7.4.

7.3.2.2 Statistical Analyses

The ability of explanatory variables to predict species composition was explored using ***redundancy analysis*** (***RDA***), a linear method of ***constrained ordination*** (ter Braak and Šmilauer 2007). Such methods determine the degree of explained variation between predictors and species composition and partition it between spatial and habitat factors. The strength of the relationship between predictors and species composition estimated determination. Results were summarized as the proportion of total explained variation assigned to spatial and habitat factors, the correlation of species to the first axis, and the strength of individual explanatory variables.

7.3.2.3 Alternative States

Studies were initially designed to describe vegetation recovery, but the data also challenge the idea that only a single early-successional community type (CT) occurs in a particular ***habitat type*** (HT). We used agglomerative clustering (see Sect. 7.3.3.1) with only environmental variables to produce HTs from a set of data (see Table 7.3). CTs were formed by classifying vascular plants and mosses; for this study, fewer CTs were recognized than in the published studies in order to improve the fit between CTs and HTs. We compared species composition within CTs and HTs using percent similarity and examined the variation of DCA scores of plots within CTs and HTs.

Table 7.3 Explanatory variables used in constrained analyses. Method used to assess variable given in parentheses. Habitat factors were used to classify habitat types (HTs).

Variable	Muddy survey	Pumice Plain survey	Abraham grid	Pumice Plain grid	Potholes
Landscape factors					
Elevation (GPS; maps)	x	x		x	x
Latitude (GPS)	x	x	x	x	x
Longitude (GPS)	x	x	x	x	x
Distance-wetland (direct)	x	x			
Distance-relicts (direct)	x	x	x	x	x
Habitat factors					
Slope (inclinometer)	x	x		x	
Aspect (compass)	x	x		x	
Insolation (calculation)	x			x	
Exposure (observation)	x	x		x	
Light (meter)				x	
Moisture (gravimetric)				x	x
pH (meter)					x
Fertility (*Lupinus*)					x
Texture (sieves)	x	x		x	x
Relict soil (observation)				x	
Topography (observation)	x	x	x	x	x
Rocks (observation)	x	x	x		
Pumice (observation)	x	x	x		
Pyroclastic (observation)		x			

7.3.3 Statistical Software

We conducted statistical analyses with Statistix 10 (Analytical Software 2013) and used PC-ORD (McCune and Mefford 2006) to provide data summaries and classifications.

7.4 Rates of Succession on Primary Surfaces

Successional rates evaluate how quickly species composition and vegetation structure change. Rates are influenced by ***competition*** (Rebele 2013), ***facilitation*** (Walker et al. 2003), nutrient inputs (Butterfield and Callaway 2013), disturbance severity (Burt and Clary 2016), isolation (Boyes et al. 2011), and ***stresses*** (Kayes et al. 2010) that limit productivity (Laborde and Thompson 2013). Stresses include infertility (Long et al. 2012), drought (Kardol et al. 2010), and truncated growing seasons (Kamijo et al. 2002). The importance of these factors varies spatially and temporally (Platt and Connell 2003). Del Moral and Magnússon (2014) compared succession rates on Mount St. Helens with Surtsey (Iceland), where seabirds imported guano that radically speeded succession. Del Moral and Chang (2015) explored successional rates on successional surfaces at Mount St. Helens in more detail and found that stresses such as drought slowed succession and that repeated disturbances such as erosion often returned vegetation to an early successional stage.

We tested three hypotheses: (1) floristic development is inhibited by stresses; (2) disturbance impedes or alters successional trajectories; and (3) isolation slows structural and floristic development.

7.4.1 Classification and Structure

Development of cover and species richness describes early succession well, but because these values soon stabilize, other criteria become more useful. To describe later succession, we report time taken to reach 90% maximum plot cover, the annual rate of increase, and trajectory complexity (Table 7.2).

7.4.1.1 Lahar Deposits

Two succession patterns developed on lahar deposits. Conifers quickly invaded plots near intact vegetation (L-D, L-E), whereas prostrate subalpine herb species dominated isolated sites (L-G), which had relatively few young conifers. Only three of seven CTs persisted. The rate of cover accretion to 90% of maximum cover declined from L-D, near the woodland, to the distant L-G. All plots transitioned through four CTs, with moderate community complexity.

7.4.1.2 Plains of Abraham

Vegetation on the ten Plains of Abraham plots remained sparse. Four of six CTs persisted, of which PA-C (dune bentgrass, [*Agrostis pallens*], Mt. Hood pussypaws [*Cistanthe umbellata*], and Parry's rush [*Juncus parryi*]) had the lowest cover. Community type PA-D was dominated by *Penstemon* and persistent herbs, whereas PA-E was dominated by *Juncus* and *Luetkea*. Community type PA-F had significant moss cover. In general, cover accumulated slowly (0.3–0.5% per year) and community complexity remained low in this stressful habitat.

7.4.1.3 Pumice Plain

The 12 Pumice Plain plots suffered the same initial disturbance, but the plots experienced different degrees of solar radiation, wind, and subsequent erosion that produced very different successional rates. Only three of nine CTs persisted. Founder CTs were dominated by grasses, well-dispersed forbs such as pearly everlasting (*Anaphalis margaritacea*), and *Lupinus*. Transitional communities featured *Penstemon*, *Salix*, and several herbs. Persistent CTs occurred at PP-F (*Agrostis*, *Lupinus*, and *Carex mertensii*), which suffered repeated water erosion, PP-G (*Lupinus*, *Penstemon*, and graminoids) in exposed sites, and PP-I on the upper, protected slopes (*Lupinus*, *Penstemon*, *Agrostis*, and mosses). All CTs reached 90% of their maximum cover in 2004 after a population explosion of *Lupinus*. Plot trajectories averaged 5.33 CTs; trajectory complexity was among the highest in the study.

7.4.1.4 Studebaker Ridge

Initial vegetation development on Studebaker Ridge was delayed by stress and isolation (del Moral 2007); in 1984, only 6 of 20 plots had any plants. Nine of 12 CTs persisted because pioneer CTs first appeared at low elevation then developed at mid-elevations. SR-L cover developed in response to fluctuations of *Lupinus*, but cover in other CTs increased more slowly and remained sparse, reaching 90% maximum cover between 2004 and 2010. SR-La cover increased by 4.33% per year, and SR-G, after a slow start, achieved 1.57% cover per year, both among the most rapid of the study. Complexity of most trajectories was high.

7.4.2 Temporal Changes in Floristic Similarity

Mean plot percent similarity (PS), determined from the floristic matrix between all years, is a complementary estimator of successional rates. Plot means were pooled into each CT. Low mean similarity reflects rapid succession (Table 7.4).

Mean similarities among Lahar CTs were low; those near woodlands were lower than isolated ones—they reached 10%

Table 7.4 Characteristics of persistent community types (CT). *Impact type* describes the main influences on successional patterns; *richness* and *cover* = final values recorded; *mean% sim* = matrix of similarity mean value (mean of all plots in CT); *years to 10%* = number of years to reach 10% similarity in matrix; Euclidean distance *d* between first and final DCA position (3 dimensions; *ΔPCA* is net change over study in the principal components analysis of the DCA scores; *PCA slope* is the regression of PCA with time (all $p > 0.001$ unless noted otherwise).

CT (N plots)	Impact Type	Richness	Cover (%)	Mean% sim	Years to 10%	*d*	ΔPCA	PCA Slope
L-D (1)	Lahar-adjacent	18.0	57.2	11.4	1990	142	2.09	0.210
L-E (1)	Lahar-adjacent	18.0	54.1	12.7	1992	195	2.24	0.260
L-G (5)	Lahar-distant	17.2	15.0	17.3	1995	73	0.66	0.048
PA-C (2)	Blast + lahar	15.5	4.5	50.5	2010	60	2.35	0.149
PA-D (4)	Blast + lahar	13.0	4.8	48.3	2010	88	2.56	0.183
PA-E (2)	Blast + lahar	13.5	6.5	43.8	2010	74	2.12	0.172
PA-F (2)	Blast + lahar	19.0	5.4	45.6	2010	81	1.84	0.135
PP-F (2)	Blast + pumice[a]	20.09	7.7	25.9	2002	93	2.42	0.137
PP-G (4)	Blast + pumice[a]	18.3	19.9	22.5	2002	121	3.27	0.135
PP-Ia (3)	Blast + pumice[a]	21.7	39.2	18.4	2002	141	3.37	0.175
PP-Ib (3)	Blast + pumice[a]	20.7	58.7	21.4	2002	152	3.68	0.190
SR-F (3)	Blast	17.3	4.3	23.8	2001	49	1.49	0.106
SR-G (1)	Blast	16.0	33.1	14.9	1993	172	3.49	0.258
SR-Ha (2)	Blast	17.0	7.9	24.6	2010	113	1.81	0.135
SR-Hb (3)	Blast	18.3	5.9	30.6	2008	111	1.74	0.121
SR-I (1)	Blast	17.0	11.5	16.5	1997	149	1.12	0.083
SR-J (1)	Blast	22.0	44.1	8.4	1990	162	2.50	0.108
SR-K (1)	Blast	16.0	7.0	22.4	2008	113	2.55	0.127
SR-La (4)	Blast	18.3	60.4	10.6	1989	118	2.20	0.111
SR-Lb (4)	Blast	15.3	21.1	10.5	1995	129	2.39	0.132

[a]Pumice here implies coarse airfall deposits (tephra).

similarity more rapidly than did those of any other site. In contrast, the mean similarity in Plains of Abraham CTs was higher than those of other CTs; no plot reached the 10% threshold. Similarity of Pumice Plain CTs was intermediate to lahars and Plains of Abraham. The Studebaker Ridge transect had variable, but low, similarities. SR-La achieved the 10% threshold rapidly because the *Lupinus* quickly improved establishment conditions (Bishop 2002). Poorly vegetated sites required longer to reach this level of similarity.

We used DCA to determine succession trajectories of all permanent plots in one analysis so that rate comparisons could be made directly (Table 7.4). The analysis revealed 4.3 species turnovers in the first DCA axis, a large overall change. Representative trajectories provide understanding of successional rates, but better understanding comes from analyses of Euclidean distances and a PCA of the trajectories, which determined the strength of floristic trends.

7.4.2.1 Lahar Deposits

The trajectories of CTs on lahar deposits changed dramatically (Fig. 7.6a). Community type L-E, with conifers, ranged over half of the entire DCA space, whereas L-G changed modestly. These changes are reflected by the overall ***Euclidean distance***, in which the proximate CTs changed considerably more than did the isolated L-G, and more so than most other CTs in this study. The general trend of successional change was reflected in PCA; its slope decreased as isolation increased from L-D to L-G.

7.4.2.2 Plains of Abraham

The DCA-1 trajectories of the Plains of Abraham CTs changed little (Fig. 7.6a), reflecting only the accumulation of species with little species turnover. Values of *d* were low, changes in PCA were relatively small, and the slope of the PCA regression reflected modest change.

7.4.2.3 Pumice Plain

The trajectories of the Pumice Plain CTs showed extraordinary change. Community type PP-Ia occurs across most of the first DCA axis (Fig. 7.6b), and the other CTs changed strikingly, largely owing to *Lupinus*. The Euclidean distance (*d*) reflected floristic change that increased with elevation; PCA showed large changes and steep slopes in each case.

7.4.2.4 Studebaker Ridge

Plots along Studebaker Ridge revealed various responses to elevation and exposure. CT SR-G changed substantially in all dimensions and had a *d* comparable to that of L-E (Fig. 7.6c). Lower, protected plots of SR-La also changed substantially, but had stabilized by year 2000. CT SR-Ha had moderate change, but SR-F, located at the highest elevations, changed least. The PCA vectors capture these dynamics well.

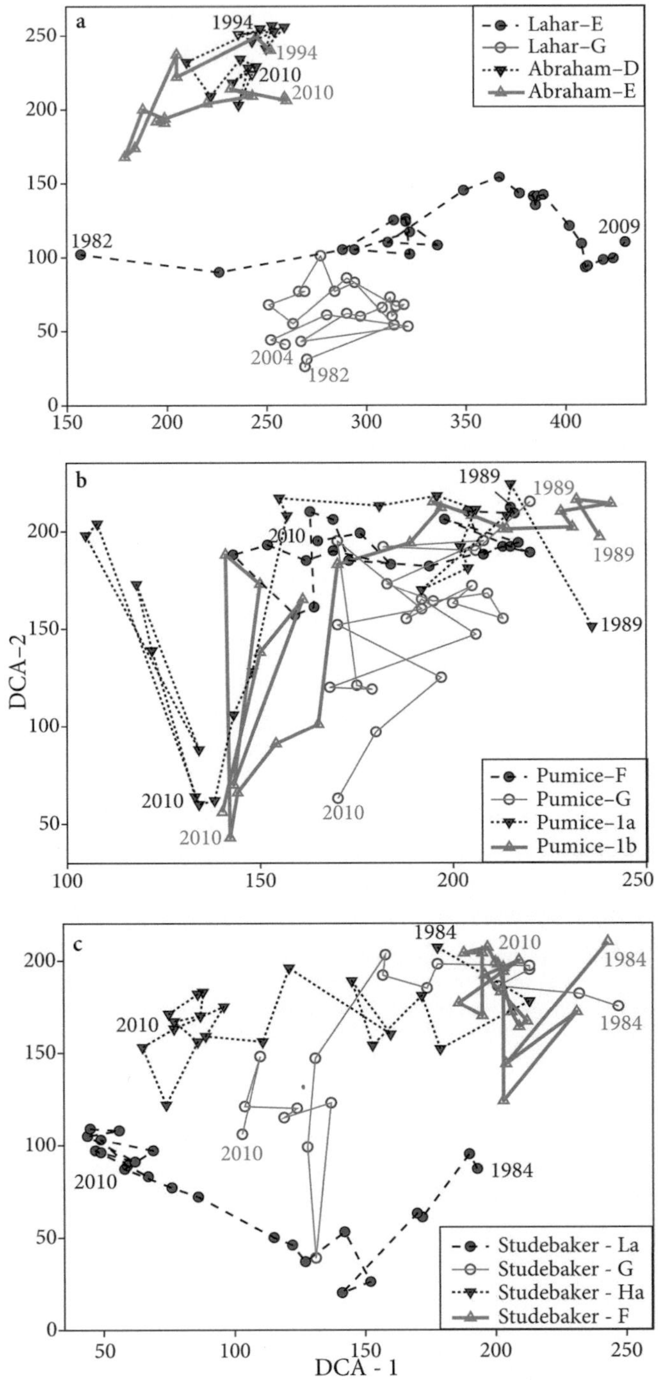

Fig. 7.6 Trajectories of community types measured by DCA: (**a**) lahars and Plains of Abraham, (**b**) Pumice Plain, (**c**) Studebaker Ridge.

7.4.3 Comparison of Measures for Determining Successional Rates

7.4.3.1 Ranking Rates

Each component of our analyses of rates of ***primary succession*** offers insight into those parameters that influenced species assembly and turnover on Mount St. Helens. Relative successional rates based on the more relevant parameters were scaled on a 5-point system to facilitate comparisons, with a value of 1 indicating rapid succession (Table 7.5).

Time to 90% maximum cover was expressed as a percentage of the particular trajectory. A relative rate score of 5 did not necessarily imply slow development, only that the vegetation continued to develop. CTs L-D and PA-E, for example, only reached the 90% cover standard in the final year of the study, yet developed at different rates. Community types that quickly reached the standard exhibited an early expansion of *Lupinus* (e.g., SR-K). The annual increment of cover to 90% is a general indicator of successional rate. Slowly developing CTs remained sparsely vegetated (e.g., L-G and most Plains of Abraham CTs).

Low trajectory complexity (H′) occurs in plots with few CTs, which suggests slow succession. High complexity, with several CTs occurring for several years each, signifies rapid succession. The few CTs found on the Plains of Abraham had long residencies and thus expressed the slowest succession. Disturbance such as water erosion enhanced the relative rate score of PP-F by partially resetting succession, providing more CTs. The protected SR-La plots changed little in DCA space after their early rapid development, but by other rate measures, this CT experienced rapid succession.

Mean matrix similarity provides an overview that mitigates the obscuring effects of nondirectional variation. Proximate lahar CTs, fertile Pumice Plain CTs (e.g., PP-I), and SR-G all changed substantially, whereas barren upper Studebaker Ridge CTs and the Plains of Abraham changed little. Mean matrix similarity is sensitive to initially low richness and cover, so patterns can be obscured and comparisons may inaccurately reflect successional rates.

The PCA supports the assessments of Euclidean distance (*d*). The CTs that changed most include L-D, PP-I, SR-G, and SR-J; those that changed least included PA-C and SR-F, which also had limited floristic turnover. The slopes of PCA curves can reflect successional rates, but they may be affected by secular fluctuations in DCA scores. By signaling trends, PCA aided the analysis in SR-G (delayed development owing to an unstable slope), L-D and L-E (developed a conifer canopy), and PP-Ib (experienced substantial facilitation effects from *Lupinus*).

Table 7.5 Rates of succession in persistent community types (CT). Rankings for *rate* ($year^{-1}$) are from 1, rapid, to 5, slow (See Tables 7.3 and 7.4 for codes to criteria).

CT	Time to 90%	Rate (y^{-1})	H′	Mean% Sim	ED	ΔPCA	PCA Slope
L-D	5	2	4	1	2	4	1
L-E	5	3	4	1	1	3	1
L-G	3	5	4	2	4	5	5
PA-C	4	5	5	5	5	3	2
PA-D	4	4	5	5	4	2	2
PA-E	5	5	5	5	4	4	2
PA-F	5	5	5	5	4	4	3
PP-F	4	4	2	4	4	3	3
PP-G	4	4	3	3	5	1	3
PP-Ia	4	2	1	2	2	1	2
PP-Ib	4	1	1	2	2	1	1
SR-F	5	5	1	3	5	5	4
SR-G	3	3	1	2	1	1	1
SR-Ha	3	4	2	3	3	4	3
SR-Hb	5	5	4	4	3	4	3
SR-I	3	4	3	2	2	5	5
SR-J	2	2	3	1	1	2	4
SR-K	1	4	2	3	3	2	3
SR-La	1	1	4	1	3	3	4
SR-Lb	3	3	2	1	3	3	3

7.4.3.2 Rate Regulation

An understanding of what controls primary successional rates facilitates ways to restore habitats (del Moral et al. 2007). Results from these several methods to estimate succession rates suggest that isolation, infertility, and drought retard succession. Communities on lahar deposits adjacent to intact vegetation became similar to these surviving woodlands, but those on sites over a hundred meters from woodlands changed less. Barren sites invaded by *Lupinus* developed quickly (Bishop et al. 2005) and demonstrated rapid succession over the 30 years of this study. In contrast, plots lacking *Lupinus* developed slowly (e.g., all CTs on the Plains of Abraham). Exposed sites, particularly those at high elevation, had short growing seasons, and were limited by drought or early frost (e.g., SR-F and SR-Hb).

7.5 Deterministic Control of Vegetation Dynamics

Succession theory once asserted that internal factors determine species composition trajectories to produce predictable, homogeneous communities (e.g., Clements 1936). This view evolved to acknowledge that existing vegetation is often only weakly directed by habitat conditions (Whittaker

1967). Turner et al. (1998) showed that trajectories are more variable during primary succession than during secondary succession because initial establishment is highly variable. Thus, erratic dispersal that allows different species to establish in similar habitats may weaken deterministic factors (Matthews and Endress 2010). Over time, the effects of deterministic factors such as fertility, moisture, and competition increase, while ***stochastic*** influences persist. Most studies of deterministic control have not produced strong statistical correlations (e.g., Brownstein et al. 2012). Both convergent trajectories and statistical links between plants and their habitats would suggest that deterministic effects are effectively molding communities, so we used the Mount St. Helens data to explore determinism during early succession.

Species composition initially was variable due to a sparse seed rain (Wood and del Moral 2000) combined with habitat stresses that produced weak correlations between environmental factors and species composition (del Moral et al. 2005). Our early studies (e.g., del Moral and Wood 1988) suggested that vegetation assembly was controlled by geographic factors; later observations suggested that the effects of habitat factors had increased and, although still feeble, were significant. Stochastic effects, competitors, and our inability to measure relevant factors combined to limit our detection of significant explanatory variables. Additionally, transient (e.g., wind storms) and contingent effects (e.g., frost, landslides) cannot be incorporated into most models, and there are no guarantees that relevant factors such as subtle disturbances (Tsuyuzaki et al. 1997) or intermittent grazing (Bishop 2002) are measurable. Here, we summarize studies on Mount St. Helens that explored vegetation–environment relations. Each sought to estimate the degree to which patterns were explicable, partition the relationship between landscape and habitat factors, and determine whether the relationship strengthened during succession (see Sect. 7.3.2.2, RDA). Results of each study are summarized in Table 7.6.

7.5.1 Muddy River Lahar Deposit

Vegetation and habitat characteristic of the upper Muddy River lahar deposit were surveyed 28 years after its formation and vegetation patterns were found to be influenced by proximity to intact forest. Vegetation varied primarily with landscape factors, yet only 31.6% of the variation was related to explanatory variables. On the lahar deposit, changes associated with elevation of 908–1320 m are associated most closely with the development of vegetation. In addition, sporadic colonization by competitive shrubs can preclude establishment of other species, forming a vegetation mosaic (Walker et al. 2006). Even when closed forests develop, much unexplained variation is likely to remain (Økland 1999) due to persistent landscape effects and competition from mats of *Racomitrium* and swards of *Lupinus* (cf. del Moral and Rozzell 2005).

Table 7.6 Summary of constrained ordination studies. *Year* is the year of sampling; *spatial* and *habitat* (%) are the fraction of the total variation by each factor; *r* is the species to explanatory variable correlation in axis 1; *Loc* location, *Ele* elevation, *Fer* fertility (estimated from *Lupinus* concentration, previous years), *Mois* soil moisture, *Iso* isolation, *Hab* habitat type, *Top* topography, *Sur* surface type, *Exp* exposure. (See Table 7.2 for variables).

				Proportion explained by factors		
Site	**Year**	**Explained variation (%)**	***r***	**Spatial (%)**	**Habitat (%)**	**Factors**
Muddy survey	2007	31.6	0.899	59.9	40.1	Ele; Loc; Iso
Pumice Plain survey	1992	14.1	0.720	85.0	15.0	Loc; Ele
	2004	29.1	0.688	60.8	39.2	Loc; Hab; Ele
Abraham grid	1990	3.4	0.261	–	–	None
	1999	10.9	0.640	42.5	57.5	Loc; Sur
	2008	12.2	0.658	38.5	61.5	Loc; Sur
Pumice Plain grid	1989	15.1	0.664	39.5	60.5	Loc; Top; Exp
	1997	23.4	0.806	43.2	56.7	Loc; Exp
	2004	30.8	0.811	44.8	55.2	Top; Loc
	2010	25.7	0.820	40.1	59.9	Exp; Top
Potholes	1993	10.2	0.597	45.4	54.6	Not significant
	1997	16.2	0.697	48.5	51.5	Loc; Ele
	2001	29.7	0.650	59.6	40.4	Ele; Loc
	2006	29.6	0.770	29.8	70.2	Fer; Mois
	2008	36.0	0.774	27.1	72.9	Fer; Mois

7.5.2 Pumice Plain Surveys

The Pumice Plain was surveyed in 1992, when 86 100-m^2 plots were sampled (del Moral et al. 1995), and in 2004, when 271 250-m^2 plots were sampled (del Moral and Lacher 2005). Geographic and geomorphic features were recorded and analyzed using ***canonical correspondence analysis*** (***CCA***). Data were reanalyzed here using RDA, which improved the fit between explanatory variables and species patterns. Total explained variance doubled and habitat factors tripled in importance during the 12-year interval between surveys, but landscape effects remained static.

7.5.3 Plains of Abraham

The Plains of Abraham grid was sampled annually from 1988 to 2010. Plant establishment appears to have been inhibited by drought stress and soil infertility, although isolation may have deterred *Lupinus* invasion. Only position, topography, and surface conditions were used in RDA, and only 3 representative years are shown in Table 7.6 (del Moral et al. 2010).

A weak relationship emerged between explanatory variables and species patterns only after 1996 when position and rills became significant. Explained variation never exceeded 13%. The best correlation to explanatory factors was 0.66, and surface factors accounted for over 60% of this variation. Both overall and habitat factors increased their explanatory power slightly.

7.5.4 Pumice Plain Grid

The grid on the eastern Pumice Plain was monitored annually from 1989 to 2010. Total explained variation maximized at 30.8% during the *Lupinus* peak of 2004, dropping to 25.7% in 2010. The covariate analyses revealed a consistent 60% of explained variation to be associated with habitat factors (Fig. 7.7), stable after 2004. Both habitat (e.g., exposure) and landscape (e.g., distance from vegetation) were important. In 2010, landscape variables accounted for 14.3% and habitat variables for 21.4%, but, owing to their statistical overlap, in combination they accounted for only 25.1%. For example, elevation combines landscape (dispersal) and habitat effects (temperature and moisture). Three habitat factors, exposure to wind (***r*** = −0.73), topography (***r*** = 0.63), and relict type (***r*** = 0.60), were significant, as were the landscape factors distance from relict sites (***r*** = −0.62) and elevation (***r*** = −0.45).

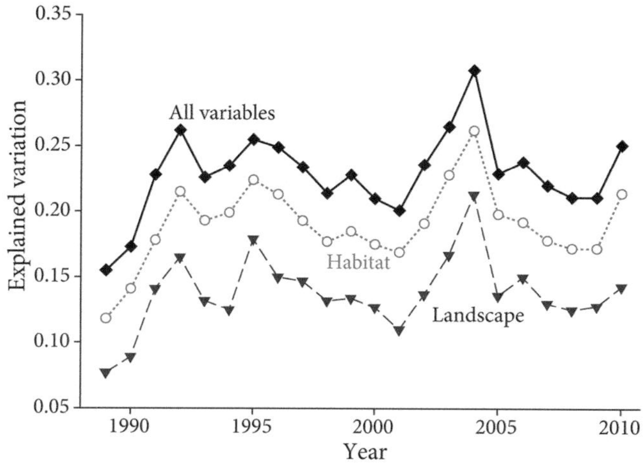

Fig. 7.7 Covariate analysis (Pumice Plain). Explained variation was divided between that explained by habitat factors and that explained by landscape factors.

7.5.5 Potholes

Redundancy analysis (RDA) of pothole data showed increasing explained variation owing to habitat conditions, whereas spatial effects remained unchanged (del Moral 2009). The correlation (***r***) increased from 0.597 to 0.774 from 1993 to 2008 as explained variation increased from 10.2% to 36.0%. The 2008 analysis explained 36% of the variation, of which 72.9% was associated with soil factors. Ephemeral species were replaced by persistent ones. Pothole vegetation became less variable as a result of deterministic factors, but the effects of chance, disturbances, and history persist.

7.6 Alternative States

Alternative states occur if two or more communities occupy one type of habitat; persistent differences are ***alternative stable states*** (Saccone et al. 2014). The evidence for alternative states resulting from stochastic factors is sparse (but see McCune and Allen 1985), yet Schröder et al. (2005) and Suding and Hobbs (2009) suggest that they result from differential dispersal (Bossuyt et al. 2005). Alternative states commonly arise from disturbances such as fire (Favier et al. 2012) or grazing (Bischoff et al. 2009), from silviculture (Beisner et al. 2003), from competition (Kefi et al. 2016), or from facilitation (Gerla and Mooij 2014).

Forest–meadow mosaics may develop on Mount St. Helens where dense *Lupinus* spp. resist conifer colonization (Titus and Bishop 2014) or after sporadic establishment by

Arctostaphylos. Distance from colonists may produce alternative states as a result of ***priority effects*** (Marteinsdottir et al. 2010), and differential disturbances often create vegetation mosaics (Efford et al. 2014). In contrast, initially variable vegetation may converge over time, as shrubs or trees come to dominate (Anthelme et al. 2007).

It is problematic to test the hypothesis that alternative states exist with the present data, because succession should require several centuries to run its course; however, it may be possible to reveal patterns consistent with the hypothesis. The essential question, do alternative *stable* states occur, is beyond our scope. Any acceptance of the alternative-state hypothesis can only be provisional, because vegetation may converge as deterministic forces strengthen; weak tendencies that suggest alternative states cannot compellingly reject the single-state hypothesis but can only make alternative states plausible. Species variation may not yet have been dampened by local factors (Sect. 7.5). To the extent that trajectories form alternative states in response to landscape conditions, stochastic factors, herbivory, disturbance, or biotic interference, they may persist.

7.6.1 Similarity and Variation in Communities and Habitats

Increasing floristic similarity within narrowly defined habitat types would imply developing deterministic control of species composition and contradict alternative states. If vegetation is strongly linked to habitat factors, then there should be a strong statistical relationship between spatial community patterns and the environment. Under such conditions, plots grouped by species composition (CTs) should be similar to plots grouped by habitat factors (HTs; Table 7.2). Even weak associations would suggest that alternative states were transient. Variation of groups defined by either species composition (CTs) or habitats (HTs), when ordered by DCA, allows a comparison of the degree to which vegetation responds to habitat control. We compared the variation of DCA scores for each plot clustered by habitat and by floristics (Table 7.7).

Table 7.7 Standard deviation of DCA scores in CTs and HT in five studies. DCA variation (SD) of DCA-1 and DCA-2 scores (SD) are the mean of the groups.

Characteristic	Muddy Survey	Pumice Survey	Plains of Abraham	Pumice Grid	Potholes
CT DCA-1 Mean SD (%)	33.64	31.48	21.92	30.76	43.90
HT DCA-1 Mean SD (%)	50.6*	53.58*	28.57*	41.61*	40.01
CT DCA-2 Mean SD (%)	27.88	34.84	22.01	23.05	30.22
HT DCA-2 Mean SD (%)	34.61*	35.63	28.26*	27.98	28.17

*HT values are significantly more variable than the comparison CT score (t-test, $p < 0.05$).

7.6.1.1 Surveys

Habitat types of the Muddy River lahar were more cohesive at lower elevations, where succession had progressed to a greater degree than at higher elevations (Fig. 7.8). Species composition of HTs at lower elevation was moderately homogeneous (58–65% similarity), whereas at the upper HTs it was less homogeneous (under 40%). Overall, there was little homogeneity of plots grouped by 10 HTs (similarity = 50.1 ± 17.1%) compared to the seven CTs (similarity = 61.5 ± 12.0%). These metrics weakly supported the possibility that alternative communities may occur.

The HTs of the Muddy River lahar deposit were distributed widely and overlapped broadly in floristic space (Fig. 7.9). Variation of plots grouped by HTs is much greater than when grouped by CTs. These patterns imply that habitats can support multiple communities.

The floristic data of the Pumice Plain survey were sorted into six communities over the entire area and classified into 13 HTs. Despite their smaller number, the communities were more cohesive, with an average of 60 ± 15.5% similarity, compared to 35.7 ± 22.1% for the HTs. There was broad overlap among the HTs in the floristic DCA space (Fig. 7.10a), and the variation of the HTs was much greater than that of the CTs.

7.6.1.2 Grids

The Plains of Abraham grid is an ideal place to find a homogeneous habitat supporting one community. However, floristic similarity of plots in five CTs (65 ± 11.6%) was significantly higher than that of plots in the ten HTs (52.5 ± 12.6%). There was a weak tendency for HTs to group within portions of the DCA graph, but overlap remained large (Fig. 7.10b), suggesting that most HTs supported several different CTs. The DCA variation in both dimensions was significantly larger for the HTs than for CTs, confirming the floristic heterogeneity of HTs.

The habitat data from the Pumice Plain grid were classified into ten HTs and the vegetation data were sorted into six CTs. The similarity of the CTs (65.3 ± 11.6%) exceeded that of the HTs (47.7 ± 16.5%), with significantly lower variation, suggesting that the habitats were supporting multiple communities. The positions of the HTs overlapped broadly in DCA space and often occurred in widely separated portions of the DCA graph (Fig. 7.10c). Variation was similar on

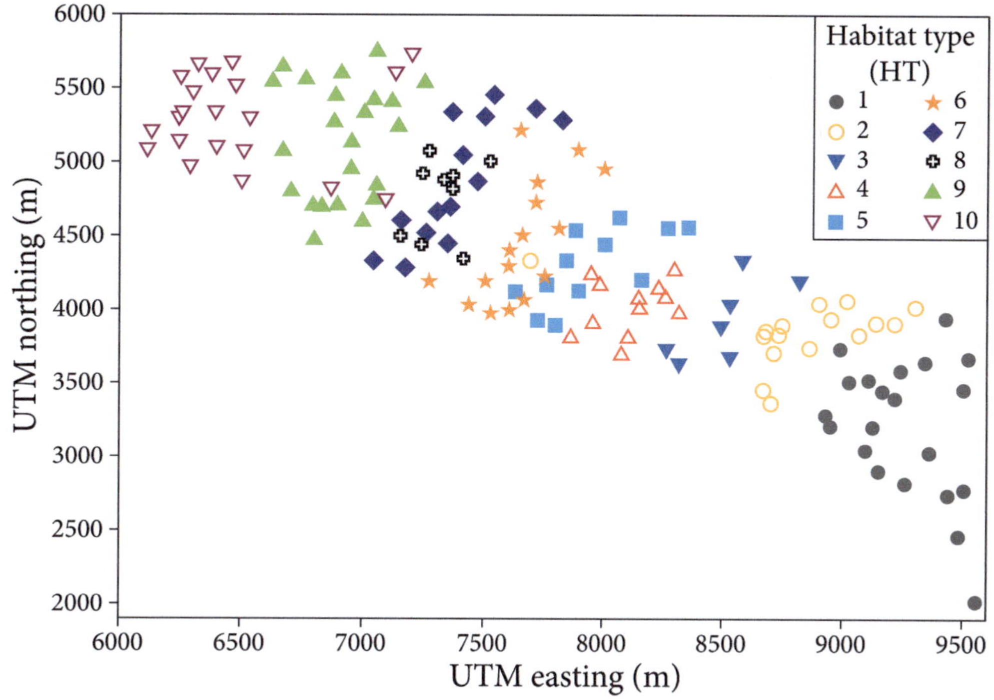

Fig. 7.8 Spatial distribution of habitat types (HT) on the Muddy River Lahar deposit.

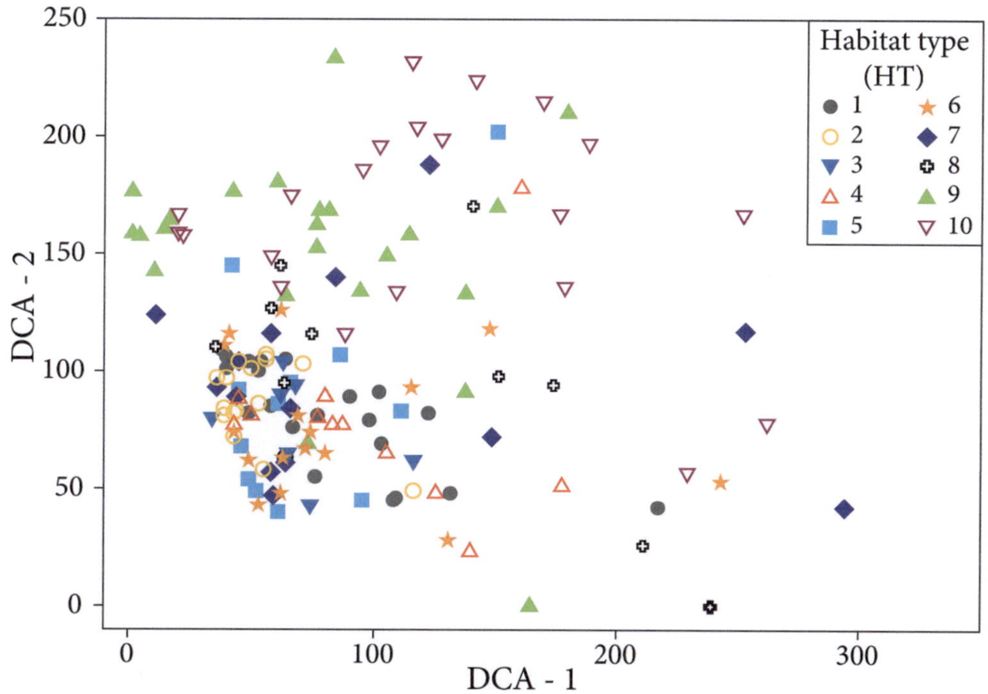

Fig. 7.9 Floristic distribution (DCA space) of habitat types (HT) on the Muddy River Lahar deposit.

Axis 2, but these data do not support a single-community-per-habitat hypothesis.

7.6.1.3 Potholes

The plot size of the 2008 pothole study was much smaller (4 m^2) than those of the other studies, and habitat measures included moisture and other soil factors measured in close proximity to the vegetation. Similarities in the five pothole CTs (62.4 ± 10.5%) and in the seven HTs (55.5 ± 13.1%) were not significantly different. The plots of the pothole study, based upon their CT membership, showed some overlap among the HTs (Fig. 7.10d), but each still ranged across much of the DCA diagram. Thus, the pothole study does not support the hypothesis of alternative CTs.

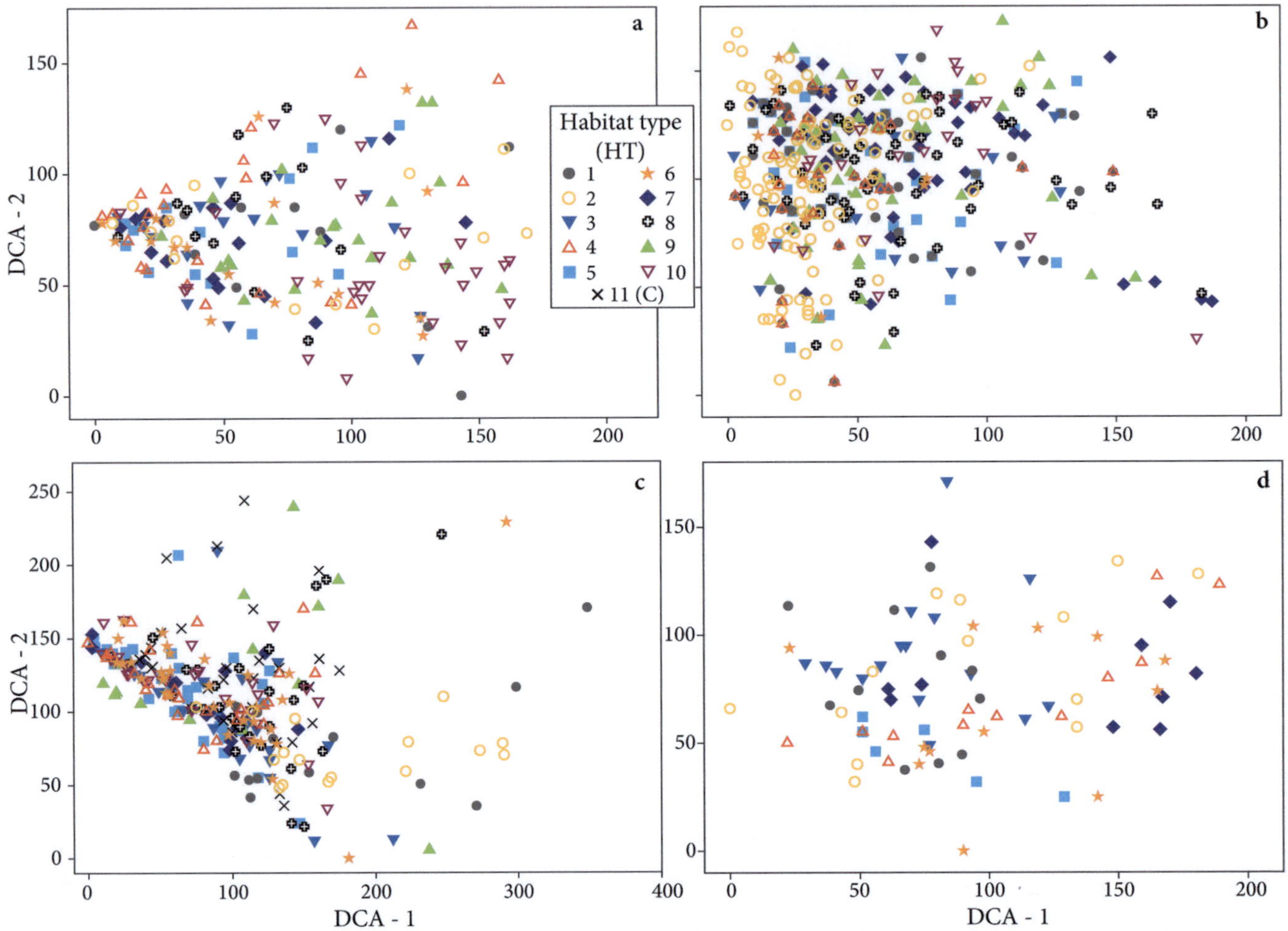

Fig. 7.10 Floristic distribution (DCA space) of habitat types (HT): (**a**) Pumice Plain survey (2004), (**b**) Plains of Abraham grid, (**c**) Pumice Plain grid, (**d**) Potholes.

7.6.1.4 Comments

The analyses of some of these sets of data weakly and provisionally support the hypothesis that homogeneous habitats can support different communities. However, analyses over a longer period may well lead to a rejection of this hypothesis. Although it is to be expected that groups based on species composition would be more homogeneous than those based on habitats, the disparities between different approaches to the question of alternative states does not exclude the possibility that similar habitats may sustain multiple community types.

7.6.2 Trajectory Patterns

If alternative states persist, species composition should not converge during succession (Walker and del Moral 2003). Because grids contain many plots in proximity, they offer a chance to observe local convergences. Two approaches, DCA and similarity changes, can be used to infer successional trajectory change. If trajectories converge in DCA space or in percent similarity, a strong effect of habitat on species composition is implied. If they ***diverge***, mechanisms associated with stochastic dispersal and contingent events are inferred.

7.6.2.1 Lahar Deposits

Variation in species composition measured by DCA on the lahar grids did not reveal ***convergence***. Some CTs converged, others diverged, and no pattern was discernible.

Similarity patterns of the lahar grids contrasted sharply. Using similarities of all plots suggested convergence on Lahar 1. When the plots were divided into three categories (adjacent to the woodland, 30–100 m, and more distant), convergence was marked. At 5-year intervals, adjacent plots increased from 48% to 53% similarity (not significant), intermediate plots from 53% to 70% ($p < 0.001$), and distant plots from 57% to 75% ($p < 0.005$). The effect appears to be due largely to the abundance of dominating conifers that masked differences in the ground layer.

Lahar 2 was divided into three elevation groups. None of these groups demonstrated similarity increases. The lack of a trend in Lahar 2 indicated that convergence was not occurring and implies that only dominant woody species can mediate convergence.

7.6.2.2 Plains of Abraham Grid

The Plains of Abraham grid occupies unremarkable, flat terrain, and yet, over duration of the study vegetation showed a complex, netlike pattern of community transitions that failed to suggest convergence (del Moral et al. 2010).

Five of ten CTs persisted. DCA scores of these CTs moved more than five half-changes along multiple trajectories with no convergence. The three common CTs were widely scattered in DCA space at the conclusion of the study. Any tendency to convergence would be evident in increasing similarity among grid plots. The grid was divided into four 100-plot blocks, and within-block similarity determined at 5-year intervals between 1990 and 2010. Similarity increased in each block only between 1990 (mean similarity = 36.1%; many low-richness plots) and 1995 (mean similarity = 57.3%). Thereafter, similarities lacked any trend. Thus, even in a homogeneous habitat, there was no convergence.

7.6.2.3 Pumice Plain Grid

Del Moral et al. (2012) studied DCA trajectories of CTs on the Pumice Plain grid. Community type trajectories changed significantly over time to similar degrees and at similar rates. Within-CT variation decreased over time in each case. When plots were clustered into eight spatial groups and five HTs, the variation of DCA scores over time declined significantly in each case. These results suggested that vegetation in a smaller area was becoming more homogeneous and might form a single community within a relatively homogeneous habitat. However, when DCA trajectories of groups based on space or habitats were compared, spatial groups from exposed sites diverged from those in protected sites. Among spatial groups in similar habitats, only 7 out of 28 (25%) comparisons showed convergence.

7.6.3 Alternative States May Exist

The present analysis does not exclude the existence of alternative states on primary landscapes of Mount St. Helens, but neither did we find evidence to contradict the single-state concept in local-scale studies. In large-scale surveys, there are some indications that alternative states *may* exist, but these remain inconclusive since key environmental factors may have gone unmeasured and environmental or biotic filters may have had insufficient time to shape vegetation to the environment. It is premature to address the question of alternative stable states. On the homogeneous Plains of Abraham grid, the CTs represent slight variations of a single community. On the more variable Pumice Plain grid, two CTs are well correlated to two HTs, and several environmental variables predicted differences among CTs. The Pothole study showed differences between CTs in several environmental variables, thus supporting the single-state hypothesis.

Although there is ample evidence in the literature that disturbance and contingent processes can produce alternative communities in similar habitats (see Sect. 7.6), the record at Mount St. Helens does not conclusively demonstrate developing alternative states. The lack of convergence to date may reflect variation in response to subtle but common characteristics of vegetation in the absence of a dominant woody canopy (cf. del Moral 1979; Dona and Galen 2006). The demonstration of a vegetation mosaic produced by factors unrelated to underlying habitats remains a challenge.

7.7 Discussion

7.7.1 Rates of Primary Succession

We evaluated factors affecting the rates of successional change. Rates generally decline with factors that restricted the growing season, including elevation, drought, and infertility. Proximity to donors accelerates early succession, but succession may be arrested when strong competitors establish quickly. Biotic effects, particularly from *Lupinus*, can drastically alter successional rates by stalling species development through competition, subsequently facilitating the development of biomass, or by changing invasion processes.

The type of initiating disturbance dictates initial responses. Lahar deposits consisted of relatively fertile pre-eruption material, so vegetation developed more rapidly than on the raw volcanic substrates. For example, interior parts of lahars developed more rapidly than did the Plains of Abraham. Subsequent erosion can reset succession and create mosaics of contrasting vegetation. Environmental stresses affect seedling establishment and biomass accumulation so low-productivity sites often demonstrate slow species accumulation (Grime and Pierce 2012).

No single criterion can fully capture successional-rate dynamics. Rates were best described by Euclidean distances between DCA scores, the PCA analysis of DCA, similarity matrix comparisons, and trajectory complexity (H′). The overall rate of change in most CTs did decelerate, but often fitfully, owing to idiosyncratic conditions (e.g., large population swings in *Lupinus*). Trajectory complexity depends upon subjective CTs so we advise caution in its use. Percent similarity of a matrix is robust, but can mask key internal dynamics. Detrended correspondence analysis and its derivatives are useful, but data with cyclic variations are poorly suited to this type of analysis.

7.7.2 Determinism

Temporal studies of the link between landscape and habitat factors are rare; seldom do they span decades (cf. Daniels et al. 2011). These studies of determinism during succession suggested principles that are applicable beyond the domain of primary succession.

Echoes of contingent events often reverberate to prevent convergence (Williamson et al. 2014). Unexplained variation dominated all of these studies, and although it may be appealing to assert that low correlations are a mere function of immature vegetation or inadequate data, tight coupling between vegetation and explanatory variables should not be expected.

Landscape studies show lower correlations between explanatory variables and species composition than do local studies, because species may be missing from suitable habitats owing to dispersal failures and because species interactions vary geographically. Among sources of unexplained variation are priority effects, the "ghosts of chaos past," to paraphrase Connell (1980). Local studies (see del Moral 1999b) improve chances to observe deterministic development because environmental factors are sampled in proximity to the sampled vegetation.

Dominant species can alter trajectories through biotic interactions, by resisting invasion or by altering microsites. Sparse seed rain and unpredictable seedling establishment enhanced variation (del Moral and Wood 1993), whereas intermittent and patchy facilitation from *Lupinus* promoted asynchronous development of fertility that generated both variable rates of succession and a vegetation mosaic (Wood and Morris 1990) that operates below the usual scale of habitat description, thus weakening correlations.

Despite many factors that impede strong deterministic ties between species and their environment, some degree of coupling can occur. Habitats that experience similar invasion histories, disturbance regimes, and isolation may develop similar trajectories, and it is reasonable to seek close ties between communities and habitats, the topic of Sect. 7.7.3.

The successful prediction of vegetation trajectories is useful to set restoration goals. The studies of determinism reported here can only suggest that links between the environment and species composition are tenuous because detailed habitat data could not be collected. Some useful principles were suggested by our results that support a view of vegetation dynamics steadfastly based on stochastic processes.

Unexplained variance in studies of plant-to-environment linkages will be largely due to contingent events, unmeasured variables, or suboptimal statistical models. Stochastic events, such as the season of major disturbances, and historical accidents, such as the paths taken by elk, can persist, so poor predictability should not be unexpected. Such variation should be embraced and recognized as important sources of ecosystem resilience. In addition, some species are more likely to influence succession than others. *Lupinus* has improved fertility to accelerate succession over longer time frames, but with temporally and spatially variable effects.

7.7.3 Alternative States

Seeking evidence for alternative states is problematic because those that may exist during early succession can be discounted by asserting that the scale of the vegetation and habitat sampling may be ill-matched or an important variable was omitted. Alternative states described here are mere suggestions, although no strong evidence demonstrated single communities in a homogeneous habitat. In most cases, only limited convergence has developed, offering at least the potential for alternative states to persist. It is likely that the large annual variation in vegetation reported in these long-term studies will decline, but that substantial unexplained variation will remain.

7.8 Conclusions

Rates, determinism, and alternative states in primary succession are each affected by random events. Dispersal from distant seed sources is intrinsically stochastic, so the suite of colonizers in a particular spot is probabilistic. Longer-lived colonists, which may require benign conditions to become established, produce consequences distinct from those produced by ephemeral species; the effects of facilitators differ from those of competitors. Over time, strong competition effects are likely to produce more homogeneous vegetation and stronger links between vegetation and habitats than will facilitators because the benefactors of facilitation are usually more variable than competitive dominants.

Successional rates are affected by the nature of the colonists and by landscape position, so they are strongly influenced by stochastic factors. Because rates at adjacent sites may differ as a result of the random colonization of different species rather than because of habitat conditions, the deterministic linkage between plants and their environment is tenuous. And because trajectories may diverge, alternative states may develop.

The story of primary plant succession on Mount St. Helens has only begun to unfold. There were many revelations during the first decades, and many more will certainly emerge. Among the many desirable future studies on Mount St. Helens would be a detailed analysis of factors that determine species composition on a local scale. Such studies would foster a better understanding of succession mechanisms and would have ample practical application, for example, in the design of restoration projects. A directed search for alternative states could also be initiated.

The study of vegetation development on Mount St. Helens initially followed a traditional approach, documenting the appearance of species in many habitats and chronicling the spread of vegetation. These studies led to new perspectives on the mechanics of primary succession that have gradually transformed the dominant view of succession to one that includes the importance of geographic factors, dispersal limitations, historical factors, and stochastic events.

Acknowledgments We thank the US National Science Foundation for funding (BSR8906544; DEB9406987, DEB0087040, DEB0541972, and OPUS grant DEB1118593) and the Mount St. Helens NVM for permission to investigate succession. Beth Brosseau, Lars Walker, and Fred Swanson made substantive comments to improve this paper, which is contribution No. 77 to the Mount St. Helens Succession Project. We dedicate this paper to the memory of David M. Wood, friend, colleague, and major contributor to the story of vegetation recovery on Mount St. Helens.

Glossary

Alternative states Two or more community types found in the same habitat.

Alternative stable states Two or more community types seeming to persist in the same habitat.

CCA Canonical correspondence analysis, constrained ordination based on DCA.

Colonization The process of arrival and establishment, both are selective of species.

Competition Negative impacts of one species on another due to use of limited resources.

Constrained ordination Method of describing species patterns in terms of environmental factors.

Convergence Occurs when two communities become increasingly similar as they mature.

DCA Detrended correspondence analysis, a method to summarize vegetation change and forming trajectories based on nonlinear assumptions.

Dispersal The process by which an organism or its reproductive units are transferred between habitats.

Divergent communities Communities that become increasingly distinct as they mature.

Facilitation Effects that promote establishment and growth of one species by another.

Euclidean distance, *d* A measure of relationship between two samples based on DCA scores: dij = square root [Σ(xik – xjk)2], i and j are values in two samples, and k = number of species over which the comparison is made.

Habitat complexity A measure of the number of CTs and their residence time calculated by the Shannon information statistic H′ = –Σ pi log pi; pi is the proportion of time occupied by each CT.

Habitat type Environmental unit based on available habitat values.

Lahar A slurry of mud and debris, here resulting from melting ice during an eruption.

Microsite Small-scale habitat (see ***safe-sites***).

Mosaic A patchwork of vegetation on the landscape.

Mycorrhizae Fungi that form mutualistic interactions with roots.

PCA Principal components analysis, a linear method to assess matrix variation.

Percent similarity (PS) A measure of relationship between two samples based on species cover: $PS_{ij} = 200\ \Sigma_{min}(x_{ik}, x_{jk})/\Sigma(x_{ik}+x_{jk})$, where min = minimum of two values xik, xjk, remaining terms as in Euclidean distance, *d*.

Permanent plots Marked sites that are repeatedly sampled.

Primary succession Ecosystem development on barren surfaces initially lacking in soil or biota.

Priority effects The consequences of establishment sequence that condition later compositional changes.

Pumice A silica-rich volcanic rock usually ejected during explosive eruptions.

Pyroclastic flow Rapidly descending current of superheated gas and tephra hugging the ground; deposits are often of very fine texture and easily eroded.

RDA Redundancy analysis, a constrained ordination based on PCA.

Relict site (vegetation) Sites protected by topography and snow that supported surviving vegetation.

Safe-site Any location that provides suitable conditions for establishment.

Stochastic Referring to a random, chance-driven process; opposed to deterministic.

Stresses Any factors that limit production.

Successional rate The change in species composition through time.

Tephrafall A rain of volcanic particles to the ground following ejection into the atmosphere by an explosive eruption. Tephra is a collective term for particles of any size, shape, or composition ejected in an explosive eruption.

Trajectory The temporal path traveled by vegetation communities, often determined by DCA.

Trajectory complexity A measure of community turnover measured by H'. See ***habitat complexity***.

References

Alvarez-Molina, L.L., M.L. Martinez, O. Perez-Maqueo, J.B. Gallego-Fernandez, and P. Flores. 2012. Richness, diversity, and rate of primary succession over 20 years in a tropical coastal dunes. *Plant Ecology* 213: 1597–1608.

Analytical Software. 2013. *Statistix 10*. Tallahassee: Analytical Software.

Anthelme, F., J.-C. Villaret, and J.-J. Brun. 2007. Shrub encroachment in the Alps gives rise to the convergence of sub-alpine communities on a regional scale. *Journal of Vegetation Science* 18: 355–362.

Baasch, A., A. Kirmer, and S. Tischew. 2012. Nine years of vegetation development in a post-mining site: Effects of spontaneous and assisted site recovery. *Journal of Applied Ecology* 49: 251–260.

Baeten, L., M. Hermy, S. Van Dael, and K. Verhyen. 2010. Unexpected understorey community development after 30 years in ancient and post-agricultural forests. *Journal of Ecology* 98: 1447–1453.

Baniya, C.B., T. Solhoy, and O.R. Vetaas. 2009. Temporal changes in species diversity and composition in abandoned fields in a trans-Himalayan landscape, Nepal. *Plant Ecology* 201: 383–399.

Beisner, B., D. Haydon, and K. Cuddington. 2003. Alternative states in ecology. *Frontiers in Ecology* 1: 376–382.

Bischoff, A., G. Warthemann, and S. Klotz. 2009. Succession of floodplain grasslands following reduction in land use intensity: The importance of environmental conditions, management, and dispersal. *Journal of Applied Ecology* 46: 241–249.

Bishop, J.G. 2002. Early primary succession on Mount St. Helens: Impact of insect herbivores on colonizing lupines. *Ecology* 83: 191–202.

Bishop, J.G., W.F. Fagan, J.D. Schade, and C.M. Crisafulli. 2005. Causes and consequences of herbivory on prairie lupine (*Lupinus lepidus*) in early primary succession. In *Ecological responses to the 1980 eruption of Mount St. Helens*, ed. V.H. Dale, F.J. Swanson, and C.M. Crisafulli, 151–162. New York: Springer.

Bossuyt, B., O. Honnay, and M. Hermy. 2005. Evidence for community assembly constraints during succession in dune slack plant communities. *Plant Ecology* 178: 201–209.

Boyes, L.J., R.M. Gunton, M.E. Griffiths, and M.J. Lawes. 2011. Causes of arrested succession in coastal dune forest. *Plant Ecology* 212: 21–32.

Brownstein, G., J.B. Steel, S. Porter, A. Gray, C. Wilson, P.G. Wilson, and J.B. Wilson. 2012. Chance in plant communities: A new approach to its measurement using the nugget from spatial autocorrelation. *Journal of Ecology* 100: 987–996.

Burt, J.W., and J.J. Clary. 2016. Initial disturbance intensity affects recovery rates and successional divergence on abandoned ski slopes. *Journal of Applied Ecology* 53: 607–615.

Butterfield, B.J., and R.M. Callaway. 2013. A functional comparative approach to facilitation and its context dependence. *Functional Ecology* 27: 907–917.

Chaideftou, E., A.S. Kallimanis, E. Bergmeier, and P. Dimopoulos. 2012. How does plant species composition change from year to year? A cases study from the herbaceous layer of a sub-Mediterranean oak woodland. *Community Ecology* 13: 88–96.

Chapman, J.L., and R.W. McEwan. 2013. Spatio-temporal dynamics of alpha- and beta-diversity across topographic gradients in the herbaceous layer of an old-growth deciduous forest. *Oikos* 122: 1679–1686.

Clements, F.E. 1936. Nature and structure of the climax. *Journal of Ecology* 24: 252–284.

Connell, J.H. 1980. Diversity and the coevolution of competitors, or the ghost of competition past. *Oikos* 35: 131–138.

Daniels, F.J.A., J.G. de Molenaar, M. Chytrý, and L. Tichy. 2011. Vegetation change in Southeast Greenland: Tasiilaq revisited after 40 years. *Applied Vegetation Science* 14: 230–241.

del Moral, R. 1979. High elevation vegetation of the Enchantment Lakes Basin. *Canadian Journal of Botany* 57: 1111–1130.

———. 1998. Early succession on lahars spawned by Mount St. Helens. *American Journal of Botany* 85: 820–828.

———. 1999a. Plant succession on pumice at Mount St. Helens, Washington. *American Midland Naturalist* 141: 101–114.

———. 1999b. Predictability of primary successional wetlands on pumice, Mount St. Helens. *Madroño* 46: 177–186.

———. 2007. Vegetation dynamics in space and time: an example from Mount St. Helens. *Journal of Vegetation Science* 18: 479–488.

———. 2009. Increasing deterministic control of primary succession on Mount St. Helens, Washington. *Journal of Vegetation Science* 20: 1145–1154.

———. 2010. Thirty years of permanent vegetation plots, Mount St. Helens, Washington. *Ecology* 91: 2185.

del Moral, R., and C.C. Chang. 2015. Multiple assessments of succession rates on Mount St. Helens. *Plant Ecology* 216: 165–176.

del Moral, R., and A.J. Eckert. 2005. Colonization of volcanic deserts from productive patches. *American Journal of Botany* 92: 27–36.

del Moral, R., and E.E. Ellis. 2004. Gradients in heterogeneity and structure on lahars, Mount St. Helens, Washington, USA. *Plant Ecology* 175: 273–286.

del Moral, R., and C.C. Jones. 2002. Early spatial development of vegetation on pumice at Mount St. Helens. *Plant Ecology* 162: 9–22.

del Moral, R., and I.L. Lacher. 2005. Vegetation patterns 25 years after the eruption of Mount St. Helens, Washington. *American Journal of Botany* 92: 1948–1956.

del Moral, R., and B. Magnússon. 2014. Surtsey and Mount St. Helens: A comparison of early succession rates. *Biogeosciences* 11: 2099–2111.

del Moral, R., and L.R. Rozzell. 2005. Effects of lupines on community structure and species association. *Plant Ecology* 180: 203–215.

del Moral, R., and D.M. Wood. 1988. Dynamics of herbaceous vegetation recovery on Mount St. Helens, Washington, USA, after a volcanic eruption. *Vegetatio* 47: 11–27.

del Moral, R., and D.M. Wood. 1993. Understanding dynamics of early succession on Mount St. Helens. *Journal of Vegetation Science* 4: 223–234.

del Moral, R., and D.M. Wood. 2012. Vegetation development on permanently established grids, Mount St. Helens (1986–2010). *Ecology* 93: 2125.

del Moral, R., J.H. Titus, and A.M. Cook. 1995. Early primary succession on Mount St. Helens, Washington, USA. *Journal of Vegetation Science* 6: 107–120.

del Moral, R., D.M. Wood, and J.H. Titus. 2005. How landscape factors affect recovery of vegetation on barren surfaces. In *Ecological responses to the 1980 eruption of Mount St. Helens*, ed. V.H. Dale, F.J. Swanson, and C.M. Crisafulli, 93–109. New York: Springer.

del Moral, R., L.R. Walker, and J.P. Bakker. 2007. Insights gained from succession for the restoration of landscape structure and function. In *Linking restoration and succession in theory and in practice*, ed. L.R. Walker, J. Walker, and R.J. Hobbs, 19–44. New York: Springer.

del Moral, R., J.E. Sandler, and C.P. Muerdter. 2009. Spatial factors affect primary succession on the Muddy River Lahar, Mount St. Helens, Washington. *Plant Ecology* 202: 177–190.

del Moral, R., J.M. Saura, and J.N. Emenegger. 2010. Primary succession trajectories on a barren plain, Mount St. Helens, Washington. *Journal of Vegetation Science* 21: 857–867.

del Moral, R., L.A. Thomason, A.C. Wenke, N. Lozanoff, and M.D. Abata. 2012. Primary succession trajectories on pumice at Mount St. Helens, Washington. *Journal of Vegetation Science* 23: 73–85.

Dona, A.J., and C. Galen. 2006. Sources of spatial and temporal heterogeneity in the colonization of an alpine krummholz environment by the weedy subalpine plant *Chamerion angustifolium* (fireweed). *Canadian Journal of Botany* 84: 933–939.

Efford, J.T., B.D. Clarkson, and R.J. Bylsma. 2014. Persistent effects of a tephra eruption (AD 1655) on tree line composition and structure, Mt Taranaki, New Zealand. *New Zealand Journal of Botany* 52: 245–261.

Favier, C., J. Aleman, L. Bremond, M.A. Dubois, V. Freycon, and J.M. Yangakola. 2012. Abrupt shifts in African savanna tree cover along a climatic gradient. *Global Ecology and Biogeography* 21: 787–797.

Freeman, J.E., and L.N. Kobziar. 2011. Tracking post-fire successional trajectories in a plant community adapted to high-severity fire. *Ecological Applications* 21: 61–74.

Fuller, R.N., and R. del Moral. 2003. The role of refugia and dispersal in primary succession on Mount St. Helens, Washington. *Journal of Vegetation Science* 14: 637–644.

Gerla, D.J., and W.M. Mooij. 2014. Alternative stable states and alternative end states of community assembly through intra- and interspecific positive and negative interactions. *Theoretical Population Biology* 96: 8–18.

Granzow-de la Cerda, I.G. Arellano, M. Brugeus, and A. Sola-Lopez. 2016. The role of distance and habitat specificity in bryophyte and perennial seed plant metacommunities in arid scrubland fragments. *Journal of Vegetation Science* 27: 414–426.

Grime, J.P., and S. Pierce. 2012. *The evolutionary strategies that shape ecosystems*. Chichester: Wiley/Blackwell.

Halpern, C.B., and M.E. Harmon. 1983. Early plant succession on the Muddy River mudflow, Mount St. Helens, Washington. *American Midland Naturalist* 110: 97–106.

Kamijo, T., K. Kitayama, A. Sugawara, S. Urushimichi, and K. Sasai. 2002. Warm temperate forest on volcano lava and ejecta. *Folia Geobotanica* 37: 71–91.

Kardol, P., D.E. Todd, P.J. Hanson, and J. Mulholland. 2010. Long-term successional forest dynamics: Species and community responses to climatic variability. *Journal of Vegetation Science* 21: 627–642.

Kayes, L.J., P.D. Anderson, and K.J. Puettmann. 2010. Vegetation succession among and within structural layers following wildfire in managed forests. *Journal of Vegetation Science* 21: 233–247.

Kefi, S., M. Holmgren, and M. Scheffer. 2016. When can positive interactions cause alternative stable states in ecosystems? *Functional Ecology* 30: 88–97.

Kovach, W.L. 1999. *MVSP, a multivariate statistics package for Windows, version 3.0*. Pentraeth: Kovach Computer Consulting Services.

Laborde, J., and K. Thompson. 2013. Colonization of limestone grasslands by woody plants: The role of seed limitation and herbivory by vertebrates. *Journal of Vegetation Science* 24: 307–319.

Lindig-Cisneros, R., S.L. Galindo-Vallejo, and S. Lara-Cabrera. 2006. Vegetation of tephra deposits 50 years after the end of the eruption of the Paricutin volcano, Mexico. *Southwestern Naturalist* 51: 455–461.

Long, W.X., X.B. Yang, and D.H. Li. 2012. Patterns of species diversity and soil nutrients along a chronosequence of vegetation recovery in Hainan Island, South China. *Ecological Restoration* 27: 561–568.

Magnússon, B., S.H. Magnússon, and S. Fridriksson. 2009. Development in plant colonization and succession on Surtsey during 1999–2008. *Surtsey Research* 12: 57–76.

Marteinsdottir, B., K. Svavarsdóttir, and T.E. Thorhallsdottir. 2010. Development of vegetation patterns in early primary succession. *Journal of Vegetation Science* 21: 531–540.

Matthews, J.W., and A.G. Endress. 2010. Rate of succession in restored wetlands and the role of site context. *Applied Vegetation Science* 13: 346–355.

McCune, B., and T.F.H. Allen. 1985. Will similar forests develop on similar sites? *Canadian Journal of Botany* 63: 367–376.

McCune, B., and D. Keon. 2002. Equations for potential annual direct incident radiation and heat load. *Journal of Vegetation Science* 13: 603–606.

McCune, B., and M.J. Mefford. 2006. *PC-ORD, multivariate analysis of ecological data, version 5.0*. Gleneden Beach: MjM Software Design.

Økland, R. 1999. On the variation explained by ordination and constrained ordination axes. *Journal of Vegetation Science* 10: 131–136.

Platt, W.J., and J.H. Connell. 2003. Natural disturbances and directional replacement of species. *Ecological Monographs* 73: 507–522.

Prach, K., P. Pyšek, and V. Jarošik. 2007. Climate and pH as determinants of vegetation succession in Central European man-made habitats. *Journal of Vegetation Science* 18: 701–710.

Rebele, F. 2013. Differential succession towards woodland along a nutrient gradient. *Applied Vegetation Science* 16: 365–378.

Saccone, P., T. Pyykkonen, A. Eskelinen, and R. Virtanen. 2014. Environmental perturbation, grazing pressure, and soil wetness jointly drive mountain tundra toward divergent alternative states. *Journal of Ecology* 102: 1661–1672.

Schröder, A., L. Persson, and A.M. De Roos. 2005. Direct experimental evidence for alternative stable states: A review. *Oikos* 110: 3–19.

Scott, A.J., and J.W. Morgan. 2012. Early life-history stages drive community reassembly in Australian old-fields. *Journal of Vegetation Science* 23: 721–731.

Sparrius, L.B., A.M. Kooijman, M.P.J.M. Riksen, and J. Sevink. 2013. Effect of geomorphology and nitrogen deposition on rate of vegetation succession in inland drift sands. *Applied Vegetation Science* 16: 379–389.

Suding, K.N., and R.J. Hobbs. 2009. Threshold models in restoration and conservation: A developing framework. *Trends in Ecology and Evolution* 24: 271–279.

Swanson, F.J., and J.J. Major. 2005. Physical events, environments, and geological-ecological interactions at Mount St. Helens: March 1980–2004. In *Ecological responses to the 1980 eruption of Mount St. Helens*, ed. V.H. Dale, F.J. Swanson, and C.M. Crisafulli, 27–44. New York: Springer.

ter Braak, C.J.F., and P. Šmilauer. 2007. CANOCO—A FORTRAN program for canonical community ordination (version 4.5). Wageningen, The Netherlands: DLO-Agricultural Mathematics Group.

Titus, J.H., and J.G. Bishop. 2014. Propagule limitation and competition with nitrogen fixers limit conifer colonization during primary succession. *Journal of Vegetation Science* 25: 990–1003.

Titus, J.H., and R. del Moral. 1998. The role of mycorrhizal fungi and microsites in primary succession on Mount St. Helens. *American Journal of Botany* 85: 370–375.

Titus, J.H., S. Whitcomb, and H.J. Pioniak. 2007. Distribution of arbuscular mycorrhizae in relation to microsites on primary successional substrates on Mount St. Helens. *Canadian Journal of Botany* 85: 941–948.

Tsuyuzaki, S., J.H. Titus, and R. del Moral. 1997. Seedling establishment patterns on the Pumice Plain, Mount St. Helens, Washington. *Journal of Vegetation Science* 8: 727–734.

Turner, M.G., W.L. Baker, C.J. Peterson, and R.K. Peet. 1998. Factors influencing succession: Lessons from large, infrequent natural disturbances. *Ecosystems* 1: 511–523.

Ujházy, K., J. Fanta, and K. Prach. 2011. Two centuries of vegetation succession in an inland sand dune area, central Netherlands. *Applied Vegetation Science* 14: 316–325.

Walker, L.R., and R. del Moral. 2003. *Primary succession and ecosystem rehabilitation*. Cambridge: Cambridge University Press.

Walker, L.R., B.D. Clarkson, W.B. Silvester, and B.R. Clarkson. 2003. Colonization dynamics and facilitative impacts of a nitrogen-fixing

shrub in primary succession. *Journal of Vegetation Science* 14: 277–290.
Walker, L.R., P.F. Bellingham, and D.A. Peltzer. 2006. Plant characteristics are poor predictors of microsite colonization during the first two years of primary succession. *Journal of Vegetation Science* 17: 397–406.
Walker, L.R., N. Hölzel, R. Marrs, R. del Moral, and K. Prach. 2014. Optimization of intervention levels in ecological restoration. *Applied Vegetation Science* 17: 187–192.
Whittaker, R.H. 1967. Gradient analysis of vegetation. *Biological Reviews* 42: 207–264.
Williamson, G.B., T.V. Bentos, J. Longworth, and R.C.G. Mesquita. 2014. Convergence and divergence in alternative successional pathways in Central Amazonia. *Plant Ecology & Diversity* 7: 341–348.
Wood, D.M., and R. del Moral. 1987. Mechanisms of early primary succession in subalpine habitats on Mount St. Helens. *Ecology* 68: 780–790.
———. 2000. Seed rain during early primary succession on Mount St. Helens, Washington. *Madroño* 47: 1–9.
Wood, D.M., and W.F. Morris. 1990. Ecological constrains to seedling establishment on the Pumice Plain, Mount St. Helens, Washington. *American Journal of Botany* 77: 1411–1418.

Plant Succession on the Mount St. Helens Debris-Avalanche Deposit and the Role of Non-native Species

8

Virginia H. Dale and Elsie M. Denton

8.1 Introduction

The debris-avalanche deposit is one of the most disturbed areas created by the 1980 eruption of Mount St. Helens (MSH), with little survival of a few plant fragments and primary succession mostly being initiated by the seeds dispersed onto the newly emplaced material. Vegetation establishment on the debris-avalanche deposit was affected by the nearly complete loss of organic matter, the nature of the deposit, local climate conditions, surviving plant life in adjacent areas, and patterns of plant colonization and growth. There was great concern about the potential for erosion of the newly emplaced volcanic material; US$2 million was spent to aerially distribute seeds of eight species (only one of which was native) over a large part of the disturbed area (Dale et al. 2005c). Seeding of the western but not the eastern portion of the deposit set up conditions for examining effects of such human interventions on plant establishment and succession. This chapter analyzes vegetation changes on the debris-avalanche deposit during the first 30 years post eruption, considers the role of non-native species, and discusses potential future vegetation patterns on the deposit. In particular, we compare measures of plant cover and species richness in permanent plots that were seeded in 1980 with plots that were not seeded to determine the 30-year effects of the seeding of mostly non-native plants on patterns on vegetation establishment.

V.H. Dale (✉)
Department of Ecology & Evolutionary Biology, The University of Tennessee, Knoxville, TN 37996, USA
e-mail: vdale@utk.edu

E.M. Denton
U.S. Department of Agriculture, Agricultural Research Service, Eastern Oregon Agricultural Research Center, 67826 OR-205-A, Burns, OR 97720, USA

8.1.1 Creation of the Debris-Avalanche Deposit

The 18 May 1980 eruption of MSH was immediately preceded by the largest debris avalanche in recorded history (Fig. 8.1). A debris avalanche occurs when an unstable slope collapses and soil, rocks, and other debris are transported downslope. Debris avalanches typically move rapidly and are common on steep volcanoes. The MSH debris avalanche was triggered by a magnitude 5.1 earthquake and followed by a laterally directed pyroclastic density current, commonly referred to as a lateral blast (hereafter, the blast PDC). About 2.8 billion m^3 of material rushed down the north side of MSH and spread over a 60-km^2 area. Later that day, earthquake-induced liquefaction of the debris deposit spawned a massive lahar that flowed down the North Fork Toutle and Toutle Rivers to the Cowlitz River (Fairchild 1985). The lahar both eroded and deposited material on the surface of the debris-avalanche deposit, with erosion exceeding deposition by about 4 million m^3 (Fairchild 1985).

Some of the debris-avalanche material was deposited in the Spirit Lake basin and the South Fork Coldwater Creek watershed, but the largest portion moved quickly down 25 km of the North Toutle River valley (Voight et al. 1981), which is the focus of this chapter. As the debris avalanche traveled, it passed through areas where the forest had already been leveled by the blast PDC. The debris-avalanche material varied in temperature, for it included blocks of glacial ice and hot rocks from the magma that had risen up into the volcano prior to the eruption. Near its end, the deposit moved through forested areas that were not otherwise affected by the eruption, toppling and burying the forest in its path.

The new environment created by the debris-avalanche deposit was mostly barren of any living organisms, viable seeds, or organic matter. Emplacements near the valley walls retained some clumps of organic material, soil, and surviving plants. Hence, the debris-avalanche deposit provided an opportunity to examine primary succession.

C.M. Crisafulli, V.H. Dale (eds.), *Ecological Responses at Mount St. Helens: Revisited 35 years after the 1980 Eruption*,
https://doi.org/10.1007/978-1-4939-7451-1_8

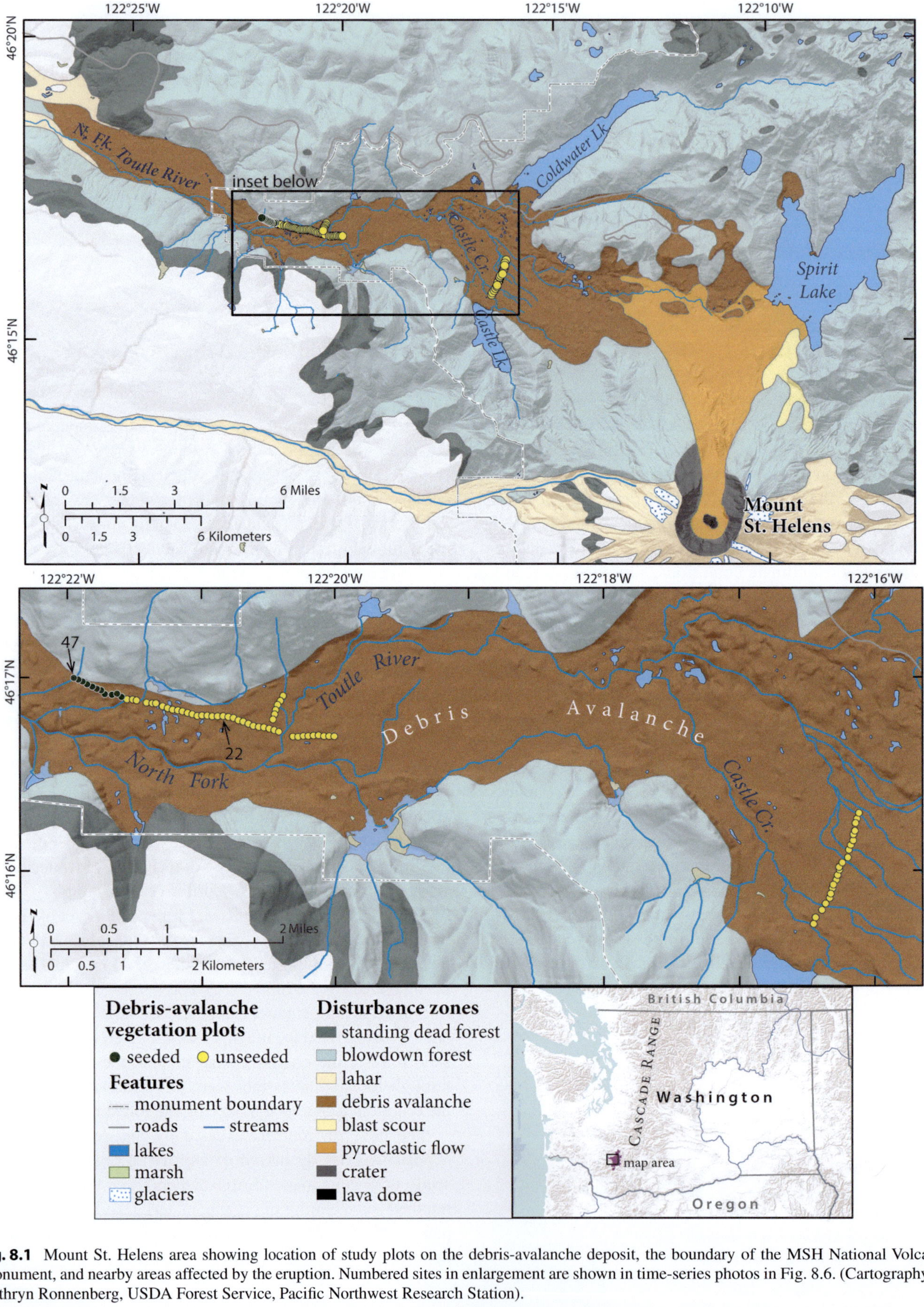

Fig. 8.1 Mount St. Helens area showing location of study plots on the debris-avalanche deposit, the boundary of the MSH National Volcanic Monument, and nearby areas affected by the eruption. Numbered sites in enlargement are shown in time-series photos in Fig. 8.6. (Cartography by Kathryn Ronnenberg, USDA Forest Service, Pacific Northwest Research Station).

8.1.2 Initial Physical and Chemical Conditions of the MSH Debris-Avalanche Deposit

The MSH debris avalanche resulted in a massive deposit composed of discrete blocks from the prior mountain in a mixture of blast-homogenized material (Glicken 1998).The heterogeneous topography of the deposit averages 45 m thick, with a maximum of 195 m (Voight et al. 1981). The terrain is relatively level where the lahar inundated the distal end of the deposit. In other locations, large mounds rise up to 50 m above the deposit surface. These hummocks are derived from lithic blocks that originated in the former interior of the volcano (Voight et al. 1981; Fairchild 1985; Glicken 1998). Many depressions formed when the deposit settled or as a result of the subsequent melting of embedded blocks of ice, creating new ponds and wetlands as these basins filled with water.

The landslide was hot, with an emplacement temperature of 100 °C in some locations (Voight et al. 1981). Ten to 12 days after the eruption, temperature varied from 68 to 98 °C measured at 1.0–1.5 m depth, with cooler temperatures occurring at a greater distance from the volcano (Banks and Hoblitt 1981).

Once it cooled, conditions on the deposit were adequate but not ideal for plant establishment and growth (Adams et al. 1986). The deposit material was poorly sorted and dominated by sand (63% by weight of the <2-mm fraction). Rocks, including all material >2 mm, constituted a variable component (36.4 ± 28.9% by weight) (Adams and Dale 1987). The sandy texture of the deposit limited moisture retention, potentially causing stressful conditions for seedling establishment and plant growth. Nitrogen levels and electrical conductivity were initially low (703 ppm NH_3 and 0.84 ± 0.71 mmho cm^{-1}, respectively). The substrate had low levels of organic matter (0.31% weight loss on ignition), a low carbon-to-nitrogen ratio (4.3:1), and was acidic, with an initial pH of 4.8 ± 0.5. For comparison, C:N ratio of forest soils in the Pacific Northwest typically ranges from 21:1 to 31:1 (Sun et al. 2004) and soils are acidic. The debris-avalanche deposit substrates had low moisture-holding capacity—saturation was 5.3 ± 0.9% per 15 atmospheres moisture (atm), where 15 atm is the water content at the wilting point and represents the point at which plant roots cannot draw any more water from the void space. Adams and Dale (1987) used lettuce (*Lactuca sativa*) to bioassay the debris-avalanche material. Seedlings were planted and well-watered under greenhouse conditions; they survived but grew little even in response to added nutrients. The limited plant growth may have been due either to soil chemistry (e.g., excessive salt) or to structural properties of the debris material (e.g., texture).

Both massive erosion and deposition have occurred on the debris-avalanche deposit since the 1980 eruption. Fluvial erosion created new channels that have grown wider and deeper over time, with channels and gullies that have steep walls (30–70°), variable incision depths (3–50 m), and widths of 3–120 m (Lehre et al. 1983; Waitt et al. 1983; Major et al. 2000). Twenty years after the eruption, erosion remained a major factor, with annual suspended sediment yield from the debris-avalanche deposit 100 times (10^4 Mg km^{-2}) typical background levels as compared to similar undisturbed rivers in the region (~10^2 Mg km^{-2}) (Major et al. 2000). Increased peak flows occurred in the autumn and winter for the first 10 years after the eruption (Major and Mark 2006).

8.1.3 Plant Propagule Survival on the Debris-Avalanche Deposit

The debris avalanche and the blast PDC removed the coniferous forests and riparian vegetation that existed in the North Fork Toutle River valley floor prior to the eruption (Adams and Adams 1982; Fairchild 1985). Adams and Dale (1987) found no seeds in assayed debris-avalanche material samples, and no seedlings germinated from test flats of deposit material. During June 1980, a search for plants was conducted across the deposit, which was found to be largely barren. No seedlings and only a few vegetatively propagating plants were found. Individual plants of 20 species survived by growing from roots or stems that had been transported in the debris-avalanche deposit and came to rest near the surface (Adams et al. 1986; Dale 1986). Most of these scattered plants were fireweed (*Chamerion angustifolium* ssp. *angustifolium*), Canada thistle (*Cirsium arvense*), and broadleaf lupine (*Lupinus latifolius*). No woody plants survived on the central portion of the debris-avalanche deposit.

Initially, there was abundant vegetation and organic debris at the terminus of the deposit, as the avalanche had scoured and transported many plants down the valley (Glicken 1998) (Fig. 8.2). However, this pile of living and dead vegetation did not influence plant reestablishment, because the Army

Fig. 8.2 As seen in summer 1980, the material at the terminus of the debris-avalanche deposit that was subsequently removed. Photo courtesy of V.H. Dale.

Corps of Engineers removed it during construction of a sediment-retention structure downstream of our study sites.

8.1.4 Patterns of Vegetation Establishment on the Debris-Avalanche Deposit

The colonists with the greatest contribution to plant cover were early-successional, wind-dispersed species that survived on the adjacent blowdown-zone hillslopes and nearby clearcuts outside the blast-PDC area (Fig. 8.1, Dale 1986). In the first 5 years after the eruption, average plant cover on the deposit was less than 1% (Dale 1986) but averaged greater than 65% by 20 years post eruption (Dale et al. 2005a). The pattern of vegetation establishment was spatially variable as related to heterogeneity of the deposit. Wetlands were areas of high plant species richness and cover, and locations near streams were susceptible to erosion, which prevented long-term plant establishment. Red alder, *Alnus rubra*—a fast-growing, early-maturing, and nitrogen-fixing tree—established quickly, had the second highest cover of any plant species by 20 years post eruption, and is predicted to be important for the next several decades (Dale et al. 2005a).

Most seeds dispersed to the deposit were transported by wind (Dale 1989). Of those seeds caught by seed traps set at 50-m intervals along the two transects on the deposit in 1981, 1982, 1983, and 1994 (1, 2, 3, and 14 years after the eruption), the number of seeds was greatest in the second year (1982), possibly because that was an unusually wet summer (Dale et al. 2005a). However, 1983 was even wetter than 1982, but fewer seeds were trapped. Heavy lupine seeds appeared in the traps for the first time in 1994. These seeds could not have been transported by the wind and likely came from plants near the traps on the deposit, evidence of the growing importance of in situ reproduction (Dale et al. 2005a).

8.1.5 Non-native Species Introduced by Natural Dispersal and by Seeding

Most of the arriving windblown seeds originated from areas adjacent to the debris-avalanche deposit that were less disturbed by the eruption. Those areas were dominated by private industrial forests, many of which had been recently cut, and those clearcuts supported an abundance of early-successional species having light, plumed seeds. In 1981, 87% of seeds trapped were native fireweed, but seeds of the three other captured species were not native: Canada thistle, field sowthistle (*Sonchus arvensis*), and woodland ragwort (*Senecio sylvaticus*) (Dale 1989). Fireweed continued to be the most common seed trapped 2, 3, and 14 years post eruption, likely because of the abundance of that species on clearcuts near the deposit and the high vagility of the seeds (Dale et al. 2005a).

In forests of the Pacific Northwest, non-native species are uncommon (Ares et al. 2009). However, 57 non-native species are considered a problem in forested ecosystems of the region (Gray et al. 2011), and over time, seeds of most of these species have dispersed onto the debris-avalanche deposit and established. Following the 1980 eruption, there was extreme apprehension about the potential effects of massive erosion of the debris-avalanche deposit and other disturbed areas, and the United States government made funds available to address these concerns. The United States Soil Conservation Service (SCS) planned to aerially seed non-native species over most of the area disturbed during the eruption in order to stabilize the recently emplaced volcanic substrates and to promote plant recovery (Dale et al. 2005c). Because many scientists objected to this management activity, which was considered a threat to research opportunities in this natural laboratory (Dale et al. 2005c), the proposed area to be seeded was reduced from 1640 to 320 km^2 in the vicinity of Mount St. Helens (Stroh and Oyler 1981). The selected species included those available from existing seed collections and were characterized by rapid germination, tolerance to a variety of environmental conditions, suitability as food or cover for wildlife, and ability to fix nitrogen via association with bacteria (Stroh and Oyler 1981). The type of volcanic disturbance and elevation determined species mixes used in specific locations across the landscape.

In the autumn of 1980, the SCS distributed seed by helicopter over about 24 km^2 on the westernmost portion of the debris-avalanche deposit and on the lahar deposit that extends downstream to the west (Figs. 8.1 and 8.3). The first author was present on the debris-avalanche deposit during the seed application and documented where the seeds fell relative to established permanent plots. The SCS used a mixture of herbaceous plants on the debris-avalanche deposit that included legumes (*Trifolium repens*, *Lotus corniculatus*, and *Melilotus officinalis*) and grasses (*Lolium perenne*, *Dactylis glomerata*, *Schedonorus arundinaceus*, *Agrostis capillaris*, and *Festuca rubra*) (Stroh and Oyler 1981). Of these species, only red fescue (*F. rubra)* is native. By September 1981, plant cover of the seeded area on the lower debris avalanche created a stark contrast with unseeded areas (Fig. 8.4) (Dale 1991). Nevertheless, large-scale erosion still occurred (Major et al. 2000).

Because the westernmost, downstream part of the debris-avalanche deposit was seeded in 1980 and the eastern part was not, the area offered an opportunity to compare successional processes occurring with and without introduction of non-native seeds. Eleven of 62 permanent plots remaining in 2010 fell in the 1980 seeded area. By

Fig. 8.3 Seeds being distributed by helicopter onto the debris-avalanche deposit in 1980. Photo courtesy of V.H. Dale.

Fig. 8.4 The initial flush of vegetation, as seen in September 1981, that grew from the non-native seeds distributed onto the debris-avalanche deposit. The abrupt line between the yellow vegetation and the gray, barren deposit marks the distinct edge of the seeded area. Photo courtesy of V.H. Dale.

1989, the seeded plots differed from those that had not received seed mix; these plots had greater cover and plant species richness but less native plant richness and greater mortality of conifer trees than the unseeded areas (Dale 1991). By 2000, 20 years post eruption, the seeded plots continued to have greater cover; overall plant richness, including non-native species, was also higher (Dale and Adams 2003). Presumably, the plant communities across our entire study area will eventually converge toward a mature coniferous forest typical of the Pacific Northwest (Franklin and Dyrness 1973), if they are not reset once more by a future eruption.

This chapter examines 30 years of vegetation data from the debris-avalanche deposit to determine if the effects of non-native seeding were still present in 2010 and whether there is evidence that these effects might be fading with time. The analysis builds upon the knowledge described above and data described below.

8.2 Methods

8.2.1 Methods of Monitoring Plant Establishment

To monitor vegetation establishment, a network of permanent plots was established in 1981 and 1982. The vegetation in these permanent plots was measured during eight summers over a 30-year period: 1981–1983, 1989, 1994, 2000, 2004, and 2010. The 250-m^2 circular plots were placed at 50-m intervals along transects between Castle and Coldwater Lakes and down the western extent of the deposit (Fig. 8.1). The plots were located on the central portion of the debris-avalanche deposit, not along the marginal facies where the debris avalanche shoved surviving plants to the edge of the valley. The initial 101 plots represented the variety of geologic and topographic conditions on the debris-avalanche deposit and distances from surviving vegetation in the adjacent landscapes. However, during the first two decades after the eruption, 39 of the unseeded plots were lost to erosion and lahars or could not be located in 1 or more years. None of the seeded plots were lost.

Ground cover and species composition are useful for monitoring the ecological integrity of ecosystems (LaPaix et al. 2009). In each sample year, presence and ground cover (vegetation up to 1 m in height) of vascular plant species, along with tree density when individual trees could be distinguished, were measured in each plot. A team that included the same botanists over the 30 years of sampling visually estimated plant cover for each species. Bråkenhielm and Qinghong (1995) found that visual estimates provide the most accurate, sensitive, and precise measure of vegetation cover. In 1982, topographic conditions of each plot were described in the field. When a tree canopy began to establish, canopy cover was recorded for all plots using a concave spherical densitometer (Peck et al. 2003).

Because abnormally high or low precipitation during parts of the 30-year observation period was likely to affect plant growth and establishment as well as erosion, annual precipitation from the nearest long-term weather station was recorded. Data have been collected since 1929 at Longview, Washington, which is about 60 km west of MSH. These data are available from the US Historical Climatology Network (Easterling et al. 1996) and the Carbon Dioxide Information Analysis Center at Oak Ridge National Laboratory. The Longview station receives less precipitation than the higher-elevation station at Spirit Lake, but the Longview precipitation record has been collected continuously before and since the eruption, whereas the Spirit Lake station was established in 1986 and thus has no pre-eruption records.

8.2.2 Methods of Data Analysis

8.2.2.1 Ground Cover and Richness

We examined changes in ground cover using repeated measures analysis of variance (ANOVA) on all plots, with seeding and years since eruption used as factors. To normalize variance, cover values were square-root transformed. Years 2 and 3 (1982 and 1983) still had such low cover that it was necessary to examine them separately from years 9 to 30. Final p-values were adjusted using a Bonferroni correction for two multiple-comparison tests (Dunn 1961). Since sampling years were not equally spaced, we used a spatial power covariance structure (Jennrich and Schluchter 1986). We tested the effects of seeding on ground plant richness using repeated measures ANOVA as well, but transformations and splitting the test over multiple years were not necessary as variance was more normally distributed. We still used a spatial power covariance structure to account for uneven distribution of sampling years.

We analyzed differences in species composition of ground vegetation using permutational multivariate analysis of variance (PERMANOVA) (Anderson 2001) with a Bray-Curtis difference matrix (Bray and Curtis 1957); seeding was included as a factor. We conducted PERMANOVA tests separately for each of the 7 years. Because the communities were not at equilibrium and multiple years elapsed between most sampling periods it was unlikely that strong similarities would exist between sampling periods; therefore including repeated measures in the PERMANOVA analysis was considered unnecessary. Final p-values were adjusted with a Bonferroni correction for seven multiple testings. As cover was very low in the early years and many plots had zero cover, we included a dummy variable with a cover value of 0.01 (the lowest possible non-zero cover value) for all plots in all years to avoid the problem of zero sums. To minimize the effect of extreme low and high cover values on the data, we applied a fourth-root transformation before analysis. Additionally, we reduced noise in the data set by removing species with a less than 5% occurrence over all years, as recommended by McCune and Grace (2002). This reduced the total number of species included in the analysis from 204 to 65 and made the data fit the assumptions for the test much better. Additionally, we ran the PERMANOVA with all 204 species to verify that using the reduced data set did not change conclusions.

To further assess the contributions of individual species to the community differences between the seeded and unseeded plots, we looked at similarity percentages using the Bray-Curtis difference matrix from the analysis of species composition (i.e., percent cover) for the ground vegetation. We applied the same dummy variable (0.01), data reduction, and transformations described above and used to examine differences in ground cover species composition. We separately examined each year for differences in ground vegetation species composition between seeded and unseeded plots, but as conclusions were not drawn across years, we did not apply a correction for multiple testing.

8.2.2.2 Trees

We examined total stem density of all tree species using repeated measures ANOVA with plot as a subject and years since eruption as the repeated factor and seeding as the treatment. Data were log-10 transformed (accounting for zeros) before analysis to improve normality of variance. To account for uneven spacing of sampling across years, we used a spatial power covariance structure.

Using PERMANOVA with a Bray-Curtis difference matrix and seeding as a factor, we also analyzed difference in stem density of tree species. Again, we ran each year as a separate test and applied a Bonferroni correction for seven multiple testings to final p-values. To account for years where no stems of any species were found within a plot, we included a dummy variable with a count of 1 (the lowest possible non-zero count value) in all plots, across all years. Before analysis, we used a log-10 transformation that accounts for the presence of zeros on the count data (McCune and Grace 2002) to reduce the influence of extreme values. Since only 14 species of trees were found in these plots across all years, we retained all tree species in the analysis.

With seeding and years since eruption as factors, we used repeated measures ANOVA on plots to examine total tree canopy cover. Variance and distribution were both normally distributed. No correction to final p-values was used, but we maintained the spatial power covariance structure for consistency with other tests even though only 2 years were included in this analysis (2004 and 2010).

8.2.2.3 Native Versus Non-native Species

Using repeated measures ANOVA, we analyzed changes in cover of non-native and native species separately. Plot was the repeated term, and we analyzed all 7 years together, with years since eruption and seeding as factors. To improve normality of variance, we used a log-10 transform accounting for zeros, along with a spatial power covariance structure to account for the lack of uniform time periods between samplings.

To perform repeated measures ANOVA, we used SAS 9.3 (SAS Institute, Cary NC). For PERMANOVA analysis, we used the Adonis package in R 3.1.1 (R Core Team 2014) and double checked using Primer 6 (PRIMER-E Ltd., Ivybridge United Kingdom). We conducted similarity percentage analyses with Primer 6.

8.3 Results

Seeded plots did not differ from unseeded plots in topographic conditions or soil particle-size distributions (Table 8.1). The 11 seeded plots contained examples of five of the six topographic categories, with only the rarest type—plots with a single high mound—not represented.

8.3.1 Ground Cover and Richness

Ground cover increased in all plots over time (Fig. 8.5). Seeding had a significant effect ($p < 0.02$) on cover (Table 8.2), as was evident from photographs (Fig. 8.6). Total cover was higher in the seeded plots than the unseeded plots in all years except year 2 ($p = 0.17$). There was a significant interaction between seeding and years since eruption in later years ($p = 0.0028$), but this result seems to be driven exclusively by year 14, when the cover in the seeded plots spiked sharply and cover in unseeded plots did not (Fig. 8.5). The year 14 spike in cover was largely a result of an increase in non-native species in seeded plots (Fig. 8.7b). The spike observed in native-species cover in the seeded plots in year 9 (Fig. 8.7a) suggests that this year may also have had a steeper increase in total cover in the seeded plots than the unseeded, but, because years 2 and 3 and 9–30 were analyzed separately, this hypothesis could not be tested directly.

Table 8.1 Mean and standard error of particle-size distribution for the >2-mm fraction of seeded and unseeded plots in August 1982.

		% boulders	% cobbles	% gravel, sand, silt, clay
Not seeded	Mean	5.8	19.4	75
	S.E.	0.6	1.6	2
Seeded	Mean	7	18.8	71.6
	S.E.	2.1	3.4	5.4

Sampling and analysis methods are described in Adams and Dale (1987).

Cover of both native and non-native species was higher in the seeded plots than the unseeded plots ($p < 0.0001$), but not significantly so in all years, as there was a year-by-treatment interaction ($p < 0.0001$) (Table 8.3). The pattern in native cover increase over time seems to be fairly consistent in seeded versus unseeded plots; the two treatments differed only in 2 years (9 and 14) ($p < 0.0001$) (Fig. 8.7a). However, non-native cover differed between seeded and unseeded plots in years 9 through 24 ($p < 0.004$). In the seeded plots, non-native cover spiked in year 14, and then remained fairly consistent, whereas in the unseeded plots, total cover slowly but consistently crept upward over time (Fig. 8.7b). The proportion of cover from non-native species was highest in the first year of sampling in unseeded plots and in year 14 in the seeded plots

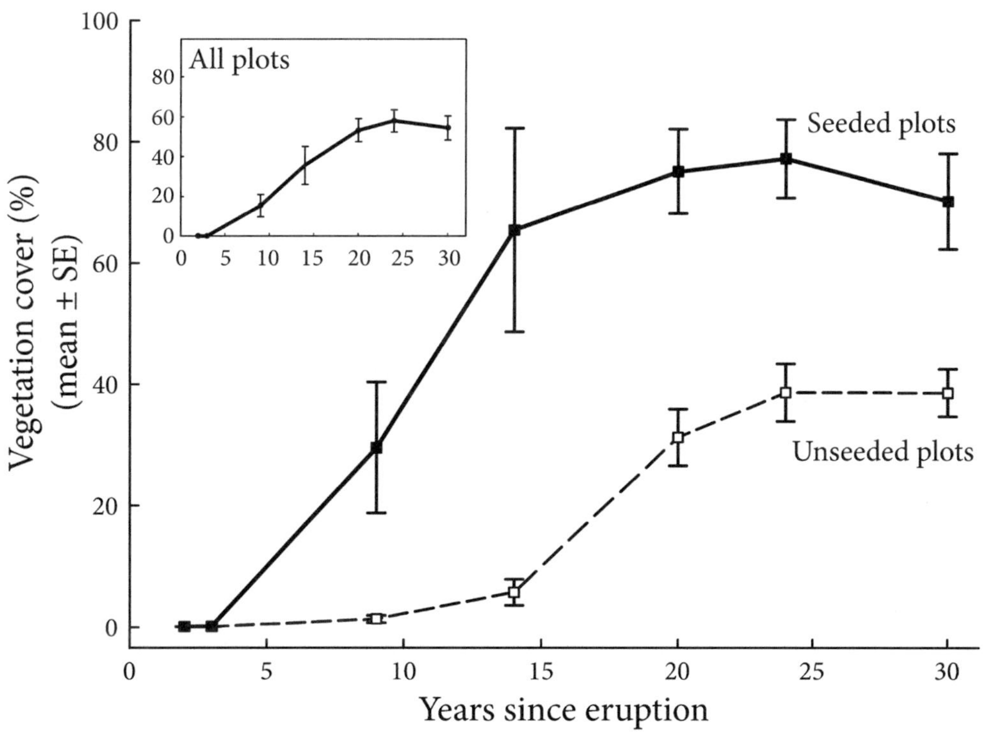

Fig. 8.5 Change in total cover over time in seeded (*solid line*) and unseeded plots (*dashed line*). *Inset* shows total cover in all plots over time on the debris-avalanche deposit. *Bars* are standard error.

Table 8.2 Repeated measures analysis on total cover of all species. Plot was the subject and years since eruption (yse) and seeded were factors. Years 2 and 3 and 9 through 30 were analyzed separately owing to variance issues. A Bonferroni correction for two multiple testings was used on final p-values.

		F value	*p*-value
Years 2–3	Seeded	7.11	0.02*
	Yse	69.68	<0.0002**
	Yse*seeded	1.85	0.36
Years 9–30	Seeded	31.64	<0.0002**
	Yse	21.31	<0.0002**
	Yse*seeded	4.59	0.0028**

* denotes significance at the 0.05 level and ** at the 0.01 level.

(Fig. 8.8). The proportion of total cover from non-native species varied over time. The contribution of non-native species to overall cover was greatest in the unseeded plots in year 14 and in the seeded plots in year 30 (Fig. 8.8 inset).

Like cover, species richness increased over time in all plots (Fig. 8.9). Species richness was higher in seeded plots compared with unseeded plots but only significantly so in years 3, 9, 24, and 30 after the eruption ($p < 0.01$). Richness on the debris-avalanche deposit as a whole has increased in every year and has followed a nearly linear trend since year 3 (Fig. 8.10). Non-native species have made up a fairly consistent proportion of all species found on the debris-avalanche deposit in all years—between 20% and 30% of total richness (Fig. 8.10, inset).

8.3.2 Species Composition

There was a significant difference in species composition comparing the seeded versus unseeded plots in all years ($p < 0.035$) (Table 8.4). Based on R^2 values, seeding explained the highest percent of difference in community composition in years 2 (15%), 9 (10%), 14 (14%), and 20 (10%), but the influence of seeding on plant community in later years was reduced, explaining only 7% of the difference between treatments in years 24 and 30. Seeding explained the lowest percent difference in ground-cover plant communities in year 3 ($R^2 = 6\%$).

In the similarity percentage analysis (Table 8.5), a total of 28 species, 19 native and 9 non-natives, were responsible for at least 50% of the divergence between seeded and unseeded plots over all sampled years. Among species contributing to differences, geophytes (perennial plants that resprout by means of underground bulbs, tubers, or corms) were the most abundant plant life form (14 total species), though annuals (5 species) were also important. Deciduous trees first appeared among the most abundant species in year 9 post eruption and increased in contribution from then onward. Only one species of hemicryptophyte (perennial plants with overwintering buds at soil level), *Hypochaeris radicata*, occurred frequently and was a large contributor in 5 of the 7 years it was present. For the first 14 years of the study, those species contributing to dissimilarity always occurred in higher abundance in the seeded plots. However, starting in year 20, some such species occurred in higher abundance in the unseeded plots (forbs *Lupinus lepidus*, *L. latifolius*, and *H. radicata* and deciduous shrubs *Salix sitchensis* and *Alnus viridis*). In year 30, *Alnus rubra* occurred in higher frequencies in the unseeded plots, even though it had occurred in higher abundance in the seeded plots in years 14–24 (Table 8.5). Among those species found in greater abundance only in the unseeded plots, all but *H. radicata* were native, but the sample size is small, so this is more likely to be random variation than a significant pattern. Also, a number of new forest understory species (e.g., *Claytonia sibirica* var. *sibirica*, *Rubus leucodermis*, and *Prunella vulgaris*) first explained community divergence in the plots in year 30, so there seems to have been a community shift toward closed-forest conditions between year 24 and 30. Overall, contribution of any one species to divergence declined over the course of the study.

8.3.3 Tree-Count Analysis and Canopy Cover

Total stem density of tree species increased through time in all years except year 14, but stem counts were not different in seeded versus unseeded plots ($p = 0.69$) (Fig. 8.11). *Alnus rubra* made up the largest fraction of stems counted in most years: year 2 (21%), 3 (1%), 9 (78%), 14 (59%), 20 (91%), 24 (89%), and 30 (92%). Additionally, no difference was found in tree species composition between the seeded and unseeded plots in any year ($p > 0.12$) (Table 8.6). Overstory canopy cover was examined only in years 24 and 30. Values did not differ between the seeded and unseeded plots ($p = 0.53$), and mean canopy density was 27%.

8.4 Discussion and Conclusions

Thirty years after the eruption of MSH, the debris-avalanche deposit has undergone extensive vegetation development. This study sought to determine whether initial differences seen in the trajectory of vegetation establishment have persisted, comparing areas that were seeded with a largely non-native plant mix in 1980 with areas that were left to undergo natural primary succession, and whether these differences are likely to continue into the future.

Fig. 8.6 Time series of photographs for plots 22 and 47. Plot 47 was seeded in 1980, whereas plot 22 was not seeded. Plot locations are labeled in Fig. 8.1. Photos taken by Virginia Dale, Bob Holland, and (in 2010) Ian Kuliasha.

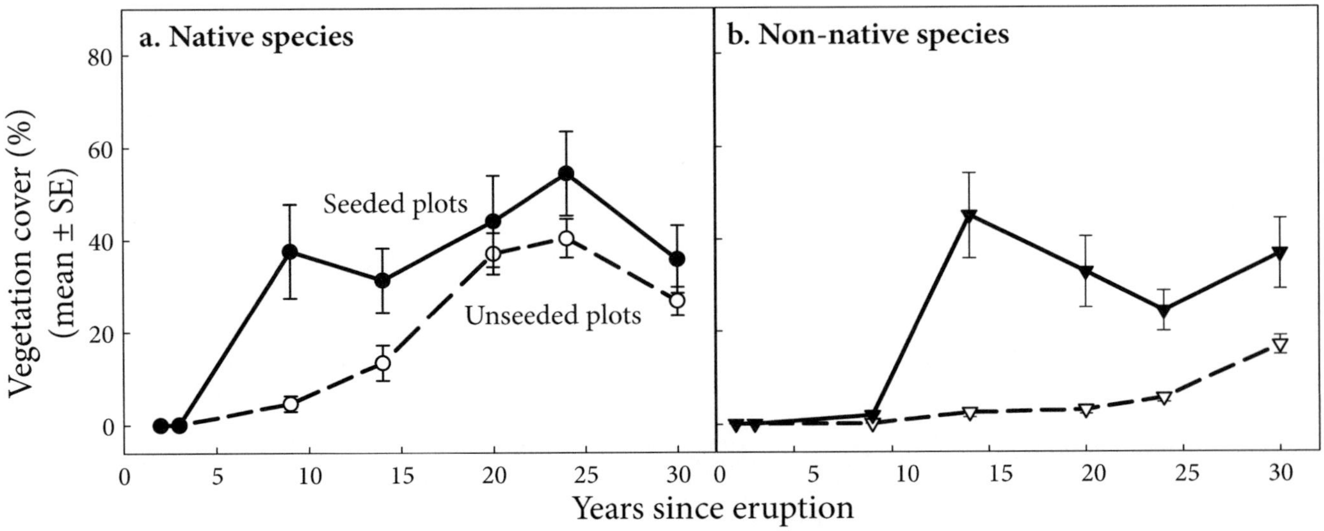

Fig. 8.7 Native (**a**) and non-native (**b**) species cover in seeded (*solid lines*) and unseeded (*dashed line*) plots over time. *Error bars* are standard error.

Table 8.3 Results of repeated measures on species richness and native and non-native cover per plot in the debris-avalanche deposit over seven sampling periods from year 2 to year 30 after the eruption. Plot was the subject and years since eruption (yse) and seeded were factors.

		F value	*p*-value
Native cover	Seeded	16.44	0.0001***
	Yse	93.03	<0.0001***
	Yse*seeded	10.45	<0.0001***
Non-native cover	Seeded	39.64	<0.0001***
	Yse	61.29	<0.0001***
	Yse*seeded	7.33	<0.0001***
Richness	Seeded	22.34	<0.0001***
	Yse	103.36	<0.0001***
	Yse*seeded	3.39	0.0029**

** denotes significance at the 0.01 level and *** at the 0.0001 level.

Total species richness on the debris-avalanche deposit increased in the 30 years since the eruption and continues to increase, as is expected for primary succession in the northwestern United States (Lichter 1998). However, cover of herbaceous vegetation plateaued at around 50% after year 20 (Fig. 8.5, inset). In future years, ground cover may even decrease as the overstory canopy closes (Halpern and Lutz 2013), altering the light regime and leading to a reduction in abundance of shade-intolerant species. The dominance of early-seral species has declined with time. Year 30 appears to have been a transition year, with species that are found in forest understories (*Claytonia sibirica* var. *sibirica*, *Rubus leucodermis*, and *Prunella vulgaris*) becoming drivers of community divergence for the first time.

Ecological effects of several volcanic debris-avalanche deposits have been studied at a number of other volcanoes: Mt. Taranaki in New Zealand (Clarkson 1990), Ksudach in Russia (Grishin et al. 1998), the Ontake volcano in Japan (Nakashizuka et al. 1993), and Mount Katmai in the state of Alaska in the United States (Griggs 1918a, b, c, 1919). However, plant recovery has not been followed at any of those sites over the decades subsequent to the initial deposit.

The aerial distribution of largely non-native seeds on a subset of plots at MSH in 1980 has had a pronounced and enduring effect on subsequent vegetation communities. The trajectory of both cover and species richness in the seeded and unseeded plots appears similar; however, total values differ. Cover and richness were higher in seeded plots throughout the entire period of this study (Fig. 8.5). Interestingly, this pattern does not seem to be driven only by higher abundance of species from the original seeding mix, as both native and non-native species had higher cover in the seeded plots (Fig. 8.7). Furthermore, seeded species never contributed more than 5% to the total cover in the seeded plots (Fig. 8.8, inset). The community analysis supports the assumption that the seeded species may not be the direct cause of the observed differences between the seeded and unseeded plots. Only two species from the original mix, *Trifolium repens* and *Festuca rubra,* are among the top contributors to community divergence (in years 20 and 30), even though ground-cover vegetation communities differed between treatments in all years. However, the presence of those seeded species may have facilitated the establishment of other non-native species, for example, by providing vegetative cover and shade. If the seeded species are not driving the community differences, it is unclear what else might be causing community divergence. There is a notable difference in the cover of non-native species in the seeded plots after 1994, but why non-native species should have increased there and not in the unseeded plots is unclear (Fig. 8.7b). At first glance, it seems tempting

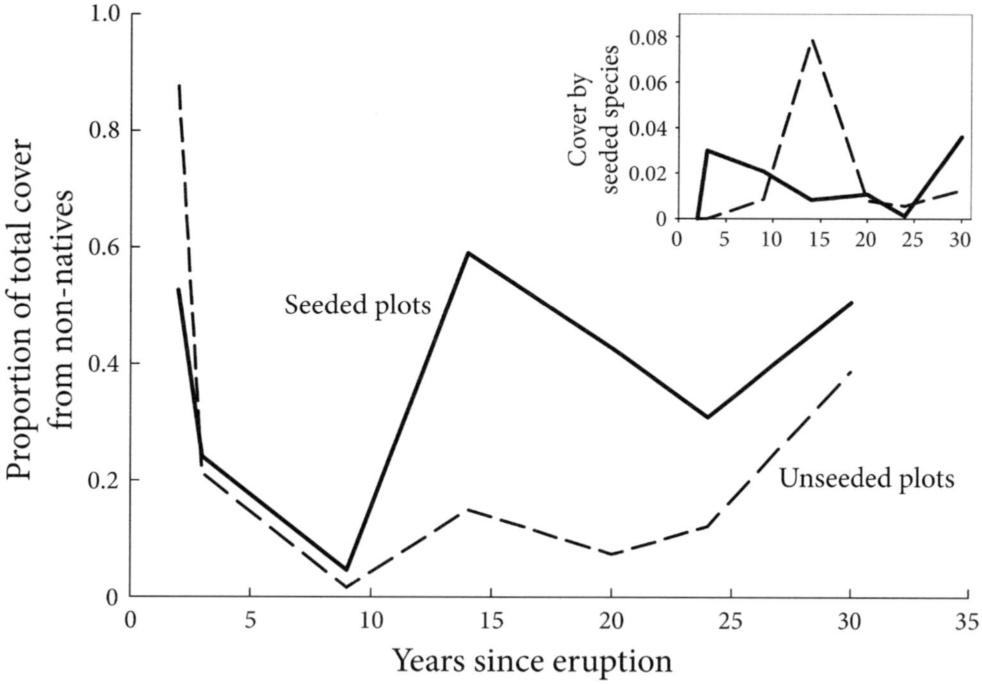

Fig. 8.8 Proportion of total cover from non-native species through time for plots that were seeded (*solid line*) and unseeded (*dashed line*). *Inset* is proportion of total cover from seeded species through time in both seeded and unseeded plots.

Fig. 8.9 Change in number of species per plot over time in seeded (*solid line*) and unseeded (*dashed line*) plots. *Bars* are standard error. *Inset* is richness per plot over time for all plots.

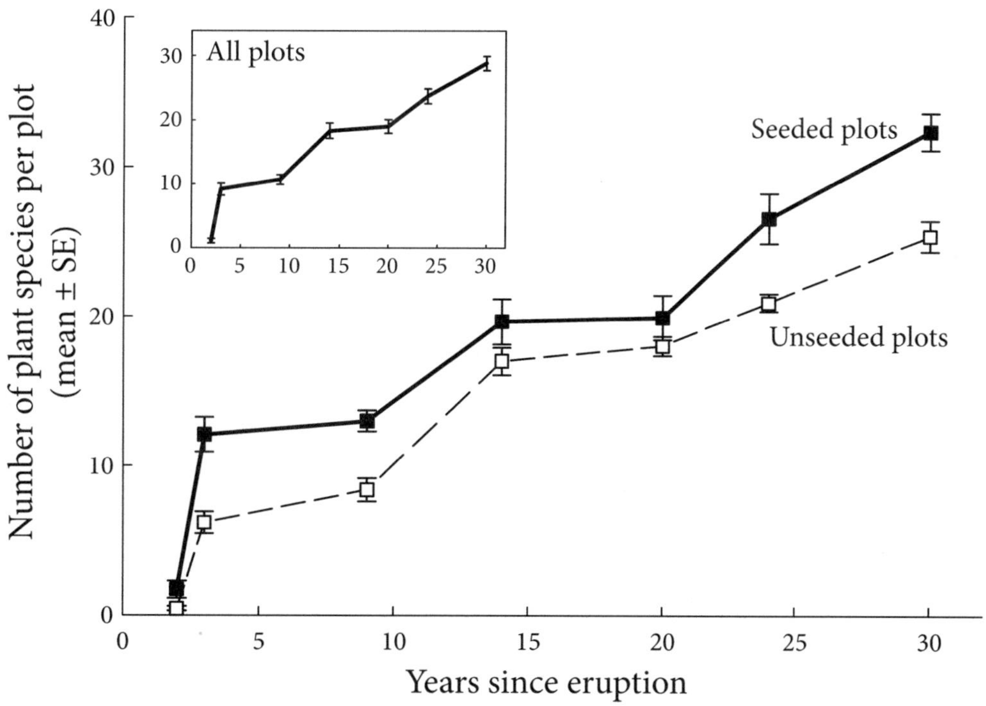

to conclude that the anomalous increase in cover is a response to climatic conditions, as the summer of 1993 was considerably wetter than average (Fig. 8.12), which can result in an increased production legacy (Sala et al. 2012). However, it is puzzling that only non-native plant cover in seeded plots would respond in this fashion, since weather events should affect all plots similarly. Certainly, MSH has a higher percentage of non-native species than the Pacific Northwest in general. Of the 107 species of non-native plants reported to occur in Cowlitz County, Washington (Washington Native

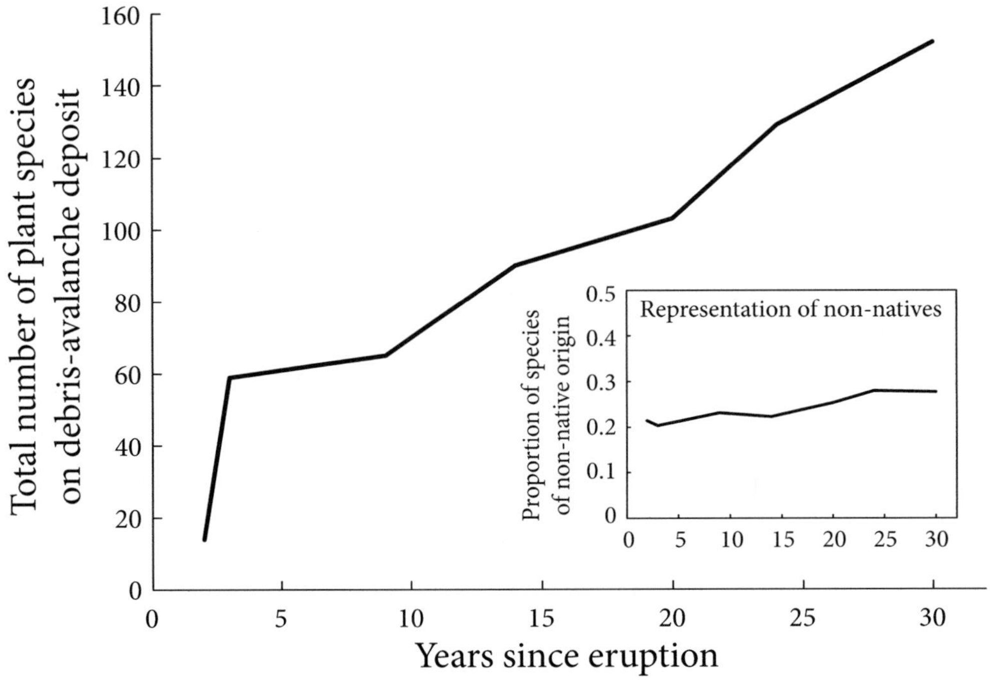

Fig. 8.10 Richness of all species on the debris-avalanche deposit over time. *Inset*: proportion of richness made up of non-native species.

Table 8.4 PERMANOVA analysis of species composition and seeded versus unseeded plots.

Year since eruption	df	Psuedo-F	R^2	Adj. p-value
2	1	10.4	0.15	0.007**
3	1	3.56	0.06	0.035*
9	1	6.38	0.10	0.007**
14	1	9.95	0.14	0.007**
20	1	6.37	0.10	0.007**
24	1	4.87	0.07	0.007**
30	1	4.47	0.07	0.014*

Test for each year was run separately, and a Bonferroni correction for seven tests was applied to final p-values. Species composition values were fourth-root transformed, and a reduced data set including only species with >5% cover in all years was used (total species = 65).
* denotes significance at the 0.05 level and ** at the 0.01 level.

Plant Society 2006), 40 have been found on the debris-avalanche deposit since the eruption, and 16 species of non-native plants have apparently arrived from outside the county. Non-native species have been arriving at a nearly constant rate and have made up 20–30% of the richness on the debris-avalanche deposit in the 30 years since the eruption (Fig. 8.10), a high proportion for the Pacific Northwest where non-native species represent about 12% of overall species richness in forests across the region (Franklin and Dyrness 1973). However, non-natives often make up a large proportion of total cover (Stapanian et al. 1998), as is also the case at MSH, particularly in the seeded plots (Fig. 8.7).

Seeding by the SCS in 1980 was based on the assumption that there would be no effect other than short-term increases in plant cover, which would result in reduced erosion. At the time, little research had been done to suggest otherwise. Because erosion from the seeded and unseeded plots was not measured, it is difficult to say whether seeding achieved its desired goal. None of the original 11 seeded plots were lost to erosion, but a number of the original 90 unseeded plots were. However, there were many more unseeded plots, and some were in the path of stream channels that eroded over time. Certainly erosion from the entire debris-avalanche deposit remained high, two orders of magnitude above background levels in the Pacific Northwest (Major et al. 2000) 20 years post eruption. Seeding after other types of disturbance has not been found to consistently reduce erosion. Post-wildfire seedings in rangelands had little effect on erosion during the first year after fire, and when they did it was only in years of high establishment (Pyke et al. 2013). Whether or not seeding achieved the stated goal of reduced erosion, the differences in cover, species richness, and community composition on the seeded versus unseeded plots continue to persist after three decades.

At least one previous study in Washington state examining plant communities and comparing erosion after seeding treatments found differences that persisted for decades, though successful soil stabilization occurred only if the seeding treatments were combined with straw and clay (Cline

Table 8.5 The results of the similarity percentage analysis on the species composition data from all years. Species that explained up to 50% of the observed difference were included. Other information not included in the test but relevant to the reported species was included as well. Species in **bold face** occurred at higher abundances in unseeded plots. *Origin* non-native or native; and *life form*: *G* geophytes, plants with buds below ground, *H* hemicryptophytes, plants with buds 0–50 cm above soil surface, *A* annual, *PE* phanerophyte, evergreen tree, *PD* phanerophyte, deciduous tree, *S* shrub.

Species	Contribution (%)	Origin	Life form
Year 2: average dissimilarity = 94.63			
Senecio sylvaticus	44.7	Non-native	A
Anaphalis margaritacea	13.1	Native	G
Year 3: average dissimilarity = 57.55			
Epilobium brachycarpum	8.1	Native	G
Chamerion angustifolium ssp. *angustifolium*	7.9	Native	G
Senecio sylvaticus	7.5	Non-native	A
Epilobium glaberrimum	7.4	Native	G
Hypochaeris radicata	7.3	Non-native	H
Anaphalis margaritacea	6.3	Native	G
Abies procera	6.0	Native	PE
Year 9: average dissimilarity = 76.49			
Lotus unifoliolatus var. *unifoliolatus*	24.1	Native	A
Agrostis exarata	7.2	Native	G
Populus balsamifera ssp. *trichocarpa*	5.7	Native	PD
Agrostis pallens	5.7	Native	G
Anaphalis margaritacea	5.3	Native	G
Holcus lanatus	4.7	Non-native	G
Year 14: average dissimilarity = 73.27			
Lotus unifoliolatus var. *unifoliolatus*	9.9	Native	A
Holcus lanatus	8.2	Non-native	G
Aira caryophyllea	7.9	Non-native	A
Hypochaeris radicata	7.8	Non-native	H
Alnus rubra	6.1	Native	PD
Vulpia myuros	5.2	Non-native	A
Epilobium glaberrimum	3.8	Native	G
Salix scouleriana	3.3	Native	PD
Year 20: average dissimilarity = 61.44			
Holcus lanatus	11.0	Non-native	G
Alnus rubra	10.4	Native	PD
Lotus unifoliolatus var. *unifoliolatus*	6.3	Native	A
Trifolium repens	5.0	Non-native	G
Agrostis pallens	4.3	Native	G
Salix sitchensis	**4.0**	**Native**	**PD**
Lupinus lepidus	**4.0**	**Native**	**G**
Hypochaeris radicata	3.7	Non-native	H
Lupinus latifolius	**3.5**	**Native**	**G**
Year 24: average dissimilarity = 59.73			
Alnus rubra	8.5	Native	PD
Holcus lanatus	7.2	Non-native	G
Hypochaeris radicata	5.0	Non-native	H
Vulpia myuros	4.3	Non-native	A
Lupinus latifolius	**3.5**	**Native**	**G**
Alnus viridis ssp. sinuata	**3.4**	**Native**	**PD**
Salix sitchensis	**3.2**	**Native**	**PD**
Lotus unifoliolatus var. *unifoliolatus*	3.1	Native	A
Lupinus lepidus	**3.0**	**Native**	**G**
Epilobium glaberrimum	2.7	Native	G
Anaphalis margaritacea	2.5	Native	G
Cytisus scoparius	2.5	Non-native	S
Digitalis purpurea	2.5	Non-native	G
Year 30: average dissimilarity = 60.89			
Epilobium glaberrimum	7.1	Native	G
Holcus lanatus	6.7	Non-native	G
Claytonia sibirica var. sibirica	6.5	Native	A
Hypochaeris radicata	**4.9**	**Non-native**	**H**
Rubus leucodermis	4.9	Native	H
Festuca rubra	3.7	Native	G
Cytisus scoparius	3.4	Non-native	S
Prunella vulgaris	3.4	Native	G
Rumex acetosella	3.1	Non-native	G
Trifolium repens	3.0	Non-native	G
Lupinus latifolius	**2.8**	**Native**	**G**
Alnus rubra	**2.8**	**Native**	**PD**

and Uresk 1979), which was not done at MSH. Furthermore, the Cline and Uresk (1979) study seeded a well-known invasive, cheatgrass (*Bromus tectorum*). Although seeding with an invasive species might be expected to produce effects that last for decades, the record from MSH demonstrates that even seeding treatments of more benign species may produce long-lasting community effects.

The difference between the seeded and unseeded areas seems unlikely to continue indefinitely. It is expected that the plant community on the debris-avalanche deposit will eventually develop into a forest typical of the Pacific Northwest (Franklin and Dyrness 1973). Results from the present study may provide evidence that this transition is

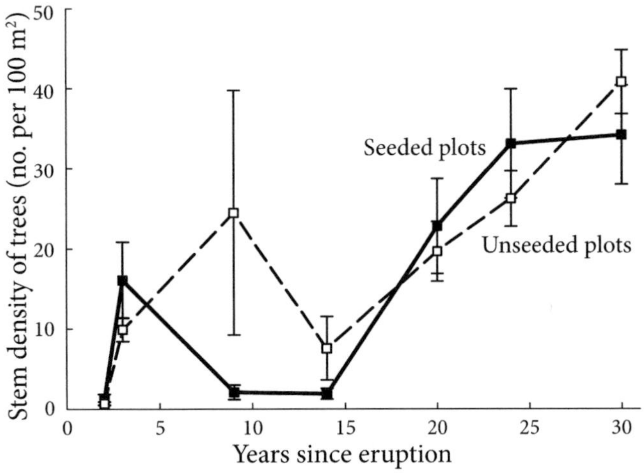

Fig. 8.11 Stem density counts of trees in seeded (*solid line*) and unseeded (*dashed line*) plots. *Error bars* are standard error.

Table 8.6 PERMANOVA analysis of tree-stem counts in seeded vs unseeded plots.

Year since eruption	df	Psuedo-F	R^2	Adj. p-value
2	1	0.46	0.01	1
3	1	0.82	0.01	1
9	1	2.03	0.03	0.91
14	1	3.49	0.06	0.12
20	1	2.51	0.04	0.34
24	1	2.96	0.05	0.2
30	1	1.72	0.03	1

Test for each year was run separately, and a Bonferroni correction for seven tests was applied to final p-values. Tree-count values were log-10 transformed before analysis to account for zeros. Data were not reduced to species with <5% occurrence as only 14 tree species were found.

Fig. 8.12 Precipitation during the growing season (1 May to 31 August) from the weather station at Longview, Washington.

already happening. Thirty years post eruption, herbaceous communities continue to differ between seeded and unseeded plots, but seeding explains a smaller percentage of the difference seen than in previous years, only 7% in year 30 post eruption as compared with a maximum of 15% in year 14. Further, trees, which will presumably become the future overstory, do not differ between seeded and unseeded plots in either stem density, species composition, or canopy cover in any year in which these metrics were sampled (Table 8.6). This finding is fascinating, in part because earlier in succession conifer tree-seedling mortality was found to be higher in the seeded plots (Dale 1991), yet that result does not seem to have had a lasting effect on the developing tree community. It should be noted that the present tree community is largely red alder (92% of all tree stems counted in year 30), not species associated with the conifer forest that is expected to eventually dominate the area.

Resource managers dealing with the pressure to establish vegetation and reduce erosion in disturbed areas are often "hamstrung by resource shortages" and consider the use of non-native seed species (Kuebbing and Simberloff 2015). Proactive management should focus on realistic short-term and long-term costs and benefits of management options. After a large disturbance, funds are often available to "fix" the problem rather than to realistically deal with the new ecosystem properties (Dale et al. 1998). Furthermore, the rarity of benefits provided by non-native species (Kuebbing and Nuñez 2015) means that those species with potential for positive effects should be identified upfront. Kuebbing and Nuñez (2015) note that positive interactions are less frequent with grasses and herbs and more common with nitrogen-fixing species and tree species. Even though nitrogen fixers were included in the mix of seeds distributed in the aftermath of the eruption of MSH, no native tree species were considered. In 1980, the first author suggested to the SCS that the native red alder (*Alnus rubra*) be considered in the mix of species seeded at MSH, as it is both a tree and a nitrogen fixer, but the SCS lamented that no seeds were available. This study revealed that red alder was a major contributor to community divergence between years 14 and 30 at MSH and was the most abundant tree in most years, a finding that is in line with earlier results from this system (Dale et al. 2005a).

Few studies using permanent plots over several decades exist that evaluate the effects of management choices on forest succession (Dale et al. 2005d; Kirby et al. 2014), and those that do mostly examine succession after logging or fire (see Halpern and Spies 1995). Thus, this study of vegetation succession at MSH over the last 30 years is a unique contribution. Although these results are of interest to the study of plant succession, they also have implications for resource management and response in the face of large disturbances. In spite of the US$2 million spent on seeding, major erosion still occurred (Major et al. 2000), and effects of the seeding on vegetation cover and plant richness are still apparent after three decades. Even so, our findings suggest that the first forest to develop will show little difference between seeded and unseeded areas.

Acknowledgments Logistic support was provided by the USDA Forest Service, the Washington State Department of Natural Resources, and Weyerhaeuser Company. Over the 30 years of the study, fieldwork was conducted with the assistance of A.B. Adams, Wendy Adams, Charles Crisafulli, Ginny Dains, D. Donohue, Margaret Evans, Howard Haemmaerle, Brad Hensley, Charlie Hensley, Asa Holland, Robert Holland, Keith Kline, Jasmine Kuliashi, Ian Kulaishi, Bridgette Nyberg, Eric Smith, Mandy Tu, and John Wallace. Over the years of this study, the research was partially funded by the National Science Foundation, the National Geographic Society, Earthwatch, and The Center for Field Research of Belmont, Massachusetts. The University of Washington Herbarium was used to verify plant species identifications. Statisticians from Colorado State University helped with the analysis: Phil Turk advised on PERMANOVA and James R. zumBrunnen advised on the repeated measures analysis. Comments of Charles Crisafulli and two anonymous reviewers on an earlier draft were helpful. Oak Ridge National Laboratory is managed by UT-Battelle, LLC, for the US Department of Energy under contract DE-AC05-00OR22725.

References

Adams, V.D., and A.B. Adams. 1982. Initial recovery of the vegetation on Mount St. Helens. In *Mount St. Helens: One year later*, ed. S.A.C. Keller, 105–113. Cheney: Eastern Washington University Press.

Adams, A.B., J.R. Wallace, J.T. Jones, and W.K. McElroy. 1986. Plant ecosystem resilience following the 1980 eruptions of Mount St. Helens, Washington. In *Mount St. Helens: Five years later*, ed. S.A.C. Keller, 182–207. Cheney: Eastern Washington University Press.

Adams, A.B., and V.H. Dale. 1987. Vegetative succession following glacial and volcanic disturbances in the Cascade Mountain Range of Washington, U.S.A. In *Mount St. Helens 1980: Botanical consequences of the explosive eruption*, ed. D.E. Bilderback, 70–147. Los Angeles: University of California Press.

Anderson, M. 2001. A new method for non-parametric multivariate analysis of variance. *Austral Ecology* 26: 32–46.

Ares, A., S.D. Berryman, and K.J. Puettmann. 2009. Understory vegetation response to thinning disturbance of varying complexity in coniferous stands. *Applied Vegetation Science* 12: 472–487.

Banks, N.G., and R.P. Hoblitt. 1981. Summary of temperature studies of 1980 deposits. In *The 1980 eruptions of Mount St. Helens, Washington*, Professional Paper 1250, ed. P.W. Lipman and D.R. Mullineaux, 295–314. Washington, DC: U.S. Geological Survey.

Bråkenhielm, S., and L. Qinghong. 1995. Comparison of field methods in vegetation monitoring. *Water, Air, and Soil Pollution* 79: 75–87.

Bray, J.R., and J.T. Curtis. 1957. An ordination of upland forest communities of southern Wisconsin. *Ecological Monographs* 27: 325–349.

Clarkson, B.D. 1990. A review of vegetation development following recent (<450 years) volcanic disturbance in North Island, New Zealand. *New Zealand Journal of Ecology* 14: 59–71.

Cline, J.F., and V.A. Uresk. 1979. Revegetation of disturbed grounds in the semi-arid climate of southcentral Washington. *Health Physics* 36: 289–294.

Dale, V.H. 1986. Plant recovery on the debris avalanche at Mount St. Helens. In *Mount St. Helens: Five years later*, ed. S.A.C. Keller, 208–214. Cheney: Eastern Washington University Press.

———. 1989. Wind dispersed seeds and plant recovery on the Mount St. Helens debris avalanche. *Canadian Journal of Botany* 67: 1434–1441.

———. 1991. The debris avalanche at Mount St. Helens: Vegetation establishment in the ten years since the eruption. *National Geographic Research and Exploration* 7: 328–341.

Dale, V.H., and W. Adams. 2003. Plant establishment fifteen years after the debris avalanche at Mount St. Helens, Washington. *Science of the Total Environment* 313: 101–113.

Dale, V.H., A. Lugo, J. MacMahon, and S. Pickett. 1998. Ecosystem management in the context of large, infrequent disturbances. *Ecosystems* 1: 546–557.

Dale, V.H., D.R. Campbell, W.M. Adams, C.M. Crisafulli, V. Dains, P.M. Frenzen, and R. Holland. 2005a. Plant succession on the Mount St. Helens debris-avalanche deposit. In *Ecological responses to the 1980 eruption of Mount St. Helens*, ed. V.H. Dale, F.J. Swanson, and C.M. Crisafulli, 59–74. New York: Springer.

Dale, V.H., J. Delgado-Acevedo, and J. MacMahon. 2005b. Effects of modern volcanic eruptions on vegetation. In *Volcanoes and the environment*, ed. J. Marti and G.G.J. Ernst, 227–249. Cambridge: Cambridge University Press.

Dale, V.H., F.J. Swanson, and C.M. Crisafulli. 2005c. Ecological perspectives on environmental management following the 1980 eruption of Mount St. Helens. In *Ecological responses to the 1980 eruption of Mount St. Helens*, ed. V.H. Dale, F.J. Swanson, and C.M. Crisafulli, 277–286. New York: Springer.

———. 2005d. Disturbance, survival, and succession: Context for understanding ecological responses to the 1980 eruption of Mount St. Helens. In *Ecological responses to the 1980 eruption of Mount St. Helens*, ed. V.H. Dale, F.J. Swanson, and C.M. Crisafulli, 3–12. New York: Springer.

Dunn, O.J. 1961. Multiple comparisons among means. *Journal of the American Statistical Association* 56 (293): 52–64.

Easterling, D.R., T.R. Karl, E.H. Mason, P.Y. Hughes, D.P. Bowman, R.C. Daniels, and T.A. Boden, eds. 1996. *United States historical climatology network (U.S. HCN) monthly temperature and precipitation data*, ORNL/CDIAC-87, NDP-019/R3. Oak Ridge: Carbon Dioxide Information Analysis Center, Oak Ridge National Laboratory.

Fairchild, L.H. 1985. Lahars at Mount St. Helens, Washington. Master's thesis. Seattle: University of Washington.

Franklin, J.F., and C.T. Dyrness. 1973. *Natural vegetation of Oregon and Washington*, General Technical Report PNW-8. Portland: Pacific Northwest Forest and Range Experiment Station. (Republished by Oregon State University Press in 1988).

Glicken, H. 1998. Rockslide-debris avalanche of May 18, 1980, Mount St. Helens volcano, Washington. *Bulletin of the Geological Survey of Japan* 49: 55–106.

Gray, A.N., K. Barndt, and S.H. Reichard. 2011. *Nonnative invasive plants of Pacific coast forests: A field guide for identification*, General Technical Report PNW-GTR-817. Portland: U.S. Department of Agriculture, Forest Service, Pacific Northwest Research Station.

Griggs, R.F. 1918a. The recovery of vegetation at Kodiak. *Ohio Journal of Science* 19: 1–57.

———. 1918b. The great hot mudflow of the Valley of 10,000 Smokes. *Ohio Journal of Science* 19: 117–142.

———. 1918c. The Valley of Ten Thousand Smokes: An account of the discovery and exploration of the most wonderful volcanic region in the world. *The National Geographic Magazine* 33: 10–68.

———. 1919. The beginnings of revegetation in Katmai Valley. *Ohio Journal of Science* 19: 318–342.

Grishin, S. Yu, R. del Moral, P. Krestov, and V.P. Verkholat. 1998. Succession following the catastrophic eruption of Ksudach volcano (Kamchatka, 1907). *Vegetatio* 127: 129–153.

Halpern, C.B., and J.A. Lutz. 2013. Canopy closure exerts weak controls on understory dynamics: A 30-year study of overstory-understory interactions. *Ecological Monographs* 83: 221–237.

Halpern, C.B., and T.A. Spies. 1995. Plant species diversity in natural and managed forests in the Pacific Northwest. *Ecological Applications* 5: 913–934.

Jennrich, R., and M. Schluchter. 1986. Unbalanced repeated-measures models with structured covariance matrices. *Biometrics* 42: 805–820.

Kirby, K.J., D.R. Bazely, E.A. Goldberg, J.E. Hall, R. Isted, S.C. Perry, and R.C. Thomas. 2014. Changes in the tree and shrub layer of Wytham Woods (southern England) 1974–2012: Local and national trends compared. *Forestry* 87: 663–673.

Kuebbing, S.E., and M.A. Nuñez. 2015. Negative, neutral, and positive interactions among nonnative plants: Patterns, processes, and management implications. *Global Change Biology* 21: 926–934.

Kuebbing, S.E., and D. Simberloff. 2015. Missing the bandwagon: Nonnative species impacts still concern managers. *NeoBiota* 25: 73–86.

LaPaix, R., B. Freedman, and D. Patriquina. 2009. Ground vegetation as an indicator of ecological integrity. *Environmental Reviews* 17: 249–265.

Lehre, A.K., B.D. Collins, and T. Dunne. 1983. Post-eruptive sediment budget for the North Fork Toutle River drainage, June 1980–June 1981. *Zeitschrift für Geomorphologie (Supplement)* 46: 143–163.

Lichter, J. 1998. Primary succession and forest development on coastal Lake Michigan sand dunes. *Ecological Monographs* 68: 487–510.

Major, J.J., and L.E. Mark. 2006. Peak flow responses to landscape disturbances caused by the cataclysmic 1980 eruption of Mount St. Helens, Washington. *Geological Society of America Bulletin* 118: 938–958.

Major, J.J., T.C. Pierson, R.L. Dinehart, and J.E. Costa. 2000. Sediment yield following severe volcanic disturbance—a two-decade perspective from Mount St. Helens. *Geology* 28: 819–822.

McCune, B., and J.B. Grace. 2002. Analysis of ecological communities. MjM Software, Gleneden Beach (www.pcord.com).

Nakashizuka, T., S. Iida, W. Suzuki, and T. Tanimoto. 1993. Seed dispersal and vegetation development on a debris avalanche on the Ontake volcano, Central Japan. *Journal of Vegetation Science* 4: 537–542.

Peck, D., J.M. Lazorchak, and D.J. Klemm, eds. 2003. *Environmental monitoring and assessment program-surface waters: Western pilot study field operations manual for wadeable streams*. Corvallis: U.S. Environmental Protection Agency, Western Ecology Division. http://www.epa.gov/emap/html/pubs/docs/groupdocs/surfwatr/field/ewwsm01.pdf.

Pyke, D.A., T.A. Wirth, and J.L. Beyers. 2013. Does seeding after wildfires in rangelands reduce erosion or invasive species? *Restoration Ecology* 21: 415–421.

R Core Team. 2014. R: A language and environment for statistical computing. R Foundation for Statistical Computing, Vienna, Austria. URL http://www.R-project.org/.

Sala, O.E., L.A. Gherardi, L. Reichmann, E. Jobbágy, and D. Peters. 2012. Legacies of precipitation fluctuations on primary production: Theory and data synthesis. *Philosophical Transactions of the Royal Society—Biological Sciences* 367: 3135–3144.

Stapanian, M.S., S.D. Sundberg, G.A. Baumgardner, and A. Liston. 1998. Alien plant species composition and associations with anthropogenic disturbance in North American forests. *Plant Ecology* 139: 49–62.

Stroh, J.R., and J.A. Oyler. 1981. *SCS seeding evaluation on Mount Saint Helens: assessment of grass–legume seedlings in the Mount Saint Helens blast zone and lower Toutle River mudflow*. Spokane: U.S. Department of Agriculture, Soil Conservation Service.

Sun, O.J., J. Campbell, B.E. Law, and V. Wolf. 2004. Dynamics of carbon stocks in soils and detritus across chronosequences of different forest types in the Pacific Northwest, USA. *Global Change Biology* 10: 1470–1481.

Swanson, F.J., and J.J. Major. 2005. Physical events, environments, and geological–ecological interactions at Mount St. Helens, March 1980–2004. In *Ecological responses to the 1980 eruption of Mount St. Helens*, ed. V.H. Dale, F.J. Swanson, and C.M. Crisafulli, 27–46. New York: Springer.

Voight, B., H. Glicken, R.J. Janda, and P.M. Douglas. 1981. Catastrophic rockslide avalanche of May 18. In *The 1980 eruptions of Mount St. Helens, Washington*, Professional Paper 1250, ed. P.W. Lipman and D.R. Mullineaux, 347–378. Washington, DC: U.S. Geological Survey.

Waitt, R.B., Jr., T.C. Pierson, N.S. MacLeod, R.J. Janda, B. Voight, and R.T. Holcomb. 1983. Eruption-triggered avalanche, flood, and lahar at Mount St. Helens—effects of winter snowpack. *Science* 221: 1394–1397.

Washington Native Plant Society. 2006. Vascular plant list: Cowlitz county. http://www.wnps.org/plant_lists/counties/cowlitz/documents/CowlitzCounty.pdf. Accessed Nov 2016.

9 The Spread of Exotic Plant Species at Mount St. Helens: The Roles of a Road, Disturbance Type, and Post-disturbance Management

Lindsey L. Karr, Charles M. Crisafulli, and Jeffrey J. Gerwing

9.1 Introduction

9.1.1 Background

The landscape surrounding Mount St. Helens was dramatically altered during and after the 18 May 1980 eruption, creating an exceptional opportunity for ecologists to study plant succession following a major natural disturbance. To protect the land and allow natural succession to proceed largely unimpeded, on 27 August 1982, Congress designated 44,550 ha as the Mount St. Helens National Volcanic Monument (Dale et al. 2005). Some areas of the monument are closed or have limited public access and are primarily managed for research, whereas other areas allow recreation and education in ways that protect geological, ecological, and cultural resources. Despite efforts to preserve the area for natural processes and scientific study, anthropogenic disturbances such as roads and management activities have augmented the invasion potential of the disturbed landscape to exotic plants.

We adopted the Natural Resources Conservation Service's definition of an exotic plant species as one that is not native to the study area (www.plants.usda.gov, May 2014). These exotic plants may become invasive when there is a lack of competition, herbivory, or environmental controls; if their new system is disturbed or otherwise altered; or simply when sufficient propagules have been introduced (Lockwood et al. 2007). Although the focus of our study is on exotic plants, some research we cite examines exotic invasive plants, which are introduced species that have established and are currently causing or "likely to cause economic or environmental harm, or harm to human health" (Clinton 1999).

Exotic species are known to have high environmental and economic costs (Wilson 1992; Pimentel et al. 2005). Exotic invasive plants can compete with native species, interrupt food webs, and alter fire regimes, decomposition, hydrology, and nutrient cycling (Vitousek et al. 1997; Mack et al. 2000). Exotic plants that successfully colonize often respond well to increased light conditions and disturbance (Newsome and Noble 1986; Parendes and Jones 2000; Gray 2007). In an extensive literature review of exotic species and disturbance, Lozon and MacIsaac (1997) found that 86% of terrestrial exotic plant species depend on disturbance for establishment. Thus, it might be expected that the conditions created by the 1980 eruption would facilitate the establishment and spread of exotic species.

It is best to control exotic plant populations while they are small, as it is often more economical and efficient than waiting for them to cause a problem (Moody and Mack 1988) and because it can be virtually impossible to eradicate an invasive species once it has established (Mooney and Drake 1989). Furthermore, it can be quite difficult to predict when and where a plant will become invasive, and the line between invasive and noninvasive is not always clear (Lockwood et al. 2007). Therefore, we looked at all exotic plant species, rather than just plants considered to be invasive.

9.1.2 Exotic Plant Concerns in Volcanic Disturbance Zones

The 1980 eruption created several disturbance zones, each with different successional starting points and potential issues related to exotic plant species. Below we describe the types and mechanisms of volcanic disturbance for three of these zones, ranging

Terms in ***bold italic face*** are defined in the glossary.

L.L. Karr (✉) • J.J. Gerwing
Environmental Sciences and Management,
Portland State University, 218 SRTC- ESM, 1719 SW 10th Ave,
PO Box 751, Portland, OR 97207, USA
e-mail: lindsey@spu.edu; jgerwing@pdx.edu

C.M. Crisafulli
U.S. Department of Agriculture, Forest Service, Pacific Northwest Research Station, Mount St. Helens National Volcanic Monument, 42218 NE Yale Bridge Rd, Amboy, WA 98601, USA
e-mail: ccrisafulli@fs.fed.us

C.M. Crisafulli, V.H. Dale (eds.), *Ecological Responses at Mount St. Helens: Revisited 35 years after the 1980 Eruption*,
https://doi.org/10.1007/978-1-4939-7451-1_9

from most to least disturbed, and identify ecological features connected to each disturbance and relevant to exotic plants.

The pyroclastic-flow zone (Fig. 9.1) was the most intensely disturbed area and was influenced by a sequence of eruption events that commenced with a gigantic rockfall avalanche as the north flank of the volcano failed and cascaded downslope, followed by a powerful, hot, gas-driven pyroclastic density current, commonly referred to as a lateral blast and herein called the ***blast PDC***. It culminated in numerous superheated, incandescent ***pyroclastic flows*** (Lipman and Mullineaux 1981). This landscape was covered by over 100 m of ***debris-avalanche*** deposits, blast-PDC deposits, and pyroclastic flows during the main 1980 eruption event and received subsequent pyroclastic flows in 1980 and several ***lahars*** afterward (Lipman and Mullineaux 1981). Lacking any remnants from the pre-eruption ecosystem, biotic response initiated under primary succession (Walker and del Moral 2003). A portion of the pyroclastic-flow zone, named the Pumice Plain, is an area of intensive scientific study, and we conducted some of our research in the eastern portion of this area (Fig. 9.2). The pyroclastic-flow deposits possess several features problematic for plant establishment and growth. These include infertility, instability, and low water-holding capacity (Halvorson et al. 1991).

The blowdown zone, the largest disturbance zone in the area influenced by the blast PDC, extends beyond the pyroclastic-flow zone (Fig. 9.1) and covers the 370-km^2 area where the volcanic forces knocked down or snapped off trees (Lipman and Mullineaux 1981).

Up to 1 m of blast-PDC deposits and infertile tephra were deposited during the eruption. However, some vegetation survived under late-lying snow or on leeward slopes, leading to numerous epicenters of survival that fostered rapid regeneration (Halpern et al. 1990). The ***tephrafall*** zone extends beyond the blowdown zone, north- and eastward of the mountain, where forest overstory species were left largely intact, but understory plant communities were dramatically altered over 1000 km^2 by various amounts of tephra ejected from the volcano. Antos and Zobel (2005) observed survival of understory species in the first years after the 1980 eruption but also found that tephrafall in forests resulted in the death of many understory plants, largely depending on microsite conditions such as snowpack, pre-existing flora, and depth of tephra. Where tephra was shallower than 4.5 cm, most herbaceous plants survived, although bryophytes did not. However, when tephra was deeper, many herbaceous plants failed to push through the mantle of tephra and perished. Furthermore, where areas were covered in deep snow under thick ash, as the snow melted, plants became plastered to the forest floor and did not survive. Except for bryophytes, species not present in the pre-eruption forest had not invaded the tephra-covered forest by the year 2000 (Antos and Zobel 2005).

9.1.3 Road Influences on Exotic Plant Spread

The 1980 eruption provided opportunities for research, education, and recreation, and roads were needed for both access and management purposes. In 1983, Windy Ridge Viewpoint opened for visitors at the end of National Forest Road 99 (NFR-99) overlooking the Pumice Plain, only 6.5 km northeast of the crater (Fig. 9.1). This road winds through the tephrafall zone and blowdown zone (Fig. 9.1) and contains several viewpoints, parking lots, restroom facilities, and trailheads, attracting 17,000–20,000 visitor vehicles each year (USFS, unpublished data, 2015). In studies of global patterns of plant invasions, Lonsdale (1999) and Usher (1988) both found that species richness of exotic plants increased with the average number of visitors in nature reserves, most likely because tourists and their vehicles increase accidental introductions.

Road construction, road management, and the presence of roads themselves disturb roadside environments (Forman et al. 2003). The construction process disturbs existing soil and vegetation (Johnston and Johnston 2004) and typically creates bare earth, which ruderal and wind-dispersed annuals often colonize first (Forman et al. 2003). Mowing is a common roadside management technique that can actually benefit some exotic plant species, especially those that grow from rosettes, because they are not killed by mowing and can grow deeper roots to withstand roadside erosion (Forman et al. 2003). Roadside trees are often thinned for increased driver visibility, but this activity can facilitate many ruderal exotic species because of the increased light (Mortensen et al. 2009). The presence of roads can increase both the sunlight exposure and temperature of roadside habitat (Parendes and Jones 2000), whereas runoff from roads can increase the moisture and nutrients (Trombulak and Frissell 2000) and change the pH of roadside habitats (Johnston and Johnston 2004). All of these conditions can aid the establishment of exotic invasive species (Davis et al. 2000).

Not only do roadsides create microhabitats for exotic species, they can also serve as a point of introduction and a dispersal corridor for seeds (Christen and Matlack 2007). Vehicles associated with tourism and maintenance can increase wind and introduce exotic seeds and vegetative propagules that attach to tires and other parts of vehicles (Lonsdale and Lane 1993). Taylor et al. (2012) found that seeds can be carried by vehicles for over 250 km but will typically detach from vehicles when wet conditions are encountered. Wind, water, birds, and other animals may also travel along roadways and further spread seeds (Flory and Clay 2009).

9.1.4 Effects of Management Activities

Some management activities at Mount St. Helens have facilitated the entry of exotic species to the post-eruption landscape, including pre-eruption clearcutting, aerial seeding and

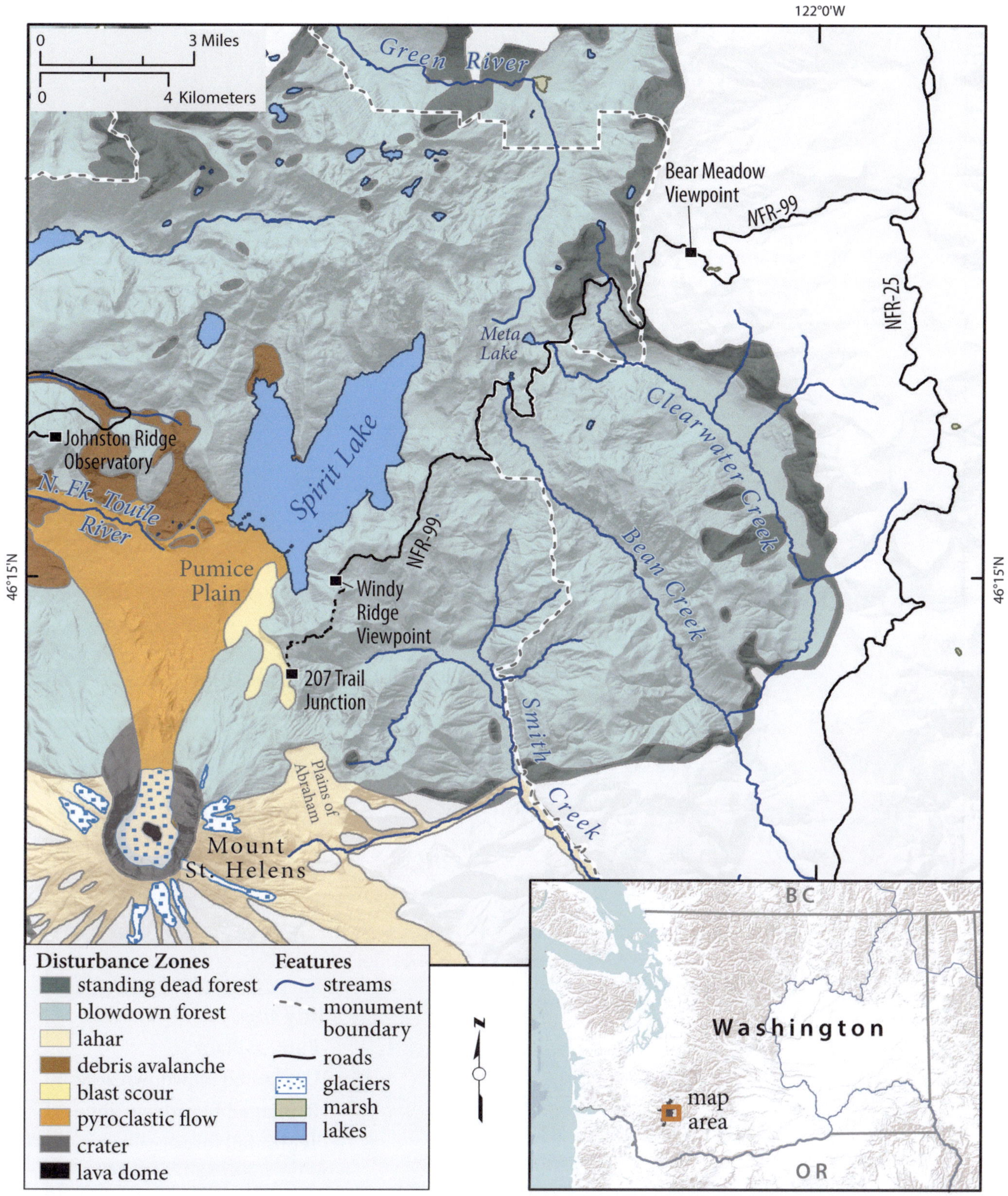

Fig. 9.1 Disturbance zones at Mount St. Helens, with viewpoints shown for context. Replanting in the blowdown zone was done only outside the monument boundary (*gray-dashed white line*). The tephrafall zone includes the area outside the standing dead forest zone to the north and east of the mountain. (Cartography by Kathryn Ronnenberg, USDA Forest Service, Pacific Northwest Research Station).

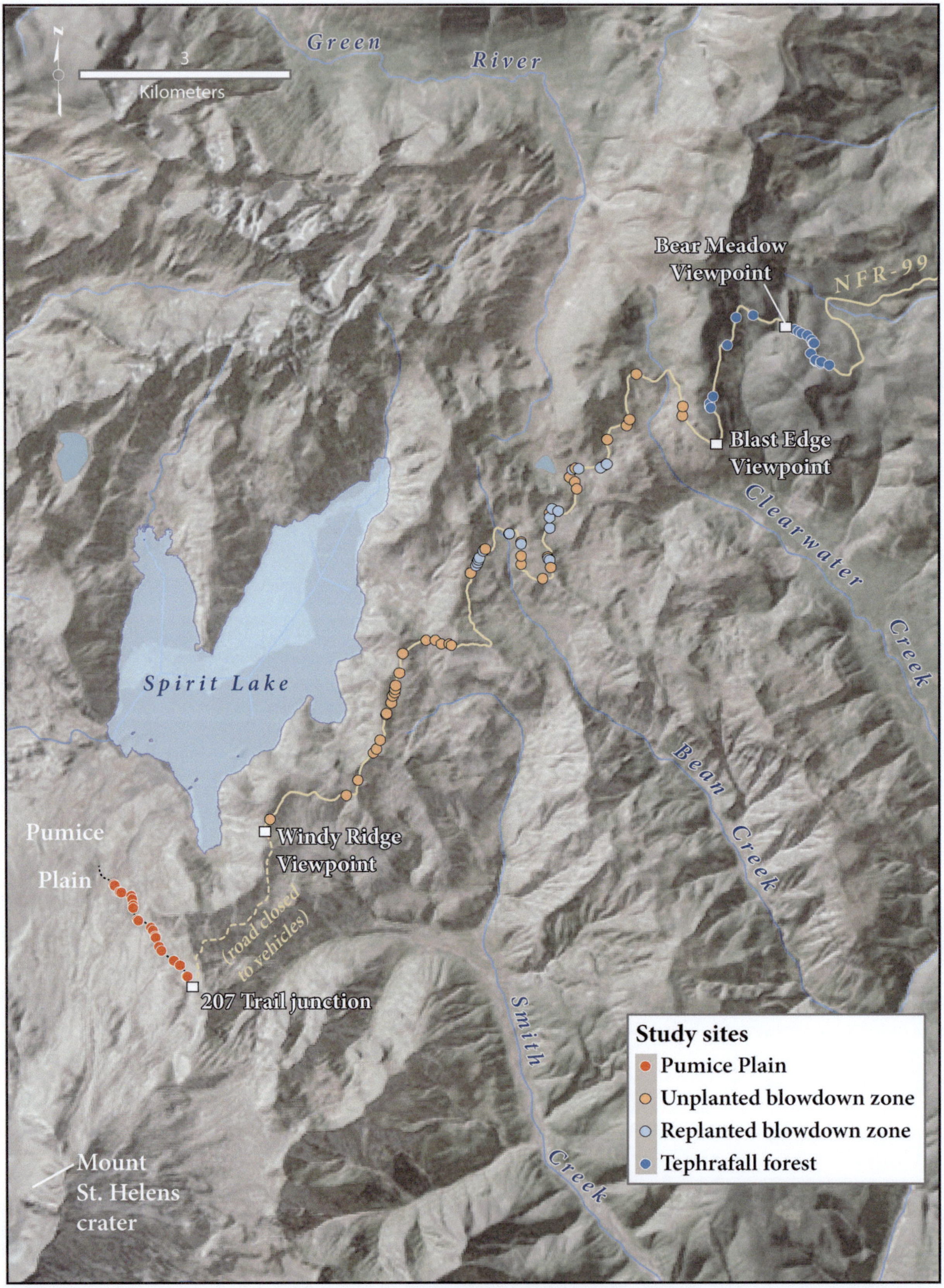

Fig. 9.2 Transect locations, showing National Forest Road (NFR) 99 and Trail 207. Transects were 70 m in length and were placed perpendicular to the road or trail at each point. (Cartography by Kelly Christiansen, USDA Forest Service, Pacific Northwest Research Station).

fertilization, salvage logging, tree replanting, bough harvesting, and tree thinning.

Pre-eruption clearcuts in the blowdown zone have increased the potential for exotic plants to establish. Many exotic plants are successful in conditions with increased light and disturbance (Parendes and Jones 2000; Gray 2005), which are characteristics of harvested forests. Gray (2005) found that exotic species were significantly more abundant in clearcuts compared to unharvested forest stands across four ecoregions in western Oregon. As Parendes (1997) noted, in a recently disturbed area, plants may reestablish from viable seed in the seedbank or from vegetative sprouting of surviving plant material. Since many seeds can remain viable in the soil for many years (Cook 1980), it follows that exotic species present in pre-eruption clearcuts could potentially invade the landscape disturbed by the 1980 eruption.

In 1980 and 1981, 8660 ha in the blowdown zone, North Fork Toutle River debris-avalanche deposit, and lower Toutle River were aerially seeded with 344 700 kg of seed, mostly exotic species, in an attempt to stabilize hillslopes, reduce erosion, and accelerate nutrient retention (Carlson et al. 1982). Carlson et al. (1982) record details of this seeding, and we give a summary of points that are relevant to our study. In the eastern portion of the blowdown zone, in the upper Clearwater and Bean Creek drainages, 3330 ha near our study area were seeded with four exotic plants (*Lolium multiflorum*, *Lolium perenne*, *Vicia villosa*, and *Trifolium subterraneum*). However, seeding only occurred in areas that had been recently clearcut and were below 1220 m in elevation. Seeding took place from 5 to 15 September 1980 at a rate of 37 kg ha^{-1} and was immediately followed by fertilizer application from 15 September to 20 October. Seeded areas were surveyed several times in the following year, and 9 months later, the seeded mixture showed 10.7% average ground cover, primarily *L. multiflorum* and *L. perenne*. However, these grasses also showed symptoms of moisture, temperature, and nutrient stress and produced few viable seeds. A refertilization was recommended but never occurred. Furthermore, Carlson et al. (1982) expected that any surviving populations of these four plants would decline rapidly.

On the North Fork Toutle River debris-avalanche deposit, about 10 km west of our study area, 4920 ha were seeded with *L. multiflorum*, *L. perenne*, *Festuca rubra*, *Phleum pratense*, *Dactylis glomerata*, *Trifolium repens*, and *Lotus corniculatus* (all exotic except *F. rubra*) at a rate of 40 kg ha^{-1} (above 853 m) and 43 kg ha^{-1} (below 853 m). Nine months later, surveys showed that this seeding was more successful than the blowdown zone sites, as the seeded areas had 27% (above 853 m) and 73.4% (below 853 m) ground cover. This ground cover consisted primarily of grasses: largely *L. multiflorum* and *L. perenne*, with some *F. rubra* and small amounts of *P. pratense*, *D. glomerata*, and one legume: *T. repens*. Although the debris-avalanche deposit is not part of our study area, seed dispersal is largely affected by prevailing winds from adjacent areas (Dale 1989), and seeds could disperse to our study area. The North Fork Toutle River debris-avalanche deposit was also fertilized after the initial seeding, but 5 months later, surveys showed that virtually none of the nitrogen and phosphorus was available for the surviving plants.

In 1993 and 1994, Titus et al. (1998) conducted a thorough inventory of vascular plants in blast-zone areas disturbed by the 1980 eruption, including the Pumice Plain, Plains of Abraham, Toutle debris-avalanche deposit, Toutle Ridge, and the crater. They found that exotic plant species comprised 57 of 341 (16.7%) plant species, with six of the initial aerially seeded exotic species found in their plots (*L. multiflorum* and *L. perenne*, *P. pratense*, *D. glomerata*, *L. corniculatus*, and *T. repens*). Twenty years after this survey, exotic species continue to influence natural succession (Dale and Adams 2003; Schoenfelder et al. 2010; Dale and Denton, Chap. 8, this volume). One example of this influence occurs on the debris-avalanche deposit area of Mount St. Helens, where sites that were heavily invaded by exotic species had 25% lower conifer survival than uninvaded sites (Dale 1991).

Along with the aerial seeding, extensive salvage logging and replanting efforts also occurred in the blast-zone forests (Foster et al. 1998). Salvage logging occurred from 1980 to 1982 in blast areas outside the Monument, as well as some areas inside the Monument before its official designation. Blast areas outside the Monument were replanted between 1981 and 1987 with Douglas-fir (*Pseudotsuga menziesii*) at elevations less than 1000 m and noble fir (*Abies procera*) at elevations greater than 1000 m, including our study area, which was replanted from 1983 to 1985 (R. Pankratz, personal communication). In July 2003, Titus and Householder (2007) surveyed blowdown areas near Meta Lake and compared unsalvaged/unplanted areas with salvaged/replanted areas. They reported that the canopy cover of trees in salvaged/replanted plots was five times higher than in unsalvaged/unplanted plots. They also found that unsalvaged/unplanted sites had greater diversity, richness, herb and shrub cover, downed woody debris, and litter depth compared to salvaged/replanted sites. Although they reported no significant differences in exotic plant cover between the two areas, exotic species were present in every 200-m^2 plot. Furthermore, the exotic species they found are plants that frequently occur in disturbed areas of the region and do not typically dominate landscapes.

Since 2006, commercial thinning and bough harvesting have occurred in replanted blowdown zone forests (R. Pankratz, personal communication), creating many canopy openings. Exotic plant cover and richness often increase in response to thinning treatments, and both Nelson et al. (2008) and Bailey et al. (1998) found significant increases of exotic plant cover and richness in thinned stands compared with unthinned stands in the Pacific Northwest. Furthermore, Bailey et al. (1998) found that roadsides adjacent to the thinned stands had an even greater increase in exotic plant abundance compared with those near unthinned stands.

The creation of the National Volcanic Monument and associated restricted areas was intended to allow succession to occur substantially unimpeded, and these restrictions may have hindered the spread of exotic plants into the area. The pyroclastic-flow zone and parts of the debris-avalanche deposit are closed to camping, off-trail hiking, dogs, horses, and all vehicles including bicycles, to minimize anthropogenic impacts on this sensitive area. Many studies have shown that hikers and vehicles can spread plant propagules (Lonsdale and Lane 1993; Pickering and Mount 2010; Taylor et al. 2012; Ansong and Pickering 2014), so these restrictions have likely reduced the introduction of exotic plants.

9.1.5 Research Objective

Understanding how both natural and anthropogenic disturbances are affecting the spread of exotic plant species across the Mount St. Helens landscape is important for both land managers and scientists (Dale and Denton, Chap. 8, this volume). The post-eruption conditions and subsequent management activities at Mount St. Helens created a variety of conditions with potentially different levels of invasibility to exotic plants. In this study, we ask how these different volcanic disturbance zones, post-disturbance management, and NFR-99 are affecting the richness, abundance, and frequency of exotic plant species at Mount St. Helens.

9.2 Methods

9.2.1 Study Area

Our study included three disturbance zones on the north and northeast sides of the volcano: pyroclastic-flow zone, blowdown zone, and tephrafall zone (Swanson and Major 2005) (Figs. 9.1 and 9.2). Conveniently, National Forest Road 99 (NFR-99) transects the blowdown and tephrafall zones, providing an opportunity to assess patterns of exotic plant spread within these zones. This 26-km paved road is associated with several vistas and trailheads and is open for tourism and recreation from June or July until snow covers roads in the autumn (USDA 2013). Our study design included 70-m transects along NFR-99 in the blowdown and tephrafall zones and 2.3 km of the Truman Trail (USFS Trail 207) in the pyroclastic-flow zone (Figs. 9.1 and 9.2).

To eliminate potential variation in plant communities related to elevation differences across roadside sites, we did not collect data along the first 6 km of road, because it has substantial elevation gain. Thus, data were collected between the 6-km mark, just east of the Bear Meadow Viewpoint, and the Windy Ridge Viewpoint (from 46°18′34″N, 122°01′46″W to 46°15′03″N, 122°08′07″W), with a modest elevation range of 220 m (1100–1320 m) across sites. Data were also collected along Trail 207, beginning at the junction with Trail 216E and heading 2.3 km northwest, between 46°13′49″N, 122°09′05″W and 46°14′33″N, 122°09′53″W (Fig. 9.2).

A portion of the pyroclastic-flow zone, named the Pumice Plain (Fig. 9.3a), covers 15 km^2 directly north of the crater and is characterized by primary succession. Two years after the eruption, isolated individuals of *Lupinus lepidus* marked the first plant colonization of the area (Halvorson et al. 1991). During our study, upland vegetation on the Pumice Plain was dominated by a few forbs and graminoids, namely, *Agrostis pallens*, *Castilleja miniata*, *Lupinus lepidus*, and *Penstemon cardwellii*, often forming dense and expansive patches, punctuated by scattered *Salix sitchensis* and *Alnus viridis* ssp. *sinuata* shrubs and occasional conifer saplings. Dissecting the upland communities were springs, seeps, and small streams that support dense riparian shrub communities of *Salix* and *Alnus*.

The blowdown zone covers 370 km^2 where overstory trees were leveled or snapped off (Swanson and Major 2005). Outside the Monument, in replanted areas of our study (Fig. 9.3d), extensive plantations of *Abies procera* now dominate the landscape (Titus and Householder 2007), with understory species commonly consisting of *Agrostis pallens*, *Anaphalis margaritacea*, *Hypochaeris radicata*, and *Vaccinium* spp., though bare substrate is also common. Canopy openings have been created by tree thinning and bough harvesting. Thinning occurred at a 4×4-m spacing (with 25% variability) in 2007, 2008, and 2010 for 8, 7, and 3 of our transects, respectively, whereas 2 transects were unthinned as of 2013. Tree boughs were harvested in 2006, 2009, and 2012 along 15, 3, and 2 of our transects, respectively, though no boughs were harvested within 15 m of the road. Furthermore, half of our transects were fertilized: 6 transects in 2005 and 4 transects in 2010 (R. Pankratz, personal communication).

Inside the Monument, in areas of the blowdown zone where trees were not replanted, diverse assemblages of graminoids, herbs, shrubs, and trees have developed (Fig. 9.3b, c). Some areas are densely vegetated, with shrub thickets dominated by *S. sitchensis* and *A. viridis* ssp. *sinuata*, often with *Anaphalis margaritacea*, *Rubus lasiococcus*, *Rubus spectabilis*, and *Vaccinium* spp. Other areas consist of more open habitat and are characterized by forbs, graminoids, and low-lying shrubs, often *Agrostis pallens*, *Anaphalis margaritacea*, *Chamerion angustifolium*, *Hypochaeris radicata*, *Penstemon cardwellii*, and *Pteridium aquilinum*. In areas that were not salvage logged, pre-eruption clearcuts are characterized by many stumps and pre-eruption forests by numerous down logs (Titus and Householder 2007).

Beyond the blowdown zone, 15 km from the crater, forests were left largely intact, but vast areas were affected by

Fig. 9.3 Four different site types. (**a**) Pumice Plain; (**b**) unplanted blowdown zone (shrub-dominated); (**c**) unplanted blowdown zone (forb-dominated); (**d**) replanted blowdown zone; (**e**) tephrafall forest. (Photos: L.L. Karr).

tephra deposits, resulting in the death of many understory plants (Antos and Zobel 2005). Forest stands today (Fig. 9.3e) are dominated by large late-seral conifer trees (*Abies amabilis*, *Tsuga mertensiana*, *T. heterophylla*, and *P. menziesii*), with saplings in the understory (Antos and Zobel 2005). During 2013, the ground cover consisted of large quantities of conifer-needle litter, and understory plants included scattered *Vaccinium* spp. and *Menziesia ferruginea*.

9.2.2 Study Design and Data Collection

Vegetation and physical site data were gathered over a 5-week period at peak biomass and flowering season, beginning in August and extending into early September of 2013. We studied three zones of disturbance intensity: pyroclastic-flow zone (Pumice Plain), blowdown zone, and tephrafall zone. Within the blowdown zone, we further discriminated between areas that were replanted with *Abies procera* and areas that were left unplanted, creating four total site types: Pumice Plain (PP), unplanted blowdown zone (UBZ), replanted blowdown zone (RBZ), and tephrafall forest (TFF) (Table 9.1). We used a random number generator and Google Earth to select random transect sites: 20 sites in each of the PP, RBZ, and TFF and 40 in the UBZ, for a total of 100 transects. The UBZ covered the largest area and had a variety of habitat, so we used a stratified random sampling method (Michalcov et al. 2011) and imagery from Google Earth to ensure data were gathered equally in both shrub-dominated and forb-dominated landscapes. We excluded sites with recent logging activity or herbicide treatment, as well as viewpoints, parking lots, culverts, and areas that were too steep or unsafe for walking. Sites were located on the ground with the aid of a handheld global positioning system (GPS).

To determine species richness, abundance, frequency, and spread of exotic species from the road into the adjacent

Table 9.1 Site types and descriptions along NFR-99 in the Gifford Pinchot National Forest.

Site type	Volcanic features	Description
Pumice Plain (pyroclastic-flow zone)	Debris-avalanche, blast-PDC deposits, pyroclastic flows, lahars	Area affected by the 1980 debris avalanche, blast PDC, pyroclastic flows (flowing volcanic hot rock and gas), and occasional lahars. Open landscape with mostly forbs, mosses, and sparse shrubs and trees. Characterized by primary succession.
Unplanted blowdown zone	Scorched and blown-down trees, tephra deposits	Area is passively managed to allow succession to occur largely unimpeded. Area where trees were blown down, snapped off, or scorched by the 1980 blast PDC. Some areas were salvage logged. Landscape consists of either open areas dominated by forbs, graminoids, and bare ground, or other areas with dense shrubs (*Alnus*, *Salix*, *Vaccinium*) and some trees.
Replanted blowdown zone	Scorched and blown-down trees, tephra deposits	Managed forest outside the National Volcanic Monument. Area where trees were blown down, snapped off, or scorched by the 1980 blast PDC. Areas inside this zone were salvage logged and replanted by the USFS with mostly *Abies procera* and are occasionally thinned or harvested for tree boughs.
Tephrafall forest	Cool tephra deposits	Forest outside the area influenced by the blast PDC, affected by cool tephra deposits (silt- to gravel-sized) from the 1980 eruption.

landscape, we recorded the presence of all vascular plant species along 100 transects, 70 m in length, positioned perpendicular to the road and greater than 20 m away from the nearest transect. We placed 1-m^2 quadrats along each transect at increasing distances from the road: 0 m, 2 m, 5 m, 10 m, 20 m, 30 m, 50 m, and 70 m (following Delgado et al. 2007). At each quadrat, we recorded all vascular ground cover plant species less than 1.5 m in height and rooted in the quadrat, with their percent cover values. Species were identified to the lowest taxonomic level possible, and percent cover was visually estimated in increasing increments as follows: less than 1% was estimated to be 0.25 or 0.5%, 1–15% was estimated at 1% increments, and 15–100% was estimated at 5% increments (adjusted from Abella et al. 2012). For consistency, one observer made these cover estimates in 92 out of 100 quadrats using a 10×10-cm square to estimate 1%, whereas, the plant cover in the remaining eight quadrats (on the Pumice Plain) was estimated by two observers trained by the main observer. Unknown plants were collected, pressed, and later identified using Hitchcock and Cronquist (1973) or Kozloff (2005). Our taxonomy was standardized according to the Integrated Taxonomic Information System online database, http://www.itis.usda.gov, and data were retrieved in April 2014. Shrub species in the genera *Salix* and *Vaccinium* were identified only to genus level, as all species in those genera in the Mount St. Helens area are native. Vascular ground cover species included herbaceous flowering plants, as well as crawling and prostrate shrubs (e.g., *Rubus lasiococcus* and *Penstemon* spp.), ferns, horsetails, and graminoids (Poaceae, Juncaceae, Cyperaceae). We also included upright woody shrub species typically less than 1.5 m in height, including member species of the genera *Arctostaphylos*, *Gaultheria*, *Menziesia*, *Pachystima*, *Ribes*, *Rubus*, and *Vaccinium*. Larger shrub species of *Acer*, *Alnus*, *Amelanchier*, *Salix*, *Sambucus*, and *Sorbus* function more as canopy and were thus excluded from ground cover and were instead treated as "canopy" species. Canopy cover was visually estimated and placed into four categories: (1) zero, (2) low (1–30%), (3) medium (31–70%), and (4) high (71–100%) (following Parendes and Jones 2000). For consistency, one observer made all canopy cover estimates, with the exception of eight transects on the Pumice Plain where canopy cover was zero. We determined elevation at the road later using data available from Google Earth.

Although we sampled 800 quadrats, each quadrat was only 1 m^2, and when considering 12 km of roadside and 2.3 km of trail, 800 m^2 is small. Realizing our resources were limited, we decided against collecting data in a few large plots in favor of a greater number of transects (100), though this would mean less total area would be sampled. In doing so, we covered a wide variety of roadside habitat.

9.2.3 Data and Statistical Analysis

We classified plant species as either native or exotic according to the Natural Resources Conservation Service (NRCS) Plants Database (www.plants.usda.gov, April 2014). To examine patterns of exotic plant frequency with respect to the road, we summed the number of quadrats in which each exotic species occurred (with frequency >3), at each distance from the road for all sites, excluding Pumice Plain sites, as they were not adjacent to a road. To characterize canopy cover, we calculated median values from four categories (zero, low, medium, high; see Sect. 9.2.2) for each site type. In order to characterize canopy cover without road or trail influence, we used canopy cover values of sites that were 70 m from the road (or trail).

We determined exotic species richness by summing the number of exotic species found in each quadrat and then calculating the mean ±1 standard error (SE) at each transect distance within each site type. We also determined native, exotic, and total plant abundance by calculating the mean percent cover ±1 SE for each transect distance, within each site type. To evaluate the influence of the road on exotic plant richness and abundance, we used the Wilcoxon Rank Sum Test to determine whether the richness and abundance of exotic plants at 0 m and 70 m were significantly different (at $\alpha < 0.05$) within each site type.

9.3 Results

9.3.1 General Flora and Exotic Species Patterns

We identified 152 plant species (Appendix 9.1). Apart from *Salix* spp. and *Vaccinium* spp., 13 additional taxa were identified only to the genus level, although some of these taxa belonged to the 152 identified species (e.g., some *Maianthemum* spp. specimens could not be distinguished between *M. racemosum* and *M. stellatum*). Such individuals could not be identified to species because they were juvenile, lacked fruits or flowers, or had been damaged by herbivory.

Twenty-two exotic species, or 14% of the total sampled flora, were recorded from the quadrats, all of which were forbs (17 species) or grasses (5 species) (Table 9.2). Fourteen of those species occurred in more than three quadrats (Fig. 9.4). Of these, ten were perennials, two were annuals (*Euphrasia officinalis*, *Mycelis muralis*), and two were either annuals or biennials (*Jacobaea vulgaris* and *Lepidium campestre*). Several species appeared to be more frequent near roadsides: *Hypochaeris radicata*, *Leucanthemum vulgare*, *Poa compressa*, *Lotus corniculatus*, *Euphrasia officinalis*, *Trifolium repens*, and *Taraxacum officinale* (Fig. 9.4). With the exception of *M. muralis* (formerly named *Lactuca muralis*), the shade tolerance of these species was intermediate to intolerant according to the NRCS (www.plants.usda.gov, May 2014). According to the Washington State Noxious Weed List (www.nwcb.wa.gov, April 2014), five of the more frequent species from our study were listed as noxious weeds: *Linaria vulgaris*, *Jacobaea vulgaris*, *Hypericum perforatum*, *Hypochaeris radicata*, and *Leucanthemum vulgare*. Three additional exotic species, *Cytisus scoparius* (shrub), *Tragopogon dubius* (forb), and *Verbascum thapsus* (forb), were observed outside of quadrats and noted in the species list (Table 9.2) but not included in analyses. The most abundant exotic species, in descending order, were *Hypochaeris radicata*, *Rumex acetosella*, *Anthoxanthum odoratum*, *Leucanthemum vulgare*, *Hypericum perforatum*, *Mycelis muralis*, and *Lotus corniculatus*. *H. radicata* was

Table 9.2 Exotic species in each site type. *PP* Pumice Plain, *UBZ* unplanted blowdown zone, *RBZ* replanted blowdown zone, *TFF* tephra-fall forest. Values are mean percent cover for all quadrats within each site type. 1 = 0–0.09% cover, 2 = 0.1 to 0.39% cover, 3 = 0.4 to 1.85% cover. The letter "W" indicates species on the Washington State Noxious Weed List (www.nwcb.wa.gov, April 2014).

	Site type				
Species name	**PP**	**UBZ**	**RBZ**	**TFF**	**Max. obs. dist.**[a]
Agrostis capillaris, colonial bentgrass			1		
Anthoxanthum odoratum, sweet vernalgrass		3	2		
Cirsium arvense (W), Canada thistle		1			
Digitalis purpurea, purple foxglove		1			
Euphrasia officinalis, eyebright		1	1		2 m
Holcus lanatus, common velvetgrass		1			
Holcus mollis, creeping velvetgrass				1	2 m
Hypericum perforatum (W), common St. Johnswort		2	2		
Hypochaeris radicata (W), hairy cat's ear	2	3	3	2	
Jacobaea vulgaris (W), tansy ragwort, stinking willie		1	1		
Lepidium campestre, field pepperweed		2	1		
Leucanthemum vulgare (W), oxeye daisy		2	3	1	
Linaria vulgaris (W), butter and eggs		1			10 m
Lotus corniculatus, bird's-foot trefoil		2	2		
Mycelis muralis, wall-lettuce		2	2	2	
Plantago lanceolata, narrowleaf plantain		1	1		10 m
Poa compressa, Canada bluegrass		1	2		10 m
Rumex acetosella, common sheep sorrel	1	3	2	1	
Rumex crispus, curly dock		1			10 m
Spergularia rubra, red sandspurry		1			
Taraxacum officinale ssp. *officinale*, common dandelion		1		1	2 m
Trifolium repens, white clover				1	5 m
Tragopogon dubius, yellow salsify[b]					2 m

(continued)

Table 9.2 (continued)

	Site type				Max. obs.
Species name	**PP**	**UBZ**	**RBZ**	**TFF**	**dist.**[a]
Cytisus scoparius, Scotch broom[b]					2 m
Verbascum thapsus, common mullein[b]					
Total count of species	2	19[c]	13	7	

[a]Roadside-only species: maximum distance observed from the road.
[b]Species were observed outside of study plots.
[c]The unplanted blowdown zone had 40 transects, whereas other zones had 20.

the most abundant exotic plant species in every site type (Table 9.2, Fig. 9.4).

Exotic species abundance was low overall, compared with native species (Fig. 9.5, Appendix 9.2). Mean exotic species abundance ranged from 0 to 6.1 (±2.6 SE) percent cover, whereas native species abundance ranged from 7.5 (±1.8 SE) to 39.7 (±6.4 SE) percent cover for the four site types and different distances from the road. The highest mean values for exotic species richness were 2.1 species (±0.24 SE) occurring in the replanted blowdown zone at 2 m from the road and 2 species (±0.18 SE) in the unplanted blowdown zone at 2 m from the road (Fig. 9.6a, b). The highest mean exotic species abundance values were 6.1% cover (±2.6 SE) and 6.0% cover (±1.6 SE) in the replanted blowdown zone at 5 m and 2 m from the road, respectively, and 5.5% cover (±0.98 SE) in the unplanted blowdown zone at 0 m from the road (Fig. 9.6e, f; Appendix 9.2).

9.3.2 Differences Among Site Types

Overall, 58% of all quadrats contained at least one exotic species. In the Pumice Plain, 26% of plots contained an exotic species, whereas in the unplanted blowdown zone, replanted blowdown zone, and tephrafall forest, 82%, 77%, and 23% of quadrats contained at least one exotic species, respectively. The two blowdown zone site types contained the greatest number of exotic species (Table 9.2). Fifteen of the 22 exotic species were found only in the blowdown zone, but just two exotic species (*Holcus mollis* and *Trifolium repens*) were unique to the tephrafall forest, and no exotic species were unique to the Pumice Plain (Table 9.2).

The four site types had different median levels of canopy cover. Both the replanted and unplanted blowdown zones are characterized by low (1–30%) canopy cover. The canopy cover in the Pumice Plain was zero, whereas the tephrafall forest had the highest median canopy cover (71–100%). Before the commercial thinning and bough harvesting of the replanted blowdown zone that occurred between 2006 and 2012, the canopy cover was higher than 2013 levels.

9.3.3 Site Type Differences Across Distances from the Road or Trail

On the Pumice Plain, where transects were perpendicular to a trail, rather than a road, the total plant cover was low compared with other sites and was dominated by native plants (Fig. 9.5d). The most abundant species were *Lupinus lepidus*, *Penstemon cardwellii*, and *Agrostis pallens*, all natives (Appendix 9.1). Only two exotic species were observed on the Pumice Plain, *Hypochaeris radicata* and *Rumex acetosella* (Table 9.2). *R. acetosella* was observed only once and *H. radicata* was observed at various distances from the trail in low cover amounts, typically less than 1%. The maximum percent cover (6%) for *H. radicata* was observed once. There was no significant difference in richness or abundance of exotic species between the trailside quadrats (0 m) and quadrats far from the trail (70 m) (Fig. 9.6d, h).

In the unplanted blowdown zone, although exotic species represented a higher percentage of the total flora than on the Pumice Plain, native species still accounted for more of the total cover than exotics (Fig. 9.5a). The most abundant species were *Anaphalis margaritacea*, *Pteridium aquilinum*, *Rubus lasiococcus*, and *Vaccinium* spp., all natives (Appendix 9.1). Nineteen exotic species were observed, with *Anthoxanthum odoratum*, *Hypochaeris radicata*, and *Rumex acetosella* being the most abundant (Table 9.2). Abundance of exotic plants was significantly higher near the road than far from the road, with 5.5% cover (±1.0 SE) at 0 m from the road and 3% cover (±0.6 SE) at 70 m from the road (Fig. 9.6e). However, exotic plant richness was not significantly different near and far from the road, with 1.8 species (±0.2 SE) at 0 m from the road and 1.4 species (±0.2 SE) at 70 m from the road (Fig. 9.6a).

In the replanted blowdown zone, native species also accounted for more of the total flora (Fig. 9.5b), with *Agrostis pallens*, *Anaphalis margaritacea*, *H. radicata*, and *Vaccinium* spp. as the most common species (Appendix 9.1). However, at 2 m and 5 m from the road, exotic plants composed a greater proportion of the total flora than at any other location in this study at 44% of total cover (±7% SE) and 43% of total cover (±7 SE), respectively (Appendix 9.2, Fig. 9.5b). Thirteen exotic plant species were identified, and *H. radicata* and *Leucanthemum vulgare* were the two most abundant species (Table 9.2). Both exotic plant richness and abundance were significantly higher near the road (0 m) than far from the road (70 m) (Fig. 9.6b, f). Exotic plant richness was 1.8 species (±0.2 SE) at 0 m from the road and 1.0 species (±0.2 SE) at 70 m from the road, whereas abundance was 4.7% cover (±1.3 SE) at 0 m from the road and 1.7% cover (±0.4 SE) at 70 m from the road (Fig. 9.6b, f).

The flora in the tephrafall forest, where median canopy cover was high, was also largely dominated by native plants

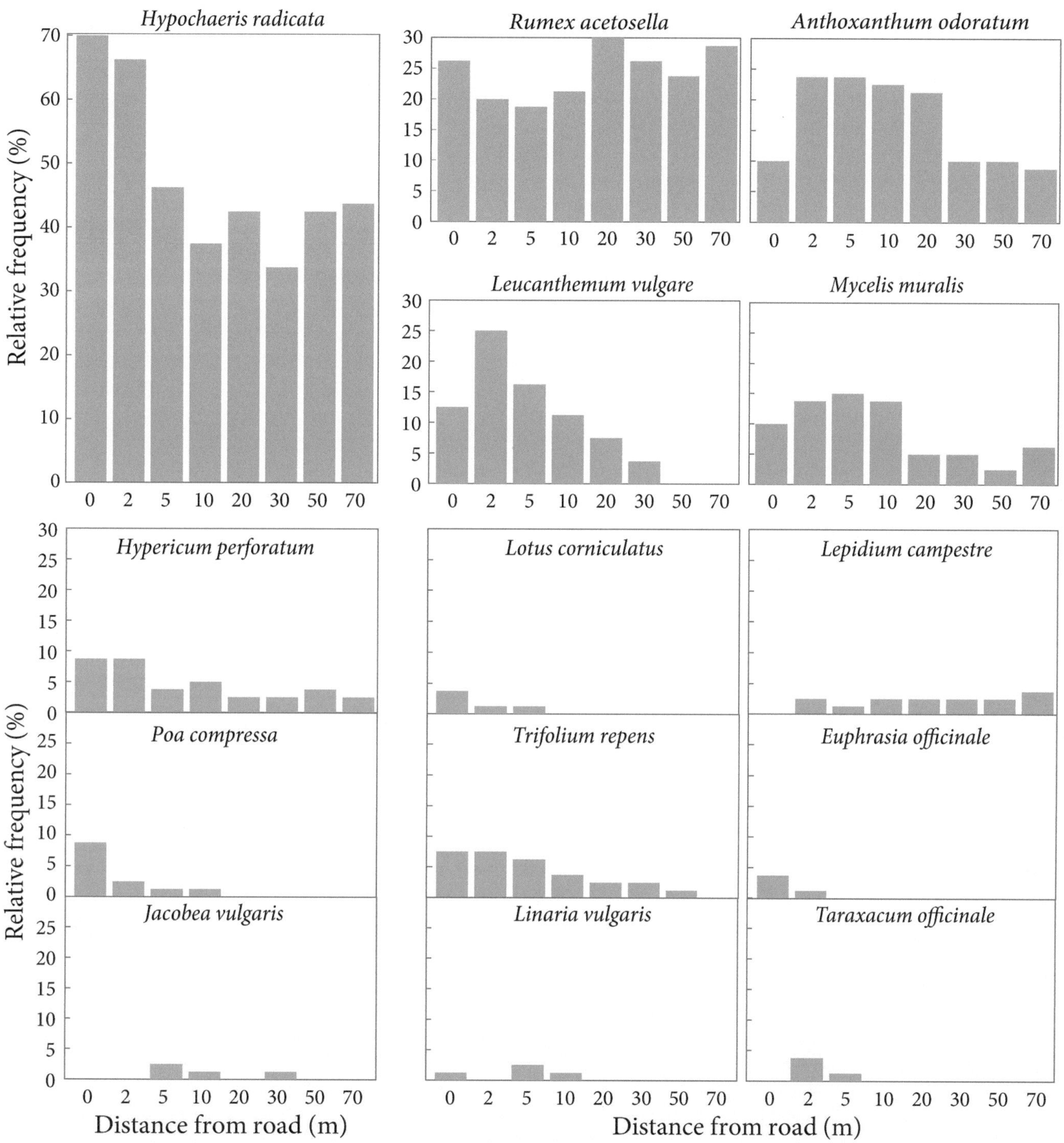

Fig. 9.4 Relative frequency graphs for exotic species occurring in more than three quadrats in the tephrafall forest, unplanted blowdown zone, and replanted blowdown zone ($n = 640$). Species are shown in order of most to least frequent.

(Fig. 9.5c), and exotic plants showed a very distinct pattern relative to distance from the road. In this zone, almost all exotic plants were restricted to the roadside, and thus, roadside quadrats were significantly higher in both exotic plant richness and abundance compared with quadrats far from the road (Fig. 9.6c, g). At distances of 10, 20, and 70 m, only one exotic plant was observed at each distance (either *H. radicata* or *Mycelis muralis*), and no exotic plants were observed at distances of 30 and 50 m (Fig. 9.5c). However, across those same distances, native plant cover ranged from 25 (±5 SE) to 32 (±7 SE) percent

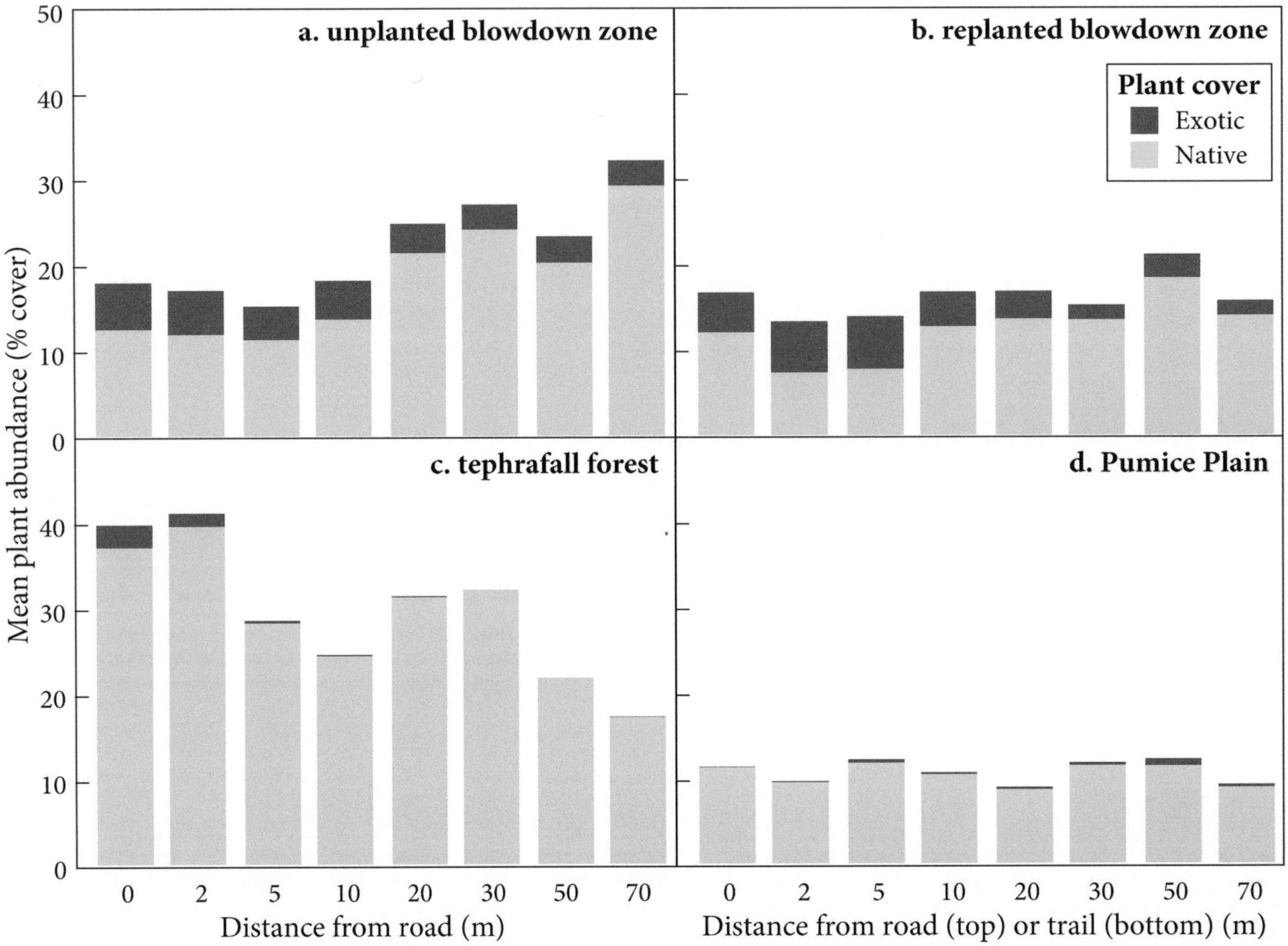

Fig. 9.5 Total plant abundance divided into native and exotic plant cover. The unplanted blowdown zone, replanted blowdown zone, and tephrafall forest are characterized by secondary succession and distances are from a road, whereas the Pumice Plain is characterized by primary succession and distances are measured from a trail. For standard errors of total abundance, see Appendix 9.2.

cover (Fig. 9.5c, Appendix 9.2). Exotic plant richness was 1.3 species (±0.2 SE) at 0 m from the road and 0.05 species (±0.05 SE) at 70 m from the road, and exotic plant abundance was 2.6% cover (±0.6 SE) at 0 m from the road and 0.05% cover (±0.05 SE) at 70 m from the road (Fig. 9.6c, g). Seven exotic plant species were observed, and the two most abundant of these species were *H. radicata* and *M. muralis* (Table 9.2). Overall, the understory was sparsely vegetated, and the most abundant plants were *Anaphalis margaritacea*, *Lupinus latifolius*, *Menziesia ferruginea*, *Vaccinium* spp., as well as *Abies amabilis* and *Tsuga mertensiana* saplings (Appendix 9.1).

9.4 Discussion

9.4.1 Low Exotic Plant Cover

Different disturbance types resulting from the 1980 eruption of Mount St. Helens created a highly variable landscape, and aerial seeding, timber management, and NFR-99 have provided potential pathways for some exotic plant species to increase in richness and abundance. Despite this potential and the increased richness and abundance of exotic plants near the road, the overall richness and abundance of exotic plants in all areas and at all distances from the road was relatively low (Fig. 9.6).

One possible explanation for the low richness and abundance of exotic plants relates to the elevation and associated climate at Mount St. Helens. In our study, the elevation ranged from 1100 to 1320 m, and thus, during the wetter months of November through April, much of the precipitation falls as snow, and our study sites may be under deep (2–3 m) snowpack for 5–6 months. The summer growing period is typically dry, and volcanic tephra substrate has poor nutrient status, is porous, and does not retain water well (Reynolds and Bliss 1986; del Moral and Wood 1993). Many of the more successful exotic plants are small-seeded and wind-dispersed, and thus, they dry out quickly and are not able to establish, whereas larger seeds are better able to withstand the stressful conditions typical at Mount St. Helens

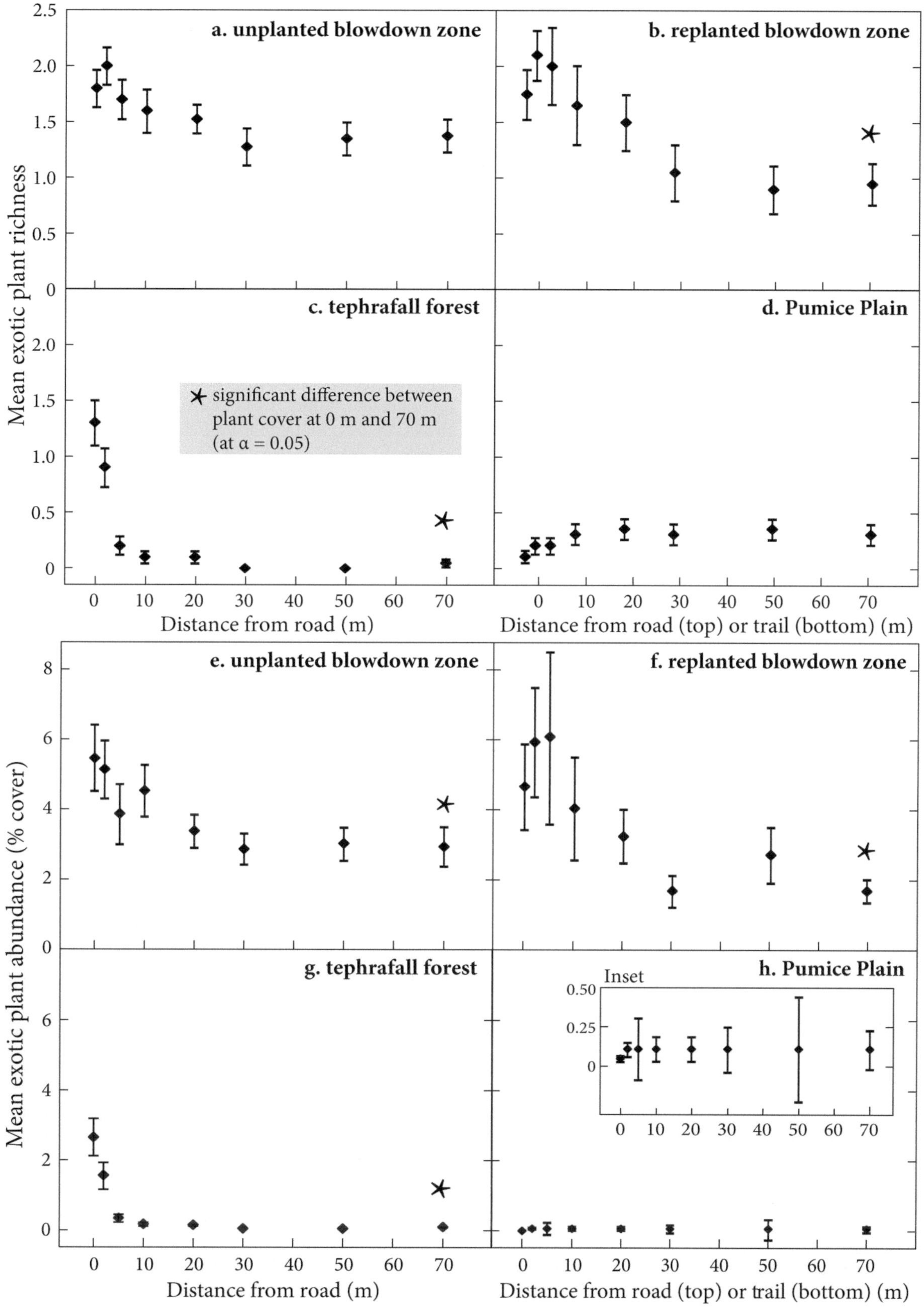

Fig. 9.6 Means ±1 standard error for (**a–d**) exotic plant richness and (**e–h**) abundance in the four site types. A star indicates significant differences at $\alpha = 0.05$ between 0 m and 70 m from the road. The unplanted blowdown zone, replanted blowdown zone, and tephrafall forest are characterized by secondary succession and distances are measured from a road, whereas the Pumice Plain is characterized by primary succession and distances are measured from a trail. *Inset* on Pumice Plain abundance graph (**h**) expands the y-axis scale to show small differences more clearly.

(Fuller and del Moral 2003). This pattern relates to Wood and del Moral's (1987) finding that many species with high dispersal traits are typically not able to establish on the stressful sites of open volcanic areas, and to Grime's (1979) conclusions that ruderal, successful dispersers do not typically establish at high elevations.

The 1980 aerial seeding was a potential source of invasive plant seeds (Carlson et al. 1982). According to Carlson et al. (1982), the four exotic seeded plants in the blowdown zone (*Lolium multiflorum*, *Lolium perenne*, *Vicia villosa*, and *Trifolium subterraneum*) had low survival rates. Furthermore, none of these species were observed in this study. However, on the North Fork Toutle River debris avalanche, survival of some of the seeded species was higher, and two of the exotic seeded species were found in our transects (*Trifolium repens* and *Lotus corniculatus*). Both species had relatively low frequency and were more common near the road. As neither of these species is wind dispersed and the seeded area is 10 km away from our study site, it is possible they were introduced by other sources such as animals, vehicles, or pre-eruption clearcutting and would have arrived in the area regardless of the seeding. Furthermore, both *T. repens* and *L. corniculatus* are common weeds in the Pacific Northwest (Hitchcock and Cronquist 1973; Gilkey and Dennis 2001).

Three species were widespread throughout our blowdown and forest transects regardless of their distance from the road: *Hypochaeris radicata*, *Rumex acetosella*, and *Anthoxanthum odoratum*. All three species are recognized internationally as invasive weeds, and *H. radicata* and *A. odoratum* are both found on every continent except Antarctica (Ortiz et al. 2008; Invasive Species Compendium November 2015). Holm et al. (1997) lists *R. acetosella* as one of the worst weeds worldwide, as it infests 45 different crops in 70 countries. All three species are common in disturbed areas (Hitchcock and Cronquist 1973), so it is not surprising they would be common at Mount St. Helens.

9.4.2 Habitat Invasibility and Roads

Our results agree with other studies that show that roads facilitate the dispersal and establishment of exotic species and that these species often spread further from the road in disturbed areas with less canopy and are restricted to the roadsides in forests (Parendes and Jones 2000; Forman et al. 2003; Hansen and Clevenger 2005; Mortensen et al. 2009; Birdsall et al. 2012). Half of our more common species were more frequent near the road: *Hypochaeris radicata*, *Leucanthemum vulgare*, *Lotus corniculatus*, *Poa compressa*, *Trifolium repens*, *Euphrasia officinalis*, and *Taraxacum officinale*. Tyser (1992) found exotic species richness to be high along a National Park road and low 100 m away from the road. Birdsall et al. (2012) found that roads contribute more to the spread of exotic species than do silvicultural treatments. Hansen and Clevenger's (2005) study showed that frequency of exotic plants decreased significantly with increasing distance from both a road and railway. Their study also showed that distance was significant only in forests and not in grasslands, suggesting that open habitats provide less of a barrier to exotic plant spread.

Six out of seven of the most common species (*H. radicata*, *R. acetosella*, *A. odoratum*, *L. vulgare*, *H. perforatum*, and *M. muralis*) have lightweight seeds and/or appendages that suggest wind as a primary dispersal mechanism (Parendes and Jones 2000). Because vegetation creates a barrier that can reduce dispersal distances (McEvoy and Cox 1987), roads may increase the distance seeds can travel, and cars may increase roadside wind (Hansen and Clevenger 2005). Seeds, especially those of grasses, can also attach to vehicles to be carried even longer distances (Lonsdale and Lane 1993).

9.4.3 Habitat Invasibility and Canopy Cover

Although some exotic species are able to establish as a result of their life-history traits (e.g., rapid growth, small seeds, persistent seedbank) (Rejmanek and Richardson 1996), many plant invasions have been linked to such habitat characteristics as canopy cover. For example, both Parks et al. (2005) and Hansen and Clevenger (2005) found that forests had significantly fewer invasive plants than grasslands. In our results, both the unplanted and replanted blowdown zones (low canopy cover) had more exotic plants than the tephrafall forest, which had high canopy cover (Fig. 9.5). However, as blowdown areas were more disturbed by the eruption than the tephrafall forest, it would be difficult to tease apart whether the increased abundance and richness of exotic plants is due to the disturbance intensity, canopy cover, or both as the two are closely related (i.e., the tephrafall forest has higher canopy cover than the blowdown zone because it only received tephrafall, which allowed many mature trees and saplings to survive, rather than the force of the blast PDC, which toppled and killed trees).

The sharp decline in exotic species richness and abundance with increasing distance from the road in the tephrafall forest is a likely edge effect. A study on *Pseudotsuga menziesii* forests of the US Pacific Northwest found that interior forest light levels were less than 15% of edge levels at midday (Chen et al. 1995). In addition to light, Chen et al. (1995) also found that edges differ in humidity, wind, soil moisture, and air temperature, and that these differences are greatly affected by aspect, season, and time of day. Another study of forest edge effect in a laurel and pine forest found that interior forest conditions appear at 6–10 m from the road (Delgado et al. 2007). These studies and others (Chen et al. 1993; Harper et al. 2005; Alignier and Deconchat 2013) show that roadside habitats are considerably different from interior forests, which suggests why quadrats at zero and 2 m from the road had up to three exotic species and quadrats greater than 10 m from the road in the tephrafall forest had essentially no exotic species (Fig. 9.6c, g).

Despite the relationship between increased exotic plant abundance and areas with greater light, some exotic species do well in shaded areas. *M. muralis* was often found under high or medium canopy cover in the blowdown zone site types (Appendix 9.1). This result is consistent with the findings of Parendes and Jones (2000), who found that five out of six of the most frequent exotic species (*H. perforatum*, *L. vulgare*, *C. vulgare*, *H. radicata*, and *T. repens*) were present significantly more often in areas with greater disturbance and greater light, although *M. muralis*, which occurred frequently, was found in both high and low light and disturbance levels.

9.4.4 The Pumice Plain

Although our roadside transect results followed an inverse pattern of higher exotic plant cover with lower canopy cover, the Pumice Plain followed the opposite pattern (e.g., low canopy cover and low exotic plant cover) (Fig. 9.5). It should be noted that the total plant abundance and richness on the Pumice Plain were also low compared with blowdown-zone site types, although the vegetation was still dominated by native plants (Fig. 9.5d); essentially only the wind-dispersed *H. radicata* was found (but see Titus et al. [1998], who found 20 species across a larger area, most of them infrequent). Despite the absence of a road and vehicles, seeds of many plants, including exotic ones, must be transported to the area by wind, water, animals, or pedestrian hikers. Even if this is true, the richness and abundance of exotic plants on the Pumice Plain remains very low (Fig. 9.6d, h).

One possible reason for the absence of exotic species is the stressful habitat conditions (discussed earlier). At the time of emplacement, the pyroclastic volcanic material also lacked nitrogen and has been slow to accrue this critically important plant nutrient (del Moral and Clampitt 1985; Halvorson et al. 1991). This deficiency greatly limits the establishment of plants that cannot supply their own nitrogen. The nitrogen-fixing *Lupinus lepidus* was among the first plants to establish on the Pumice Plain (Wood and del Moral 1988); it also has large seeds and is able to quickly put down roots to survive in the harsh landscape (Braatne and Bliss 1999). *Hypochaeris radicata* appeared on the Pumice Plain soon after the 1980 eruption (Schoenfelder et al. 2010) and was the most common exotic species in 1993–1994, according to Titus et al. (1998). Despite *H. radicata*'s high dispersal rates and rapid nitrogen-uptake abilities that allow it to take advantage of small nitrogen pulses (Schoenfelder et al. 2010), its seedling survival remains low (del Moral and Wood 1993).

Other possible reasons for the lower abundance of exotic plants on the Pumice Plain may be that it was severely disturbed, lacks a road, and receives fewer visitors. The 1980 eruption removed or buried all vegetation on the Pumice Plain, and it is also isolated from other areas of vegetation (del Moral and Bliss 1993; Tsuyuzaki et al. 1997). Yet on the nearby debris-avalanche deposit, Dale (1989) did not find that distance to surviving vegetation was related to number of plants that became established. The Pumice Plain also lacks a road, and Tyser's (1992) study showed that hiking trails contribute less to exotic plant spread than do roads. Furthermore, visitors must hike 3.2 km from Windy Ridge before reaching the Pumice Plain, thus greatly reducing both the number of visitors and potential of hitchhiking seeds.

9.5 Conclusions

This study showed that quadrats nearer to the road in the tephrafall forest, unplanted blowdown zone, and replanted blowdown zone exhibited higher exotic plant abundance than quadrats far from the road (Fig. 9.6e–g). The low light regime of the forest canopy in the tephrafall forest might have provided an efficient barrier to exotic plant spread from the road. In contrast, exotic species were more prevalent in the unplanted and replanted blowdown zones where there was less canopy cover and increased disturbance intensity. Nonetheless, 33 years after the eruption, exotic species were not strongly represented in the flora of any of these areas. As succession proceeds, barring other large disturbances, the blowdown zone will likely see an increase in canopy cover as trees establish and mature. However, continued monitoring is warranted to document whether invasion patterns change as soils and other environmental conditions alter under a developing canopy.

Acknowledgments Field data collection assistance was provided by Jessica Camp, Sara Copp, Abi Groskopf, and Kate Rhodes. Bill Becker, John Bishop, Sara Copp, Tim Elder, Ed Guerrant, Richard Halse (and the Oregon State University Herbarium), Dominic Maize, Ashley Smithers, Duncan Thomas, and Anthony Wenke provided assistance with plant identifications. We thank Sara Copp, Mat Dorfman, Yangdong Pan, and Jill VanWinkle for statistical consultation. We also thank Robert "Rocky" Pankratz for providing US Forest Service data regarding planting, thinning, and bough harvesting. LLK thanks the Mount St. Helens Institute for providing living arrangements while in the field. The USFS Pacific Northwest Research Station provided funding for CMC. This work was supported by the National Science Foundation. LTREB Program DEB-0614538 to CMC.

Glossary

Blast pyroclastic density current In the case of the Mount St. Helens 1980 eruption, failure of the volcano's north flank unroofed pressurized magma and superheated water. Rapid exsolution of magmatic gases and conversion of superheated water to steam produced a laterally directed blast, which formed a density current that flowed across rugged topography. The current contained fragmented rock debris as well as shattered forest material.

Debris avalanche A rapid granular flow of an unsaturated or partly saturated mixture of volcanic rock particles (± ice) and water, initiated by the gravitational collapse and disintegration of part of a volcanic edifice. Debris avalanches differ from debris flows in that they are not water-saturated. Although debris avalanches commonly occur in association with eruptions, they can also occur during periods when a volcano is dormant.

Lahar An Indonesian term for a rapid granular flow of a fully saturated mixture of volcanic rock particles (± ice), water, and commonly woody debris. A lahar that has ≥50% solids by volume is termed a *debris flow*; one that has roughly 10–50% solids by volume is termed a *hyperconcentrated flow*. Flow type can evolve with time and distance along a flow path as sediment is entrained or deposited.

Pyroclastic flow Rapid flow of a dense, dry mixture of hot (commonly >700 °C) solid particles, gases, and air that moves across the ground surface, often following landscape typography. Flows are generally gravity driven but may be accelerated initially by impulsive lateral forces of directed volcanic explosions. Flows typically move at high velocity (up to several hundred km/h).

Tephrafall A rain of volcanic particles to the ground following ejection into the atmosphere by an explosive eruption. Tephra is a collective term for particles of any size, shape, or composition ejected in an explosive eruption.

Appendices

Appendix 9.1 List of all species found and their mean percent cover (%). *PP* Pumice Plain, *UBZ* unplanted blowdown zone, *RBZ* replanted blowdown zone, *TFF* tephrafall forest. *E* following species name designates exotic species.

	% Mean cover			
Species name	**PP**	**UBZ**	**RBZ**	**TFF**
Abies amabilis		0.156		0.984
Abies procera	0.019	0.008	0.058	
Acer circinatum				0.013
Acer glabrum		0.003		
Achillea millefolium	0.108	0.298	0.141	0.006
Achlys triphylla				0.063
Acmispon parviflorus (Lotus micranthus)		0.181	0.030	
Agoseris aurantiaca/ grandiflora	0.003	0.003		
Agoseris aurantiaca		0.009	0.006	0.050
Agoseris grandiflora	0.003	0.022		
Agrostis spp.			0.013	
Agrostis capillaris (E)			0.003	
Agrostis exarata		0.028		
Agrostis pallens	1.597	0.801	1.148	0.716
Agrostis scabra	0.144	0.034	0.020	0.013
Alnus viridis ssp. *sinuata*	0.002	0.001	0.013	
Anaphalis margaritacea	0.153	1.814	1.322	3.623
Anemone deltoidea		0.003		
Anthoxanthum odoratum (E)		0.443	0.336	
Arctostaphylos uva-ursi	0.441			
Arnica cordifolia				0.025
Aruncus dioicus	0.002			0.288
Asteraceae		0.006		
Athyrium filix-femina		0.130	0.050	0.375
Bromus spp.				0.025
Bromus carinatus		0.034	0.003	
Bromus vulgaris		0.003		
Calamagrostis canadensis		0.028	0.450	
Campanula scouleri		0.098		0.091
Carex spectabilis/mertensii	0.009			
Carex spp.	0.047	0.002	0.028	0.031
Carex lenticularis		0.234		
Carex mertensii	0.059	0.210	0.016	0.025
Carex pachystachya	0.094	0.395	0.166	0.031
Carex preslii	0.013			
Carex spectabilis	0.045			
Castilleja miniata	0.508	0.316	0.234	0.019
Chamerion angustifolium	0.041	0.227	0.184	0.177
Chimaphila menziesii/ umbellata				0.003
Chimaphila menziesii			0.003	0.006

(continued)

Appendix 9.1 (continued)

	% Mean cover			
Species name	**PP**	**UBZ**	**RBZ**	**TFF**
Chimaphila umbellata		0.006	0.044	0.097
Cinna latifolia		0.009		
Circaea alpina				0.006
Cirsium arvense (E)		0.028		
Cirsium edule			0.513	
Cistanthe umbellata	0.014			
Claytonia sibirica				0.025
Clintonia uniflora				0.088
Collomia heterophylla		0.009		
Corallorhiza maculata				0.002
Deschampsia spp.		0.006		
Dicentra formosa				0.013
Digitalis purpurea (E)		0.006		
Elymus glaucus		0.728	0.122	0.044
Epilobium spp.		0.002		0.013
Epilobium anagallidifolium	0.009	0.157	0.130	0.214
Epilobium brachycarpum	0.003	0.108	0.030	
Epilobium ciliatum			0.006	
Epilobium glaberrimum	0.003			
Epilobium hornemannii		0.002		
Epilobium howellii	0.006			
Epilobium minutum	0.016	0.195	0.139	
Equisetum arvense		0.014	0.006	
Eriogonum nudum		0.146	0.006	
Eriogonum pyrolifolium		0.002	0.044	
Eriophyllum lanatum			0.006	
Eucephalus ledophyllus		0.013	0.009	
Euphrasia officinalis (E)		0.005	0.008	
Festuca spp.		0.002		0.256
Festuca rubra		0.025		0.236
Fragaria vesca		0.088	0.288	
Fragaria virginiana	0.002	0.036	0.006	0.075
Galium triflorum				0.050
Gaultheria humifusa				0.056
Gaultheria ovatifolia		0.006		0.270
Gaultheria shallon		0.006		
Glyceria striata		0.003	0.094	
Grass spp.			0.009	
Gymnocarpium dryopteris				0.075
Hieracium albiflorum	0.217	0.487	0.634	0.589
Holcus lanatus (E)		0.003		
Holcus mollis (E)				0.019
Hypericum perforatum (E)		0.322	0.322	
Hypericum scouleri			0.003	
Hypochaeris radicata (E)	0.298	1.470	1.803	0.294
Jacobaea vulgaris (E)		0.003	0.022	
Juncus spp.		0.006	0.003	
Juncus mertensianus	0.008			
Juncus parryi	0.316	0.086	0.042	
Lepidium campestre (E)		0.100	0.003	
Leucanthemum vulgare (E)		0.258	0.438	0.031
Linaria vulgaris (E)		0.041		
Lomatium martindalei		0.002		
Lotus corniculatus (E)		0.110	0.198	
Luetkea pectinata	0.139			
Lupinus latifolius	0.175	0.466	0.261	2.902
Lupinus lepidus	2.897		0.019	
Luzula spp.			0.003	0.006
Luzula parviflora	0.092	0.267	0.075	0.050
Maianthemum racemosum/ stellatum		0.003		0.002
Maianthemum racemosum (Smilacina racemosa)		0.003	0.006	0.063
Maianthemum stellatum (Smilacina stellata)				0.041
Menziesia ferruginea		0.125	0.063	2.044
Mertensia paniculata				0.156
Micranthes ferruginea (Saxifraga ferruginea)	0.013	0.022		
Microsteris gracilis		0.023	0.005	
Mitella pentandra		0.016		
Montia parvifolia		0.011	0.003	
Mycelis muralis (E)		0.177	0.113	0.150
Nothochelone nemorosa		0.412	0.567	0.019
Orthilia secunda				0.038
Osmorhiza berteroi				0.053
Osmorhiza occidentalis				0.006
Paxistima myrsinites		0.047		
Penstemon cardwellii	2.531	1.036	0.402	0.058
Penstemon serrulatus	0.036	0.067	0.019	0.013
Phalaris arundinacea		0.036	0.006	
Phacelia nemoralis	0.056	0.285	0.402	
Pinus monticola			0.044	0.006
Plantago lanceolata (E)		0.003	0.006	
Poa spp.		0.031		
Poa compressa (E)		0.030	0.197	
Poa palustris			0.025	
Poa pratensis		0.033	0.013	
Polygonum douglasii		0.004		
Polygonum minimum	0.150	0.063	0.233	
Polystichum munitum		0.066	0.059	0.100
Populus trichocarpa	0.044	0.016		
Pseudotsuga menziesii	0.533		0.006	0.022
Pteridium aquilinum		1.363	0.344	0.531
Pyrola asarifolia var. *purpurea*				0.019
Pyrola picta				0.013
Ranunculus uncinatus				0.006

(continued)

Appendix 9.1 (continued)

Species name	% Mean cover PP	UBZ	RBZ	TFF
Ribes acerifolium (Ribes howellii)		0.006		
Ribes lacustre	0.002	0.163		0.063
Rubus lasiococcus		1.323	0.617	0.391
Rubus parviflorus		0.475		0.300
Rubus spectabilis		0.831	0.063	0.031
Rubus ursinus		0.128		
Rumex acetosella (E)	0.002	0.900	0.320	0.025
Rumex crispus (E)		0.002		
Salix spp.		0.002		0.003
Sambucus racemosa		0.003		
Sedum oreganum		0.005	0.025	
Senecio triangularis				0.013
Shrub spp.		0.001		
Sorbus scopulina		0.013		0.005
Sorbus sitchensis		0.006	0.006	
Spergularia rubra (E)		0.007		
Stellaria crispa		0.159	0.006	0.019
Taraxacum officinale ssp. *officinale* (E)		0.013		0.050
Tellima grandiflora			0.025	
Tiarella trifoliata var. *unifoliata*				0.122
Trautvetteria caroliniensis				0.025
Trifolium repens (E)				0.020
Trillium ovatum		0.003	0.006	0.014
Trisetum spicatum		0.003		
Tsuga heterophylla	0.050	0.002		0.544
Tsuga mertensiana		0.003		2.356
Vaccinium spp.		3.495	3.266	10.364
Valeriana sitchensis				0.025
Vancouveria hexandra				0.006
Veratrum viride		0.003		
Veronica americana		0.006		
Viola glabella				0.014
Viola palustris		0.006		
Viola sempervirens				0.002
Total	10.897	22.14	16.355	29.700

Appendix 9.2 Mean abundance values (percent cover) for plant species at each distance from the road (or from the trail, for the Pumice Plain). Native and exotic values do not always add to 100% owing to occasional unknown species.

Distance from road/trail (m)	Total plant percent cover ±1 std. error (%)	Native plant percent cover (%) and percent of total cover	Exotic plant percent cover (%) and percent of total cover
Unplanted blowdown zone			
0	18.2 ± 2.4	12.6 (69%)	5.5 (30%)
2	17.3 ± 1.6	12.0 (70%)	5.2 (30%)
5	15.4 ± 2.5	11.4 (74%)	3.9 (25%)
10	18.4 ± 2.0	13.8 (75%)	4.5 (25%)
20	24.9 ± 3.5	21.5 (86%)	3.4 (14%)
30	27.2 ± 5.2	24.3 (89%)	2.9 (11%)
50	23.4 ± 3.6	20.4 (87%)	3.1 (13%)
70	32.3 ± 6.8	29.4 (91%)	3.0 (9%)
Planted blowdown zone			
0	16.9 ± 3.0	12.2 (72%)	4.7 (28%)
2	13.5 ± 2.5	7.5 (56%)	6.0 (44%)
5	14.1 ± 2.6	7.9 (56%)	6.1 (43%)
10	16.9 ± 3.4	12.9 (76%)	4.1 (24%)
20	17.0 ± 4.5	13.7 (81%)	3.3 (19%)
30	15.4 ± 4.0	13.6 (89%)	1.7 (11%)
50	21.3 ± 6.1	18.6 (87%)	2.7 (13%)
70	15.8 ± 4.4	14.1 (89%)	1.7 (11%)
Tephrafall forest			
0	39.8 ± 5.4	37.2 (93%)	2.6 (7%)
2	41.2 ± 6.5	39.7 (96%)	1.5 (4%)
5	28.7 ± 7.8	28.4 (99%)	0.3 (1%)
10	24.8 ± 5.4	24.6 (99%)	0.1 (1%)
20	31.5 ± 6.7	31.4 (100%)	0.1 (0%)
30	32.2 ± 6.4	32.2 (100%)	0 (0%)
50	22.0 ± 7.3	22.0 (100%)	0 (0%)
70	17.4 ± 5.0	17.3 (100%)	0.1 (0%)
Pumice Plain			
0	11.5 ± 1.4	11.5 (100%)	0.03 (0%)
2	9.8 ± 2.1	9.7 (99%)	0.1 (1%)
5	12.3 ± 2.0	11.9 (97%)	0.4 (3%)
10	10.8 ± 1.7	10.6 (98%)	0.2 (2%)
20	9.1 ± 1.0	8.8 (97%)	0.3 (3%)
30	11.9 ± 3.0	11.6 (97%)	0.3 (3%)
50	12.4 ± 3.3	11.6 (94%)	0.8 (6%)
70	9.4 ± 1.6	9.1 (97%)	0.3 (3%)

References

Abella, S.R., E.C. Engel, J.D. Springer, and W.W. Covington. 2012. Relationships of exotic plant communities with native vegetation, environmental factors, disturbance, and landscape ecosystems of *Pinus ponderosa* forests, USA. *Forest Ecology and Management* 271: 65–74.

Alignier, A., and M. Deconchat. 2013. Patterns of forest vegetation responses to edge effect as revealed by a continuous approach. *Annals of Forest Science* 70: 601–609.

Ansong, M., and C. Pickering. 2014. Weed seeds on clothing: A global review. *Journal of Environmental Management* 144: 203–211.

Antos, J.A., and D.B. Zobel. 2005. Plant responses in forests of the tephra-fall zone. In *Ecological responses to the 1980 eruption of Mount St. Helens*, ed. V.H. Dale, F.J. Swanson, and C.M. Crisafulli, 47–58. New York: Springer.

Bailey, J.D., C. Mayrsohn, P.S. Doescher, E.St. Pierre, and J.C. Tappeiner. 1998. Understory vegetation in old and young Douglas-fir forests of western Oregon. *Forest Ecology and Management* 112: 289–302.

Birdsall, J.L., W. McCaughey, and J.B. Runyon. 2012. Roads impact the distribution of noxious weeds more than restoration treatments in a lodgepole pine forest in Montana, U.S.A. *Restoration Ecology* 20: 517–523.

Braatne, J.H., and L.C. Bliss. 1999. Comparative physiological of lupines colonizing early successional habitats on Mount St. Helens. *Ecology* 80: 891–907.

Carlson, J.R., J.R. Stroh, J.A. Oyler, and F. Reckendorf. 1982. Erosion control revegetation on land damaged by the 1980 Mt. St. Helens eruption. In *Proceedings: High-altitude revegetation workshop No. 5*, Information Series No. 48, ed. R.L. Cuany and J. Etra, 174–189. Fort Collins: Colorado Water Resources Research Institute, Colorado State University. http://hermes.cde.state.co.us/drupal/islandora/object/co:3830.

Chen, J., J.F. Franklin, and T.A. Spies. 1993. Contrasting microclimates among clearcut, edge, and interior of old-growth Douglas-fir forest. *Agricultural and Forest Meteorology* 63: 219–237.

———. 1995. Growing season microclimatic gradients from clearcut edges into old-growth Douglas-fir forests. *Ecological Applications* 5: 74–86.

Christen, C.C., and G.R. Matlack. 2007. The habitat and conduit functions of roads of three invasive plant species. *Biological Invasions* 11: 453–465.

Clinton, W. 1999. U.S. presidential executive order 13112 re: Invasive species. February 3, 1999. *Federal Register* 64: 6183–6186.

Cook, R. 1980. The biology of seeds in the soil. In *Demography and evolution in plant populations*, ed. O.T. Solbrig, 107–130. Berkeley: University of California Press.

Dale, V.H. 1989. Wind dispersed seeds and plant recovery on the Mount St. Helens debris avalanche. *Canadian Journal of Botany* 67: 1434–1441.

———. 1991. Revegetation of Mount St. Helens debris avalanche 10 years post eruption. *National Geographic Research and Exploration* 7: 328–341.

Dale, V.H., and W.M. Adams. 2003. Plant reestablishment 15 years after the debris avalanche at Mount St. Helens, Washington. *Science of the Total Environment* 313 (1–3): 101–113.

Dale, V.H., C.M. Crisafulli, and F.J. Swanson. 2005. 25 years of ecological change at Mount St. Helens. *Science* 308: 961–962.

Davis, M.A., J.P. Grime, and K. Thompson. 2000. Fluctuating resources in plant communities: A general theory of invasibility. *Journal of Ecology* 88: 528–534.

del Moral, R., and L.C. Bliss. 1993. Mechanisms of primary succession: Insights resulting from the eruption of Mount St. Helens. *Advances in Ecological Restoration* 24: 1–66.

del Moral, R., and C.A. Clampitt. 1985. Growth of native plant species on recent volcanic substrates form Mount St. Helens. *American Midland Naturalist* 114: 374–383.

del Moral, R., and D.M. Wood. 1993. Early primary succession on a barren volcanic plain at Mount St. Helens, Washington. *American Journal of Botany* 80: 981–991.

Delgado, J.D., N.L. Arroyo, J.R. Arevalo, and J.M. Fernandez-Palacios. 2007. Edge effects of roads on temperature, light, canopy cover, and canopy height in laurel and pine forests (Tenerife, Canary Islands). *Landscape and Urban Planning* 81: 328–340.

Flory, S.L., and K. Clay. 2009. Effects of roads and forest successional age on experimental plant invasions. *Biological Conservation* 142: 2531–2537.

Forman, R.T.T., D. Sperling, J.A. Bissonette, A.P. Clevenger, C.D. Cutshall, V.H. Dale, L. Fahrig, R. France, C.R. Goldman, K. Heanue, J.A. Jones, F.J. Swanson, T. Turrentine, and T.C. Winter. 2003. *Road ecology: Science and solutions*. Washington, DC: Island Press.

Foster, D.R., D.H. Knight, and J.F. Franklin. 1998. Landscape patterns and legacies resulting from large, infrequent forest disturbances. *Ecosystems* 1: 497–510.

Fuller, R.N., and R. del Moral. 2003. The role of refugia and dispersal in primary succession on Mount St. Helens, Washington. *Journal of Vegetation Science* 14: 637–644.

Gilkey, H.M., and L.J. Dennis. 2001. *Handbook of Northwestern plants*. Corvallis: Oregon State University Press.

Gray, A.N. 2005. Eight nonnative plants in western Oregon forests: Associations with environment and management. *Environmental Monitoring and Assessments* 100: 109–127.

Gray, A. 2007. Distribution and abundance of invasive plants in Pacific Northwest forests. In *Meeting the challenge: Invasive plants in Pacific Northwest ecosystems*. General Technical Report PNW-GTR-694, ed. T.B. Harrington and S.H. Reichard, 143–148. Portland: U.S. Department of Agriculture, Forest Service, Pacific Northwest Research Station.

Grime, J.P. 1979. *Plant strategies and vegetation processes*. New York: Wiley.

Halpern, C.B., P.M. Frenzen, J.E. Means, and J.F. Franklin. 1990. Plant succession in areas of scorched and blown-down forest after the 1980 eruption of Mount St. Helens, Washington. *Journal of Vegetation Science* 1: 181–194.

Halvorson, J.J., J.L. Smith, and E.H. Franz. 1991. Lupine influence on soil carbon, nitrogen and microbial activity in developing ecosystems at Mount St. Helens. *Oecologia* 87: 162–170.

Hansen, M.J., and A.P. Clevenger. 2005. The influence of disturbance and habitat on the presence of non-native plant species along transport corridors. *Biological Conservation* 125: 249–259.

Harper, K.A., S.E. Macdonald, P.J. Burton, J.Q. Chen, K.D. Brosofske, S.C. Saunders, E.S. Euskirchen, D. Roberts, M.S. Jaiteh, and P.A. Esseen. 2005. Edge influence on forest structure and composition in fragmented landscapes. *Conservation Biology* 19: 768–782.

Hitchcock, C.L., and A. Cronquist. 1973. *Flora of the Pacific Northwest*. Seattle: University of Washington Press.

Holm, L., J. Doll, E. Holm, J. Pancho, and J. Herberger. 1997. *World weeds: Natural histories and distribution*. New York: Wiley.

Johnston, F.M., and S.W. Johnston. 2004. Impacts of road disturbance on soil properties and on exotic plant occurrence in subalpine areas of the Australian Alps. *Arctic Antarctic and Alpine Research* 36: 201–207.

Kozloff, E.N. 2005. *Plants of Western Oregon, Washington and British Columbia*. Portland: Timber Press.

Lipman, P.W., and D.R. Mullineaux, eds. 1981. *The 1980 Eruptions of Mount St. Helens, Washington*. Professional Paper 1250. Washington, DC: U.S. Geological Survey.

Lockwood, J.L., M.F. Hoopes, and M.P. Marchetti. 2007. *Invasion ecology*. Malden: Blackwell Publishing.

Lonsdale, W.M. 1999. Global patterns of plant Invasions and the concept of invasibility. *Ecology* 80: 1552–1536.

Lonsdale, W.M., and A.M. Lane. 1993. Tourist vehicles as vectors of weed seeds in Kakadu National Park, Northern Australia. *Biological Conservation* 69: 277–283.

Lozon, J.D., and H.J. MacIsaac. 1997. Biological invasions: Are they dependent on disturbance? *Environmental Review* 5: 131–141.

Mack, R.N., D. Simberloff, W.M. Lonsdale, H. Evans, M. Clout, and F.A. Bazzaz. 2000. Biotic invasions: Causes, epidemiology, global consequences and control. *Ecological Applications* 10: 689–710.

McEvoy, P.B., and C.S. Cox. 1987. Wind dispersal distances in dimorphic achenes of ragwort, *Senecio jacobaea*. *Ecology* 68: 2006–2015.

Michalcov, D., S. Lvoncik, M. Chytry, and O. Hajek. 2011. Bias in vegetation databases? A comparison of stratified-random and preferential sampling. *Journal of Vegetation Science* 22: 281–291.

Moody, M.E., and R.N. Mack. 1988. Controlling the spread of plant invasions: The importance of nascent foci. *Journal of Applied Ecology* 25: 1009–1021.

Mooney, H.A., and J.A. Drake. 1989. Biological invasions: A SCOPE program overview. In *Biological invasions: A global perspective*, ed. J.A. Drake, H.A. Mooney, F. Di Castri, R.H. Groves, F.J. Kruger, M. Rejmanek, and M. Williamson, 491–506. Chicheser: Wiley.

Mortensen, D.A., E.S.J. Rauschert, A.N. Nord, and B.P. Jones. 2009. Forest roads facilitate the spread of invasive plants. *Invasive Plant Science and Management* 2: 191–199.

Nelson, C.R., C.B. Halpern, and J.K. Agee. 2008. Thinning and burning result in low-level invasion by nonnative plants but neutral effect on natives. *Ecological Applications* 18: 762–770.

Newsome, A.E., and I.R. Noble. 1986. Ecological and physiological characters of invading species. In *Ecology of biological invasions*, ed. R.H. Groves and J.J. Burton, 1–20. Cambridge: Cambridge University Press.

Ortiz, M.Â., K. Tremetsberger, A. Terrab, T.F. Stuessy, J.L. García-Castaño, E. Urtubey, C.M. Baeza, C.F. Ruas, P.E. Gibbs, and S. Talavera. 2008. Phylogeography of the invasive weed *Hypochaeris radicata* (Asteraceae): From Moroccan origin to worldwide introduced populations. *Molecular Ecology* 17: 3654–3667.

Parendes, L.A. 1997. Spatial patterns of invasion by exotic plants in a forested landscape. Doctoral dissertation. Corvallis: Oregon State University.

Parendes, L.A., and J.A. Jones. 2000. Role of light availability and dispersal in exotic plant invasion along roads and streams in the H. J. Andrews Experimental Forest, Oregon. *Conservation Biology* 14: 64–75.

Parks, C.G., S.R. Radosevich, B.A. Endress, B.J. Naylor, D. Anzinger, L.J. Rew, B.D. Maxwell, and K.A. Dwire. 2005. Natural and land-use history of the northwest mountain regions (USA) in relation to patterns of plant invasions. *Perspectives in Plant Ecology Systems* 7: 137–158.

Pickering, C., and A. Mount. 2010. Do tourists disperse weed seed? A global review of unintentional human-mediated terrestrial seed dispersal on clothing, vehicles and horses. *Journal of Sustainable Tourism* 18: 239–256.

Pimentel, D., R. Zuniga, and D. Morrison. 2005. Update on the environmental and economic costs associated with alien-invasive species in the United States. *Ecological Economics* 52: 273–288.

Rejmanek, M., and D.M. Richardson. 1996. What attributes make some plant species more invasive? *Ecology* 77: 1655–1661.

Reynolds, G.D., and L.C. Bliss. 1986. Microenvironmental investigations of tephra covered surfaces at Mount St. Helens. In *Mount St. Helens: Five years later*, ed. S.A.C. Keller, 147–152. Cheney: Eastern Washington University.

Schoenfelder, A.C., J.G. Bishop, H.M. Martinson, and W.F. Fagan. 2010. Resource use efficiency and community effects of invasive *Hypochaeris radicata* (Asteraceae) during primary succession. *American Journal of Botany* 97: 1772–1779.

Swanson, F.J., and J.J. Major. 2005. Physical events, environments, and geological-ecological interactions at Mount St. Helens: March 1980–2004. In *Ecological responses to the 1980 eruption of Mount St. Helens*, ed. V.H. Dale, F.J. Swanson, and C.M. Crisafulli, 27–44. New York: Springer.

Taylor, K., T. Brummer, M.L. Taper, A. Wing, and L.J. Rew. 2012. Human-mediated long-distance dispersal: An empirical evaluation of seed dispersal by vehicles. *Biodiversity Research* 18: 942–951.

Titus, J.H., and E. Householder. 2007. Salvage logging and replanting reduce understory cover and richness compared to unsalvaged-unplanted sites at Mount St. Helens, Washington. *Western North American Naturalist* 67: 219–231.

Titus, J.H., S. Moore, M. Arnot, and P. Titus. 1998. Inventory of the vascular flora of the blast zone, Mount St. Helens, Washington. *Madroño* 45: 146–161.

Trombulak, S.C., and C.A. Frissell. 2000. Review of ecological effects of roads on terrestrial and aquatic communities. *Conservation Biology* 14: 18–30.

Tsuyuzaki, S., J.H. Titus, and R. del Moral. 1997. Seedling establishment patterns on the Pumice Plain, Mount St. Helens, Washington. *Journal of Vegetation Science* 8: 727–734.

Tyser, R.W. 1992. Alien flora in grasslands adjacent to road and trail corridors in Glacier National Park, Montana (U.S.A.). *Conservation Biology* 6: 253–262.

U.S. Department of Agriculture (USDA). 2013. Mount St. Helens National Volcanic Monument. Accessed 14 May 2013. http://www.fs.usda.gov/mountsthelens.

Usher, M.B. 1988. Biological invasions of nature reserves: A search for generalisations. *Biological Conservation* 44: 119–135.

Vitousek, P.M., C.M. D'Antonio, L.L. Loope, M. Rejmanek, and R. Westbrooks. 1997. Introduced species: A significant component of human-caused global change. *New Zealand Journal of Ecology* 21: 1–16.

Walker, L.R., and R. del Moral. 2003. *Primary succession and ecosystem rehabilitation*. Cambridge: Cambridge University Press.

Wilson, E.O. 1992. *The diversity of life*. Cambridge, MA: Belknap Press.

Wood, D.M., and R. del Moral. 1987. Mechanisms of early primary succession in subalpine habitats on Mount St. Helens. *Ecology* 68: 780–790.

———. 1988. Colonizing plants on the Pumice Plains, Mount St. Helens, Washington. *American Journal of Botany* 75: 1228–1237.

Lichen Community Development Along a Volcanic Disturbance Gradient at Mount St. Helens

10

Peter R. Nelson, Bruce McCune, Tim Wheeler, Linda H. Geiser, and Charles M. Crisafulli

10.1 Introduction

Many studies have documented vascular plant community responses to volcanic eruptions (e.g., Whittaker et al. 1989; Grishin et al. 1996; Tsuyuzaki and Hase 2005; Magnússon et al. 2009), and some are specific to the plant reestablishment subsequent to the 1980 eruption of Mount St. Helens (Antos and Zobel 2005; Dale et al. 2005; del Moral and Wood 2012). Nearly all these studies either omit lichens, mention them as a monolithic group with mosses (e.g., Beard 1945, 1976; Tsuyuzaki and Hase 2005), or focus exclusively on lava flows (e.g., Skottsberg 1941; Jackson and Keller 1970; Jackson 1971; Kurina and Vitousek 1999, 2001).

P.R. Nelson (✉)
Department of Botany and Plant Pathology, Oregon State University, 2082 Cordley Hall, Corvallis, OR 97331, USA

Arts and Sciences Division, University of Maine-Fort Kent, 23 University Drive, Fort Kent, ME 04743, USA
e-mail: peter.nelson@maine.edu

B. McCune
Department of Botany and Plant Pathology, Oregon State University, 2082 Cordley Hall, Corvallis, OR 97331, USA
e-mail: mccuneb@oregonstate.edu

T. Wheeler
6606 Jocko Canyon Road, Arlee 59821, MT, USA
e-mail: timothybwheeler@gmail.com

L.H. Geiser
U.S. Department of Agriculture, Forest Service, Pacific Northwest Region, Air Resources Management Program, Corvallis 97331, OR, USA

U.S. Department of Agriculture, Forest Service, Water Fish Wildlife and Rare Plants, Sidney R. Yates Federal Building, 201 14th Street, SW Suite 3SC, Mailstop 1121, Washington, DC 20250, USA
e-mail: lgeiser@fs.fed.us

C.M. Crisafulli
U.S. Department of Agriculture, Forest Service, Pacific Northwest Research Station, Mount St. Helens National Volcanic Monument, 42218 NE Yale Bridge Rd, Amboy, WA 98601, USA
e-mail: ccrisafulli@fs.fed.us

Lichens are symbiotic composite organisms composed of a fungus (ascomycete or rarely basidiomycete), plus an alga or a cyanobacterium, or occasionally all three. Lichens can be important components of the post-eruption vegetation recovery because of their ability to fix nitrogen, which is absent or at low levels in new volcanic deposits (Kurina and Vitousek 1999, 2001). Moreover, lichens can stabilize soil and inorganic substrates and weather primary surfaces, such as lava, preparing the way for colonization by vascular plants (Jackson and Keller 1970). Lichens can also greatly increase diversity measures of communities when included in vegetation studies and provide a more holistic view of vegetation establishment on volcanic surfaces. Lichens are also particularly important to conditions in wet montane temperate forests, which were widespread at Mount St. Helens before the 1980 eruption, where they contributed significantly to epiphytic biomass (Berryman and McCune 2006), biodiversity (Peck and McCune 1997), and nitrogen (Pike 1978). For these reasons, it is important to examine lichen community development after the 1980 eruption of Mount St. Helens. Our goals are to describe and interpret patterns of lichen species richness, abundance, and community composition in the context of the characteristics of the different volcanic disturbances created by a single eruption. Our work on lichens complements numerous other taxa that have been studied at the volcano (Dale et al. 2005).

Early on 18 May 1980, a large earthquake hastened a mass downslope movement of the summit and north slope of Mount St. Helens, creating a massive rockfall avalanche, which displaced Spirit Lake to the north, and, upon turning west, buried and filled much of the North Fork Toutle River drainage. Within seconds of the avalanche, a powerful laterally directed pyroclastic density current (PDC, commonly called the lateral blast but herein called the blast PDC), of superheated water vapor, volcanic gases, rocks, and entrained forest debris uprooted, shattered, and removed trees in a nearly 180° swath centered to the north and extended outward for 25 km from the crater (Waitt 1981).

C.M. Crisafulli, V.H. Dale (eds.), *Ecological Responses at Mount St. Helens: Revisited 35 years after the 1980 Eruption*,
https://doi.org/10.1007/978-1-4939-7451-1_10

As the blast PDC slowed, trees were knocked down but not removed, creating a large zone of toppled forest (blowdown forest zone). The blast PDC further slowed and cooled, creating an area where trees remained standing but were scorched (standing dead forest zone). Beyond the standing dead zone and in protected pockets within the zone influenced by the blast PDC were areas that experienced only cool air-fall deposits of ash and popcorn-size pumice (tephrafall zone). Beginning about noon on 18 May and lasting for several hours, numerous pyroclastic flows surged from the crater, creating a vast, sterile area later referred to as the Pumice Plain. Lahars generated from melting snow and ice flowed down the river channels draining the eruption, causing severe impacts on pre-eruption ecosystems. Many shallow and less forceful lahars traveled down the cone, causing minor damage to vegetation (Foxworthy and Hill 1982) (Fig. 10.1). For our sampling purposes, we divided the eruption-influenced landscape into seven strata that correspond to the disturbance zones used by other ecologists (Dale et al. 2005): (1) tephrafall (not shown on map), (2) standing dead, (3) blowdown, (4) tree removal (not shown on map), (5) lahar, (6) debris-avalanche deposit, and (7) pyroclastic flow (Fig. 10.1).

Lichens avoid competition with larger, faster-growing vascular plants by living in high-stress environments. Compared to vascular plants, most lichens are more toler-

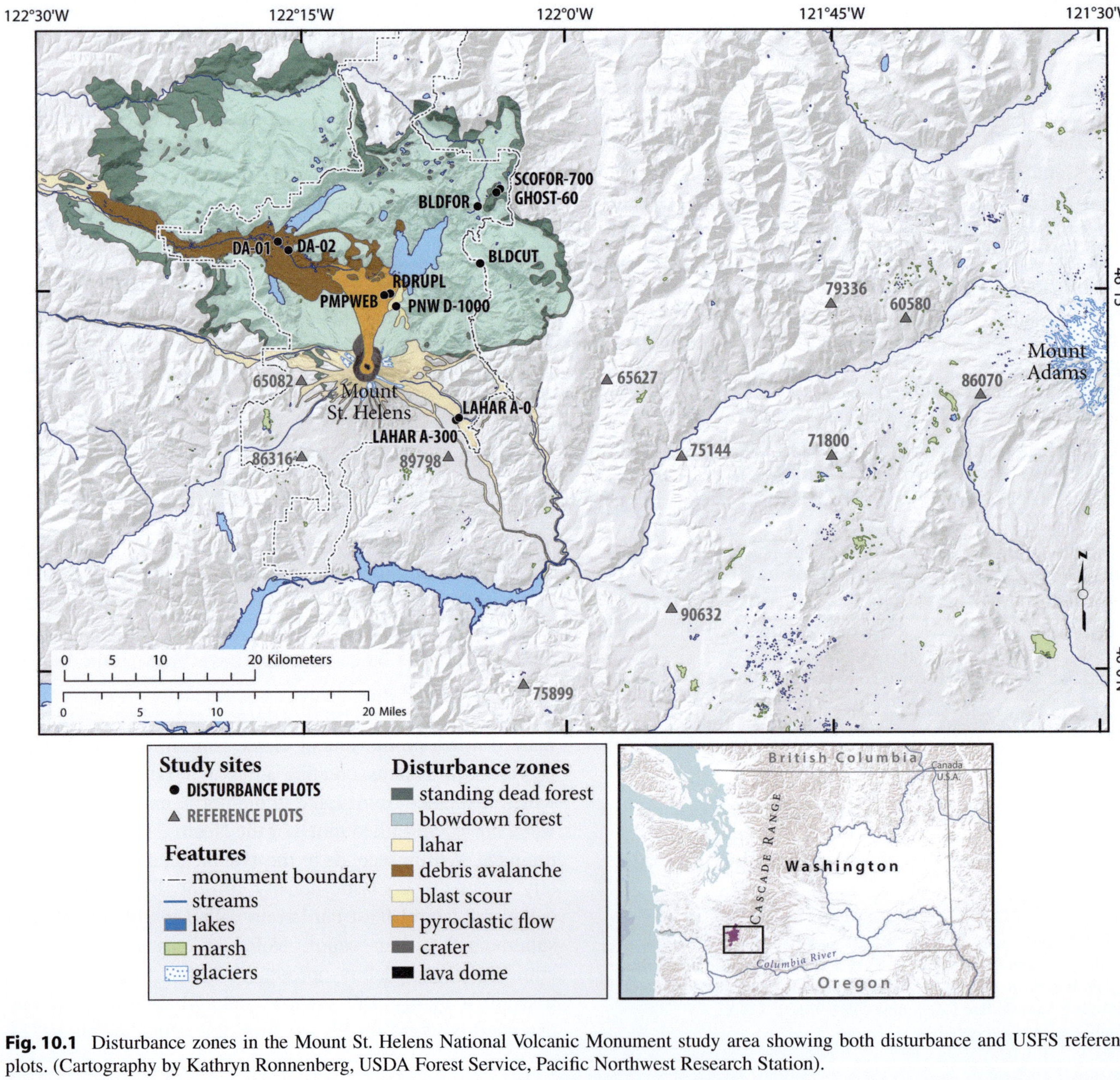

Fig. 10.1 Disturbance zones in the Mount St. Helens National Volcanic Monument study area showing both disturbance and USFS reference plots. (Cartography by Kathryn Ronnenberg, USDA Forest Service, Pacific Northwest Research Station).

ant of temperature fluctuations, desiccation, and nutrient deficiency (Grime 1977). They also derive their nutrition from air- and water-borne particles, largely independent of the surfaces they inhabit, making them less dependent on the nutrient availability of their substrates. Owing to these traits, lichens are often the first vegetation to colonize after disturbances produce conditions for primary succession, such as following volcanic eruptions (Kitayama et al. 1995).

Most previous studies on the response of the lichen community to volcanic disturbance have focused on lava flows (e.g., Skottsberg 1941; Jackson 1971; Bjarnason 1991; Kitayama et al. 1995; Sipman and Raus 1995). Lava flows are only one of the many types of volcanic disturbance; at Mount St. Helens, they were confined only to the crater after the 1980 eruption and thus were not present in significant areas. Hence, prior studies lend little insight into the lichen community responses to the types of disturbances that did occur over extensive areas around Mount St. Helens. Only one prior study in Russia looked at the lichen response to a variety of volcanic disturbance types; they described vegetation responses along a gradient of tephra depths 90 years post eruption (Grishin et al. 1996). We add to that work by sampling lichen communities across a variety of volcanic disturbance types, using the 1980 eruption of Mount St. Helens as a case study.

10.2 Methods

We report lichen abundance and richness for 11 disturbed and 11 reference plots, which are described below (Table 10.1 and Fig. 10.1). In the autumn of 2012, we sampled 11 plots across seven disturbance zones, with plot location being constrained by road and trail access. To increase understanding of ecological responses of a number of taxonomic groups to the

Table 10.1 Epiphytic lichen species richness and abundance data, for disturbance zone and reference plots, sorted by richness. Reference plot plant associations are derived from USDA Forest Service Region 6 protocols. Abundance is measured on a unitless linear scale calculated as the sum of transformed counts of species in four abundance classes (see Sect. 10.2). Disturbance plots were measured in autumn 2012, and most reference plots were measured in 1996–1997; two reference plots were measured in 2005. Only the first five digits of the reference plot numbers appear in Fig. 10.1.

		Epiphytic lichens		Non-epiphytic lichens		
Plot	Disturbance zone	Species richness	Abundance	Species richness	Abundance	Elevation (m)
DA-01	Debris-avalanche deposit	23	0.626	20	3.584	769
DA-02	Debris-avalanche deposit	21	2.784	7	0.133	767
BLDFOR	Blowdown	15	0.501	7	0.151	1122
BLDCUT	Blowdown	15	0.573	3	0.111	1050
LAHAR A-0	Lahar	14	3.821	9	0.243	929
GHOST-60	Tephrafall	13	2.452	6	0.141	1158
LAHAR A-300	Lahar	11	0.461	4	0.031	919
PNW D-1000	Pyroclastic flow	10	0.037	11	0.722	1125
SCOFOR-700	Standing dead	7	0.610	5	0.410	1153
PMPWEB	Pyroclastic flow	0	0	8	0.332	1125
RDRUPL	Tree removal	0	0	7	0.340	1165
	Reference plot plant association					
75144 53 59	Douglas-fir/vine maple/salal	26	9.710	–	–	1707
79336 53 59	Western hemlock/vine maple/vanilla leaf	25	7.522	–	–	1494
86070 53 59	Silver fir/big huckleberry/bear grass	24	6.666	–	–	1402
90635 53 59	Silver fir/salal	23	1.427	–	–	792
71800 53 59	Silver fir/coolwort foamflower	23	8.834	–	–	1036
65627 53 59	Western hemlock/vanilla leaf	23	2.993	–	–	701
65082 53 15	Silver fir/big huckleberry/beadlily	22	15.601	–	–	1158
89798 53 59	Silver fir/Alaska huckleberry	20	1.451	–	–	853
86316 53 15	Silver fir/big huckleberry/bear grass	18	9.621	–	–	853
60580 53 59	Silver fir/vanilla leaf-beadlily	18	1.143	–	–	1402
75899 53 59	Western hemlock/vanilla leaf	18	8.010	–	–	823

1980 eruption of Mount St. Helens, we took advantage of pre-existing studies and superimposed our lichen plots (Fig. 10.1) in the tephrafall, standing dead, blowdown, tree-removal, and pyroclastic-flow zones on long-term mammal, arthropod, and mycorrhizae study sites (Allen et al. 2005; Crisafulli et al. 2005a; Parmenter et al. 2005). Our sites in the debris-avalanche deposit were on and adjacent to long-term amphibian study sites (Crisafulli et al. 2005b), and our lahar sites were located on long-term vegetation plots (Frenzen et al. 2005). On each circular plot (34.7-m radius), we used the Forest Inventory and Analysis (FIA) lichen communities indicator survey method (McCune et al. 1997), modified to include lichens that were located less than 0.5 m high on trees or shrubs and on soil, rocks, and logs. We did not collect species that were clearly identified in the field. Collected vouchers were identified using standard light microscopy, chemical spot tests, and thin-layer chromatography methods following the nomenclature of Esslinger (2012), except for some macrolichen genera classified following the nomenclature in McCune and Geiser (2009). Voucher specimens reside at the Oregon State University herbarium (herbarium code OSC) and the Mount St. Helens National Volcanic Monument. The same observer (first author) searched each plot area for a minimum of 30 min until either 15 min had elapsed without finding a new species or 2 h had elapsed. Each lichen species was assigned one of four abundance classes, based on the number of thalli (spatially separate and distinct individuals): 1 = 1–3 thalli; 2 = 4–10 thalli; 3 = >10 thalli, covering less than half of the available substrate in the plot; and 4 = more than half available substrate in the plot covered by the species. Plots ranged from 767 to 1161 m in elevation, one or two plots per disturbance zone (Fig. 10.1, Table 10.1). Vegetation on the plots ranged from sparse graminoid and shrub cover in the blast area to regenerating young conifers in the blowdown zone to old-growth coniferous forest in the tephrafall zone (Table 10.2).

Table 10.2 Description of vegetation status in 2012 for plots by disturbance zone. Plot short names are in parentheses. See Sect. 10.2 for citations of studies from which these vegetation descriptions were obtained.

Disturbance	Vegetation description
Tephrafall forest (GHOST-60)	Late-seral (300- to 400-year-old) forest dominated by mature Pacific silver fir (*Abies amabilis*) and mountain hemlock (*Tsuga mertensiana*), with few very large Douglas-fir (*Pseudotsuga menziesii*). Understory composed of huckleberries (*Vaccinium membranaceum, V. ovalifolium*) and other ericaceous shrubs. Sparse ground layer with prince's pine (*Chimaphila umbellata, C. menziesii*) and wintergreens (*Pyrola* spp).
Standing dead forest (SCOFOR)	Numerous standing dead large conifers. Saplings of mountain hemlock (*Tsuga mertensiana*) and Pacific silver fir (*Abies amabilis*). Abundant ericaceous shrubs (*Vaccinium membranaceum, V. ovalifolium*, and *Menziesia ferruginea*) and a mix of herbs (*Chamerion angustifolium, Anaphalis margaritacea, Hieracium albiflorum, Chimaphilia umbellata*, and *Pyrola* spp).
Blowdown forest (BLDFOR)	Pole- and sapling-sized Pacific silver fir (*Abies amabilis*), mountain hemlock (*Tsuga mertensiana*), and few Douglas-fir (*Pseudotsuga menziesii*). Complex understory of forest understory shrubs, huckleberries (*Vaccinium membranaceum* and *V. ovalifolium*), and sun-requiring species (*Salix sitchensis, S. scouleriana*, and *Alnus viridis*). Abundant early-successional forbs and graminoids (*Elymus glaucus, Agrostis* spp., and *Carex mertensii*).
Blowdown clearcut (BLDCUT)	Widely scattered noble fir (*Abies procera*) and Douglas-fir (*Pseudotsuga menziesii*). Dense thickets of willow (*Salix sitchensis* and *S. scouleriana*) interspersed with large openings. Sparse forb and grasses (<10% cover). Moss (*Racomitrium canescens*) abundant in some portions of plot.
Tree removal (RDRUPL)	Early-successional herb-dominated community with scattered shrubs (*Salix sitchensis*) and a few Douglas-fir (*Pseudotsuga menziesii*) saplings. Dominant herbs include *Lupinus lepidus, Castilleja miniata, Hieracium albiflorum*, and *Agrostis* spp.). Total plant cover is 40–50%.
Pyroclastic flow (PMPWEB)	Sparse early-successional herb-dominated community. Total plant cover is 20–30%, composed of mixed grasses (*Agrostis* spp., *Elymus glaucus*, and *Festuca* spp.), sedges (*Carex* spp.), and forbs (e.g., *Lupinus lepidus, Castilleja miniata*, and *Hieracium albiflorum*). The shrubs, *Salix sitchensis* and *Alnus viridis*, are widely scattered.
Pyroclastic flow (PNW D-1000)	Mostly barren early-successional herb-dominated community. Total plant cover <10% and composed of widely scattered forbs (*Lupinus lepidus* and *Agrostis* spp.), with widely spaced willow (*Salix sitchensis*) and small conifers (*Abies procera* and *Pseudotsuga menziesii*). Hummocky topography with numerous large boulders.
Debris-avalanche deposit (DA-01)	Early-successional grass-dominated community. Total plant cover is 60–70%.
Debris-avalanche deposit (DA-02)	Young, closed-canopy red alder (*Alnus rubra*) forest with dense herbaceous understory of grasses, forbs (*Galium* spp., *Cerastium* spp., and *Montia sibirica*), and fern (*Polystichum munitum*).
Lahar (A-0)	Mixed mature conifer forest of Douglas-fir (*Pseudotsuga menziesii*), western hemlock (*Tsuga heterophylla*), and western white pine (*Pinus monticolus*). Understory with few vascular plants but high cover of moss (*Racomitrium canescens*).
Lahar (A-300)	Young conifer forest dominated by Douglas-fir (*Pseudotsuga menziesii*) and lodgepole pine (*Pinus contorta*) pole trees and saplings. Tree interspaces with high moss (*Racomitrium canescens*) cover and numerous patches of manzanita (*Arctostaphylos uva-ursi*).

To augment our disturbance plot data, we also conducted searches of all substrates (trees, shrubs, logs, stumps, rocks, soil) near each plot for lichen taxa not captured in the plot survey. Three of the coauthors searched for crustose lichens (lichens adhering strongly to the substrate) or other macrolichens outside the plot using their expertise to guide their search to areas likely to yield additional species, ranging up to 1000 m away from the plot. No searches for incidental lichens (meaning those not collected or observed on the plot in the time-constrained search) were conducted on the debris avalanche deposit because those plots were sampled a week later than the rest of the fieldwork and only two authors were available.

In order to compare our observations in the MSH disturbance zones to reference areas relatively undisturbed by the eruption, we acquired lichen community data surveyed on the Forest Inventory and Analysis (FIA) 5.4-km grid by the US Forest Service (USFS) Air Resource Management Program. We randomly selected 11 plots on the Gifford Pinchot National Forest from the USFS data set, constrained to an elevation range (701–1707 m) similar to the 11 plots sampled in the disturbance zone, at a distance of between 4 and 45 km from the summit of MSH. One of these reference plots was located in the Mt. Adams Wilderness, whereas the rest were located in the Monument. Reference plots (Fig. 10.1 and Table 10.1) reflected a range of forest structure and composition in the region, including stands of young, dense Douglas-fir (*Pseudotsuga menziesii*), 50- to 100-year-old mixed conifers, 100- to 200-year-old middle- to high-elevation fir (*Abies amabilis* and *A. procera*), and old-growth Douglas-fir or western hemlock (*Tsuga heterophylla*). Plant associations for each plot are given in Table 10.1, excerpted from Region 6 US Forest Service protocols (USDA 2014). We sampled these plots in the same fashion as the plots in the volcanically disturbed areas but restricted surveys to epiphytic lichens (those growing on trees or shrubs) greater than 0.5 m above ground level or litterfall obviously from the canopy. Reference plots were sampled by Forest Service staff between 1996 and 2005.

To describe differences between lichen communities of the disturbance zones, we used nonmetric multidimensional scaling (NMS) ordination in PC-ORD (McCune and Mefford 2011) with a Sørensen distance measure and "slow and thorough" autopilot settings. We calculated species richness (total number of species), evenness, and Simpson's and Shannon's diversity indices to compare lichen communities among disturbance zones. To compare total lichen abundance among disturbance zones, we treated abundance categories as logarithmically scaled. We transformed all abundances to the units of the highest abundance class (4) by dividing the total number of entries in lower abundance classes by a multiple of 10. We divided the total number of entries in class 3 by 10, class 2 by 100, and class 1 by 1000. We then summed the transformed abundances across all macrolichens present on a plot for a total lichen abundance presented on a relative, unitless scale. We used the highest abundance for each lichen species if it was found more than once in a disturbance zone. We calculated species richness and abundance for epiphytic lichens on disturbance plots to compare to reference plots where only epiphytic lichens species were surveyed.

10.3 Results

On plots in the eruption disturbance zones, we found 75 macrolichen species, 11 of which occurred on at least half the plots and 27 that were found only once (see Fig. 10.2 for images of some lichen taxa found at MSH). The 11 species that occurred on at least half the plots were *Alectoria sarmentosa*, *Cetraria chlorophylla*, *Cladonia fimbriata*, *C. transcendens*, *C. verruculosa*, *Hypogymnia inactiva*, *H. physodes*, *Nodobryoria oregana*, *Parmelia sulcata*, *Peltigera didactyla*, and *Platismatia glauca* (Table 10.3). Plot macrolichen richness (epiphytic and non-epiphytic richness combined) in the disturbance zones ranged from 7 to 43 species (Table 10.1). The debris-avalanche deposit had the highest average macrolichen richness (36 species), and the blowdown, lahar, and tephrafall zones had moderate macrolichen richness (19–20 species). The pyroclastic-flow and standing dead zones had moderate to low macrolichen richness (12–15 species), whereas the tree-removal zone had the lowest macrolichen richness (7 species) (Table 10.1). Macrolichen richness (epiphytic and non-epiphytic richness combined) on disturbance plots was negatively related to elevation ($R^2 = 0.56$) and positively related to total macrolichen abundance ($R^2 = 0.60$).

Total macrolichen abundance (epiphytic and non-epiphytic combined) inside the disturbance zones ranged from 0.34 to 4.21 units, and epiphytic lichen abundance ranged from zero in the tree-removal zone to 3.821 in one of the lahar plots (Table 10.1). Total lichen abundance was highest on the debris-avalanche deposit but also very high in the lahar plot that was on the edge of the deposit where trees were not killed and in the tephrafall plots. The pyroclastic-flow and blowdown zone plots had moderate total lichen abundance, but much lower than plots with high lichen abundance. The plots in the standing dead forest, tree-removal zone, and middle of the lahar had the lowest total lichen abundance but were only slightly lower than the pyroclastic-flow and blowdown zone plots (Table 10.1).

Epiphytic lichen abundance in the disturbance zones was highest in the lahar plot on the edge of the deposit, the plot in the tephrafall zone, and the debris-avalanche deposit plot, where a red alder (*Alnus rubra*) patch had established. The blowdown zone plot, other lahar plot, and standing dead zone plot all had much lower epiphytic lichen abundance, and the pyroclastic-flow and tree-removal zones had few to no trees, although one plot in the pyroclastic-flow zone had some epiphytic species that had colonized a boulder.

Fig. 10.2 Common or distinctive lichens of each disturbance zone: (**a**) *Cladonia fimbriata*, (**b**) *Placopsis lambii*, (**c**) *Platismatia stenophylla*, (**d**) *Sphaerophorus tuckermanii*, (**e**) *Peltigera rufescens*, and (**f**) *Lecidea plana*. Photos by Tim Wheeler.

Fifty-seven epiphytic macrolichen species were found on the 11 randomly selected plots outside the eruption zone, 30 species of which were also found in the disturbance plots (Table 10.4). Twenty of these lichens occurred on more than half the plots outside the disturbance zone, including *Alectoria sarmentosa*, *Bryoria glabra*, *Cetraria chlorophylla*, *C. orbata*, *Hypogymnia apinnata*, *H. enteromorpha*, *H. hultenii*, *H. imshaugii*, *H. inactiva*, *H. occidentalis*, *H. physodes*, *H. tubulosa*, *H. wilfiana*, *Nodobryoria oregana*, *Parmelia hygrophila*, *Parmeliopsis hyperopta*, *Platismatia glauca*, *P. herrei*, *P. stenophylla*, and *Sphaerophorus globosus* (which was split into *S. tuckermanii* and *S. venerabilis* after the plots were sampled) (Table 10.4). Species richness for epiphytic macrolichens on the reference plots ranged from 19 to 26 species (Table 10.1) and was weakly positively related to elevation (R^2 = 0.24). Epiphytic macrolichen abundance on plots outside the disturbance zone ranged from 1.143 to 15.601 units (Table 10.1) but was unrelated to species richness (R^2 = 0.07).

Incidental lichens found in the disturbance zone included 107 lichen species, 30 of which were also found on the plots in the disturbance zone (Table 10.5). Most of the additional incidental lichens not recorded on the plots were crustose species that are excluded from the Forest Inventory and Analysis lichen survey protocol and could have therefore occurred in both the disturbance and reference plots. Interesting incidental taxa included *Lecidea plana*, a small crustose species with black apothecia and a sparse thallus, which was common growing on pumice stones in areas that were otherwise devoid of vegetation (Fig. 10.2f). *Rhizocarpon* (ten species) was the most diverse saxicolous crustose genus, followed by *Lecanora* (seven species) and *Lecidea* (six species), all of which were found mostly on andesitic boulders in the Pumice Plain created by the pyroclastic flow (Table 10.5). We observed a surprising number of normally epiphytic lichen species colonizing andesitic boulders on the Pumice Plain (site PNWD1000), including *Alectoria sarmentosa*, *Cetraria chlorophylla*, *C. platyphylla*, *Platismatia glauca*, *Hypogymnia inactiva*, *H. physodes*, *Nodobryoria oregana*, *Platismatia glauca*, *P. stenophylla*, and *Ramalina farinacea* (Table 10.3).

The nonmetric multidimensional scaling (NMS) ordination of plots (using all macrolichens found on disturbance plots) in species space represented 88% of the community variation in the 75 lichen species on 11 plots on two axes (p = 0.04, final instability <0.00001, final stress = 6.4, 250 runs with real and randomized data). Axis 1 explained most of the variation (67%), representing a gradient of forested plots (negative axis scores) to treeless or sparsely treed plots (positive axis scores). Axis 2 explained 21% of the variation, representing a gradient of low-elevation, species-rich plots to higher-elevation plots with fewer lichen species (Fig. 10.3).

10.4 Discussion

10.4.1 Patterns in Lichen Abundance

Epiphytic macrolichen species richness was similar between plots inside and outside the disturbance zones (75 species in the disturbance plots and 59 species in the reference plots; Table 10.1), although the composition was quite different (Tables 10.2 and 10.3). Lichens of mature coniferous forests, such as *Lobaria oregana* and *Sphaerophorus tuckermanii*, were frequent in the reference plots (Table 10.4) but were missing from most disturbed plots, except on large relict trees in the tephrafall zone plots (Table 10.3). Colonization and recruitment by conifers in the blowdown zone may have allowed some forest lichen species to colonize, and many survived in the relict forest patch of the tephrafall zone

Table 10.3 Lichen species found in eruption disturbance zone plots, sorted alphabetically, with abundances within zone in the following categories: 1 = 1–3 thalli, 2 = 4–10 thalli, 3 = >10 thalli but less than half of plot being covered, 4 = more than half of available substrate contains the species. A thallus constitutes a single individual. **Bold face** indicates species found on more than half of the disturbance zone plots sampled.

	Disturbance zone						
Lichen species	Pyroclastic flow	Debris avalanche	Tree removal	Lahar	Blowdown	Standing dead	Tephrafall
Alectoria sarmentosa	**1**			**4**	**2**	**3**	**4**
Bryoria friabilis				2		3	
Bryoria fuscescens				3			2
Bryoria pseudofuscescens					3		
Candelariella pacifica		2			2		
Cetraria chlorophylla	**1**	**2**		**2**	**3**		
Cetraria orbata		2			3		
Cetraria platyphylla	1						
Cladonia bellidiflora				2			
Cladonia borealis	1						
Cladonia cariosa		2	2				
Cladonia carneola				2	2	2	2
Cladonia chlorophaea group	2	2	2		2		3
Cladonia coniocraea			2	2	2		
Cladonia cornuta		1				3	
Cladonia ecmocyna				3			
Cladonia fimbriata	**3**	**3**	**3**	**2**	**2**	**3**	**2**
Cladonia macrophyllodes				1			
Cladonia rangiferina				1			
Cladonia squamosa		2			1		1
Cladonia transcendens	**2**			**3**	**3**	**3**	**2**
Cladonia umbricola					2	3	2
Cladonia verruculosa	3	4	3	2			
Evernia prunastri		4		2			
Hypogymnia enteromorpha		2		3	1		2
Hypogymnia hultenii							1
Hypogymnia imshaugii		1		3	3	3	
Hypogymnia inactiva	**1**	**2**		**3**	**2**		**3**
Hypogymnia physodes	**2**	**3**		**3**	**3**	**3**	
Hypogymnia tubulosa		3		2	3		
Hypotrachyna sinuosa		2					
Leptogium lichenoides							
Leptogium palmatum		2					
Lobaria oregana							2
Melanelixia subaurifera		3					
Melanohalea exasperatula					2		
Melanohalea multispora		2					
Menegazzia subsimilis							
Montanelia disjuncta		1					
Nodobryoria oregana	**1**			**3**	**2**		**3**
Parmelia hygrophila		3			1		
Parmelia sulcata	**1**	**4**			**2**		**2**
Parmeliopsis hyperopta		1		3		3	3
Peltigera collina		3					
Peltigera didactyla	**3**	**3**	**3**		**1**		
Peltigera mebranacea		3					
Peltigera rufescens		4					
Peltigera venosa		2					
Physcia adscendans		2					

(continued)

Table 10.3 (continued)

	Disturbance zone						
Lichen species	Pyroclastic flow	Debris avalanche	Tree removal	Lahar	Blowdown	Standing dead	Tephrafall
Physcia aipolia		1					
Physcia tenella	1	2					
Physconia enteroxantha		1					
Platismatia glauca	**2**	**2**		**4**	**3**	**3**	**4**
Platismatia herrei				3			3
Platismatia norvegica							1
Platismatia stenophylla	2			4	2	2	
Polychidium muscicola		1					
Protopannaria pezizoides		1					
Ramalina dilacerata		2					
Ramalina farinacea		3		2	2		
Sphaerophorus tuckermanii							2
Stereocaulon tomentosum	2	4		2			
Stereocaulon vesuvianum	2	2		2			
Umbilicaria cylindrica	2						
Umbilicaria hyperborea	3	1					
Um. hyperborea var. *hyperborea*	3	2					
Umbilicaria polyphylla	3	2					
Umbilicaria torrefacta	3	2					
Umbilicaria vellea s. lat.		1					
Usnea cornuta		3					
Usnea flavocardia		2					
Usnea scabrata				2	2		
Usnea sp.		2					
Xanthoria candelaria		3					
Xanthoria polycarpa	1	3					

(Table 10.3). However, structural and climatic conditions that produce epiphytic lichen diversity, such as gaps, multiple-aged tree cohorts, and a mix of deciduous and coniferous trees (Neitlich and McCune 1997), have yet to develop in the 32 years since the eruption.

Both the lahar and the tephrafall plots had abundant epiphytic lichen communities, with prolific *Alectoria sarmentosa* and *Platismatia glauca* (Table 10.3). These species are two of the most common and abundant lichens outside the disturbance zone (Table 10.4). However, epiphytic lichen abundance was an order of magnitude higher in the reference plots as compared to plots in the disturbance zones (Table 10.1). Forest structure conducive to abundant epiphytic lichen growth has not yet developed within the disturbance zones studied.

Terricolous lichens (lichens living on the ground or in the soil) accounted for the highest lichen abundance in the disturbance zones, especially on the debris-avalanche deposit; species included *Cladonia verruculosa*, *Peltigera rufescens*, and *Stereocaulon tomentosum*, although a red alder patch on the debris-avalanche deposit plot also had the abundant epiphytes *Evernia prunastri* and *Parmelia sulcata* (Table 10.3). The abundance of lichens on the lahar and tephrafall sites represents legacy lichen communities that likely survived the eruption, whereas the high lichen abundance on the debris-avalanche deposit is attributable to extensive cover of a few terricolous, disturbance-adapted species that colonized since the eruption. However, most of the disturbance plots had low lichen abundance. We hypothesize that the mixture of substrates on the debris-avalanche deposit is more conducive to lichen proliferation than the very droughty pumice of the pyroclastic-flow or blowdown zones.

10.4.2 Patterns of Lichen Species Richness Among Disturbance Zones

Substrate diversity and variable substrate stability were the likely drivers of the increased species diversity on the debris-avalanche deposit, the most species-rich disturbance zone. Substrates on the debris-avalanche deposit plots included boulders, mineral and organic soils, down wood, and a few small patches of conifers, deciduous trees, and shrubs. On the debris-avalanche deposit, there were ground-dwelling species commonly found in disturbed areas, such as *Cladonia verruculosa*, *Peltigera didactyla*, and *Leptogium palmatum*,

Table 10.4 Epiphytic lichen species found on reference plots outside eruption disturbance zone, sorted alphabetically, with abundances in the following: 1 = 1–3 thalli, 2 = 4–10 thalli, 3 = >10 thalli but less than half of plot being covered, 4 = more than half of available substrate contains the species. A thallus constitutes a single individual. **Bold face** indicates species found on more than half of the reference plots sampled.

	Reference plot numbers										
Lichen species	75899 53 59	90632 53 59	86316 53 15	89789 53 59	75144 53 59	71800 53 59	65082 53 15	65627 53 59	86070 53 59	79336 53 59	60580 53 59
Alectoria imshaugii									1		
***Alectoria sarmentosa*[a]**	**4**	**3**	**4**	**3**	**4**	**4**	**4**	**3**	**4**	**4**	**3**
Bryoria capillaris				2		2	4			3	
Bryoria fremontii											3
Bryoria friabilis[a]			4				4		2	4	
Bryoria fuscescens[a]							4			4	
Bryoria glabra			**4**	**3**	**3**	**4**	**3**		**3**		**2**
Bryoria pseudofuscescens[a]					3						
Bryoria trichodes				3			4	3	3		
***Cetraria chlorophylla*[a]**	**3**	**3**		**3**	**3**	**3**	**4**	**1**	**2**	**3**	**3**
***Cetraria orbata*[a,b]**		**2**			**3**	**3**		**2**	**2**	**3**	
Cetraria platyphylla[a]			3			3				3	3
Cladonia coniocraea[a]							3				
Cladonia fimbriata[a]									3		
Cladonia macilenta									2		
Cladonia norvegica								2			
Cladonia ochrochlora					3			1			1
Cladonia squamosa[a]			2		4						
Cladonia transcendens[a]								3			1
Cladonia umbricola[a]						2					
Evernia prunastri[a]		1									
Hypogymnia apinnata	**4**	**3**	**1**	**2**	**3**	**4**	**4**		**4**	**3**	**3**
***Hypogymnia enteromorpha*[a]**	**3**	**3**	**4**	**3**	**4**	**4**	**4**	**3**		**3**	
***Hypogymnia hultenii*[a]**	**2**	**3**		**3**	**2**		**4**	**1**			
***Hypogymnia imshaugii*[a]**		**3**		**3**		**4**		**3**		**3**	**3**
***Hypogymnia inactiva*[a]**	**4**	**3**	**4**	**3**	**4**	**3**	**4**	**3**	**2**	**3**	**2**
Hypogymnia occidentalis	**3**	**1**				**4**			**4**	**1**	**3**
Hypogymnia oceanica						1			1		
Hypogymnia physodes[a]	3	3	3	3	3	1		2	1	2	
Hypogymnia rugosa	3			1							
***Hypogymnia tubulosa*[a]**		**3**		**3**	**3**			**2**		**3**	**3**
Hypogymnia wilfiana	**3**	**3**				**3**		**2**	**3**	**3**	
Hypotrachyna sinuosa[1]		1			3						
Kaernefeltia merrillii									1		
Letharia vulpina										1	
Lobaria oregana			4		3		3				
Melanelixia glabratula[a,c]		1			3						
Melanohalea subelegantula									1		
Menegazzia terebrata					3						
Nodobryoria abbreviata	3									3	3
***Nodobryoria oregana*[a]**		**2**	**4**	**3**		**3**	**4**		**4**	**4**	**2**
***Parmelia hygrophylia*[a]**	**3**			**2**	**3**	**1**	**3**	**2**	**3**	**3**	**2**
Parmelia pseudosulcata					3						
Parmelia sulcata[1]		3	2	3	3						
Parmeliopsis ambigua						1					
***Parmeliopsis hyperopta*[a]**	**4**	**1**	**4**	**2**	**4**	**4**	**4**	**4**	**4**	**4**	**3**
Physcia aipolia[a]		1									
***Platismatia glauca*[1]**	**4**	**3**	**4**	**3**	**4**	**4**	**4**	**4**	**4**	**4**	**3**

(continued)

Table 10.4 (continued)

	Reference plot numbers										
Lichen species	**75899 53 59**	**90632 53 59**	**86316 53 15**	**89789 53 59**	**75144 53 59**	**71800 53 59**	**65082 53 15**	**65627 53 59**	**86070 53 59**	**79336 53 59**	**60580 53 59**
Platismatia herrei	**4**	**3**	**3**		**4**	**3**	**3**	**3**	**3**	**3**	
Platismata norvegica	3						3				
Platismata stenophylla[a]	**4**	**3**	**3**	**3**	**3**	**3**	**4**	**3**	**2**	**3**	
Sphaerophorus globosus[a,b]	**3**	**1**	**3**	**2**	**4**		**4**	**3**			
Tuckermannopsis subalpina									1	3	
Usnea chaetophora							1	2			
Usnea filipendula			3			2		2			
Usnea scabrata[a]								2			
Usnea subfloridana					3						

[a]Species also found on plots in the disturbance zone (Table 10.3).
[b]*Melanelixia glabratula* was likely misidentified as *M. subaurifera*.
[c]*Sphaerophorus globosus* represents a combination of *S. venerabilis* and *S. tuckermanii*, which were described after these data were collected.

Table 10.5 All lichen species, including crustose lichens, found in incidental surveys in the areas immediately surrounding disturbance zone plots, sorted alphabetically. x = present. **Bold face** denotes species also found on disturbance zone plots. Note that no incidental collections were made on the debris-avalanche deposit.

	Disturbance zone					
Lichen species	**Pyroclastic flow**	**Tree removal**	**Lahar**	**Blowdown**	**Standing dead**	**Refugium**
Acarospora fuscata		x	x			
Acarospora oreophila			x			
Acarospora veronensis			x			
Agyrium rufum		x				
Alectoria sarmentosa	**x**		**x**			
Arthropyrenia punctiformis	x			x		
Biatora flavopunctata				x		
Biatora sp.				x	x	
Bryoria glabra			x			
Bryoria sp.						
Buellia aethalea	x		x			
Buellia stigmatea	x					
Candelaria "cascadensis"			x			
Cetraria chlorophylla			**x**			
Cetraria orbata			**x**			
Cetraria platyphylla			**x**		**x**	
Cladonia carneola	**x**		**x**			
Cladonia coniocraea	**x**					
Cladonia gracilis		**x**				
Cladonia macrophyllodes		**x**				
Cladonia ochrochlora			**x**			
Cladonia verruculosa		**x**	**x**			
Elixia flexella				x		
Evernia prunastri	**x**		**x**			
Fuscopannaria cyanolepra			x			
Hypogymnia enteromorpha			**x**			
Hypogymnia hultenii			**x**			
Hypogymnia imshaugii	**x**		**x**		**x**	
Hypogymnia inactiva	**x**		**x**			
Hypogymnia metaphysodes	x		x			
Hypogymnia occidentalis			x			
Hypogymnia physodes	x		x		x	

(continued)

Table 10.5 (continued)

	Disturbance zone					
Lichen species	**Pyroclastic flow**	**Tree removal**	**Lahar**	**Blowdown**	**Standing dead**	**Refugium**
Hypogymnia rugosa				x		
Hypogymnia tubulosa	x		x			
Lecanora argentea	x					
Lecanora carpinea	x					
Lecanora circumborealis	x	x		x	x	
Lecanora intricata	x					
Lecanora polytropa	x		x			
Lecanora schizochromatica	x	x				
Lecanora zosterae	x					
Lecidea atrobrunnea	x	x				
Lecidea auriculata	x					
Lecidea lapicida		x				
Lecidea paddensis			x		x	
Lecidea plana	x	x	x			
Lecidea sp.	x	x				
Lecidea tesselata		x				
Lepraria borealis				x		
Lepraria pacifica					x	
Letharia vulpina	x					
Melanelixia glabratula				**x**		
Melanohalea exasperatula	**x**					
Melanohalea subelegantula				x		
Micarea sp.					x	
Miriquidica atrofulva				x		
Mycoblastus affinis						x
Mycoblastus caesius					x	
Mycoblastus sanguinarius					x	x
Myriospora scabrida	x					
Nodobryoria oregana		**x**		**x**		
Parmelia hygrophila	**x**					
Parmelia sulcata	**x**			**x**		
Peltigera membranacea	**x**					
Peltigera rufescens	**x**					
Pertusaria mccroryae					x	
Pertusaria ophthalmiza					x	
Physcia aipolia				**x**		
Physcia tenella				x		
Placopsis lambii	x	x		x		
Placynthiella dasaea					x	
Platismatia glauca				x		
Platismatia stenophylla				x		
Polysporina simplex	x					
Porpidia thomsonii				x		
Pseudephebe pubescens						x
Ramalina farinacea				**x**		
Rhizocarpon distinctum				x		
Rhizocarpon geminatum				x		x
Rhizocarpon geographicum	x	x				
Rhizocarpon grande	x					
Rhizocarpon macrosporum	x	x				
Rhizocarpon polycarpon	x					

(continued)

Table 10.5 (continued)

Lichen species	Disturbance zone					
	Pyroclastic flow	Tree removal	Lahar	Blowdown	Standing dead	Refugium
Rhizocarpon postumum				x		
Rhizocarpon reductum				x		
Rhizocarpon riparium			x			
Rhizocarpon sphaerosporum				x		
Rinodina pyrina	x					
Staurothele aff. *fissa*				x		
Stereocaulon nivale			x			
Stereocaulon tomentosum	**x**		**x**			
Stereocaulon vesuvianum	**x**	**x**			**x**	
Thelomma occidentale	x					
Trapelia sp.	x					
Umbilicaria sp.		x				
Umbilicaria decussate	x					
Umbilicaria hyperborea	**x**	**x**	**x**			**x**
Um. hyperborea* var. *radicicula	**x**					
Umbilicaria polyphylla	x					x
Umbilicaria torrefacta			x			x
Umbilicaria virginis		x				
Usnea sp.				x		
Usnea subfloridana				x		
Verrucaria no spore				x		
Xanthoria candelaria	x			x		
Xanthoria polycarpa				x		
Xylographa trunciseda					x	

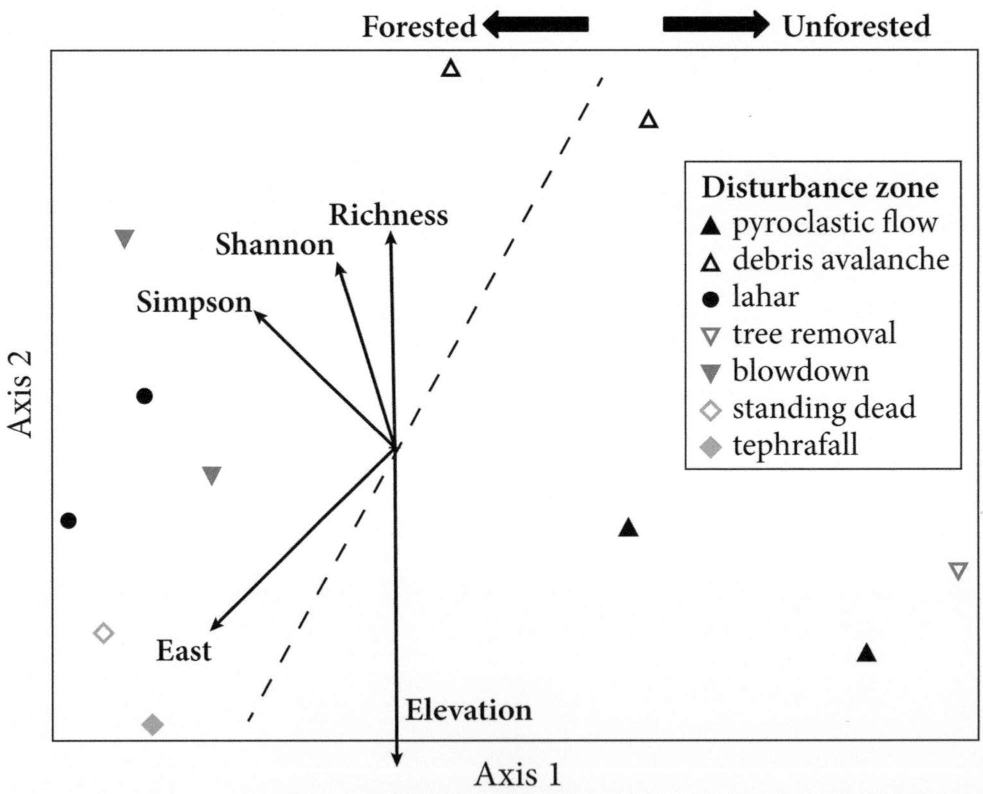

Fig. 10.3 Ordination of plots from the disturbance zone in species space. Each symbol is a plot coded by disturbance type. Plots closer to each other have more similar lichen communities. Vectors point in the increasing direction of assumed relationship to Shannon's and Simpson diversity indices. *Dotted line* and *thick black* arrows show the main gradient thought to be represented by axis 1, from forested to nonforested plots.

as well as rock-dwelling species, such as four *Umbilicaria* species. The epiphytes *Hypotrachyna sinuosa*, *Menegazzia subsimilis*, and *Physcia aipolia* were found on the red alders in the debris-avalanche deposit plots, and the scattered, small Douglas-fir trees also provided substrate for a separate group of lichens most common on conifers, including several species of *Hypogymnia* and *Cetraria* plus *Nodobryoria oregana*. The combination of all these substrates contributed to the development of a relatively diverse lichen community on the debris-avalanche deposit (Tables 10.1 and 10.2). However, the debris-avalanche deposit plots were at the lowest elevation, confounding the effect of the disturbance on species richness with elevation (Fig. 10.3). Interestingly, lichen species richness exhibited a weak negative relationship with elevation in the reference plots, opposite to the relationship between lichen species richness and elevation in the disturbance area. This implies that disturbance type is mainly responsible for lichen species richness in the eruption zone, not factors related to elevation.

We observed abundant *Niphotrichum elongatum*, a mat-forming moss, on the lahar plots, some places in the pyroclastic-flow zone, and on the debris-avalanche deposit. We suspect these mats inhibited lichen establishment, partially explaining the lower diversity on the lahar and pyroclastic-flow plots. Despite the passage of 32 years since the eruption, most of the ground surface in the tree-removal and pyroclastic-flow zones was still pumice, which is very unstable and droughty. This was also the likely cause of the low lichen diversity in many disturbance zones. Lichens apparently cannot colonize pumice, except in small patches, despite their desiccation tolerance. The exception was *Lecidea plana*, which embedded itself in and on the individual pumice pebbles. Habitat heterogeneity within the pyroclastic-flow zone, mainly boulders of more stable rock types, gave other lichens the opportunity to colonize and persist. Colonization of these boulders by saxicolous genera (such as *Umbilicaria*) and by epiphytic taxa increased lichen species richness at one of the pyroclastic-flow sites. We suspect that birds transported epiphytic lichen propagules to some prominent andesitic boulders in the pyroclastic-flow sites that served as perches.

At the tephrafall site, lichen species associated with old-growth forest survived the eruption, including the dispersal-limited *Lobaria oregana* (Denison 1979), which was moderately abundant. Forests for miles around the site were toppled or singed, greatly reducing the number of propagules likely to reach disturbed areas. Many old-growth-associated lichens have very large propagules, which decrease their dispersal distance. Increased distance and large dispersal propagules greatly reduce the chance of colonization of the remnant patch of old-growth forest in the tephrafall zone. Other old-growth-associated lichen species that were present, such as *Platismatia herrei* and *Sphaerophorus tuckermanii* (sensu Pike et al. 1975), lend support to the notion that the forest patch in the tephrafall zone served as a refuge for lichens.

In the study most comparable to our Mount St. Helens work, researchers sampled the vegetation community, including lichens, in disturbance zones created by the 1907 eruption of the Ksudach volcano in Kamchatka, Russia (Grishin et al. 1996). The Ksudach eruption lacked the diverse and spatially extensive disturbance features generated by the 1980 Mount St. Helens eruption and was studied nearly 90 years post eruption. Nonetheless, the researchers stratified their sampling among the different disturbance types created by a gradient of tephra-deposit thickness, in which they found that lichens, mostly *Stereocaulon grande*, were most abundant in areas with pumice deposits greater than 100 cm deep. Lichen diversity was highest in forest patches that survived the eruption where tephra deposits were shallow (Grishin et al. 1996). However, they did not sample epiphytic lichens, preventing direct comparisons with our species richness estimates. Nonetheless, the diversity of substrates likely present at the sites with shallow tephra deposits (wood, older outcrops, new tephra, organic soils) probably increased species richness in the Ksudach eruption in the same way as on our plots on the debris-avalanche deposit.

10.5 Conclusions

Lichen abundance and community composition varied among disturbance zones. The sampled area of the debris-avalanche deposit had the highest total and epiphytic lichen species richness, likely as a result of habitat heterogeneity that provided substrates for many different lichen species. The plots in the tree-removal zone had the lowest species richness because most of the surface was covered with pumice, which may have been initially too unstable and droughty for lichens to establish or persist. Some old-growth-associated lichens survived in forest patches that experienced only heavy tephrafall. Compared with lichen communities outside the eruption disturbance zones, many epiphytic lichen species have yet to colonize trees inside the disturbance area. Total lichen abundance peaked in the sample plots on the debris-avalanche deposit relative to other disturbance zones because there are many different textures of substrates available, which provide a diversity of habitats for lichens to colonize. Lichen abundance peaked on one lahar plot and in the tephrafall zone owing to the survival of abundant epiphytic lichens. Our results are a starting point to better understand the variable responses of different groups of lichens to a variety of volcanic disturbances produced by a single eruption.

Acknowledgments This work was the outcome of a foray convened by the Northwest Lichenologist Guild and hosted by the USDA Forest

Service, Pacific Northwest Research Station, and the Mount St. Helens Institute. Support for CMC was provided by the US Forest Service-Pacific Northwest Research Station and the US National Science Foundation LTREB Program (DEB-0614538).

References

Allen, M.F., C.M. Crisafulli, S.J. Morris, L.M. Egerton-Warburton, J.A. MacMahon, and J.A. Trappe. 2005. Mycorrhizae and Mount St. Helens: Story of a symbiosis. In *Ecological responses to the 1980 eruption of Mount St. Helens*, ed. V.H. Dale, F.J. Swanson, and C.M. Crisafulli, 221–232. New York: Springer.

Antos, J.A., and D.B. Zobel. 2005. Plant responses in forests of the tephra-fall zone. In *Ecological responses to the 1980 eruption of Mount St. Helens*, ed. V.H. Dale, F.J. Swanson, and C.M. Crisafulli, 47–58. New York: Springer.

Beard, J.S. 1945. The progress of plant succession on the Soufrière of St. Vincent. *Journal of Ecology* 33: 1–9.

———. 1976. The progress of plant succession on the Soufrière of St. Vincent: Observations in 1972. *Vegetatio* 31: 69–77.

Berryman, S., and B. McCune. 2006. Estimating epiphytic macrolichen biomass from topography, stand structure and lichen community data. *Journal of Vegetation Science* 17: 157–170.

Bjarnason, Á.H. 1991. Vegetation on lava fields in the Hekla area, Iceland. PhD dissertation. Uppsala: Uppsala University.

Crisafulli, C.M., J.A. MacMahon, and R.R. Parmenter. 2005a. Small-mammal survival and colonization on the Mount St. Helens volcano: 1980–2002. In *Ecological responses to the 1980 eruption of Mount St. Helens*, ed. V.H. Dale, F.J. Swanson, and C.M. Crisafulli, 199–202. New York: Springer.

Crisafulli, C.M., L.S. Trippe, C.P. Hawkins, and J.A. MacMahon. 2005b. Amphibian responses to the 1980 eruption of Mount St. Helens. In *Ecological responses to the 1980 eruption of Mount St. Helens*, ed. V.H. Dale, F.J. Swanson, and C.M. Crisafulli, 183–198. New York: Springer.

Dale, V.H., F.J. Swanson, and C.M. Crisafulli. 2005. Disturbance, survival, and succession: Understanding ecological responses to the 1980 eruption of Mount St. Helens. In *Ecological responses to the 1980 eruption of Mount St. Helens*, ed. V.H. Dale, F.J. Swanson, and C.M. Crisafulli, 3–11. New York: Springer.

del Moral, R., and D.M. Wood. 2012. Vegetation development on permanently established grids, Mount St. Helens (1986–2010). *Ecology* 93: 2125–2125.

Denison, W.C. 1979. *Lobaria oregana*, a nitrogen-fixing lichen in old-growth Douglas-fir forests. In *Symbiotic nitrogen fixation in the management of temperate forests: Proceedings of a workshop held April 2–5, 1979*, ed. J.C. Gordon, C.T. Wheeler, and D.A. Perry, 266–275. Corvallis: Forest Research Laboratory, Oregon State University.

Esslinger, T.L. 2012. A cumulative checklist for the lichen-forming, lichenicolous and allied fungi of the continental United States and Canada. North Dakota State University. Version #18. Accessed Dec 2012. http://www.ndsu.edu/pubweb/~esslinge/chcklst/chcklst7-18.htm

Foxworthy, B.L., and M. Hill. 1982. *Volcanic eruptions of 1980 at Mount St. Helens: The first 100 days*, Professional Paper 1249. Washington, DC: U.S. Geological Survey.

Frenzen, P.M., K.S. Hadley, J.J. Major, M.H. Weber, J.F. Franklin, J.H. Hardison III, and S.M. Stanton. 2005. Geomorphic change and vegetation development on the muddy river mudflow deposit. In *Ecological responses to the 1980 eruption of Mount St. Helens*, ed. V.H. Dale, F.J. Swanson, and C.M. Crisafulli, 75–91. New York: Springer.

Grime, J. 1977. Evidence for the existence of three primary strategies in plants and its relevance to ecological and evolutionary theory. *American Naturalist* 111: 1169–1194.

Grishin, S.Y., R. Moral, P.V. Krestov, and V.P. Verkholat. 1996. Succession following the catastrophic eruption of Ksudach volcano (Kamchatka, 1907). *Plant Ecology* 127: 129–153.

Jackson, T.A. 1971. A study of the ecology of pioneer lichens, mosses, and algae on recent Hawaiian lava flows. *Pacific Science* 1: 22–32.

Jackson, T.A., and W.D. Keller. 1970. A comparative study of the role of lichens and 'inorganic' processes in the chemical weathering of recent Hawaiian lava flows. *American Journal of Science* 269: 446–466.

Kitayama, K., D. Mueller-Dombois, and P.M. Vitousek. 1995. Primary succession of Hawaiian montane rain forest on a chronosequence of eight lava flows. *Journal of Vegetation Science* 6: 211–222.

Kurina, L.M., and P.M. Vitousek. 1999. Controls over the accumulation and decline of a nitrogen-fixing lichen, *Stereocaulon vulcani*, on young Hawaiian lava flows. *Journal of Ecology* 87: 784–799.

———. 2001. Nitrogen fixation rates of *Stereocaulon vulcani* on young Hawaiian lava flows. *Biogeochemistry* 55: 179–194.

Magnússon, B., S.H. Magnússon, and S. Fridriksson. 2009. Developments in plant colonization and succession on Surtsey during 1999–2008. *Surtsey Research* 12: 57–76.

McCune, B., and L. Geiser. 2009. *Macrolichens of the Pacific Northwest*. Corvallis: Oregon State University Press.

McCune, B., and M.J. Mefford. 2011. *PC-ORD version 6.0. Multivariate analysis of ecological data, users guide*. Gleneden Beach: MjM Software Design.

McCune, B., J.P. Dey, J.L.E. Peck, D. Cassell, K. Heiman, S. Will-Wolf, and P.N. Neitlich. 1997. Repeatability of community data: species richness versus gradient scores in large-scale lichen studies. *The Bryologist* 100: 40–46.

Neitlich, P.N., and B. McCune. 1997. Hotspots of epiphytic lichen diversity in two young managed forests. *Conservation Biology* 11: 172–182.

Parmenter, R.R., C.M. Crisafulli, N. Korbe, G. Parsons, M. Edgar, and J.A. MacMahon. 2005. Posteruption arthropod succession on the Mount St. Helens volcano: The ground-dwelling beetle fauna (Coleoptera). In *Ecological responses to the 1980 eruption of Mount St. Helens*, ed. V.H. Dale, F.J. Swanson, and C.M. Crisafulli, 139–150. New York: Springer.

Peck, J.E., and B. McCune. 1997. Remnant trees and canopy lichen communities in western Oregon: A retrospective approach. *Ecological Applications* 7: 1181–1187.

Pike, L.H. 1978. The importance of epiphytic lichens in mineral cycling. *The Bryologist* 81: 247.

Pike, L.H., W.C. Denison, D.M. Tracy, M.A. Sherwood, and F.M. Rhoades. 1975. Floristic survey of epiphytic lichens and bryophytes growing on old-growth conifers in western Oregon. *The Bryologist* 78: 389–402.

Sipman, H.J.M., and T. Raus. 1995. Lichen observations from Santorini (Greece). *Bibliotheca Lichenologica* 57: 409–428.

Skottsberg, C. 1941. Plant succession on recent lava flows in the island of Hawaii: Goteborgs Kungl. *Goteborgs Kungliga Vetenskaps och Vitterhets-Samhället Handlingar, Sjatte Foljden, Series B* 1: 8. 32 pages.

Tsuyuzaki, S., and A. Hase. 2005. Plant community dynamics on the volcano Mount Koma, northern Japan, after the 1996 eruption. *Folia Geobotanica* 40: 319–330.

U.S. Department of Agriculture, Forest Service [USDA]. 2014. Field Sampled Vegetation (FSVeg), common stand exam field guides, region 6 field guide, appendix E existing vegetation references and codes. Accessed 31 Oct 2016. http://www.fs.fed.us/nrm/documents/fsveg/cse_user_guides/R6FG_appendix.pdf.

Waitt, R.B., Jr. 1981. Devastating pyroclastic density flow and attendant airfall of May 18: Stratigraphy and sedimentology of deposits. In *The 1980 eruptions of Mount St. Helens, Washington*, Professional Paper 1250, ed. P.W. Lipman and D.R. Mullineaux, 439–458. Washington, DC: U.S. Geological Survey.

Whittaker, R.J., M.B. Bush, and K. Richards. 1989. Plant recolonization and vegetation succession on the Krakatau Islands, Indonesia. *Ecological Monographs* 59: 59–123.

11 Succession and Mycorrhizae on Mount St. Helens

Michael F. Allen, Matthew R. O'Neill, Charles M. Crisafulli, and James A. MacMahon

11.1 Introduction

The eruption of Mount St. Helens in May 1980 is a clear example of a large, infrequent disturbance (Dale et al. 1998). But what does that mean? Biologically, most large disturbances encompass many plant communities and soil types and have variable intensity within the scope of the single event (volcanoes, hurricanes, and landslides are examples). Living and dead organisms persist after disturbance, varying in type, abundance, and spatial distribution. These are termed *biological legacies*. Legacies come in many forms and may include survival of trees, persistence of plant propagules, presence of topsoil with its myriad microorganisms, or even survival of animals in protected patches or burrowed below the soil surface. All of these legacy types and scales can occur within a single large disturbance. These legacies may strongly influence post-disturbance ecological processes, and subsequent small perturbations that disrupt the structure and functioning can both facilitate and retard successional processes within a disturbance matrix. From a landscape perspective, all of these factors interact to create a complex array of communities whose trajectories and patterns dictate the structure of the resulting landscape.

Much of the theory of ecological succession, beginning with Cowles (1899) and Cooper (1923), has been developed using space-for-time substitutions in comparing chronosequences, where an underlying assumption is that a patch advances along a linear trajectory and that differences among sites are due to the time since disturbance. The eruption of Mount St. Helens presented a unique opportunity to study differences in successional dynamics among locations altered to varying degrees by the same event, with its multiplicity of disturbance types and intensities. Furthermore, the protection of the core research area in the Mount St. Helens National Monument and the initial and continued support for research activity have provided a novel chance to independently observe multiple sites for over three decades, with the opportunity to continue observations into the foreseeable future.

The conceptual framework of Clements (1916), as modified by MacMahon (1981) (Fig. 11.1), is still useful for describing successional processes that are critical to understanding key features of the changing landscape of today and to predicting tomorrow's patterns. Specifically, succession is a sequence of organisms affected by processes that occur post-disturbance (or post-*nudation*). These organisms include both *survivors* (also known as *residuals*) and *immigrants*, organisms and propagules that arrive at the site and undergo *ecesis*, the process of establishment. *Biotic interactions*, such as competition, select successful individuals from the invading pool of organisms. Together, these processes result in a *stabilization* (or, as per Clements [1916], *climax*) phase.

In the Pacific Northwest, forest communities change continually, even across multi-century time scales (e.g., Franklin et al. 2002). Tree establishment may occur across multi-decadal scales (Poage et al. 2009; Freund et al. 2014), and basal area, root biomass, and soil organic matter likely never

M.F. Allen (✉)
Departments of Plant Pathology and Microbiology, Center for Conservation Biology, University of California-Riverside, 1900 University Avenue, Riverside, California, CA 92521-0334, USA
e-mail: michael.allen@ucr.edu

M.R. O'Neill
Department of Biology, University of California-Riverside, 1900 University Avenue, Riverside, CA 92521, USA
e-mail: monei003@ucr.edu

C.M. Crisafulli
U.S. Department of Agriculture, Forest Service, Pacific Northwest Research Station, Mount St. Helens National Volcanic Monument, 42218 NE Yale Bridge Rd., Amboy, Washington, DC 98601, USA
e-mail: ccrisafulli@fs.fed.us

J.A. MacMahon
Department of Biology and Ecology Center, Utah State University, 1688 Willow Way, Logan, UT 84341, USA
e-mail: jim.macmahon@usu.edu

C.M. Crisafulli, V.H. Dale (eds.), *Ecological Responses at Mount St. Helens: Revisited 35 years after the 1980 Eruption*,
https://doi.org/10.1007/978-1-4939-7451-1_11

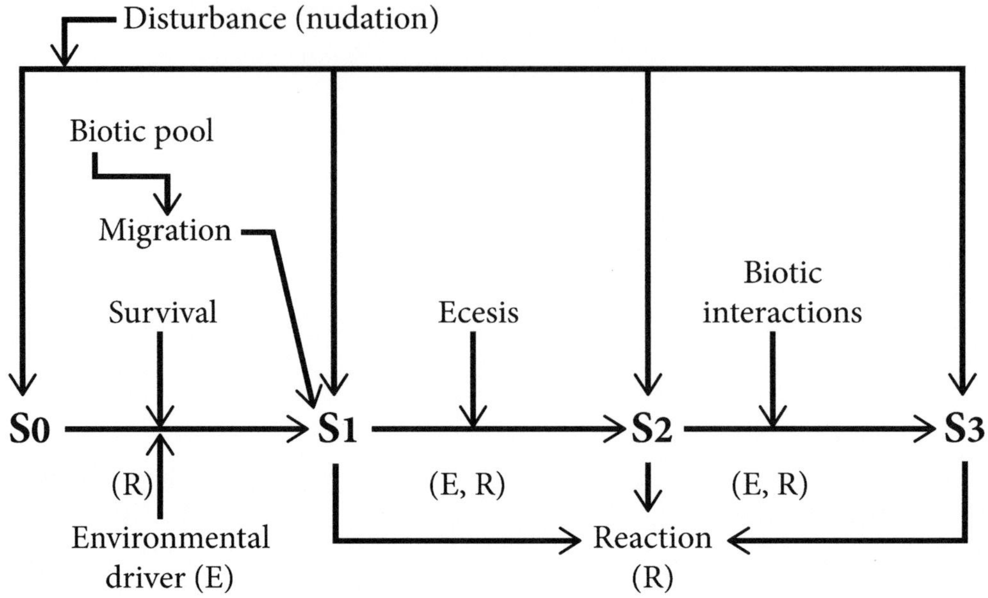

Fig. 11.1 Successional model for evaluating the roles of different processes, derived and redrawn from MacMahon (1981). S refers to the successional state of the system. S0 represents baseline conditions following disturbance, before organisms are present; S1 includes the total pool of surviving and invading organisms before they start growing; S2, after organisms establish (ecesis), is largely a function of soils and climate; and S3 results after biotic interactions such as competition. E refers to environmental drivers and R to reaction of organisms to these actions.

reach equilibrium (Clemmensen et al. 2013). Thus, depending on the intervals between stand-replacement disturbances, a "climax" may never be reached, but a dynamic equilibrium may be established (Edmonds 1982).

In part, the changing nature of forests over time is due to the shifting composition of organisms and to complex interactions with symbiotic organisms. Thus, symbiotic microbes offer a unique opportunity to understand the intricacies of succession. Our focal organisms are the mycorrhizal fungi (MF). A *mycorrhiza* is a mutualistic association between plants and fungi, localized in the plant root. These fungi form a symbiosis with plants whereby the fungus enters a root and its mycelium extends outward from the root surface into the soil, where an expanding hyphal network absorbs nutrients and water. Nutrients are then exchanged with the plant for carbon. Mycorrhizae can be formed between nearly all phyla of fungi and almost all land plants, with only a few exceptions (Allen 1991). MF indirectly affect other microorganisms, such as nitrogen-fixing bacteria, by providing the plant with the phosphorus, magnesium, or iron required for nitrogen fixation while simultaneously competing with the same microorganisms for the plant's carbon (Allen and Allen 2017).

In some locations, MF can survive a large disturbance. If soil remains intact, buried soils can retain their biological viability. Spores of MF and plant seeds persisted in soil below the buried tephra for between 5 and 7 years on Mount St. Helens (Allen et al. 2005). Upon exposure and renewed contact between plants and MF, new mycorrhizal infections are readily established.

In other cases, MF must reinvade (immigrate) into the disturbed area. MF disperse by a variety of mechanisms. Some serve as a food source for animals, and in the process, spores are dispersed as the animals traverse their environment. Spores on the soil surface interact with the physical environment, and may disperse by wind, depending on the size and shape of the spores, the intensity and structure of wind turbulence, and the humidity of the air (Allen et al. 1993).

Once dispersed, an immigrating mycorrhizal propagule is subject to conditions specific to the place where it lands in the disturbed landscape. This means that an individual organism is sensitive to legacy effects, defined by the particular spatial location across various topographic and disturbance-intensity gradients within the disturbance area. Once established, the hyphae of an individual MF connect with other individuals of compatible strains (anastomosis) to form a dikaryotic mycelial network that can expand as much as several meters, connecting multiple plants. This process is part of ecesis (establishment).

Although MF have similar functions in different ecosystems—that is, soil resources are exchanged by the fungus for carbon (C) from the plant—variations on this theme are numerous. The role of mycorrhizae in succession has been investigated for decades (Allen 1991), but most studies focus on how mycorrhizal fungi alter plant communities (e.g., van der Heijden and Sanders 2002). In many classic studies of mycorrhizae and succession, the first colonizing plants tend to be *nonmycotrophic*, that is, they never form mycorrhizae (see, e.g., Allen and Allen 1980; Janos 1980). Mycorrhizal plants and their associated MF comprise the next wave of colonists. The plant community then shifts from facultative to obligate mycorrhizal status. Plant colonization at Mount St. Helens never behaved in that predictable manner. Beginning with extended work on Mount St. Helens, a new perspective has arisen, focusing on the simultaneous interactions among fungi, plants, animals, and environmental change in the process of succession. On Mount St. Helens, many of the colonizing species (e.g., the trees) are mycorrhizal obligates (i.e., they require MF to survive and reproduce in nature), and all other plants are facultatively mycorrhizal (they form mycorrhizae if both the plant and fungus are present). There are no nonmycotrophic higher plant species at Mount St. Helens (see discussion in Sect. 11.4.1). This pattern represents a paradigm shift in studying the role of mycorrhizae in succession.

11.2 Mycorrhizae on Mount St. Helens

Mycorrhizae are the most abundant and have the greatest biomass of mutualistic associations in terrestrial ecosystems. The fungus is an obligate dependent upon its host for carbon. To obtain that carbon, it penetrates the root cortex and grows intra- or intercellularly (Fig. 11.2). Using up to 30% of the net carbon fixed by the plant, the fungus explores the soil for nutrients and water. Because of the small size of the individual hypha and the large extent of the mycelial network, MF access molecules of nutrients, which the hyphae transport and exchange with the plant for the host's carbon. Many of the categorizations of mycorrhizae have changed over the years during which symbiosis has been studied at Mount St. Helens. Because every known type of mycorrhiza has been found at Mount St. Helens, understanding mycorrhizal dynamics at the volcano helps us understand mycorrhizal functioning at a global scale.

A brief overview of the types of mycorrhizae found on Mount St. Helens is given below; for more detail, see van der Heijden and Sanders (2002). There are five basic types of mycorrhizae on Mount St. Helens—arbuscular mycorrhizae (AM), ectomycorrhizae (EM), ericoid mycorrhizae, arbutoid

Fig. 11.2 Mycorrhizae on Mount St. Helens from the Smith Creek site. Shown are (**a**) arbuscular mycorrhizae with arbuscules, hyphae, and vesicles within a *Lupinus latifolius*; (**b**) arbuscular mycorrhizal fungal hyphae extending out from roots; (**c**) an ectomycorrhizal root tip and extending hyphae from roots of an *Abies amabilis*; and (**d**) black ectomycorrhizal fungi on *A. amabilis*. Photos: M.F. Allen.

mycorrhizae, and orchid mycorrhizae. In the AM, the most primitive and most common type of mycorrhiza, hyphae of fungi of the division Glomeromycota, penetrate the cell walls of the inner cortical cells (but not the membrane) of plant rootlets. Within the cell walls, nutrients from the fungi are exchanged for carbon from the host across the fungal membrane, fungal wall, intracellular space, and plant membrane. Most plants form AM, known in the literature at the time of our earlier publications as vesicular-arbuscular mycorrhizae (VAM). Although not all MF form vesicles (members of the Gigasporales do not), neither do they always form arbuscules (e.g., in maples, *Acer* spp.). However, the AM label remains. AM fungi (AMF) can infect seedlings of taxa that are often not considered to be hosts for AM. For example, on Mount St. Helens and elsewhere in the southern Washington Cascades, AMF were found in Pinaceae (pine family) (Allen, unpublished; see also Smith et al. 1998), *Lupinus* (Allen and MacMahon 1988), *Equisetum* (horsetails), and *Carex* (sedges) (Allen, unpublished observations).

Mycorrhizae of Pinales (conifers) and many angiosperm trees (e.g., Fagaceae, beeches, and oaks) do not penetrate the plant cell walls but form a mantle covering the tips of fine roots. In this relationship, the MF penetrate the fine roots and form an interweaving (hartig) net between cortical cells that enables nutrients from the fungus to be exchanged for carbon from the host plant. These networks are called ectomycorrhizae (EM). EM fungi are members of the genus *Endogone* (in the Zygomycota) or two other phyla of fungi, the ascomycetes and basidiomycetes. Some plants are switchers that can form both AM and EM, including *Alnus*, *Salix*, and *Populus* (alders, willows, and cottonwoods), all found at Mount St. Helens.

Most AM and EM fungi are generalist taxa with respect to host specificity (Molina et al. 1992; Allen et al. 1995), although many exhibit preferential relationships. However, some EM fungi and other mycorrhizal types have MF that form highly specialized partnerships.

At Mount St. Helens, three other mycorrhizal types can be found. Ericoid mycorrhizae form between fungi in the ascomycetes and plants in the Ericaceae (heath family; subfamilies Vaccinioideae [huckleberries] and Ericoideae [heathers and rhododendrons]), in which the MF also penetrate the cortical cells. Arbutoid mycorrhizae form between basidiomycetes and plants in the subfamilies Arbutoideae (manzanitas, kinnikinnick) and Monotropoideae (Indian pipe). This type of mycorrhiza is often considered simply a variant of an EM but with a less-well-developed mantle and with penetration of cortical cells by the MF. Finally, orchids (Orchidaceae) form their own type of mycorrhiza with basidiomycete fungi in the Cantharellales and Tulasnellales.

A mycorrhiza improves the ability of the host plant to acquire soil nutrients and water because the hyphae radiate out into the substrate and penetrate small pores and organic matter fragments that plant roots cannot access. In the case of AM, the individual hypha grows approximately 0.5–1 mm, then bifurcates, and repeats, approximately daily (Allen et al. 2003; Hernandez and Allen 2013). This bifurcating network, termed a *mycelium*, produces approximately 130 cm of hyphae for each penetration point, with a penetration point every 1–5 mm of fine root. EM hyphal architecture is not as regular as AM (Allen et al. 2003), but *sporocarps* can be found up to 20 m distant from a host plant. Large networks of mycelia link plants in a "common mycorrhizal network" (CMN) (Bledsoe et al. 2014), regulating the allocation of resources among individual plants. There are both upper and lower limits of soil nutrient concentrations that constrain mycorrhizal functioning (Fig. 11.3). This concentration threshold is an extensively documented part of mycorrhizal dynamics. Mycorrhizal activity peaks when nutrient concentrations are low because hyphae extend beyond root systems, increasing the soil volume explored. Above those concentrations, the plant roots themselves readily acquire nutrients and will actively reject the mycorrhiza, as the fungus becomes a carbon drain. However, at extremely low levels of nutrient nitrogen (N) or phosphorus (P) availability, the fungal mycelium becomes nutrient limited; there is simply an inadequate amount of P or N to support fungal biomass and simultaneously transport nutrients to the host plant (Treseder and Allen 2002). This conceptual framework is of critical importance in understanding the dynamics of microbial symbioses and succession on Mount St. Helens.

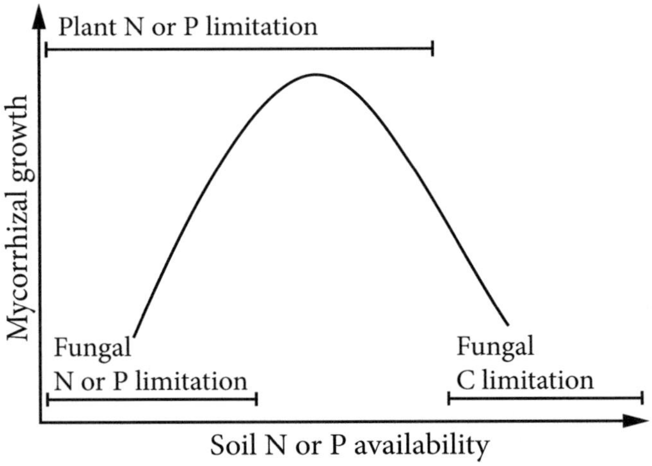

Fig. 11.3 The response of mycorrhizae to soil and plant nutrient limitations (derived from Treseder and Allen 2002). At extremely low nutrient N or P availability, plants and fungi compete for nutrients, and as the fungus is an obligate upon the plant for carbon, it cannot survive and grow. At intermediate levels, plants are nutrient limited but can provide carbon to increase nutrient uptake. At high nutrient levels (such as artificial fertilization), the plant no longer needs the fungal nutrients and rejects the mycorrhizal partner.

Mycorrhizal dynamics occur in two predominant phases that are often confusing in the literature. The first process is colonization. Many authors use the term colonization to refer to the actual amount of infection of an individual root or root system. Because we are concerned with moving plant and fungal inoculum across space, we distinguish the immigration process (colonization) from the symbiosis formation (infection). Infection can occur from residual propagules (spores or infected root fragments) that survived the eruption or from immigrating propagules.

Infection is the actual penetration of a root by a fungus and the establishment of the symbiosis. Nicholson (1960) described three factors regulating infection: compatible host genotype, edaphic factors (particularly energy, nutrients, and water), and the "activity" of the microbe (the active inoculum density, viability, and spatial distance from infecting point). Garrett (1970) termed these three factors "inoculum potential." Edaphic factors play a large role in the infection process, based on nutrients and water availability (Fig. 11.3). These factors will be discussed in Sect. 11.4. Inoculum density, viability, and spatial pattern are also key to initiating new infections.

At sites where there are no residuals, colonization depends upon immigration of spores or transport of infected root fragments to the site. Immigration occurs via two predominant transportation processes: wind and animals. Wind can move propagules up to several km from source populations (e.g., Allen 1988). Dispersal depends on wind intensity, direction, and landscape characteristics (Allen et al. 1989). This mechanism of dispersal has been documented for EM fungal propagules at scales of several km (Allen 1987). Animals can also transport propagules, externally or internally, and this appears to be a dominant mechanism of dispersal at Mount St. Helens (Allen 1988), as discussed in greater detail in Sect. 11.4.

Following colonization, hyphae grow from spores or previously infected root segments toward newly formed root segments. This local colonization can include plant-to-plant infection (forming the common mycorrhizal network). Only primary roots can be infected, so colonization occurs as hyphae grow across the patch. Mosse (1977) found that AM hyphae could infect new plants across crop rows up to a 20-cm distance during a growing season. We have observed new pine seedlings of many species at many sites being infected by *Pisolithus tinctorius* from neighboring trees across tens of meters of mine spoils (Allen, unpublished observations).

We previously studied the mechanisms whereby mycorrhizae are dispersed and established in a suite of patches across a successional landscape and how symbiosis, in turn, regulated the plant composition of that patch (Allen et al. 2005). Subsequently, we have looked at how our initial patches changed and how the plant communities across the entire Mount St. Helens disturbance area have developed. Our goal in this work is to describe how the mechanisms of reinvasion and establishment repeated themselves across Mount St. Helens, creating the complex landscape that we observe today.

11.3 Pre-eruption Vegetation, Disturbance Processes, and Their Legacies

Our study sites are located in areas that represented three basic types of pre-eruption ecological systems, each capable of providing different types and amounts of biological legacies: (1) old-growth forests, with trees often exceeding a meter in diameter; (2) timber-harvest areas (hereafter referred to as clearcuts) that were dominated by deciduous shrubs or herbaceous vegetation; and (3) alpine meadows supporting herbaceous vegetation. Overlying these pre-eruption vegetation conditions were three zones of volcanic disturbance (Fig. 11.4) from the eruption on 18 May 1980: the pyroclastic-flow, blowdown, and tephrafall zones (Swanson and Major 2005). The interaction of the pre-eruption vegetation conditions and the type and severity of eruption processes resulted in varying post-eruption conditions that affected the process of succession through the survival, immigration, and establishment of mycorrhizal fungi in the three disturbance zones studied.

11.3.1 Pre-eruption Vegetation

Both mature forest and previously harvested forest areas (clearcuts) existed in our study area before the eruption. These two extremes had vastly different architectures and supported different biological communities before the eruption.

Late-seral forests present around the mountain represent the Pacific silver fir (*Abies amabilis*), mountain hemlock (*Tsuga mertensiana*), and western hemlock (*T. heterophylla*) zones (Franklin and Dyrness 1973), each with its own codominant and subdominant tree species, along with saplings and a variety of ericaceous shrubs in the understories (Swanson et al. 2005). All of the tree species in these three forest types are exclusively EM (e.g., Allen et al. 1995). Most carbon fixation occurs in the upper portions of the canopies (Segura et al. 1994). The coarse and tannin-rich roots provide little in the way of food resources for fossorial mammals but do form EM with truffle fungi. These fungi may make up a major fraction of the diet of a number of animals, such as northern flying squirrels (*Glaucomys sabrinus*) and southern red-backed voles (*Myodes [=Clethrionomys] gapperi*), and even deer (*Odocoileus* spp.) (Maser et al. 1978).

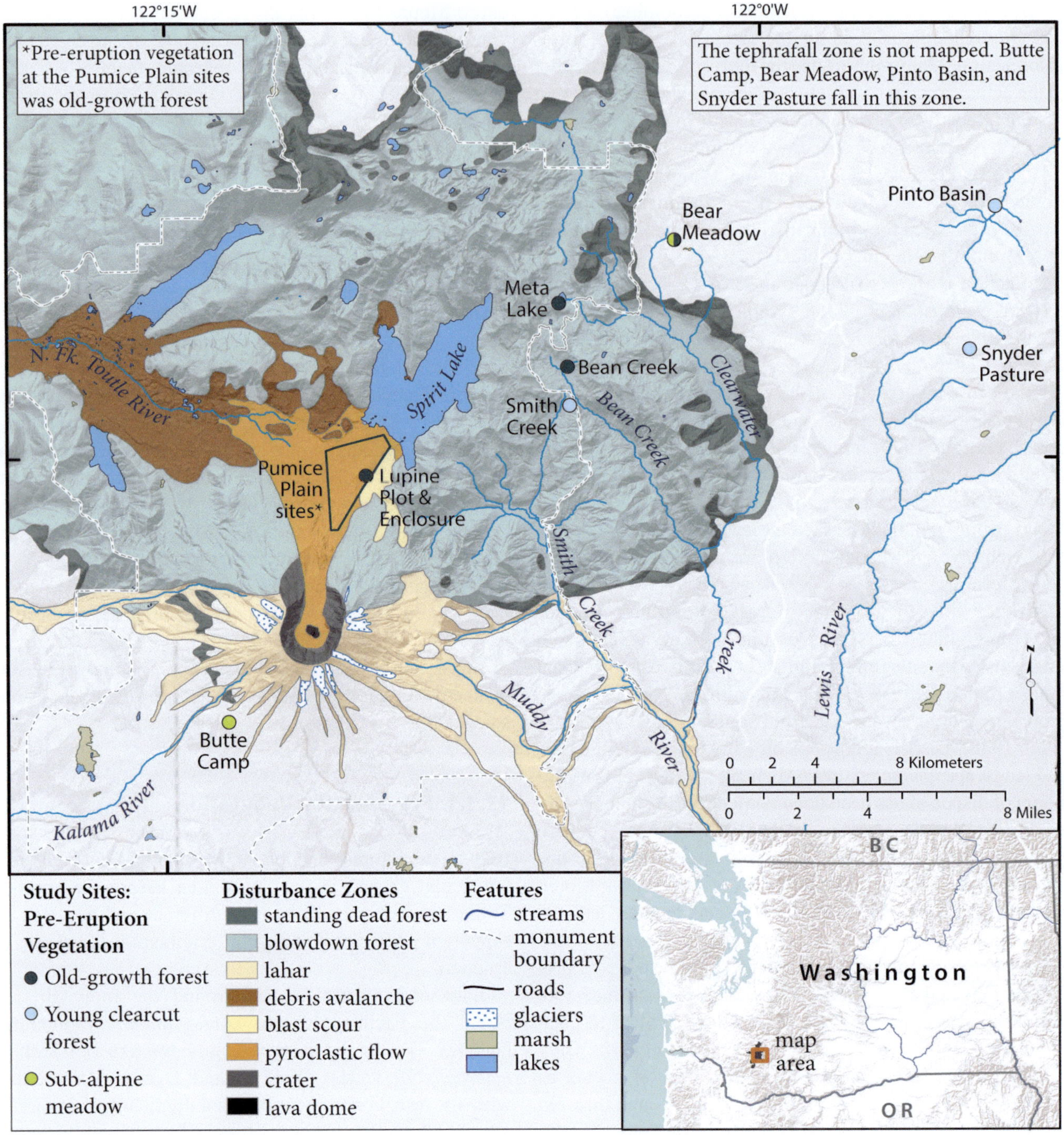

Fig. 11.4 Locations around Mount St. Helens of study sites for mycorrhizae and small-mammal survival and recolonization. (Cartography by Kathryn Ronnenberg, USDA Forest Service, Pacific Northwest Research Station).

In contrast to mature forests, young, pre-canopy-closure clearcuts are early-successional communities with a high diversity and abundant standing crop of highly palatable grasses and forbs, as well as shrubs, hardwood and conifer seedlings, and arthropods. These young clearcuts supported a wide variety of animals, from diverse assemblages of birds and small mammals to large species such as elk (*Cervus elaphus*), deer, and American black bears (*Ursus americanus*).

Natural meadows in subalpine areas near tree line consist of AM forbs and grasses, with high densities of northern pocket gophers (*Thomomys talpoides*) that continually churn the soil through their burrowing activities, effectively retarding tree production.

11.3.2 Eruption Processes and Effects on Biota

On 18 May 1980, Mount St. Helens underwent a highly energetic explosive eruption that involved a complex set of geophysical processes that grossly transformed a large segment of Cascadian terrain (570 km^2) over a 12-h period. Briefly, the event was initiated by an enormous rockslide avalanche that allowed the release of a powerful, horizontally directed, pyroclastic density current (commonly referred to as the lateral blast and hereafter as the blast PDC) that removed, toppled, or seared vegetation. The tremendous heat associated with the blast melted glaciers and snow fields, spawning volcanic mudflows (lahars) down all drainages on the volcano. Emanating from the vent, pumice and ash formed a large eruptive column that was carried by prevailing winds to the northeast, creating a tephrafall zone. Finally, searing-hot pyroclastic flows surged from the crater, creating a vast, sterile area immediately north of the volcano. In the aftermath, several structurally distinct disturbance zones were present that dictated the types and amounts of biophysical legacies remaining from the pre-eruption ecosystems—a result of different volcanic processes and their associated levels of intensity (Fig. 11.5). In our study, we consider sites disturbed by just three of the primary volcanic disturbance processes that occurred during the eruption: pyroclastic flows, the blast PDC, and tephrafall.

Our study area in the pyroclastic-flow zone, the Pumice Plain, is a ~10-km^2 area immediately north of the mountain edifice, bounded to the north by Spirit Lake. The Pumice Plain (Fig. 11.4) was formed by a series of pyroclastic flows, 0.25–40-m thick, that capped the preceding blast PDC and debris-avalanche deposits. The deposit is composed of low-density, golf-ball- to orange-sized dacite pumice that is highly erodible. The superhot (up to 800 °C) pyroclastic flows obliterated all life from this area, thus setting the stage for primary succession.

Our study sites in the area influenced by the blast PDC lie in the blowdown zone, where trees were leveled or snapped off and the former forest floor was covered by blast-PDC deposits up to 50 cm deep, then topped with 10–50 cm of pumiceous tephra. In areas that had been forested before the eruption, the downed tree boles remained on site and were coated by these deposits.

Fig. 11.5 An overview of the three differing conditions resulting from the 1980 eruption. Shown are Pinto Basin (**a**) with just tephrafall; Smith Creek (**b**) and Meta Lake (**c**) in the blowdown zone; and the Pumice Plain (**d**) with pyroclastic-flow material where plant roots had no access to old soil. Photos: M.F. Allen.

Our third focal area of investigation was in the tephrafall zone, which lay beyond the range of the blast PDC. Cool, popcorn-sized pumice and ash (particle size <2 mm) were carried on the wind from the eruptive column above the volcano and deposited toward the northeast, to depths of about 20–30 cm in our study area. Further distant from the volcano, the eruption deposited a layer of fine ash as far east as Nebraska, ~1800 km away.

11.3.3 Post-eruption Biological Legacies and Processes Across Disturbance Zones

Following the eruption, there were two dominant types of small, post-eruption perturbations: water erosion and animal activities, particularly burrowing by northern pocket gophers that mixed tephra with buried pre-eruption soil. Additional perturbations were mediated by other animals, most notably trampling and herbivory by elk.

The most severely disturbed area in our study is the Pumice Plain, in the pyroclastic-flow zone, where the blast force, extreme heat, and cumulative depth of debris-avalanche, blast-PDC, and pyroclastic-flow deposits left no living or dead biophysical legacies from the pre-eruption ecosystem. Therefore, all reestablishment on the Pumice Plain came from migrants.

In contrast, there is a strong legacy of survival throughout the blowdown zone. Smaller trees were bent under the weight of the ash, and shrubs and herbaceous vegetation were buried but often not killed. Many of these plants or their propagules survived, as did many animals living in sheltered habitats (e.g., logs, soil, lakes) or protected beneath layers of snow or ice (Franklin et al. 1985; Crisafulli et al. 2005). Although larger mammals did not survive in the blowdown zone, fossorial animals, particularly northern pocket gophers, often persisted through the eruption (MacMahon et al. 1989; Crisafulli et al. 2005). When they emerged, they brought buried soils back to the surface, along with both chemical resources and biological soil communities. We observed nematodes, collembolans, and saprotrophic and mycorrhizal fungi in sites that had previously been clearcut (e.g., Allen et al. 1984). The underlying soil may well have been the source of rhizobia (bacteria) that colonized lupines in these areas.

In the tephrafall zone, forest overstory trees, saplings, and robust shrubs survived, but short-statured shrubs, herbs, and bryophytes were decimated (Antos and Zobel 2005), and these communities have remained severely altered for 30 years. Animal survival was also much higher in the tephrafall zone as compared with the other disturbance zones studied (Franklin et al. 1985). Fossorial rodents, insects, and most of the belowground community elements survived (Allen et al. 1984).

Subalpine meadow vegetation experienced high survival at our Butte Camp site, despite being subjected to 15-cm-thick tephrafall deposit of pumice lapilli and ash (Allen et al. 1984). Here, surviving gophers and ants mixed the new tephra with buried meadow soil, producing an almost perfect potting mix for plants consisting of aerating material (pumice) and organic matter brought up from below the tephra (Allen et al. 1984). On steep slopes, rain and snowmelt water eroded the tephra, creating rills and gullies that exposed buried soil and releasing buried perennial plants that sent up new shoots. In these areas, surviving small-mammal and insect populations rapidly rebounded.

11.3.4 Signature of Prior Eruptions

Past eruptions have left evidence that forces us to carefully examine the current patterns of succession. The forests present around Mount St. Helens grew atop an organic layer that covers the tephrafall from an earlier eruption in 1842–1843 (Yamaguchi and Lawrence 1993). In the interval between eruptions, roots and EM proliferate through the new soils. Organic compounds leach downward from the surface organic layers, turning deeper tephra layers a darker, yellow-to-brown color. Although we do not yet know the exact chemistry of the leachate, deeper, coarse roots penetrate through these tephra layers but proliferate laterally only in the old organic layers, not within the tephra layers. This layering is repeated at least a meter deep through multiple cycles of eruption and forest reestablishment (Allen et al. 2005).

One needs only to look at the numerous road cuts east of Mount St. Helens to see clear evidence of the "ghost of eruptions past," where earlier eruptions deposited successive tephra layers, each laid down on top of an organic soil layer that developed as forest grew between volcanic episodes. As a specific case, one of the Goat Rocks eruptions (possibly the explosive 1800 eruption) left a deep tephra layer (>1 m) in a road cut between Meta and Ryan Lakes (Fig. 11.6). No roots were observed in this tephra layer in either 2010 or 2012. All plant roots and MF were found in the finer-textured, organic soils that developed prior to the 1800 Goat Rocks eruption (below the tephra layer) or above the Goat Rocks tephra in the organic layer preceding the 1980 eruption.

Given the repeated eruptive activity at Mount St. Helens over past millennia, the biota has contended with repeated volcanic disturbances and has developed resilience to these perturbations (Adams et al. 1987). Mycorrhizal fungi, their host symbiont plants, and their vectors were clearly lost from the pyroclastic-flow zone, but surviving organisms were observed in patches where topographic features or snow cover protected animals, plants, and microbes along the edge of the pyroclastic flow. The frequency and abundance of

Fig. 11.6 The ash layers at a blowdown site between Meta and Ryan Lakes. Observable features include alternating layers of organic soil and tephra from multiple eruptions (see Sect. 11.3.4). Roots and mycorrhizae proliferate in the organic layers, rapidly penetrating the ash layers without spreading out laterally. Organic materials leach downward from the organic layers, coloring the tephra brown. Photo: Michael F. Allen.

survivors increased with distance from the volcano as the intensity of disturbance diminished (Dale et al. 2005). Although there were areas where all life perished in the eruption, the vast majority of the landscape disturbed by the blast PDC inherited numerous biophysical legacies that strongly influenced the pace and pattern of succession (Crisafulli et al. 2005). Variation in the types, amounts, and spatial distribution of legacies, among other factors, led to a number of successional trajectories (Dale et al. 2005).

11.4 Post-eruption Dynamics of Individual Patches: Immigration and Ecesis

Three sets of post-eruption templates are critical to our discussion here: the buried soils and surviving animals, plants, and their propagules, from sites previously covered by mature forests; the different biological legacies from sites that supported clearcut, early-successional or young-forest communities or natural alpine meadows; and the blank slate of sites where all such legacies were eliminated (Fig. 11.5). Mycorrhizal development and plant succession followed different trajectories in each of these three types of sites.

11.4.1 Mycotrophy and Colonizing Plants

A generalized model of succession for most disturbance studies has revolved around perturbation of a relatively nutrient-enriched soil, as depicted in old-field succession studies. In this framework, the initial invading plants tend to be nonmycotrophic, that is, not forming mycorrhizae, and to use soils enriched in nutrients from the dead plants and microbes (e.g., Allen and Allen 1980; Janos 1980). In oceanic sand dune succession, nonmycotrophic plants initially colonize (Nicolson 1960), but this substrate is enriched with high concentrations of nutrients delivered by high tides (Allen and Allen 1991).

Early succession at Mount St. Helens showed an intriguingly different suite of patterns. Across all disturbance zones, a common feature is the high silica content of volcanic products (60–65% by mass) (MacMahon and Warner 1984). In 1980–1981, both N and P were at the low end of detection capability. In this substrate with virtually no organic matter but high silica and pH 6.7, bicarbonate-extractable P was 0.5 mg kg^{-1} and NO_3-N was 0.9 mg kg^{-1} (MacMahon and Warner 1984). These values are extremely low compared with available P of 26 and NO_3-N of 10.9 mg kg^{-1}, respectively, in the organic soil (pH = 4.9) buried below the fresh volcanic deposits.

In the tephrafall and blowdown zones, the volcanic deposits were relatively thin (15–30 and 40–80 cm, respectively) layers overlying a buried soil rich in nutrients and organic matter. In many cases, plants were able to access these soils (see Sects. 11.3.2 and 11.3.4). No nonmycotrophic plants were observed, but many of the plants that were observed are facultatively mycotrophic, that is, they form mycorrhizae when inoculum is present but can also grow and reproduce without mycorrhizae. Plants that are facultatively mycotrophic, such as *Chamerion angustifolium* and *Anaphalis margaritacea*, are common to the three disturbance types studied.

Lupines (*Lupinus* spp.) are keystone species that exert a crucial influence on succession at Mount St. Helens. Lupines have been considered as either nonmycotrophic or facultatively mycotrophic plants. By 1982, in the tephrafall and blowdown zones, *L. latifolius* was mycorrhizal when growing where inoculum was present in the form of spores or old infected root fragments (Allen et al. 1984). In old soil exposed by erosion of overlying deposits or by gopher burrowing, both *Lupinus latifolius* and *L. lepidus* formed nitrogen-fixing nodules (an association with the bacterium, *Rhizobia*) as well as AM, whereas in the tephra, with no inoculum, both species formed nodules but neither had AM. No evidence of growth enhancements with AMF has been documented; however, the mycorrhizal plants had higher N and P concentrations and a higher conductance for water and CO_2, evidence for greater rates of photosynthesis (Allen et al. 1992). In 1982, small patches of *L. lepidus* were found on the Pumice Plain. None of these plants were mycorrhizal, but all were heavily nodulated, forming large, red nodules indicative of O_2-rich leghemoglobin and active N_2 fixation. To test for conditions under which colonization of mycorrhizal inoculum occurred, we placed a northern pocket gopher in an enclosure around an individual *Lupinus* plant at our Smith Creek blowdown site (*L. latifolius*) and on the Pumice Plain (*L. lepidus*) (Fig. 11.7). The gophers remained for 24 h and then were returned to point of capture. These gophers presumably transported inoculum, perhaps passed in their fecal pellets, as by 1 year later, the enclosed plants were inoculated with AM but the surrounding individuals outside the enclosures were not (Allen and MacMahon 1988) (Fig. 11.8).

Fig. 11.7 The original gopher study plot on the Pumice Plain. In 1982 a gopher, trapped at Pinto Basin, was placed in an enclosure around a lupine for 24 h (**a**) and then returned to point of capture. The gopher inoculated this plot (Allen and MacMahon 1988), presumably with AM spores deposited in its droppings. (**b**) This patch has remained mycorrhizal through 2015 (Allen, personal observations). Photos: Michael F. Allen.

11.4.2 Succession in Study Plots

A key element in the successional process was the simultaneous invasion of plants and MF, under environmental factors in which plants can establish, infection can occur, and plants and MF can grow and reproduce.

11.4.2.1 Tephrafall Zone

Tephra provided few nutrients and minimal water-holding capacity (MacMahon and Warner 1984; Allen, unpublished data). Even by 2015, tephra deposited by eruptions from both the 1800s and 1980 supported roots and mycelial networks only poorly at best.

In the old-growth forests surrounding Bear Meadow (Fig. 11.4), the surviving overstory trees persisted throughout the 30-year duration of our study. A new organic layer has developed on top of the 1980 tephra layer. With few fossorial rodents mixing the soil, roots and fungi proliferated throughout the organic layers, but only a few coarse taproots penetrated through the tephra layer without lateral branching and then ramified in the deeper, pre-eruption organic layers (Allen et al. 2005). EMF hyphae expanded out from the colonized root tips. By 2012, achlorophyllous plants became common in the old-growth forest as an organic horizon accumulated above the 1980 tephra layer. Achlorophyllous plants in the Ericaceae (heath family) (Fig. 11.9a) form an ericoid or an arbutoid symbiosis with known EM fungi, extracting carbon from its MF, which had ultimately been captured from the nearby tree (e.g., Zimmer et al. 2007). The fungi of achlorophyllous orchids (Fig. 11.9b), in contrast, may parasitize neighboring trees with their MF.

Mixing of old soil and tephra by fossorial rodents provided the plant-growth medium that was still apparent in 2000 in the Butte Camp meadow and in the formerly clearcut openings that were buried by deep tephrafall (Bear Meadow, Snyder Pasture, Pinto Basin). In 2000, the mycorrhizae at Butte Camp meadow were indistinguishable from pre-eruption levels except that new tephra (20 years old) had been mixed into the soils by extensive gopher bioturbation. Most of the former clearcuts are now young forests with a high diversity of fungal associations, including both hypogeous and epigeous EM fungi.

11.4.2.2 Blowdown Zone

Prior to the eruption, the region that would become the blowdown zone consisted of a mix of timber, meadows, and clearcut areas. The blast PDC toppled, sheared, and killed all trees, except in some locations where late-lying snowpack

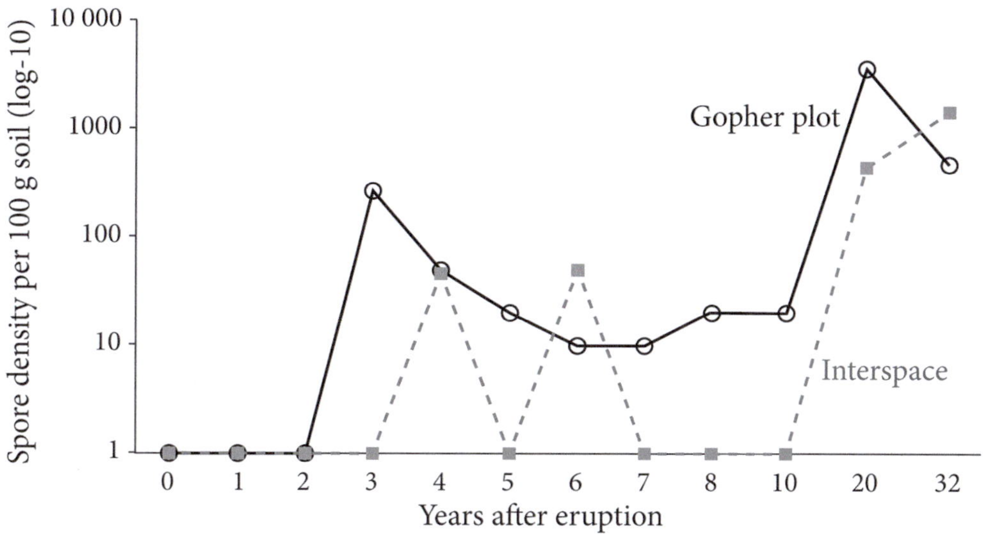

Fig. 11.8 Density of arbuscular mycorrhizal fungal spores at the Pumice Plain site since the eruption through 2012. For 24 h in 1982, a gopher was enclosed with the lupine plant shown in Fig. 11.7a. The lupine plant was found to be mycorrhizal in 1983 (Allen and MacMahon 1988). At that time no other plants on the Pumice Plain were found to be mycorrhizal. For the first decade, AM fungal spores tended to occur at higher densities under the original inoculated plant and the vegetation that subsequently developed around it (Fig. 11.7b).

Fig. 11.9 The last group of plants reestablishing at Mount St. Helens, the achlorophyllous plants, including (**a**) Indian pipe (*Monotropa hypopitys*) and (**b**) orchids (*Corallorhiza maculata*). In both cases, the plant depends upon its mycorrhizal fungus for both nutrients and carbon, and the fungus derives its carbon from the adjacent trees. Photos: M.F. Allen.

was deep enough to protect saplings. These patches of surviving saplings (*A. amabilis* and *Tsuga* spp.) and their MF persisted through our 2015 sampling. Different AM and EM inoculum survived in these patches, affecting the matrix of diverse vegetation that could be seen by 1984.

One site, Smith Creek (Fig. 11.4), is especially interesting. This pre-eruption clearcut site was severely disturbed by the powerful blast PDC. The site received about 30 cm of lithic blast deposit that was overtopped with pumiceous tephra of lapilli and ash ~30–40-cm thick. This site was not actively managed by planting young conifers, but instead was managed passively, allowing natural succession to proceed. On steep slopes, water erosion cut through the tephra deposits during the first few post-eruption years, exposing the former, buried soil. Here, numerous residual early-seral herbaceous and woody (*Salix* spp.) plants sprouted from perennial rootstocks, which had preexisting AM fungal associations. AM fungal spores were observed in these erosional features (rills and gullies) but not in adjacent tephra (Allen 1987). At numerous locations across the valley walls, patches where pocket gophers mixed tephra with old soils could be observed. *Lupinus latifolius*, *Anaphalis margaritacea*, *Chamerion angustifolium*, *Hieracium albiflorum*, and many other AM plants were observed and remained dominant in 2015. Interestingly, chronic gopher bioturbation has maintained the early-successional, herbaceous status of this site. In the areas where the erosional tephra was >1 m deep, *L. latifolius* established with extensive root systems and N-fixing nodules. These plants and adjacent seedlings of *L. latifolius* and many other species became AM by the mid-1980s (Allen et al. 1992).

Mountain hemlock and Pacific silver fir saplings in the vicinity of Meta Lake (Fig. 11.4) survived the 1980 eruption because they were protected beneath snow and remained rooted in the old soil. Their roots were EM. As surface organic matter accumulated from dead and decaying senesced plants, fungi, and other organic inputs (e.g., dead arthropods, lichens), a new EM/organic profile developed atop the tephra. Again, a layered profile emerged, consisting of a root/EM/organic-matter layer of old soil, a tephra layer with taproots running through it but not branching, and a new surface root/EM/organic soil layer.

Pocket gophers survived at the Bean Creek blowdown site (Fig. 11.4), mixing buried soil with tephra, forming patches of excellent growth medium, and producing high rates of AM plant growth. Gophers concentrated their burrowing activities in areas of abundant preferred forage, creating areas of composite mounds from which spores dispersed laterally between 1982 and 1984 (Allen 1987). In 1983, the site was planted with 2-year-old nursery-grown conifer seedlings, likely introducing at least one EM fungus, *Pisolithus tinctorius*, to the site. Other conifer seedlings were augur-planted into the buried soil with EM inoculum. The diversity of EM fungi at this site has continued to increase. During a 2012 foray, *Melanogaster* sp., a basidiomycete truffle, was found.

11.4.2.3 Pyroclastic Flow

The Pumice Plain, our only primary successional site, represents an entirely new surface that has base material derived from the debris avalanche up to 200 m thick, overtopped with blast-PDC deposits, and then capped with numerous pyroclastic-flow deposits that collectively are several meters thick. The searing heat of the pyroclastic flows (up to 800 °C) sterilized these pumice and ash deposits. On this necessarily inorganic substrate, it is not surprising that nitrogen was the primary nutrient limiting plant growth, below detectable levels when the pyroclastic material was assayed (MacMahon and Warner 1984). Among initial colonizers of this area were the prairie lupine (*Lupinus lepidus*), symbiotic with *Rhizobia* bacteria capable of fixing N_2 in root nodules.

We hypothesize that the Pumice Plain soils would initially fall at the extreme low-nutrient end of the mycorrhizal functioning curve (Fig. 11.3). However, carbon produced by plants could provide resources allowing for successful establishment and persistence of MF, even if the plants showed little or no growth increment. This pattern also fits other facultatively mycotrophic plants on Mount St. Helens, as discussed in Sect. 11.4.1 (Titus and del Moral 1998).

Over time, the large (golf-ball-sized) pumice in the top 5 cm of deposits has weathered to a smaller grain size, primarily sand (0.0625–2.0 mm). By 1997, the root, leaf, and microbial litter formed an organic-rich patch around each established lupine that could hold nutrients, including fixed N_2 (Halvorson and Smith 2009). These patches supported a wide variety of forbs (Allen et al. 1992; Titus et al. 2007).

The gopher enclosure patch continued to increase in C and N over time (Halvorson et al. 2005; Halvorson and Smith 2009; Halvorson et al., Chap. 5, this volume) and continued to support an array of mycorrhizal forbs (Fig. 11.7). It also continued to attract animals that transport an increasing diversity of AM fungi to the patch. The initial experimental gopher treatment inoculated the lupine with *Glomus macrocarpum*. After two decades, Allen et al. (2005) reported *Acaulospora morrowiae*, *Ac.* sp. 1, *Glomus aggregatum*, *Gl. claroideum*, *Gl. geosporum*, *Gl. leptotichum*, *Gl. spurcum*, *Gl. tenue*, *Sclerocystis coremioides*, and *Scutellospora calospora* in this plot. By 2012, the spore counts of the interspace between lupine mounds equaled those of the mounds, a sign that the initial difference in inoculum between the experimental mound and its immediate vicinity appears to have largely disappeared (Fig. 11.8 graph).

11.4.3 Interactions and Scaling

Mammals that can transport inoculum, ranging in size from shrews (*Sorex* spp.) to elk, rapidly dispersed across the Pumice Plain (Fig. 11.10), following willows growing along stream courses and attracted to scattered lupine patches (MacMahon et al. 1989; Crisafulli et al. 2005; Crisafulli et al., Chap. 15, this volume). Small mammals also migrated among microsites favorable for plants and mycorrhizal fungal establishment. By 1985, lupine patches surrounding the original Pumice Plain enclosure treatment plot (described in Sect. 11.4.1) had AM fungi associated with local rodent

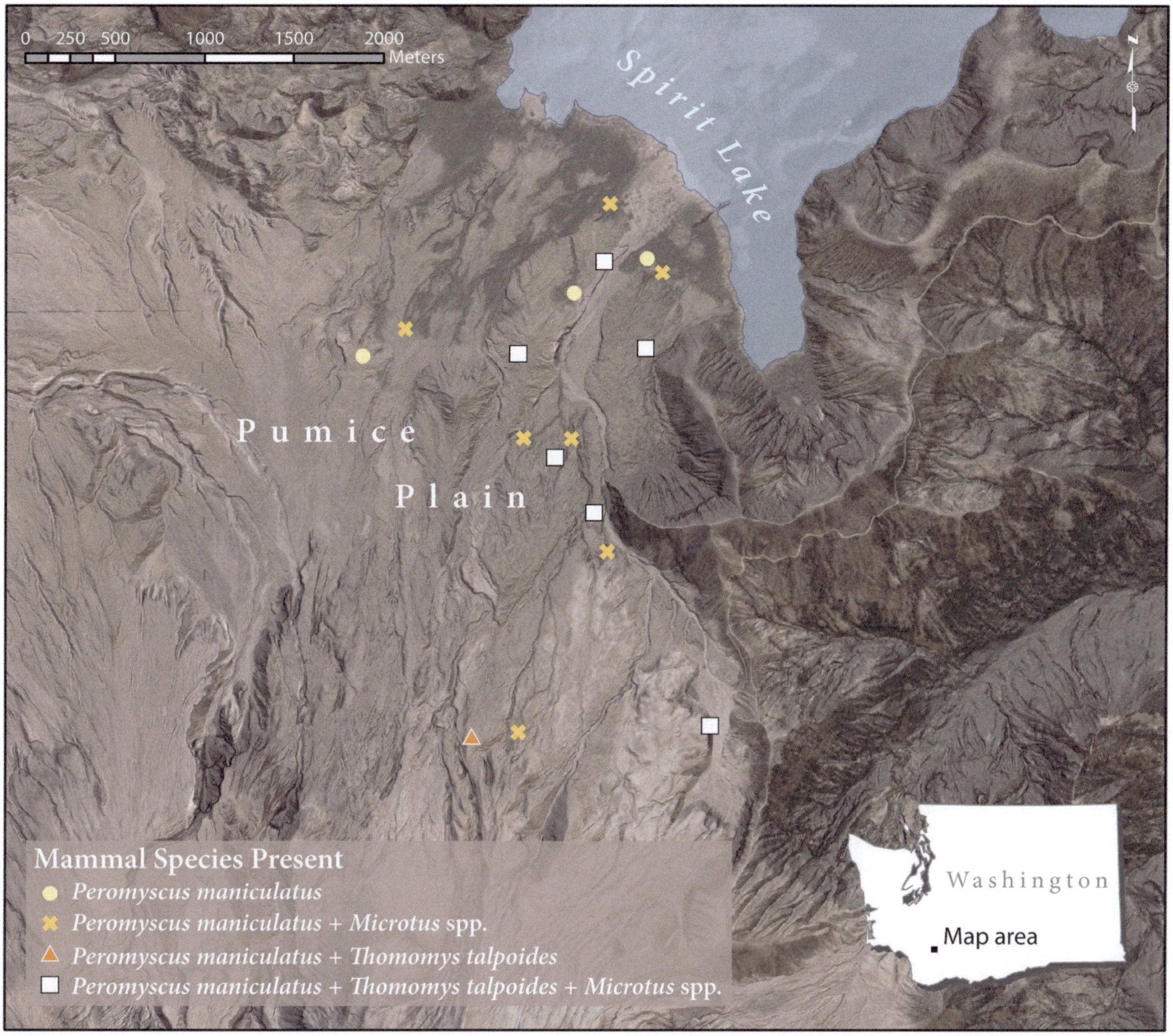

Fig. 11.10 Migration of small mammals across the successional landscape following the eruption of Mount St. Helens (see Crisafulli et al., Chap. 15, this volume). In all cases, AM fungi were found approximately 1 year following the invasion of these mammals. Shown are the distributions of *Peromyscus maniculatus* alone, *Peromyscus maniculatus* with *Microtus* spp., *Peromyscus maniculatus* plus *Thomomys talpoides*, and all three taxa. (Cartography by Kelly Christiansen, USDA Forest Service, Pacific Northwest Research Station).

dispersal (Allen 1987). By 2003, Titus et al. (2007) found that most sampled microsites had AMF, including a variety of taxa such as *Acaulospora delicata*, *A. koskei*, *A. mellea*, two additional undescribed species of *Acaulospora*, *Archaeospora trappei*, *Gl. intraradices*, *Gl. luteum*, *Gl. microaggregatum*, *Gl. tortuosum*, *Paraglomus occultum*, and an undescribed *Glomus* sp. AZ123.

By 1985, the first EM trees had established in protected areas such as the lee sides of rocks or ridges (Allen 1987). The majority of coniferous species present on the north slope of Mount St. Helens prior to eruption could be found by 2000, but conifers occurred only at low densities, scattered across the Pumice Plain (Birchfield 2011).

Willows (*Salix* spp.) and alders (*Alnus* spp.) are genera that have member species that form AM and EM. By growing along streams, these plants form corridors for animal movement and, in turn, dispersal of AM fungi. All willows sampled in 2015 were AM, with 25% (±16%) of root segments infected. In any place where woody plants emerge, the vegetation architecture facilitates trapping of airborne fungal spores (Allen 1987), so EM fungi may also be migrating along stream corridors. We have observed *Laccaria bicolor*, a common EM fungus in the Pacific Northwest, in riparian alder stands.

The most recognizable pattern emerging from our study sites is that succession is not necessarily directional. In many of our study plots, plant species have come and gone and returned again, and only where shrubs have colonized mesic sites have we seen a substantial shift in plant community structure. Individual EM tree seedlings have colonized and disappeared, but the mycorrhizal fungi remain and are potentially available to develop symbioses with new seedlings. Over time, the substrate of each patch weathers to a finer texture, and organic-matter inputs from plant litter and fungal organic residues, such as glomalin and chitin, accumulate (Halvorson et al., Chap. 5, this volume). Each patch undergoes succession at its own rate and in its own pattern. But as each patch grows, it eventually becomes a magnet for animals, such as elk, that strongly influence site conditions through herbivory, trampling, and additional nutrient inputs. When coupled with periodic drought or wet conditions, which alter the competitive relationships among plants and their mycorrhizal partners, a dynamic trajectory of succession emerges.

11.5 Complex Structure of a Successional Landscape

Mount St. Helens has been an extremely important study area for understanding successional processes (Crisafulli et al. 2015; Swanson and Crisafulli, Chap. 16, this volume). The model of Clements (1916), as modified by MacMahon (1981) (Fig. 11.1), represents all of the steps that happen within a small spatial unit (a few m^2). Ecologists often extrapolate to a broader disturbance and make the assumption that when two areas are different, the difference is due to time since disturbance. This assumption is often crucial for management purposes or for assessing external stressors. However, as we can see at Mount St. Helens, differences in composition or structure between two locations can also be due to legacies, distance from sources of propagules, or landscape structure. Because Mount St. Helens represents a single large disturbance, when two patches differ, it is due to the distinctive ecological phenomena that drive those patches.

Within a few years of the eruption, patterns emerged that foreshadowed differences in the rate, patterns, and processes of succession that would unfold across disturbance zones over the following decades. Initial conditions of the vegetation and the specific types and intensities of volcanic disturbance would have a lasting impression throughout our 30-year study, but ecological differences among the disturbance zones did lessen, particular during the third post-eruption decade.

The Pumice Plain is the most obvious area where the effects of the eruption remain. Here, succession processes were initialized exclusively by colonizers, occurred at a much slower rate, and appear to be more stochastic when compared with the blowdown and tephrafall disturbance zones. As facultatively mycorrhizal plants establish in favorable microsites (Titus et al. 2007), these patches attract animals and provide hosts for MF. The primary vector for MF dispersal was mammals, especially small mammals such as gophers and voles. These rodents disperse across the landscape, moving from one patch of plants to another (Fig. 11.10), not along a migration front.

MF do not remain only where the initial inoculum was deposited. In 1985, we found that wind dispersed inoculum locally from gopher mounds that rose high enough above the surrounding substrate to allow entrainment of spores. Most AMF spores are too heavy for long-distance dispersal by wind (Allen 1987), but tree patches could provide inoculum for epigeous EMF dispersal (Allen et al. 1993).

By 2010, gopher activity was widespread across the lower Pumice Plain. Gopher mounds continued to be focal points for AMF inoculum in newly colonized areas. However, by 2012, the interspace areas contained as many spores as the gopher mounds (Fig. 11.8), but the soil surrounding the mounds was highly patchy. The mean spore count of the surrounding soil equaled that of the gopher-mound soil, but displayed more variability, with both higher and lower counts. The relative importance of this patchiness varies with topography and plant-species composition. As patches develop at different rates in different locations, they create a complex array of communities across the landscape.

From the initial inoculation enclosure study to the natural inoculation a year later, animals and wind have continued to move AMF and EMF inoculum across the Mount St. Helens landscape as the fungi reinvaded the primary substrate. Titus et al. (2007) found that by 2003, AMF were widely distributed across the Pumice Plain. Our continuing observations show that all of the AM plants studied were facultatively mycorrhizal and that each of those patches continues to build up inoculum. Individual tree seedlings became inoculated with EM by 1985 (Allen 1987). Although most seedlings germinated, grew for a few years and then died, some persist. All these surviving trees form EM, although the EMF species composition remains to be analyzed. Our observations demonstrate that, 35 years after the eruption, mycorrhizal fungi continue to follow the plant patches across and upslope on the Pumice Plain and to infill between colonizing patches. As of 2015, all mycorrhizal types (AM, EM, ericoid, orchid, monotropoid) that were present prior to the eruption have colonized and are continuing to spread.

In contrast to the Pumice Plain, millions of residual plants emerged from blowdown sites, grew, created patches, and eventually spread and coalesced. The rapid development of plant resources fueled a positive feedback between the plants and animals, such as the northern pocket gopher. Residual plants provided food that facilitated the growth of gopher populations, which in turn increased the rate of bioturbation, leading to additional improvements in edaphic conditions and increased the concentrations of spores on the tephra surface. However, these processes were spatially patchy in the landscape during the first two post-eruption decades, and large gaps existed between loci of activity.

In some portions of the blowdown zone, the vertical and horizontal structure of patches of plants, mycorrhizae, and soil are beginning to resemble the pre-eruption conditions. The young, planted conifers present in clearcuts shortly after the eruption are now young forests, and the vertical layering resulting from past eruptions and cycles of reestablishment can be readily seen in the aggrading forest soils.

In the tephrafall zone, linkages among plants, animals, and fungi established quickly and appeared to be much more ephemeral than in the blowdown zone and Pumice Plain. This was probably caused by the relatively thin deposits (<30 cm) at our tephrafall-zone study sites, as well as the high-level survivorship for MF, plants, and animals. However, the suite of processes differed dramatically based on the pre-eruption plant community condition. Pre-eruption clearcut sites followed the pattern outlined above and stood in sharp contrast to the old-growth forest sites where groundcover vegetation survivorship was low, fossorial mammals few, bioturbation and mixing of tephra and buried soil infrequent, all of which retarded the rate of understory regeneration.

11.6 Space-for-Time Substitutions as a Research Model

As discussed in Sect. 11.1 for studies of succession more generally, the role of mycorrhizae has largely been studied using space-for-time substitution (e.g., Allen and Allen 1980; Janos 1980; Helm and Allen 1995). However, that model has proved to be of limited use in understanding the complexities of landscape-scale recovery. The different legacies of survival of plants, animals, and fungi set the stage for an intricate pattern of both primary and secondary succession. Mount St. Helens has demonstrated that primary succession is largely a process of interacting patch dynamics, dependent upon migrations and establishment of plants, rodents, and mycorrhizal fungi. Secondary succession at Mount St. Helens is also patchy but depends more upon the depth of the tephra layer and the underlying topography. Some animals and plants that invaded the primary successional area (Pumice Plain) immigrated from surviving patches of the secondary-successional areas. Many of these areas inherited legacies from the meadows, clearcuts, and forests that predated the eruption.

During the first 15 years following the eruption, we documented the mechanisms whereby mycorrhizal fungi interacted with animals and plants to establish successional patches scattered across the landscape. During the subsequent 20 years, we have observed that plants and animals disperse across the newly formed landscape. The mechanisms of succession were the same, but the legacies and spatial topography, not only the passage of time, dictated the rate of recovery. Succession at Mount St. Helens is not dominated by time since disturbance but rather by convoluted and often contradictory biotic and environmental interactions through time. MF were vectored by animals and wind in unpredictable patterns, but continue to expand across this complex landscape, depending upon individuals and patches of plants that simultaneously migrated and germinated. The incredible biological diversity, even in an area formerly dominated by coniferous forest, demonstrates the intricate web of interactions among numerous species, integrating dispersal, food webs, ecesis, competition, and reproduction. If any one of these interactions was absent, the existing, dynamic pattern of succession on Mount St. Helens would not have followed the multiple trajectories that we see expressed on the landscape 35 years later.

References

Adams, A.B., V.H. Dale, A.R. Kruckeberg, and E. Smith. 1987. Plant survival, growth form and regeneration following the May 18, 1980, eruption of Mount St. Helens, Washington. *Northwest Science* 61: 160–170.

Allen, E.B., and M.F. Allen. 1980. Natural re-establishment of vesicular-arbuscular mycorrhizae following stripmine reclamation in Wyoming. *Journal of Applied Ecology* 17: 139–147.

Allen, E.B., M.F. Allen, D.J. Helm, J.M. Trappe, R. Molina, and E. Rincon. 1995. Patterns and regulation of arbuscular and ectomycorrhizal plant and fungal diversity. *Plant and Soil* 170: 47–62.

Allen, M.F. 1987. Re-establishment of mycorrhizas on Mount St. Helens: Migration vectors. *Transactions of the British Mycological Society* 88: 413–417.

———. 1988. Re-establishment of VA mycorrhizae following severe disturbance: Comparative patch dynamics of a shrub desert and subalpine volcano. *Proceedings of the Royal Society of Edinburgh* 94B: 63–71.

———. 1991. *The ecology of mycorrhizae*. New York: Cambridge University Press.

Allen, M.F., and E.B. Allen. 1991. Mycorrhizae and plant community development: Mechanisms and patterns. In *The fungal community*, ed. G.C. Carroll and D.T. Wicklow, 455–479. New York: Marcel Dekker.

———. 2017. Mycorrhizal mediation of soil fertility amidst nitrogen eutrophication and climate change. Chapter 12. In *Mycorrhizal mediation of soil: Fertility, structure, and carbon storage*, ed. N.C. Johnson, C. Gehring, and J. Jansa, 213–231. Amsterdam: Elsevier Press.

Allen, M.F., and J.A. MacMahon. 1988. Direct VA mycorrhizal inoculation of colonizing plants by pocket gophers (*Thomomys talpoides*) on Mount St. Helens. *Mycologia* 80: 754–756.

Allen, M.F., J.A. MacMahon, and D.C. Andersen. 1984. Reestablishment of Endogonaceae on Mount St. Helens: Survival of residuals. *Mycologia* 76: 1031–1038.

Allen, M.F., L.E. Hipps, and G.L. Wooldridge. 1989. Wind dispersal and subsequent establishment of VA mycorrhizal fungi across a successional arid landscape. *Landscape Ecology* 2: 165–171.

Allen, M.F., C. Crisafulli, C.F. Friese, and S. Jeakins. 1992. Reformation of mycorrhizal symbioses on Mount St. Helens, 1980–1990: Interactions of rodents and mycorrhizal fungi. *Mycological Research* 96: 447–453.

Allen, M.F., E.B. Allen, C.N. Dahm, and F.S. Edwards. 1993. Preservation of biological diversity in mycorrhizal fungi: Importance and human impacts. In *International symposium on human impacts on self-recruiting populations*, ed. G. Sundnes, 81–108. Trondheim: The Royal Norwegian Academy of Sciences.

Allen, M.F., W. Swenson, J.I. Querejeta, L.M. Egerton-Warburton, and K.K. Treseder. 2003. Ecology of mycorrhizae: A conceptual framework for complex interactions among plants and fungi. *Annual Review of Phytopathology* 41: 271–303.

Allen, M.F., C.M. Crisafulli, S.J. Morris, L.M. Egerton-Warburton, J.A. MacMahon, and J.M. Trappe. 2005. Mycorrhizae and Mount St. Helens: Story of a symbiosis. In *Ecological responses to the 1980 eruption of Mount St. Helens*, ed. V.H. Dale, F.J. Swanson, and C.M. Crisafulli, 221–232. New York: Springer.

Antos, J.A., and D.B. Zobel. 2005. Plant responses in forests of the tephra-fall zone. In *Ecological responses to the 1980 eruption of Mount St. Helens*, ed. V.H. Dale, F.J. Swanson, and C.M. Crisafulli, 47–58. New York: Springer.

Birchfield, M.K. 2011. Thirty years of conifer establishment in volcanic primary succession at Mount St. Helens: Patterns and factors affecting establishment. M.S. thesis. Pullman: Washington State University.

Bledsoe, C.S., M.F. Allen, and D. Southworth. 2014. Beyond mutualism: Complex mycorrhizal interactions. *Progress in Botany* 75: 311–334.

Clements, F.E. 1916. *Plant succession: An analysis of the development of vegetation*. Publication no. 242. Washington, DC: Carnegie Institution.

Clemmensen, K.E., A. Bahr, O. Ovaskainen, A. Dahlberg, A. Ekbald, H. Wallander, J. Stenlid, R.D. Finlay, D.A. Wardle, and B.D. Lindahl. 2013. Roots and associated fungi drive long-term carbon sequestration in boreal forest. *Science* 339: 1615–1618.

Cooper, W.S. 1923. The recent ecological history of Glacier Bay, Alaska: The present vegetation cycle. *Ecology* 4: 223–246.

Cowles, H.C. 1899. The ecological relations of vegetation on the sand dunes of Lake Michigan. *Botanical Gazette* 27: 95–117. 167–202, 281–308, 361–391.

Crisafulli, C.M., J.A. MacMahon, and R.R. Parmenter. 2005. Small-mammal survival and colonization on the Mount St. Helens volcano: 1980–2002. In *Ecological responses to the 1980 eruption of Mount St. Helens*, ed. V.H. Dale, F.J. Swanson, and C.M. Crisafulli, 199–218. New York: Springer.

Crisafulli, C.M., F.J. Swanson, J.J. Halvorson, and B. Clarkson. 2015. Volcano ecology: Disturbance characteristics and assembly of biological communities. In *Encyclopedia of volcanoes*, ed. H. Sigurdsson, B. Houghton, S. McNutt, H. Rymer, and J. Stix, 2nd ed., 1265–1284. London: Elsevier Publishing.

Dale, V.H., A. Lugo, J. MacMahon, and S.T.A. Pickett. 1998. Ecosystem management in the context of large, infrequent disturbances. *Ecosystems* 1: 546–557.

Dale, V.H., F.J. Swanson, and C.M. Crisafulli. 2005. Disturbance, survival, and succession: Understanding ecological responses to the 1980 eruption of Mount St. Helens. In *Ecological responses to the 1980 eruption of Mount St. Helens*, ed. V.H. Dale, F.J. Swanson, and C.M. Crisafulli, 3–12. New York: Springer.

Edmonds, R.L., ed. 1982. *Analysis of coniferous forest ecosystems in the western United States*, US/IBP Synthesis Series 14. Stroudsburg: Hutchinson Ross Publishing Company.

Franklin, J.F., and C.T. Dyrness. 1973. *Natural vegetation of Oregon and Washington*. General Technical Report PNW-8. Portland: U.S. Department of Agriculture, Forest Service, Pacific Northwest Forest and Range Experiment Station.

Franklin, J.F., J.A. MacMahon, F.J. Swanson, and J.R. Sedell. 1985. Ecosystem responses to the eruption of Mount St. Helens. *National Geographic Research* 1: 198–216.

Franklin, J.F., T.A. Spies, R. Van Pelt, A.B. Carey, D.A. Thornburgh, D.R. Berg, D.B. Lindenmayer, M.E. Harmon, W.S. Keeton, D.C. Shaw, K. Bible, and J.Q. Chen. 2002. Disturbances and structural development of natural forest ecosystems with silvicultural implications, using Douglas-fir forests as an example. *Forest Ecology and Management* 155: 399–423.

Freund, J.A., J.F. Franklin, A.J. Larson, and J.A. Lutz. 2014. Multi-decadal establishment for single-cohort Douglas-fir forests. *Canadian Journal of Forest Research* 44: 1068–1078.

Garrett, S.D. 1970. *Pathogenic root-infecting fungi*. Cambridge: Cambridge University Press.

Halvorson, J.J., and J.L. Smith. 2009. Carbon and nitrogen accumulation and microbial activity in Mount St. Helens pyroclastic substrates after 25 years. *Plant and Soil* 315: 211–228.

Halvorson, J.J., J.L. Smith, and A.C. Kennedy. 2005. Lupine effects on soil development and function during early primary succession at Mount St. Helens. In *Ecological responses to the 1980 eruption of Mount St. Helens*, ed. V.H. Dale, F.J. Swanson, and C.M. Crisafulli, 243–254. New York: Springer.

Helm, D.J., and E.B. Allen. 1995. Vegetation chronosequence near Exit Glacier, Kenai Fjords National Park, Alaska. *Arctic and Alpine Research* 27: 246–257.

Hernandez, R.R., and M.F. Allen. 2013. Diurnal patterns of productivity of arbuscular mycorrhizal fungi revealed with soil ecosystem observatory. *New Phytologist* 200: 547–557.

Janos, D.P. 1980. Mycorrhizae influence tropical succession. *Biotropica* 12: 56–64.

MacMahon, J.A. 1981. Successional processes: Comparisons among biomes with special reference to probable roles of and influences on animals. In *Forest succession, concepts and application*, ed. D.C. West, H.H. Shugart, and D.B. Botkin, 277–304. New York: Springer-Verlag.

MacMahon, J.A., and N. Warner. 1984. Dispersal of mycorrhizal fungi: Processes and agents. In *VA mycorrhizae and reclamation of arid and semi-arid lands*, Scientific Report Number SA1261, ed. S.E. Williams and M.F. Allen, 28–41. Laramie: Wyoming Agricultural Experiment Station.

MacMahon, J.A., R.R. Parmenter, K.A. Johnson, and C.M. Crisafulli. 1989. Small mammal recolonization on the Mount St. Helens volcano: 1980–1987. *American Midland Naturalist* 122: 365–387.

Maser, C., J.M. Trappe, and R.A. Nussbaum. 1978. Fungal–small mammal interrelationships with emphasis on Oregon coniferous forest. *Ecology* 59: 799–809.

Molina, R., H. Massicotte, and J.M. Trappe. 1992. Specificity phenomena in mycorrhizal symbiosis: Community-ecological consequences and practical implications. In *Mycorrhizal functioning: An integral plant-fungal process*, ed. M.F. Allen, 357–423. New York: Chapman & Hall.

Mosse, B. 1977. Plant growth responses to vesicular-arbuscular mycorrhiza. X. Responses of Stylosanthes and maize to inoculation in unsterilized soil. *New Phytologist* 78: 277–288.

Nicolson, T.H. 1960. Mycorrhiza in the Gramineae. II. Development in different habitats, particularly sand dunes. *Transactions of the British Mycological Society* 43: 132–145.

Poage, N.J., P.J. Weisberg, P.C. Impara, J.C. Tappeiner, and T.S. Sensenig. 2009. Influences of climate, fire, and topography on contemporary age structure patterns of Douglas-fir at 205 old forest sites in western Oregon. *Canadian Journal of Forest Research* 39: 1518–1530.

Segura, G., L.B. Brubaker, J.F. Franklin, T.M. Hinckley, D.A. MacGuire, and G. Wright. 1994. Recent mortality and decline in mature Abies amabilis: The interaction between site factors and tephra deposition from Mount St. Helens. *Canadian Journal of Forest Research* 24: 1112–1122.

Smith, J.E., K.A. Johnson, and E. Cazares. 1998. Vesicular mycorrhizal colonization of seedlings of Pinaceae and Betulaceae after spore inoculation with *Glomus intraradices*. *Mycorrhiza* 7: 279–285.

Swanson, F.J., and J.J. Major. 2005. Physical events, environments, and geological ecological interactions at Mount St. Helens: March 1980–2004. In *Ecological responses to the 1980 eruption of Mount St. Helens*, ed. V.H. Dale, F.J. Swanson, and C.M. Crisafulli, 27–44. New York: Springer.

Swanson, F.J., C.M. Crisafulli, and D.K. Yamaguchi. 2005. Geological and ecological settings of Mount St. Helens before May 18, 1980. In *Ecological responses to the 1980 eruption of Mount St. Helens*, ed. V.H. Dale, F.J. Swanson, and C.M. Crisafulli, 13–26. New York: Springer.

Titus, J.H., and R. del Moral. 1998. Vesicular-arbuscular mycorrhizae influence Mount St. Helens pioneer species in greenhouse experiments. *Oikos* 81: 495–510.

Titus, J.H., S. Whitcomb, and H.J. Pitoniak. 2007. Distribution of arbuscular mycorrhizae in relation to microsites on primary successional substrates on Mount St. Helens. *Canadian Journal of Botany* 85: 941–948.

Treseder, K.K., and M.F. Allen. 2002. Direct nitrogen and phosphorus limitation of arbuscular mycorrhizal fungi: A model and field test. *New Phytologist* 155: 507–515.

van der Heijden, M.G.A., and I.R. Sanders. 2002. *Mycorrhizal ecology*. New York: Springer-Verlag.

Yamaguchi, D.K., and D.B. Lawrence. 1993. Tree-ring evidence for 1842–1843 eruptive activity at the Goat Rocks dome, Mount St. Helens, Washington. *Bulletin of Volcanology* 55: 264–272.

Zimmer, K., N.A. Hynson, G. Gebauer, E.B. Allen, M.F. Allen, and D.J. Read. 2007. Wide geographical and ecological distribution of nitrogen and carbon gains from fungi in pyroloids and monotropoids (Ericaeae) and in orchids. *New Phytologist* 175: 166–175.

12 Primary Succession on the Mount St. Helens Volcano: Ground Beetle (Coleoptera: Carabidae) Community Assembly and Species Turnover, 1980–2010

Robert R. Parmenter, Charles M. Crisafulli, Tara E. Blackman, Cheryl A. Parmenter, Gary L. Parsons, Danny Shpeley, and James A. MacMahon

12.1 Introduction

The eruption of the Mount St. Helens volcano on 18 May 1980 created a landscape mosaic of disturbances. A rock fall-debris avalanche, pyroclastic flows, and lahars created new land surfaces of pumice and other volcanic rock; other areas were influenced by a powerful blast pyroclastic density current (commonly referred to as a lateral blast, herein called the blast PDC) that leveled forest in a large swath, creating the blowdown forest zone, heat-killed standing trees in the standing dead forest zone, and landscapes blanketed beneath varying depths of lapilli and ash in the tephrafall zone (Fig. 12.1). Within the pyroclastic-flow zone, where no legacies of the former forest persisted, ecological responses have followed patterns of primary succession, beginning with mineral substrate of volcanic materials such as juvenile dacite pumice and assorted older volcanic lithics of varying dimensions. However, even as the main eruption event slackened, an aerial bombardment of flying and wind-borne invertebrates immigrated to the newly created landscape, providing a significant source of organic materials and nutrients (Edwards 1986a, b; Edwards et al. 1986; Edwards and Sugg 1993; Sugg and Edwards 1998). Some new species searched for this type of microhabitat, but most were "accidental" arrivals, having been blown into the volcanic landscape and unable to escape to undisturbed habitats. Hence, the vast majority of insects arriving on the barren Pumice Plain in the initial post-eruption years required unavailable habitats and quickly perished. However, over time, ambulatory species arrived in the area, dispersing from adjacent source populations less disturbed by the eruption and even from distant undisturbed areas. During this time, the mineral surface began to weather, soils began to form, and cobble- and gravel-sized particles changed to finer-textured materials. Concomitantly, a small but increasing proportion of organic matter arrived by wind (loess) or was created by the

R.R. Parmenter (✉)
National Park Service, Valles Caldera National Preserve, P.O. Box 359, 090 Villa Louis Martin, Jemez Springs, NM 87025, USA

Museum of Southwestern Biology, Department of Biology, University of New Mexico, Albuquerque, NM 87131, USA
e-mail: Robert_Parmenter@nps.gov

C.M. Crisafulli
U.S. Department of Agriculture, Forest Service, Pacific Northwest Research Station, Mount St. Helens National Volcanic Monument, 42218 NE Yale Bridge Rd., Amboy, WA 98601, USA
e-mail: ccrisafulli@fs.fed.us

T.E. Blackman
U.S. Department of Agriculture, Forest Service, Pacific Northwest Research Station, Mount St. Helens National Volcanic Monument, 42218 NE Yale Bridge Rd., Amboy, WA 98601, USA
e-mail: tarablackman@gmail.com

C.A. Parmenter
Museum of Southwestern Biology, Department of Biology, University of New Mexico, Albuquerque, NM 87131, USA
e-mail: cparment@unm.edu

G.L. Parsons
Department of Entomology, A.J. Cook Arthropod Research Collection, Michigan State University, 288 Farm Lane, Rm 243 Natural Sciences Bldg., East Lansing, MI 48824, USA
e-mail: parsons@msu.edu

D. Shpeley
Department of Biological Sciences, University of Alberta, CW-405 Biological Sciences Building, Edmonton, AB T6G2E9, Canada
e-mail: dshpeley@ualberta.ca

J.A. MacMahon
Department of Biology and the Ecology Center, Utah State University, 1688 Willow Way, Logan, UT 84341, USA
e-mail: jim.macmahon@usu.edu

C.M. Crisafulli, V.H. Dale (eds.), *Ecological Responses at Mount St. Helens: Revisited 35 years after the 1980 Eruption*,
https://doi.org/10.1007/978-1-4939-7451-1_12

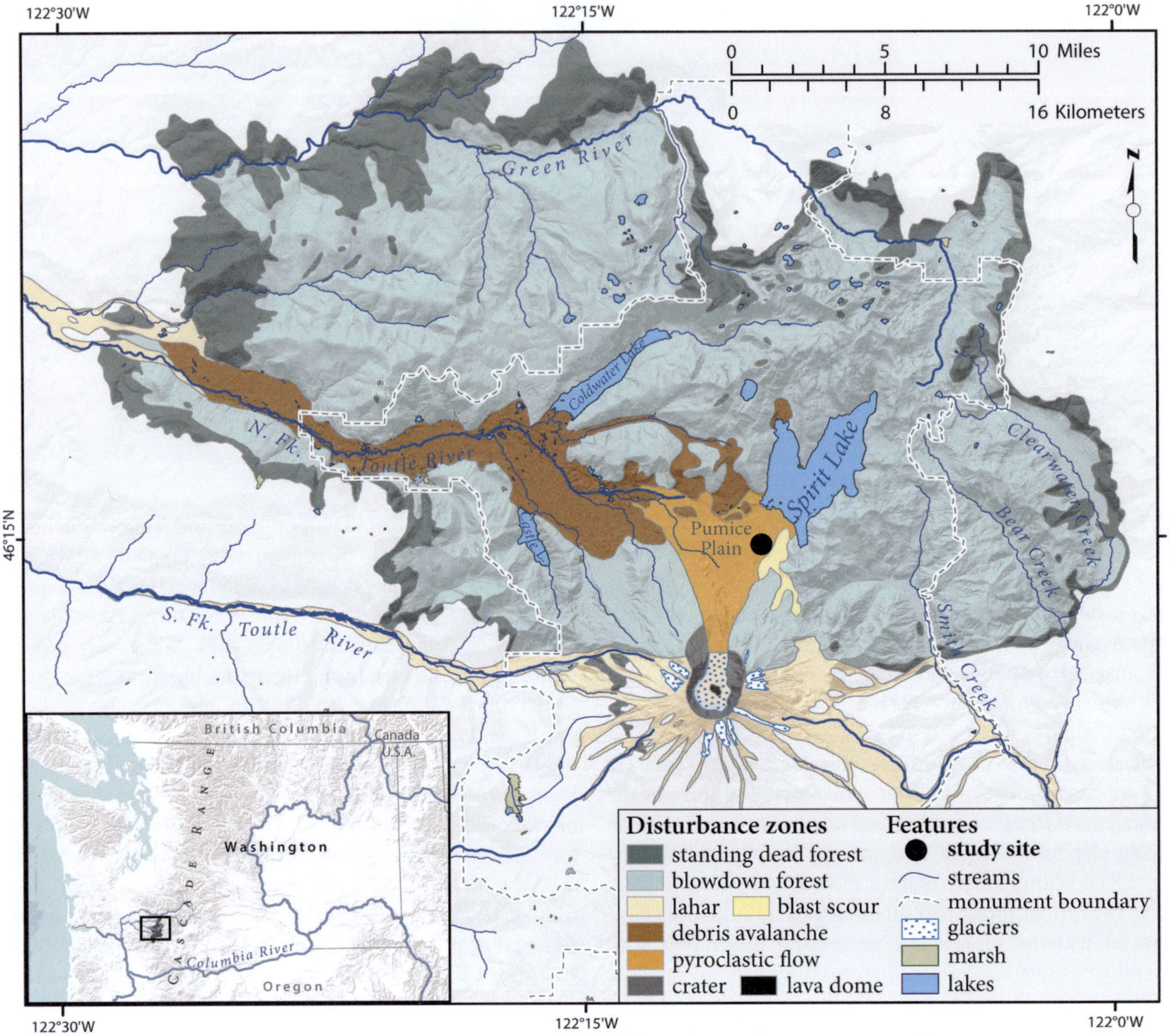

Fig. 12.1 The Mount St. Helens volcano, showing the Pumice Plain study area south of Spirit Lake. (Cartography by Kathryn Ronnenberg, USDA Forest Service, Pacific Northwest Research Station).

early-colonizing flora and fauna. In 2010, 30 years post eruption, primary successional processes have transformed much of the pyroclastic-flow zone into a montane meadow dominated by forbs and grasses, with scattered thickets of shrubs and young trees (Fig. 12.2, bottom).

The newly created Pumice Plain north of the Mount St. Helens' crater provided an opportunity for studying primary successional processes. As described in Parmenter et al. (2005) for Coleoptera assemblages during 1987–2000, these processes followed relay successional patterns (Egler 1954), with a sequential replacement of species through time driven by temporal changes in both abiotic and biotic factors. Initial successful colonists (species that established reproducing populations) to the Pumice Plain included scavengers and predator species able to take advantage of the "rain" of arthropods falling either on the Pumice Plain or onto the surface of Spirit Lake and later washed ashore (Sugg and Edwards 1998). Species that were well adapted to the gravelly habitats of the Pumice Plain prospered in the prey-rich environment, becoming numerically dominant during early post-eruption years. During soil formation, vegetation established (Dale et al. 2005a), facilitating the colonization of herbivorous arthropods. As microhabitats transformed from gravelly barrens to montane meadow, early colonist beetle species became less abundant, and in some cases disappeared altogether on monitored study plots, but still persisted in the region in ever-shrinking suitable microsites. With the arrival of additional insect prey species, predatory beetle species also shifted, leading to the establishment of new predator populations. Interspecific relationships of predation, competition, herbivory,

Fig. 12.2 Time series photographs of the Pumice Plain study site. (**a**) Pumice Plain (foreground) and Spirit Lake in 1982. Note lack of vegetation and pumice-gravel surface. (**b**) Same area in 2010. Note nearly complete ground coverage of herbaceous vegetation with much reduced presence of pumice gravel.

and disease-parasite-parasitoid interactions concomitant with dynamic abiotic processes (climate, soil weathering) contributed to each species' likelihood of establishment (Dale et al. 2005b). These patterns of arthropod succession observed on the Pumice Plain mimic other volcanoes worldwide (see Elizalde 2014 for a recent review), especially during the initial periods of primary succession (e.g., Hodkinson et al. 2002).

To examine and illustrate the patterns of beetle species turnover during primary succession, we have selected the family Carabidae, commonly known as ground beetles. Ground beetles are common inhabitants of forests, meadows, and riparian zones of the Pacific Northwest and are often found in abundant numbers and species-rich assemblages. Carabid beetle communities worldwide have been extensively studied as indicators of environmental conditions and ecosystem disturbances, such as forestry (Pearce and Venier 2006; Niemelä et al. 2007; Blanchet et al. 2013), agriculture (Cole et al. 2012; Rusch et al. 2013), livestock grazing (Dennis et al. 1997, 1998), mining (Parmenter and MacMahon 1987; Dunger 1989), wildland and prescribed fire (Sasal et al. 2010, Elia et al. 2012, Iglay et al. 2012), windstorms (Gandhi et al. 2008), polder construction (Meijer 1989), natural successional processes (Schirmel and Buchholz 2011), and urbanization (Martinson and Raupp 2013). As such, ground beetle assemblage dynamics across time should provide insights into rates of changing successional conditions on the Pumice Plain and provide interpretive results applicable to future changes in other taxa that may respond more slowly to environmental change.

The overall goal of this report is to document the temporal and compositional changes of the ground beetle assemblage over the 30-year period following the 1980 eruption, with particular attention to the patterns of ground beetle species accumulation versus species turnover. This study has the advantage of examining species-specific changes on a single site for three decades, compared with the many, more common, successional "snapshot" studies assessing biotic communities at a single point in time across an array of disturbed sites of varying ages (substituting "space for time" to illustrate long-term patterns). The specific objective of our study is to assess the continuing primary successional patterns of Carabidae on the Pumice Plain for the decades of 1980–2010 and determine if the relay succession of species is still continuing, and if so, at what rate of species gains, losses, and turnover. In addition, we ask what natural-history characteristics (body size, trophic level, dispersal traits [flying or ambulatory], microhabitat tolerances, etc.) might explain each species' colonization success, population expansion, and where applicable, population collapse and extirpation.

12.2 Study Area and Methods

The Pumice Plain study area is located immediately north of the volcano and south of Spirit Lake and covers approximately 10 km^2 of the original 15 km^2 pyroclastic flow that swept northward from the volcano into Spirit Lake basin in May 1980 (Swanson and Major 2005); the nearest undisturbed forested sites are 12.5 km to the east, 12.8 km to the west, 11.9 km to the north, and 7.4 km to the south (on the other side of the volcano) (Fig. 12.1). Prior to the eruption, the site had been an old-growth forest dominated by Pacific silver fir (*Abies amabilis*). However, the collapse of the north flank of the volcano at the beginning of the eruption on 18 May 1980, followed by a massive pyroclastic density current culminating in numerous pyroclastic flows, completely displaced all biota and buried the site in up to 100 m of gravelly sand, pumice, rock, and ash at temperatures of 300–730 °C (Banks and Hoblitt 1981; Foxworthy and Hill 1982; Swanson and Major 2005). As a result, subsequent biotic colonization and establishment over the following three decades constituted a primary successional process.

Our specific sampling site was located approximately 10 km north of the volcano on an intact pyroclastic-flow surface (N 46°14′ 51″, W 122°09′55″, elevation 1160 m; Fig. 12.1). We sampled ground-dwelling arthropod assemblages with ten pitfall traps placed along a line transect at 10- to 15-m intervals across the study area. Traps consisted of plastic cups (9.5 cm diameter, 11 cm deep, half-filled with propylene glycol) and were opened in the spring (late May to mid-June, depending on time of snowmelt), left open continuously between sample collection times (four to five times per year, generally 3 weeks apart) and closed in autumn at the end of the growing season (October–November). This sampling effort throughout the growing season (snowmelt through summer to early winter) enabled us to assess and incorporate ground beetles with different seasonal phenologies and peak activity periods into our annual analyses.

Carabid beetle data from the early post-eruption years of 1983–1985 were collected by Drs. P.M. Sugg and J. S. Edwards at the University of Washington and originally published in Dr. Sugg's dissertation (Sugg 1989); sampling by the present authors was conducted each year during 1987, 1990, 1995, 2000, 2005, and 2010. Of the multiple sites sampled by Sugg and Edwards on the Pumice Plain, the specific site that most closely matched our site was the Spirit Lake site ("S.L." in Table 3.1, p. 42, in Sugg 1989). This site was both geographically very close to our study site and similar in elevation, topography, and surface substrates. Dr. Sugg deployed a total of 20 pitfall traps at this site (6.35-cm diameter cups, partially filled with ethylene glycol, set at 10-m intervals) and sampled from mid-May to mid-October each year; carabid beetle data for this site can be found in Sugg (1989: see Table 4.6, pp. 90–93).

Arthropods were preserved in 70% ethanol, transported to the laboratory, sorted, identified, and counted. Species identifications were made by D. Shpeley (University of Alberta) and G. Parsons (Michigan State University). Carabid beetle voucher specimens from the authors' samples are archived in the Division of Arthropods, Museum of Southwestern Biology, University of New Mexico, Albuquerque, and with the USDA Forest Service's Pacific Northwest Research Station in Amboy, Washington. Voucher specimens collected by Sugg and Edwards were deposited in the collections of the California Academy of Sciences, San Francisco.

To analyze multiyear successional patterns of the ground beetle assemblage, we calculated species-specific, annual catch per unit effort as the number of individuals per species divided by the number of traps and trapping days (i.e., days traps were open). We compared beetle assemblages across years using Jaccard's similarity index (*S*; Jaccard 1908) that produced percentage similarities of species composition between years calculated as:

$$S_{1,2} = \left[a / (a + b + c)\right] \times 100$$

where a = the number of species common to both years 1 and 2, b = the number of species unique to year 1, and c = the number of species unique to year 2. We assembled these paired similarity indices into a resemblance matrix for each sample year (1983–2010) to evaluate patterns of similarity within the study area through time. To evaluate patterns of beetle species succession on the study sites, we plotted the mean catch per unit effort over time (years).

To quantify species gain, loss, and turnover rates during succession, we used the rate metrics described in Anderson (2007). Three basic metrics were calculated: species gain rate (proportion of new species between sample periods), species loss rate (proportion of species lost between sample periods), and species turnover rate (proportional changes between sample periods). Species gain rate between sample periods (G_p) was calculated as:

$$G_p = G / \left[(St_1 + St_2) / 2\right]$$

where G = number of new species gained during the time period and St_1 and St_2 are the number of species present at the beginning (t_1) and end (t_2) of the sample time period. In calculating species gain rate in any given year, we included not only newly arrived species but also species that had previously occupied the site but had temporarily disappeared and then reappeared in subsequent years.

Species loss rate per time period (L_p) was calculated as:

$$L_p = L / \left[(St_1 + St_2) / 2\right]$$

where L = number of species lost during the sample time period. Finally, species turnover rate per time period (T_p) was calculated as:

$$T_p = (G + L) / (St_1 + St_2)$$

We calculated carabid beetle species gain rate, loss rate, and turnover rate for each time interval between samples; to calculate *annual* gain, loss, and turnover rates, we divided rate values by the number of years between samples. We assumed that species richness on the Pumice Plain was zero immediately after the 1980 eruption. Annual rate values were then plotted on a timeline (1980–2010), and linear or nonlinear regressions (exponential, polynomial, logarithmic, power function) were used to assess successional patterns (selected regression models had the best fit indicated by maximum coefficient of determination). We then compared these patterns with the generalized successional patterns discussed in Anderson (2007).

From published sources (Hatch 1953; Parsons et al. 1991; Ball 2001; Ball and Bousquet 2001; Larochelle and Larivière 2003; Pearson et al. 2006; Erwin 2007), we collected natural-history attributes for each species, including known preferred habitats (riparian, lakeshore, meadow, forest), preferred soil type (texture), trophic role as adults and larvae (predator, parasitoid, omnivore, herbivore, granivore), activity periods (diurnal, nocturnal), seasonal activity (year-round, growing season, spring, spring-summer, summer, summer-autumn, autumn), dispersal capability (flight, ambulatory), and body size (length). We analyzed successional patterns of natural-history attributes using Pearson correlations for numbers of species, percentage of species, and percentage of total number of individual beetles collected in each sample year.

12.3 Results

12.3.1 The Ground Beetle Assemblage on the Pumice Plain

Over the 30 years of this study, 51 species of carabid beetles (20,202 individuals) were collected on the Pumice Plain site (Tables 12.1 and 12.2). Few species were very common, but 15 species, represented by >90 individuals each, comprised 98.8% of the total sample. Six species had 10–54 specimens collected, but the remaining 30 species were relatively uncommon, with <10 individuals sampled (10 of these 30 species were represented by only a single specimen (Table 12.1)). [Note: two species of *Trachypachus* have been included in our analyses, although their family-level classification is still uncertain (some authors place them in their own family, Trachypachidae, whereas others include them in Carabidae; see Ball [2001]); given the similarity of their natural histories with those of ground beetles, we have included them here.]

12.3.2 Ground Beetle Population Patterns During Colonization

As noted above, most species (30 of 51 species) were rare, defined as being represented by only a few individuals (<10) or single occurrences. These rarer species apparently failed to establish sustainable populations on the study site, although they may have existed in different nearby microhabitats. The more abundant species (>10 individuals) followed a pattern of "boom and bust" population dynamics. For example, *Trachypachus* spp. became established in large numbers shortly after the eruption during 1983–1985 but had nearly disappeared by 1990 (Fig. 12.3). In contrast, *Apristus constrictus* was present in low numbers in the early 1980s, and then this species increased rapidly in 1985 and 1987, before declining by 1990, and was not observed in later years (Fig. 12.3). The tiger beetle, *Cicindela depressula*, was first detected in 1987 (five specimens), peaked in abundance in 1990 and 1995, and became much less common in subsequent years. The caterpillar hunter, *Calosoma tepidum* (a wingless, ambulatory species), arrived on the Pumice Plain in 1995 and displayed a rapid population growth by 2000, after which it declined in numbers but remained in the assemblage through 2010 (Fig. 12.3). The first few *Calathus fuscipes* were observed in 2000, and this species exhibited a rapid population increase by 2005, after which the population crashed to low levels by 2010; however, a closely related species (*C. ruficollis*) failed to establish a population despite a few colonizing individuals (Fig. 12.3). The genus *Amara*, represented by four species, was present in small numbers during the first 20 years after the eruption; in 2005, two species (*A. idahoana* and *A. obesa*) displayed large population increases, followed by a moderate decline in numbers by 2010 (Fig. 12.3).

Similar patterns to *Amara* were exhibited by the five species in the genus *Harpalus* (Fig. 12.4). Only *H. nigritarsus* attained large population sizes (97 specimens) and, except for 1987, has consistently occupied the Pumice Plain since 1983 (Fig. 12.4). The other four species were relatively uncommon (≤4 specimens), with one species (*H. affinis*) being recorded up to 1990 and the other three present from 1990 onward.

The *Bembidion* species group, represented by 16 species in the genera *Bembidion*, *Amerizus*, and *Lionepha*, also followed the "boom and bust" population patterns on the Pumice Plain (Fig. 12.5). This diverse group included some of the early colonists, but each species appeared for only relatively short periods of time. Of the 14 species recorded on the study site during 1983–1990, 11 species were gone after 1990. Even *B. planatum*, the most abundant beetle recorded in the entire study (Table 12.1), had disappeared from the study site by 2000 (Fig. 12.5). In 2010, only two species of this group remained.

12.3.3 Successional Turnover Patterns of Ground Beetles

The ground beetle assemblage displayed continuous additions and losses of species during sampling years (Table 12.2). Of the 13 species of carabids collected in 1983, only two species (*A. idahoana* and *H. nigritarsis*) were still present on the study site in 2010. Calculated values of Jaccard's percentage similarity coefficients (Table 12.3) showed that the maximum species similarity occurred in 1983–1984 (56%). By 2000, species similarity to the 1983 assemblage was reduced to 8% (Table 12.3). In general, Jaccard's similarity coefficients, including number of species in common, declined as temporal periods between samples increased (Table 12.3).

Calculations of annual species gains, losses, and turnover (Anderson 2007) illustrated the patterns of continuous gains and losses of ground beetle species (Fig. 12.6). Annual species gain (G_p) was initially high, going from zero species

Table 12.1 List of ground beetle species (Carabidae), numbers of specimens collected, and natural-history attributes (see text); Pumice Plain site Mount St. Helens 1983–2010.

Species	Number collected	Body size (mm)	Preferred habitat	Soil preferences	Trophic role (adult/larva)	Daily activity	Adult activity seasons	Movement (dispersal) traits
Agonoleptus conjunctus (Say)	1	3.4	Meadow	Sand/gravel	Predator/predator	Both	Year-round	Flier, strong runner
Agonum cupripenne (Say)	9	8.5	Meadow	Sand/gravel	Predator/predator	Diurnal	Year-round	Flier, moderate runner
Amara conflata LeConte[a]	1	8.8	Meadow	General	Predator/predator	Diurnal	Spring–autumn	Flier, moderate runner
Amara idahoana (Casey)[b]	220	6.2	Meadow	General	Granivore/predator	Diurnal	Summer	Flier, moderate runner
Amara obesa (Say)	744	10.5	Meadow	Sand/gravel	Predator/predator	Nocturnal	Spring–autumn	Flier, moderate runner
Amara sp.[a]	1	–	Unknown	Unknown	Unknown	Unknown	Unknown	Flier, moderate runner
Amerizus oblongulus (Mannerheim)[a]	1	4.4	Forest	Fine organic	Predator/predator	Nocturnal	Spring–summer	Ambulatory, moderate runner
Amerizus spectabilis (Mannerheim)[a]	4	6.1	Forest	Fine organic	Predator/predator	Nocturnal	Spring–summer	Flier, moderate runner
Anisodactylus binotatus (Fabricius)	1	10.7	Meadow	General	Predator/predator	Nocturnal	Spring–autumn	Flier, moderate runner
Apristus constrictus Casey[b]	248	3.6	Riparian/lakeshore	Sand/gravel	Predator/predator	Diurnal	Spring–summer	Strong flier, strong runner
Bembidion (*Peryphus*) sp.	9	4.9	Unknown	Unknown	Predator/predator	Unknown	Summer–autumn	Flier, moderate runner
Bembidion (*Plataphus*) sp. #1	1	4.9	Unknown	Unknown	Predator/predator	Unknown	Summer	Flier, moderate runner
Bembidion (*Plataphus*) sp. #2	3	4.9	Unknown	Unknown	Predator/predator	Unknown	Summer	Flier, moderate runner
Bembidion breve Motschulsky[a]	5	4.9	Meadow	General	Predator/predator	Unknown	Summer	Flier, moderate runner
Bembidion curtulatum Casey	2	3.5	Riparian/lakeshore	Sand/gravel	Predator/predator	Nocturnal	Summer	Flier, moderate runner
Bembidion improvidens Casey[b]	3125	3.7	Meadow	General	Predator/predator	Nocturnal	Summer	Flier, moderate runner
Bembidion iridescens (LeConte)[a]	3	4.1	Riparian/lakeshore	Fine organic	Predator/predator	Nocturnal	Spring–autumn	Flier, moderate runner
Bembidion obscurellum (Motschulsky)[b]	548	4.6	Riparian/lakeshore	General	Predator/predator	Nocturnal	Growing season	Strong flier, moderate runner
Bembidion planatum (LeConte)[b]	8922	7.7	Riparian/lakeshore	Sand/gravel	Predator/predator	Nocturnal	Growing season	Strong flier, moderate runner
Bembidion quadrimaculatum dubitans (LeConte)	13	3.2	Riparian/lakeshore	General	Predator/predator	Unknown	Spring–summer	Flier, moderate runner
Bembidion recticolle LeConte[a]	10	3.2	Riparian/lakeshore	Sand/gravel	Predator/predator	Nocturnal	Spring–summer	Flier, moderate runner
Bembidion rupicola (Kirby)[a]	9	4.9	Riparian/lakeshore	Sand/gravel	Predator/predator	Nocturnal	Growing season	Strong flier, moderate runner
Bembidion sp. (*incertum* grp.)[a]	8	5.0	Unknown	Unknown	Predator/predator	Unknown	Unknown	Flier, moderate runner
Calathus fuscipes (Goeze)	834	11.2	Meadow	General	Predator/predator	Nocturnal	Growing season	Flier, moderate runner
Calathus ruficollis Dejean	6	8.5	General – meadow and Forest	Sand/gravel	Predator/predator	Nocturnal	Growing season	Ambulatory, moderate runner
Calosoma tepidum LeConte	352	19.5	Meadow	General	Predator/predator	Both	Spring–summer	Ambulatory, moderate runner
Carabus taedatus Fabricius	3	21.5	General – meadow and Forest	Sand/gravel	Predator/predator	Nocturnal	Growing season	Ambulatory, moderate runner
Cicindela depressula Casey	1875	14.0	Meadow	Sand/gravel	Predator/predator	Diurnal	Growing season	Strong flier, strong runner
Cicindela longilabris Say	6	13.5	Meadow	Sand/gravel	Predator/predator	Diurnal	Growing season	Flier, strong runner

Cymindis seriata Hatch	30	8.4	Meadow	Sand/gravel	Predator/predator	Unknown	Autumn	Ambulatory, moderate runner
Harpalus affinis (Schrank)[b]	4	10.5	General – meadow and Forest	Sand/gravel	Omnivore/predator	Diurnal	Year-round	Strong flier, moderate runner
Harpalus caliginosus (Fabricius)	2	22.0	General – meadow and Forest	Sand/gravel	Omnivore/predator	Both	Growing season	Strong flier, moderate runner
Harpalus cautus Dejean	1	8.8	General – meadow and Forest	Sand/gravel	Omnivore/predator	Nocturnal	Growing season	Flier, moderate runner
Harpalus cordifer Notman	1	8.5	Forest	Fine organic	Omnivore/predator	Nocturnal	Growing season	Flier, moderate runner
Harpalus nigritarsis Sahlberg	97	11.5	Meadow	Sand/gravel	Omnivore/predator	Both	Growing season	Strong flier, moderate runner
Lebia perita Casey[a]	1	5.5	Meadow	General	Predator/parasitoid	Unknown	Summer–autumn	Flier, moderate runner
Lebia viridis Say[a]	1	5.0	Meadow	General	Predator/parasitoid	Diurnal	Year-round	Strong flier, moderate runner
Lionepha erasa (LeConte)	92	3.5	Riparian/lakeshore	Sand/gravel	Predator/predator	Nocturnal	Spring–autumn	Flier, moderate runner
Nebria eschscholtzi Ménétriés	41	10.5	Riparian/lakeshore	Sand/gravel	Predator/predator	Nocturnal	Summer	Strong flier, strong runner
Nebria mannerheimii Fischer von Waldheim[b]	2162	11.7	Riparian/lakeshore	Sand/gravel	Predator/predator	Nocturnal	Growing season	Strong flier, strong runner
Nebria sahlbergii Fischer von Waldheim[a]	2	9.0	Riparian/lakeshore	Sand/gravel	Predator/predator	Nocturnal	Spring–summer	Strong flier, strong runner
Nipponobebria virescens (Horn)[a]	3	10.5	Meadow	General	Predator/predator	Nocturnal	Growing season	Strong flier, strong runner
Notiophilus sylvaticus Dejean	8	4.7	General – meadow and Forest	Sand/gravel	Predator/predator	Diurnal	Growing season	Flier, strong runner
Omophron ovale Horn	5	5.5	Riparian/lakeshore	General	Predator/predator	Nocturnal	Growing season	Flier, strong runner
Opisthius richardsoni Kirby[b]	256	9.5	Riparian/lakeshore	General	Predator/predator	Diurnal	Summer–autumn	Flier, strong runner
Pterostichus adstrictus Eschscholtz[b]	340	10.7	General – meadow and Forest	Sand/gravel	Predator/predator	Nocturnal	Growing season	Flier, moderate runner
Scaphinotus angusticollis (Mannerheim)	2	20.0	General – meadow and Forest	Fine organic	Predator/predator	Nocturnal	Growing season	Ambulatory, moderate runner
Sericoda obsoleta (Say)[a]	3	6.1	General – meadow and Forest	Fine organic	Predator/predator	Both	Growing season	Strong flier, moderate runner
Trachypachus inermis Motschulsky[b,c]	123	4.8	Meadow	Loam/clay	Predator/predator	Diurnal	Spring–summer	Flier, strong runner
Trachypachus slevini van Dyke[b,c]	54	7.0	Riparian/lakeshore	General	Predator/predator	Diurnal	Summer–autumn	Flier, strong runner
Trechus obtusus Erichson[b]	10	4.0	Meadow	General	Predator/predator	Nocturnal	Growing season	Flier, moderate runner

[a]Data from Sugg (1989).
[b]Recorded by Sugg in 1983–1985 and by authors in 1987–2010.
[c]Possibly classified in family Trachypachidae (see text).

Table 12.2 List and numbers of ground beetles (Carabidae), species richness, and total number of individuals collected from the Pumice Plain of Mount St. Helens, by year of occurrence (shaded boxes).

	Year of occurrence (number of individuals collected)								
Carabidae species	**1983**	**1984**	**1985**	**1987**	**1990**	**1995**	**2000**	**2005**	**2010**
Amara conflata	1								
Amerizus oblongulus	1								
Sericoda obsoletum	1	1	1						
Bembidion rupicola	4	2	3						
Harpalus affinis	1		1						
Opisthius richardsoni	2	81	172		1				
Apristus constrictus	5	5	175	49	14				
Bembidion improvidens	177	229	2514	112	93				
Trachypachus inermis	26	5	86	4	1		1		
Bembidion obscurellum	177	18	61	1	4	287			
Bembidion planatum	137	1612	7063	63	46	1			
Harpalus nigritarsis	3	3	1		26	5	21	23	15
Amara idahoana	3	1	5	3	6	4	35	134	29
Bembidion incertum grp.		3	5						
Bembidion iridescens		1	2						
Trechus obtusus		1	7	1				1	
Pterostichus adstrictus		1	4	1		151	139	44	
Nebria mannerheimii		1	1	140	1270	666	57	14	13
Amara sp.			1						
Bembidion breve			5						
Bembidion recticolle			10						
Amerizus spectabilis			4						
Lebia perita			1						
Lebia viridis			1						
Nebria sahlbergii			2						
Nippononebria virescens			3						
Trachypachus slevini			19	34	1				
Nebria eschscholtzi			18	18		3	2		
Calathus ruficollis			1					4	1
Amara obesa			5	3			41	432	263
Bembidion (Plataphus) sp. #2				3					
Bembidion curtulatum				2					
Lionepha erasa				1		5	43	25	18
Cicindela depressula				5	662	971	197	35	5
Bembidion (Plataphus) sp. #1					1				
Omophron ovale					5				
Notiophilus sylvaticus					1	7			
Harpalus cordifer						1			
Calosoma tepidum						7	295	26	24
Agonum cupripenne							9		
Anisodactylus binotatus							1		
Agonoleptus conjunctus							1		
Bembidion quadrimaculatum							12	1	
Calathus fuscipes							8	801	25
Bembidion (Peryphus) sp.							1		8
Harpalus caliginosus								2	
Scaphinotus angusticollis								2	
Carabus taedatus									3
Cicindela longilabris									6
Cymindis seriata									30
Harpalus cautus									1
Total species	13	15	28	16	15	12	16	14	14
Total number of individuals	538	1964	10,171	440	2133	2108	863	1544	441

No. individuals captured per trap day

Species	1983	1984	1985	1987	1990	1995	2000	2005	2010
Trachypachus inermis	0.0088	0.0019	0.0228	0.0047	0.0007		0.0007		
Trachypachus slevini			0.0050	0.0395	0.0007				
Apristus constrictus	0.0017	0.0019	0.0463	0.0570	0.0101				
Cicindela depressula				0.0058	**0.4763**	**0.6346**	**0.1459**	0.0280	0.0048
Calosoma tepidum						0.0046	**0.2185**	0.0208	0.0229
Calathus ruficolus			0.0003					0.0032	0.0010
Calathus fuscipes							0.0059	**0.6408**	0.0238
Amara conflata	0.0003								
Amara sp.			0.0003						
Amara idahoana	0.0010	0.0004	0.0013	0.0035	0.0043	0.0026	0.0259	**0.1072**	0.0276
Amara obesa			0.0013	0.0035			0.0304	**0.3456**	**0.2505**

Capture rate ≤0.0009 0.0010–0.0099 0.0100–0.0999 0.1000–0.9999

White = not captured in that year

Fig. 12.3 Abundances of common ground beetle species (Carabidae) by year on the Pumice Plain of Mount St. Helens, illustrating the "boom and bust" population dynamics of each species through the successional process.

Number of individuals captured per trap day

Species	1983	1984	1985	1987	1990	1995	2000	2005	2010
Harpalus nigritarsis	0.0010	0.0011	0.0003		**0.0187**	0.0033	**0.0156**	**0.0184**	**0.0143**
Harpalus affinis	0.0003		0.0003		0.0014				
Harpalus cordifer						0.0007			
Harpalus caliginosus								0.0016	
Harpalus cautus									0.0010

Capture rate ≤0.0009 0.0010–0.0099 0.0100–0.0999

White = not captured in that year

Fig. 12.4 Abundances of ground beetle *Harpalus* species (Carabidae) by year on the Pumice Plain of Mount St. Helens.

immediately following the eruption to 13 species by 1983, but declined from 1983 to 1984, followed by a large species gain from 1984 to 1985 (Fig. 12.6a). Coupled with the later years (1987–2010), overall rates of annual species gain showed a significant negative exponential pattern from 1980 to 2010. However, a regression analysis using only the period of 1987–2010 revealed no significant correlation ($G_p = -0.0018Yr + 0.11$, $r = 0.56$, $df = 4$, $P > 0.05$), indicating that annual species gain rates stabilized after 1985.

Annual species loss rate (L_p) between sample periods varied from 0% during 1980–1983 and 1984–1985 to 36% per year between 1985 and 1987, but no significant trend in species loss patterns occurred over the 30 years of the study (Fig. 12.6b); however, using the period 1987–2010, annual species loss rate did show a significant negative trend ($L_p = 0.342Yr^{-0.538}$, $r = 0.96$, $df = 4$, $P < 0.01$), indicating that annual species loss rates were decreasing through time after 1987 during the same time period that annual gain rates were stabilizing. Overall, from 1990 to 2010, annual species loss rates averaged 9.5% per year between sample periods, and species gain rates averaged 8.8% per year during the same time period.

Number of individuals captured per trap day

Species	1983	1984	1985	1987	1990	1995	2000	2005	2010
Amerizus oblongatus	0.0003								
Bembidion rupicola	0.0014	0.0008	0.0008						
Bembidion planatum	0.0463	0.6051	1.8685	0.0733	0.0331	0.0007			
Bembidion obscurellum	0.0598	0.0068	0.0161	0.0012	0.0023	0.1876			
Bembidion improvidens	0.0598	0.0860	0.6651	0.1302	0.0663				
Bembidion iridescens		0.0004	0.0005						
Bembidion breve		0.0011	0.0013						
Amerizus spectabilis			0.0011						
Bembidion recticolle			0.0026						
Lionepha erasa				0.0012		0.0033	0.0313	0.0200	0.0171
Bembidion curtulatum				0.0023					
B. (Plataphus) sp. 2				0.0035					
B. (Plataphus) sp. 1					0.0007				
B. quadrimaculatum							0.0089	0.0008	
B. (Peryphus) sp.							0.0007		0.0076

Capture rate

≤0.0009 0.0010–0.0099 0.0100–0.0999 0.1000–0.9999 ≥1.0000

White = not captured in that year

Fig. 12.5 Abundances of the ground beetle *Bembidion* species group (*Bembidion*, *Amerizus*, and *Lionepha*; Carabidae) by year on the Pumice Plain of Mount St. Helens.

Table 12.3 Jaccard's (1906) percentage similarity coefficients (upper right portion of matrix) and numbers of common species (lower left portion of matrix) among years for ground beetle species (Carabidae) on the Pumice Plain of Mount St. Helens.

	1983	1984	1985	1987	1990	1995	2000	2005	2010
1983		56	35	26	47	19	12	8	8
1984	10		54	41	41	30	19	21	12
1985	11	15		38	34	21	19	20	14
1987	6	9	12		41	40	33	30	20
1990	9	9	11	9		35	19	16	16
1995	4	6	7	8	7		40	37	30
2000	3	5	7	8	5	8		50	43
2005	2	5	7	7	4	7	10		47
2010	2	3	5	5	4	6	9	9	

Annual turnover rate (T_p) also showed an initial high during 1980–1983 and displayed a significant negative trend through 2010 (Fig. 12.6c). The best-fit power function indicated a rapid rate of change in the early post-eruption years, followed by a stabilizing trend after 1990. Annual turnover rate averaged 9.2% per year from 1990 to 2010.

12.3.4 Temporal Patterns of Natural-History Attributes

For the great majority of the ground beetle species, we found published information on their natural-history characteristics, including preferred habitats, preferred soil type, their

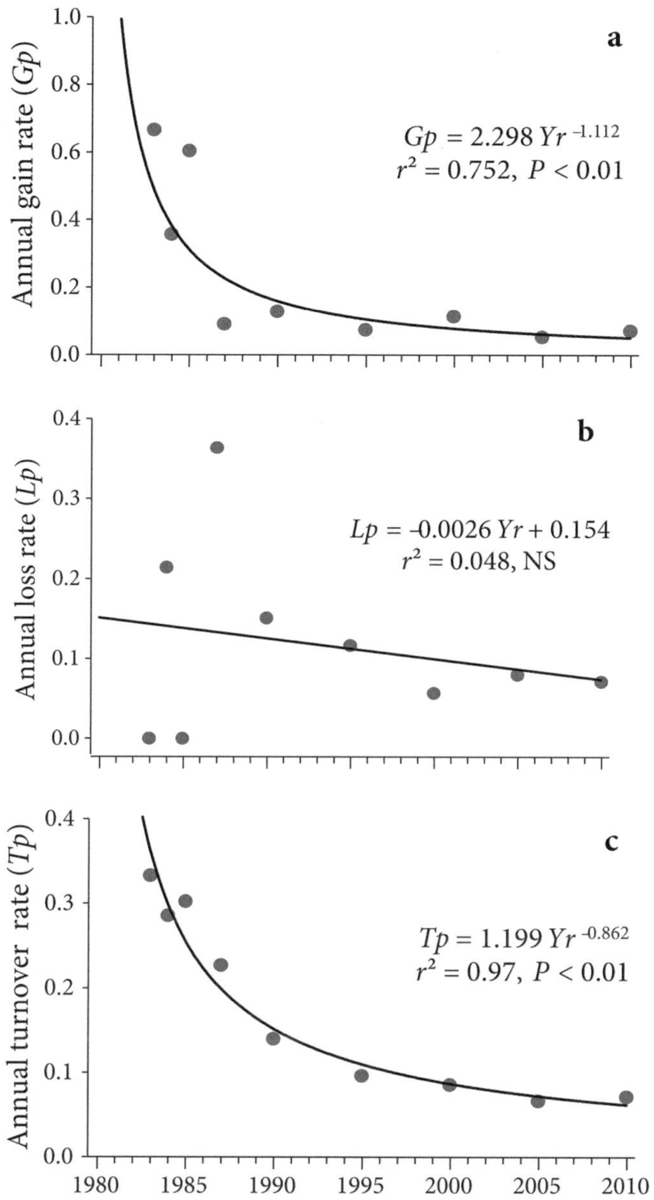

Fig. 12.6 Successional patterns of ground beetle (Carabidae) species gain rates (**a**), loss rates (**b**), and turnover rates (**c**) on the Pumice Plain of the Mount St. Helens volcano.

trophic role as both adults and larvae, dispersal capabilities, activity periods on a daily and seasonal basis, and average body sizes (Table 12.1). Analyses of these attributes of the ground beetle assemblages from 1983 through 2010 yielded a number of significant trends. First, the early post-eruption years on the Pumice Plain were dominated by species that preferred riparian/lakeshore environments and open meadow habitats (Fig. 12.7, left). Through time, as the Pumice Plain was colonized by meadow vegetation, species that preferred riparian/lakeshore habitats became significantly less common (Pearson's $r = -0.79$, $df = 7$, $P < 0.02$); this was mirrored by reductions in the percentage of species (Pearson's $r = -0.88$, $df = 7$, $P < 0.01$) and the total number of individuals (Pearson's $r = -0.95$, $df = 7$, $P < 0.01$). Although the absolute number of ground beetle species preferring meadow habitats did not significantly increase, there were significant increases in the percentage of meadow species present (Pearson's $r = 0.79$, $df = 7$, $P < 0.02$) and the total number of meadow individuals (Pearson's $r = 0.96$, $df = 7$, $P < 0.01$). In 1984, over 85% of the ground beetle individuals were specialists for riparian and lakeshore habitats, but by 2010, more than 90% of the ground beetles on the Pumice Plain were meadow-dwelling individuals.

The majority of ground beetle species on the Pumice Plain either preferred sand/gravel soils or were soil generalists (normally inhabiting a wide variety of soil types). Only a few species (and a very small number of individuals) typical of forested or shaded habitats with clay/loams and fine organic soils were observed on the Pumice Plain, and these species did not establish reproducing populations (Fig. 12.7, center). Through time, even these species significantly diminished, both in total number of species (Pearson's $r = -0.77$, $df = 7$, $P < 0.05$) and in number of individuals (Pearson's $r = -0.72$, $df = 7$, $P < 0.05$). Concomitantly, the percentage of species preferring sand/gravel soils significantly increased over time (Pearson's $r = 0.82$, $df = 7$, $P < 0.01$).

The trophic roles of ground beetles were heavily dominated by predators as both adults and soil-dwelling larvae (Fig. 12.7, right). A small number of species (in the genera *Amara* and *Harpalus*) were omnivorous and granivorous (Table 12.1). Although these species were consistently present in all years, their numbers of individuals were very small for the first two decades after the eruption. As vegetation increased, there was a significant increase in the numbers of both omnivorous individuals (Pearson's $r = 0.86$, $df = 7$, $P < 0.01$) and granivorous individuals (Pearson's $r = 0.88$, $df = 7$, $P < 0.01$); however, in 2010, over 90% of total individuals were still predators as adults and larvae.

Successional changes in dispersal capabilities of ground beetle species were also observed. All but one of the ground beetle species collected on the Pumice Plain during 1983–2000 were flight capable, although these flying species were moderate to strong runners; only a single species was ambulatory (wingless). After 2000, the number of obligate ambulatory species significantly increased (Fig. 12.8, left), both as total species (Pearson's $r = 0.84$, $df = 7$, $P < 0.01$) and as the percentage of species present (Pearson's $r = 0.86$, $df = 7$, $P < 0.01$).

Diel ground beetle activity patterns on the Pumice Plain were mixed between diurnal and nocturnal species, with some species being active during day and night (Fig. 12.8, center). In 1983, the total numbers of individual beetles were dominated by diurnal species, but

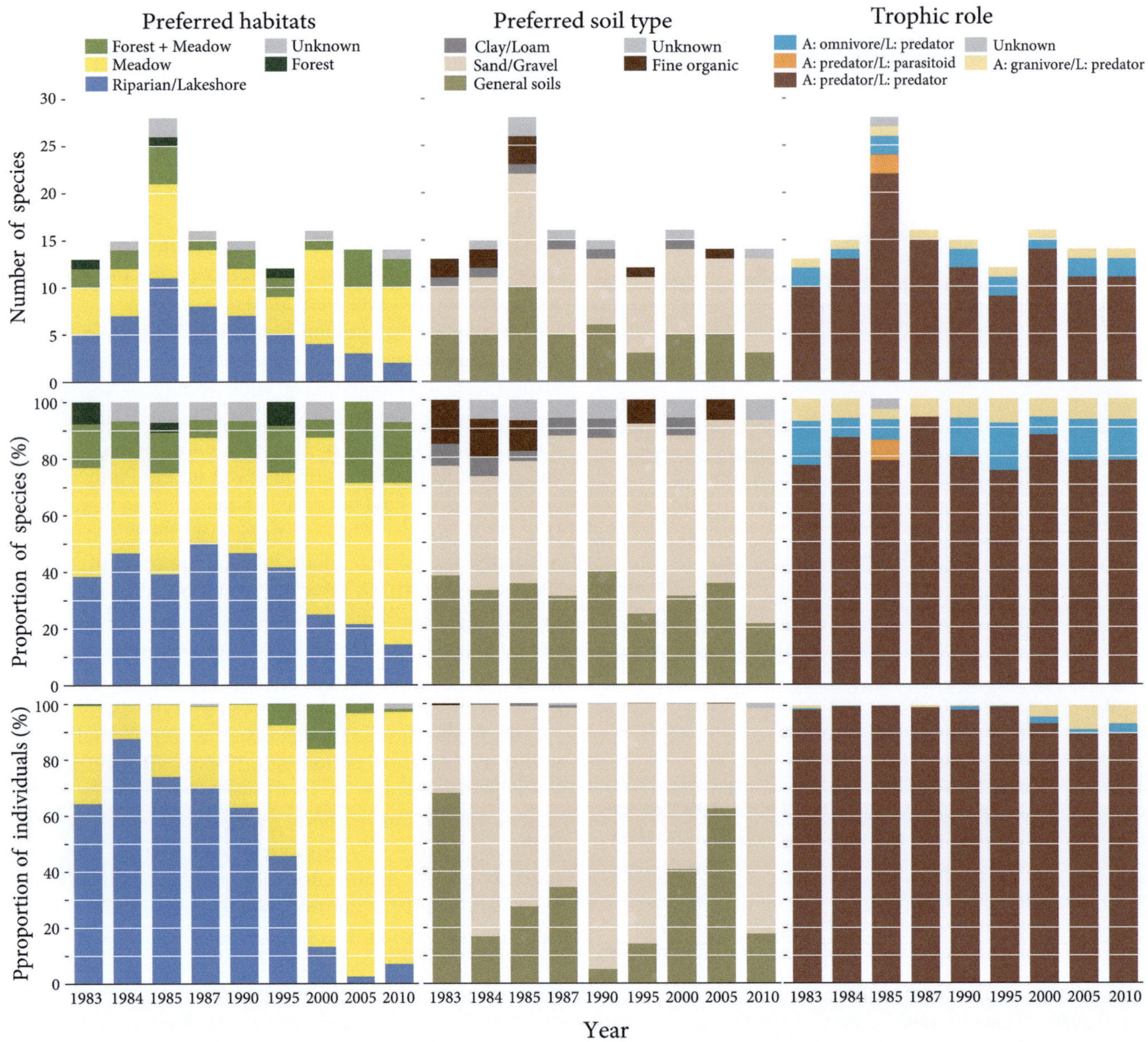

Fig. 12.7 Successional changes in ground beetle (Carabidae) natural-history traits from literature sources (see text for citations) for 1983–2010 on the Pumice Plain of the Mount St. Helens volcano. *Left* = preferred habitats; *center* = preferred soils; *right* = trophic role. Each column of graphs reports actual numbers of species (*top*), percentage of species (*middle*), and percentage of total individuals (*bottom*).

this switched to dominance by nocturnal individuals and species in 1984. Nocturnal species and individuals were more common than diurnal species (except 1990) in most years through 2010.

Ground beetles on the Pumice Plain exhibited a wide range of seasonal activity patterns, but most species and individuals were active throughout the growing season (spring-summer-autumn; Fig. 12.8, right). Between 20% and 40% of the species in a given year were limited to particular seasons; the overall successional pattern showed a significant increase in the percentage of species active during the entire growing season (Pearson's $r = 0.70$, $df = 7$, $P < 0.05$). Only one or two species in a given year were active in all sample periods, and these species were represented by only a few individuals.

Finally, an analysis of ground beetle body size through time found significant increases in body length from 1983 to 2010 for both average lengths across species and for lengths of total numbers of individuals (Fig. 12.9). In 1983, the mean species body length was 6.6 mm, whereas in 2010 it had increased linearly to 11.0 mm. Mean body lengths of total individuals also significantly increased from 7.4 mm in 1983 but peaked in 2000 at 13.9 mm before declining to 10.4 in 2010.

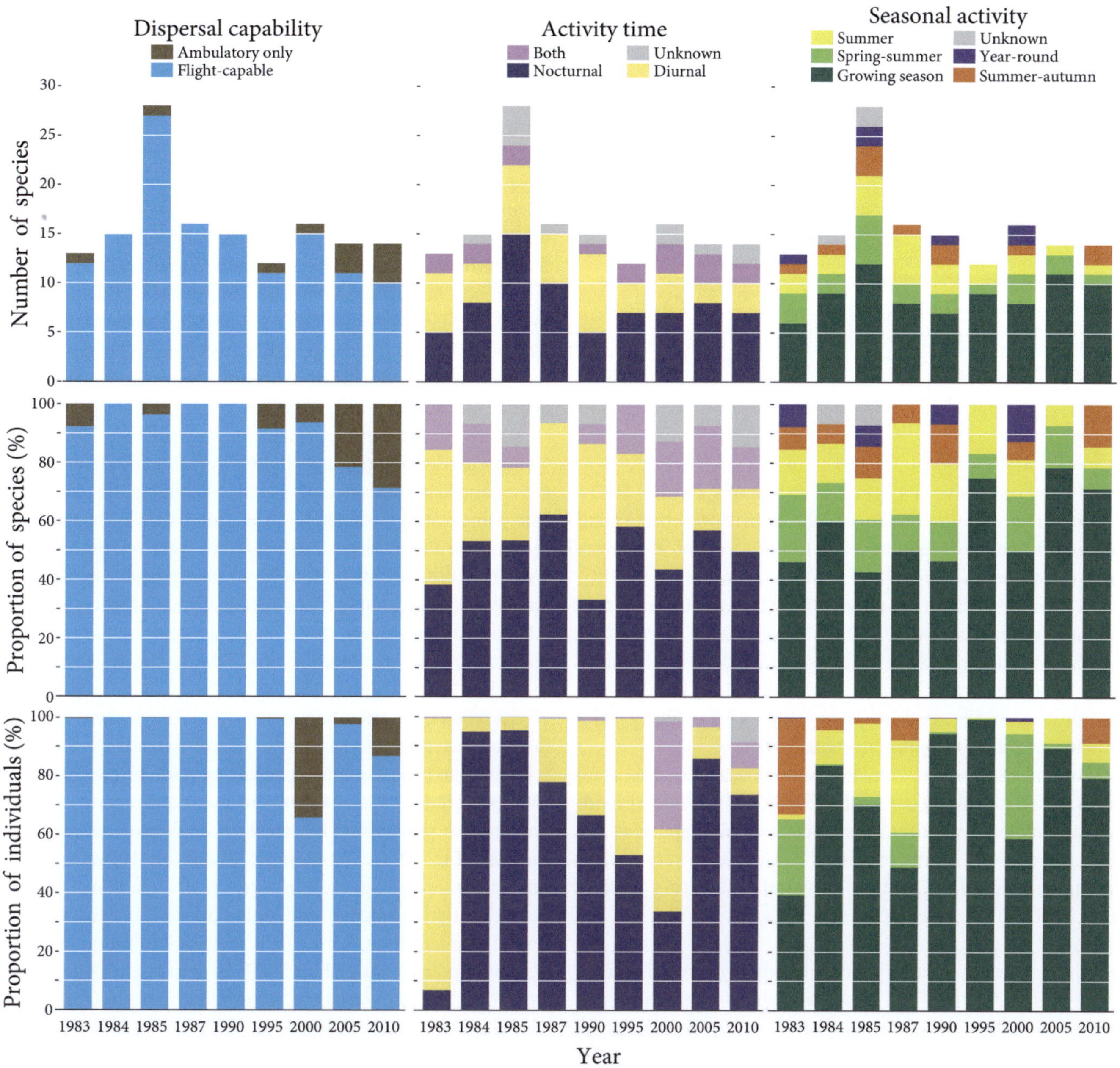

Fig. 12.8 Successional changes in ground beetle (Carabidae) natural-history traits from literature sources (see text for citations) for 1983–2010 on the Pumice Plain of the Mount St. Helens volcano. *Left* = dispersal traits; *center* = daily activity period; *right* = seasonal activity period. Each column of graphs reports actual numbers of species (*top*), percentage of species (*middle*), and percentage of total individuals (*bottom*).

12.4 Discussion

The ground beetle assemblage proved to be a sensitive indicator of primary successional processes on the Pumice Plain. Carabid species surrounding Mount St. Helens provided a continuous influx of potential colonists, and rapidly changing microenvironments and biotic/abiotic resources led to a progression of different resident species, as demonstrated by species turnover. However, unlike many other faunal and floral generalists repopulating the Pumice Plain, ground beetles expressed sensitivity through species loss as microhabitats changed during successional progression. Such "relay" succession of ground beetle species also provided a fine-scale temporal illustration of the rapid succession of microhabitats on the Pumice Plain.

Other inventories of carabid beetles on Mount St. Helens volcano observed the formation of a species-rich assemblage quickly following the 1980 eruption. Edwards (1986a, b), Edwards et al. (1986), and Edwards and Sugg (2005)

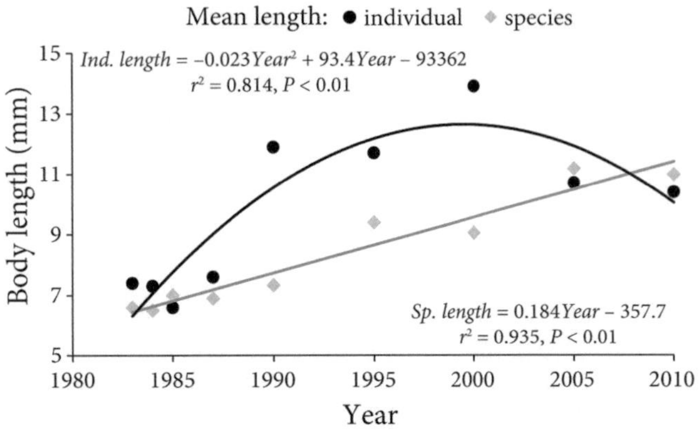

Fig. 12.9 Successional pattern of body size of ground beetles (Carabidae) on the Pumice Plain of the Mount St. Helens volcano, 1983–2000.

conducted extensive surveys for arthropods during 1981–1985 across the pyroclastic-flow deposits (incorporating many different microhabitats) and found a total of 35 species of ground beetles. Of these, 13 species were found in sufficient numbers with larvae present to verify established reproducing populations. Adams and Leffler (1986) reported 11 carabid species from the debris-avalanche deposit and adjacent blowdown zone sites in the North Toutle River valley (northwest of the Pumice Plain, Fig. 12.1) during only 10 days of pitfall trapping in August 1984.

Ground beetle natural histories emphasize the important roles of soil and the ground surface characteristics in providing microhabitat requirements of each carabid species. The soil microhabitat of the Pumice Plain changed rapidly following the eruption owing to physical weathering (Hoffer and Hoffer 1986; Reynolds and Bliss 1986; Wagner and Walker 1986; Halvorson et al., Chap. 5, this volume), the continuous "rain" of arthropods and other organic materials (Edwards and Sugg 1993), development of vegetation, and animal activity (e.g., bioturbation of pocket gophers [*Thomomys talpoides*] and concentration of nutrients by ants). The original post-eruption surfaces on the Pumice Plain consisted of rocks and loose tephra/pumice gravels, infused with minor amounts of fine volcanic ash and sand similar to riverbanks (gravel bars) or small "beaches" along a mountain lakeshore. The initial colonizing ground beetle species (see Table 12.2 and Fig. 12.7) inhabited the interstitial spaces among the gravels, likely feeding on detritus and/or preying on the aerial fallout of arthropods. However, these species also had to contend with fine ash from the eruption. Arthropods exposed to the fine dust from volcanic ash (particles of <40 to 1400 μ; Hoffer et al. 1982) suffered debilitating and lethal injuries primarily owing to increased water losses from cuticular abrasion and loss of saliva during increased grooming activities, as well as obstruction of spiracle valves and blockages in the digestive system (Edwards and Schwartz 1981; Masciocchi et al. 2012; Fernández-Arhex et al. 2013).

Most species of ground beetles in the early post-eruption years were strong fliers and could easily have flown in from nearby undisturbed habitats and landed on or near the study site (Fig. 12.8); in later years, ambulatory (wingless) species began arriving "on foot" (e.g., *Calosoma tepidum*, *Scaphinotus angusticollis*). The majority of these species were predators (Fig. 12.7) and, in the early years, would have taken advantage of the large numbers and wide variety of arthropods falling onto the site (Sugg 1989; Edwards and Sugg 1993). Other species (e.g., in the genera *Amara* and *Harpalus*) were at least partially herbivorous, occasionally feeding on seeds or other plant parts; however, even in these genera, the soil-dwelling larvae were all predators.

As the pumice and tephra gravels weathered to sand and smaller particles, the interstitial rock spaces filled, creating a pavement-like surface and greatly reducing the extent of rocky interstitial microsites. These microsites still existed on the Pumice Plain during our study but were restricted to the western portion of the plain where lobate pyroclastic flows of coarse texture (cobbles) persisted. Ground beetle species adapted to more open, drier, sandy ground, and compacted soils were able to colonize this habitat, but those species requiring moist, gravelly microhabitats were extirpated. Herbaceous vegetation began to establish, providing food for herbivorous arthropod species. These "islands of fertility" with live and dead plant materials provided new resources for both predatory and herbivorous ground beetle species, and species preadapted for meadow environments began to increase (Fig. 12.7). For example, by the late 1990s, caterpillars (Lepidoptera larvae) had reached high densities and were defoliating many of the lupine plants (*Lupinus lepidus*) on the Pumice Plain, with a peak in leaf defoliation and root destruction in 2000 (Bishop 2002; Bishop et al. 2005, 2010). In 1995, a few individuals of "caterpillar hunters" (*Calosoma tepidum*) had been recorded. By 2000, *C. tepidum* populations had numerically exploded and subsequently leveled off at lower numbers through 2010. In this example, the presence of suitable herbivorous prey species presumably facilitated the establishment and persistence of this predatory ground beetle.

The addition of a vegetation cover on the Pumice Plain not only provided direct food resources (seeds and leaf material) and indirect food resources (via supporting herbivorous prey species) to the ground beetle assemblage but also provided greater shelter from both predators and abiotic variables under the developing plant architecture. With a greater range of thermal and predator cover opportunities, one would hypothesize that the average body sizes of the ground beetle assemblage would increase, as a result of larger-bodied species surviving on the Pumice Plain (Fig. 12.9). Prior to establishment of plants (and burrowing rodents as well), shelter space was limited to the small spaces among gravels and pieces of pumice, and small body size would have been well suited for survival in such a barren environment. But with growth of clumps of forbs, grasses, shrubs, and colonizing trees (with concomitant accumulations of plant leaf and stem litter), the breadth of favorable habitat clearly increased. Although definitive experiments to determine differential survivorship of carabid beetles with varying body lengths have yet to be undertaken, the increase in the body size of the ground beetle assemblage appears to be associated with this plant-based facilitation of increasing habitat complexity.

The temporal dynamics of ground beetle species gains, losses, and turnover rates were quantified with metrics of species presence/absence through time. As noted by Anderson (2007), successional theory predicts that the rates of species change during succession should decline through time but that "harsh abiotic conditions or spatial isolation of the community appear to substantially alter temporal patterns in rates of successional change" (Anderson 2007: 780). In the case of ground beetle species dynamics on the Pumice Plain (a harsh and somewhat isolated environment), the patterns indicated high values for species gain and turnover, with highly variable loss rates, immediately after the eruption. However, in the subsequent years (1987–2010), the rates of annual species gain remained constant instead of declining (see Fig. 12.6a). Annual species loss rates and turnover rates did gradually decline, consistent with other successional studies. Yet this pattern contrasts sharply with observed patterns of species change for several major faunal taxa on the Pumice Plain (e.g., amphibians, birds, small mammals). For these groups, species numbers had generally continued to accumulate over the 30 years since the 1980 eruption, experiencing minimal species losses once a species becomes established with reproducing populations (MacMahon et al. 1989; Crisafulli et al. 2005a, b; Crisafulli et al., Chap. 15, this volume). Although these vertebrate taxa displayed species-gain dynamics comparable to those of ground beetles, the species loss rates and turnover rates were due entirely to rarer species that "wink in and out" but had not established breeding populations (see Crisafulli et al., Chap. 15, this volume). In contrast, numerous carabid beetle species through time successfully colonized and established breeding populations and then went locally extinct as successional habitat changes occurred. These results indicate that vertebrate taxa were less sensitive to the observed biotic and abiotic changes in the Pumice Plain environment and therefore changed more slowly through time in terms of loss and turnover rates of successfully established species.

Future changes in the ground beetle assemblage on the Pumice Plain will likely continue to include recruitment of additional species and losses of existing species. Arthropod sampling in forests and clearcuts of the Mount St. Helens region has produced a list of 20 additional carabid species that have not yet been recorded on the Pumice Plain, and there are undoubtedly others (R.R. Parmenter, unpublished data). A number of ground beetle species are generally found only in old-growth forest stands, and therefore their establishment on the Pumice Plain may be decades away. Continued monitoring of the ground beetle assemblage, in concert with other faunal and floral taxa, will undoubtedly yield additional insights into the rates and directions of successional change and community reassembly in the post-eruption Mount St. Helens landscape.

Acknowledgments Support for the field sampling and laboratory analyses during this study was provided by the National Science Foundation (BSR 84-0721, DBI-9820318 and LTREB Program DEB-0614538); the USDA Forest Service Pacific Northwest Research Station; the Museum of Southwestern Biology at the University of New Mexico, Albuquerque; the Department of Entomology, Michigan State University; the Department of Biological Sciences, University of Alberta, Edmonton; and the Ecology Center, Utah State University, Logan. We greatly appreciate the comments by three anonymous reviewers that led to improvements in this chapter. We thank the Gifford Pinchot National Forest and Mount St. Helens National Volcanic Monument for access to our study sites. We also thank all the students and researchers who have participated in the project over the past 30 years, and we are grateful for the insights and inspiration provided by the late Professor John S. Edwards, University of Washington.

References

Adams, A.B., and S. Leffler. 1986. Insect recolonization of the northwest sector of the Mount St. Helens blast zone. In *Mount St. Helens: Five years later*, ed. S.A.K. Keller, 307–324. Spokane: Eastern Washington University Press.

Anderson, K.J. 2007. Temporal patterns in rates of community change during succession. *American Naturalist* 169: 780–793.

Ball, G.E. 2001. Trachypachidae C.G. Thomson, 1857. In *American beetles, volume 1. Archostemata, Myxophaga, Adephaga, Polyphaga: Staphyliniformia*, ed. R.H. Arnett Jr. and M.C. Thomas, 144–146. New York: CRC Press.

Ball, G.E., and Y. Bousquet. 2001. Carabidae Latreille, 1810. In *American beetles, volume 1. Archostemata, Myxophaga, Adephaga, Polyphaga: Staphyliniformia*, ed. R.H. Arnett Jr. and M.C. Thomas, 32–132. New York: CRC Press.

Banks, N.G., and R.P. Hoblitt. 1981. Summary of temperature studies of 1980 deposits. In *The 1980 eruptions of Mount St. Helens, Washington*. Professional Paper 1250, ed. P.W. Lipman and D.R. Mullineaux, 295–313. Washington, DC: U.S. Geological Survey.

Bishop, J.G. 2002. Early primary succession on Mount St. Helens: Impact of insect herbivores on colonizing lupines. *Ecology* 83: 191–202.

Bishop, J.G., W.F. Fagan, J.D. Schade, and C.M. Crisafulli. 2005. Causes and consequences of herbivory on prairie lupine (*Lupinus lepidus*) in early primary succession. In *Ecological responses to the 1980 eruption of Mount St. Helens*, ed. V.H. Dale, F.J. Swanson, and C.M. Crisafulli, 151–161. New York: Springer.

Bishop, J.G., N.B. O'Hara, J.H. Titus, J.L. Apple, R.A. Gill, and L. Wynn. 2010. N-P co-limitation of primary production and response of arthropods to N and P in early primary succession on

Mount St. Helens volcano. *PLoS ONE* 5 (10): e13598. https://doi.org/10.1371/journal.pone.0013598.

Blanchet, F.G., J.A.C. Bergeron, J.R. Spence, and H. Fangliang. 2013. Landscape effects of disturbance, habitat heterogeneity and spatial autocorrelation for a ground beetle (Carabidae) assemblage in mature boreal forest. *Ecography* 36: 636–647.

Cole, L.J., S. Brocklehurst, D.A. Elston, and D.I. McCracken. 2012. Riparian field margins: Can they enhance the functional structure of ground beetle (Coleoptera: Carabidae) assemblages in intensively managed grassland landscapes? *Journal of Applied Ecology* 49: 1384–1395.

Crisafulli, C.M., J.A. MacMahon, and R.R. Parmenter. 2005a. Small-mammal survival and colonization on the Mount St. Helens volcano: 1980–2002. In *Ecological responses to the 1980 eruption of Mount St. Helens*, ed. V.H. Dale, F.J. Swanson, and C.M. Crisafulli, 199–218. New York: Springer.

Crisafulli, C.M., L.S. Trippe, C.P. Hawkins, and J.A. MacMahon. 2005b. Amphibian responses to the 1980 eruption of Mount St. Helens. In *Ecological responses to the 1980 eruption of Mount St. Helens*, ed. V.H. Dale, F.J. Swanson, and C.M. Crisafulli, 183–197. New York: Springer.

Dale, V.H., D.R. Campbell, W.M. Adams, C.M. Crisafulli, V.I. Dains, P.M. Frenzen, and R.F. Holland. 2005a. Plant succession on the Mount St. Helens debris-avalanche deposit. In *Ecological responses to the 1980 eruption of Mount St. Helens*, ed. V.H. Dale, F.J. Swanson, and C.M. Crisafulli, 59–73. New York: Springer.

Dale, V.H., F.J. Swanson, and C.M. Crisafulli, eds. 2005b. *Ecological responses to the 1980 eruption of Mount St. Helens*. New York: Springer.

Dennis, P., M.R. Young, C.L. Howard, and I.J. Gordon. 1997. The response of epigeal beetles (Col.: Carabidae, Staphylinidae) to varied grazing regimes on upland *Nardus stricta* grasslands. *Journal of Applied Ecology* 34: 433–443.

Dennis, P., M.R. Young, and I.J. Gordon. 1998. Distribution and abundance of small insects and arachnids in relation to structural heterogeneity of grazed, indigenous grasslands. *Ecological Entomology* 23: 253–264.

Dunger, W. 1989. The return of soil fauna to coal mined areas in the German Democratic Republic. In *Animals in primary succession*, ed. J.D. Majer, 307–337. New York: Cambridge University Press.

Edwards, J.S. 1986a. Arthropods as pioneers: recolonization of the blast zone on Mount St. Helens. *Northwest Environmental Journal* 2: 263–273.

———. 1986b. Derelicts of dispersal: Arthropod fallout on Pacific Northwest volcanoes. In *Insect flight: Dispersal and migration*, ed. W. Danthanarayana, 196–203. New York: Springer-Verlag.

Edwards, J.S., and L.M. Schwartz. 1981. Mount St. Helens ash: A natural insecticide. *Canadian Journal of Zoology* 59: 714–715.

Edwards, J.S., and P.M. Sugg. 1993. Arthropod fallout as a resource in the recolonization of Mount St. Helens. *Ecology* 74: 954–958.

———. 2005. Arthropods as pioneers in the regeneration of life on the pyroclastic-flow deposits of Mount St. Helens. In *Ecological recovery of Mount St. Helens after the 1980 eruption*, ed. V.H. Dale, F. Swanson, and C.M. Crisafulli, 127–138. New York: Springer.

Edwards, J.S., R.L. Crawford, P.M. Sugg, and M.A. Peterson. 1986. Arthropod recolonization in the blast zone of Mount St. Helens. In *Mount St. Helens: Five years later*, ed. S.A.K. Keller, 329–333. Spokane: Eastern Washington University Press.

Egler, F.E. 1954. Vegetation science concepts. I. Initial floristics composition. A factor in old-field vegetation development. *Vegetatio* 4: 412–417.

Elia, M., R. Lafortezza, E. Tarasco, G. Colangelo, and G. Sanesi. 2012. The spatial and temporal effects of fire on insect abundance in Mediterranean forest ecosystems. *Forest Ecology and Management* 263: 262–267.

Elizalde, L. 2014. Volcanism and arthropods: A review. *Ecología Austral* 24: 3–16.

Erwin, T.L. 2007. *A treatise on the western hemisphere Caraboidea (Coleoptera). Their classification, distributions, and ways of life*, Trachypachidae, Carabidae—Nebriiformes 1. Vol. 1. Sofia: Pensoft Publishers.

Fernández-Arhex, V., M. Buteler, M.E. Amadio, A. Enriquez, T. Stadler, G. Becker, and O. Bruzzone. 2013. The effects of volcanic ash from Puyehue–Caulle range eruption on the survival of *Dichroplus vittigerum* (Orthoptera: Acrididae). *Florida Entomologist* 96: 286–288.

Foxworthy, B.L., and M. Hill. 1982. *Volcanic eruptions of 1980 at Mount St. Helens: The first 100 days*, Professional Paper 1249. Washington, DC: U.S. Geological Survey.

Gandhi, K.J.K., D.W. Gilmore, S.A. Katovich, W.J. Mattson, J.C. Zasada, and S.J. Seybold. 2008. Catastrophic windstorm and fuel-reduction treatments alter ground beetle (Coleoptera: Carabidae) assemblages in a North American sub-boreal forest. *Forest Ecology and Management* 256: 1104–1123.

Hatch, M.H. 1953. *The beetles of the Pacific Northwest. Part I. Introduction and adephaga*. Seattle: University of Washington Press.

Hodkinson, I.D., N.R. Webb, and S.J. Coulson. 2002. Primary community assembly on land – The missing stages: Why are the heterotrophic organisms always there first? *Journal of Ecology* 90: 569–577.

Hoffer, J.M., and R.L. Hoffer. 1986. Chemical composition of the May 18, 1980 Mount St. Helens tephra and the underlying soil: Changes from July 1980 to May 1981. In *Mount St. Helens: Five years later*, ed. S.A.K. Keller, 108–111. Spokane: Eastern Washington University Press.

Hoffer, J.M., R.L. Hoffer, and W.M. Shannon. 1982. Preliminary analyses of the composition and texture May 18, 1980, Mount St. Helens tephra. In *Mount St. Helens: One year later*, ed. S.A.C. Keller, 21–25. Spokane: Eastern Washington University Press.

Iglay, R.B., D.A. Miller, B.D. Leopold, and G. Wang. 2012. Carabid beetle response to prescribed fire and herbicide in intensively managed, mid-rotation pine stands in Mississippi. *Forest Ecology and Management* 281: 41–47.

Jaccard, P. 1908. Nouvelles recherches sur la distribution florale. *Bulletin de la Societé Vandoise de Science Naturelle* 44: 223–270.

Larochelle, A., and M.-C. Larivière. 2003. *A natural history of the ground-beetles (Coleoptera: Carabidae) of America North of Mexico*. Sofia: Pensoft Publishers.

MacMahon, J.A., R.R. Parmenter, K.A. Johnson, and C.M. Crisafulli. 1989. Small mammal recolonization on the Mount St. Helens volcano: 1980–1987. *American Midland Naturalist* 122: 365–387.

Martinson, H.M., and M.J. Raupp. 2013. A meta-analysis of the effects of urbanization on ground beetle communities. *Ecosphere* 4 (5): 60. https://doi.org/10.1890/ES12-00262.1.

Masciocchi, M., A.J. Pereira, M.V. Lantschner, and J.C. Corley. 2012. Of volcanoes and insects: the impact of the Puyehue–Cordón Caulle ash fall on populations of invasive social wasps, *Vespula* spp. *Ecological Research* 28: 199–205.

Meijer, J. 1989. Sixteen years of fauna invasion and succession in the Lauwerszeepolder. In *Animals in primary succession*, ed. J.D. Majer, 339–369. New York: Cambridge University Press.

Niemelä, J., M. Koivula, and D.J. Kotze. 2007. The effects of forestry on carabid beetles (Coleoptera: Carabidae) in boreal forests. *Journal of Insect Conservation* 11: 5–18.

Parmenter, R.R., and J.A. MacMahon. 1987. Early successional patterns of arthropod recolonization on reclaimed strip mines in southwestern Wyoming: The ground-dwelling beetle fauna (Coleoptera). *Environmental Entomology* 16: 168–177.

Parmenter, R.R., C.M. Crisafulli, N. Korbe, G. Parsons, M. Edgar, and J.A. MacMahon. 2005. Post-eruption arthropod succession on the Mount St. Helens volcano: The ground-dwelling beetle fauna (Coleoptera). In *Ecological recovery of Mount St. Helens after the 1980 eruption*, ed. V.H. Dale, F. Swanson, and C.M. Crisafulli, 139–150. New York: Springer.

Parsons, G.L., G. Cassis, A.R. Moldenke, J.D. Lattin, N.H. Anderson, J.C. Miller, P. Hammond, and T.D. Schowalter. 1991. *Invertebrates of the H.J. Andrews Experimental Forest, Western Cascade Range, Oregon. V: An annotated list of insects and other arthropods*, General Technical Report PNW-GTR-290. Portland: U.S. Department of Agriculture, Forest Service, Pacific Northwest Research Station.

Pearce, J.L., and L.A. Venier. 2006. The use of ground beetles (Coleoptera: Carabidae) and spiders (Araneae) as bioindicators of sustainable forest management: A review. *Ecological Indicators* 6: 780–793.

Pearson, D.L., C.B. Knisley, and C.J. Kazilek. 2006. *A field guide to the tiger beetles of the United States and Canada*. New York: Oxford University Press.

Reynolds, G.D., and L.C. Bliss. 1986. Microenvironmental investigations of tephra covered surfaces at Mount St. Helens. In *Mount St. Helens: Five years later*, ed. S.A.K. Keller, 147–152. Spokane: Eastern Washington University Press.

Rusch, A., R. Bommarco, P. Chiverton, S. Oberg, H. Wallin, S. Wiktelius, and B. Ekbom. 2013. Response of ground beetle (Coleoptera, Carabidae) communities to changes in agricultural policies in Sweden over two decades. *Agriculture, Ecosystems and Environment* 176: 63–69.

Sasal, Y., E. Raffaele, and A.G. Farji-Brener. 2010. Succession of ground dwelling beetle assemblages after fire in three habitat types in the Andean forest of NW Patagonia, Argentina. *Journal of Insect Science* 10: 1–17.

Schirmel, J., and S. Buchholz. 2011. Response of carabid beetles (Coleoptera: Carabidae) and spiders (Araneae) to coastal heathland succession. *Biodiversity and Conservation* 20: 1469–1482.

Sugg, P.M. 1989. Arthropod populations at Mount St. Helens: survival and revival. Doctoral dissertation. Seattle: University of Washington.

Sugg, P.M., and J.S. Edwards. 1998. Pioneer aeolian community development on pyroclastic flows after the eruption of Mount St. Helens, Washington, U.S.A. *Arctic and Alpine Research* 30: 400–407.

Swanson, F.J., and J.J. Major. 2005. Physical events, environments, and geological-ecological interactions at Mount St. Helens: March 1980–2004. In *Ecological recovery of Mount St. Helens after the 1980 eruption*, ed. V.H. Dale, F. Swanson, and C.M. Crisafulli, 27–44. New York: Springer.

Wagner, R., and R.B. Walker. 1986. Mineral nutrient availability in some Mount St. Helens surface samples. In *Mount St. Helens: Five years later*, ed. S.A.K. Keller, 153–162. Spokane: Eastern Washington University Press.

13 Diversity of Large-Bodied Macroinvertebrates in Ponds Created on the Debris-Avalanche Deposit Following the 1980 Eruption of Mount St. Helens

Shannon M. Claeson, Charles M. Crisafulli, and William J. Gerth

13.1 Introduction

One of the most notable aspects of the 1980 eruption of Mount St. Helens was the enormous rockslide-debris avalanche from the cone of the volcano that immediately preceded the blast pyroclastic density current (blast PDC), commonly referred to as the lateral blast (Swanson and Major 2005). The majority of the 2.5-km^3 deposit of unconsolidated rocky debris came to rest in the North Fork Toutle River Valley (Fig. 13.1) and created a large complex of fishless ponds varying in size and hydroperiod (the duration that a pond basin holds water). Initial monitoring of these ponds focused on the invasion and reproductive biology of amphibians (Karlstrom 1986; Crisafulli et al. 2005). During the summers of 2003–2005, aquatic amphibians in many of the ponds were surveyed using funnel traps (also known as minnow traps). The macroinvertebrates captured in these traps were also collected for identification and community analysis. We recognize that funnel-trapping is not the standard method for sampling aquatic macroinvertebrates in lentic habitats because traps are biased against small-sized individuals (e.g., Diptera, oligochaetes, zooplankton) and less-mobile taxa (e.g., Ephemeroptera, Trichoptera) and may not accurately estimate taxa richness throughout the pond (Turner and Trexler 1997). However, this data set is unique in the literature, primarily owing to the rarity of large-scale disturbances that create lentic water bodies. It is even more unusual to have aquatic macroinvertebrate samples from 97 ponds in the same ecoregion that were all created at the same time, and from a geologic perspective, quite recently.

Previous investigations of aquatic invertebrate responses to volcanism have focused on macroinvertebrates in rivers and streams (Anderson and Wisseman 1987; Hawkins 1988; Meyerhoff 1991; Anderson 1992; Dorava and Milner 1999; Collier 2002; Miserendino et al. 2012) or on zooplankton communities in lakes and ponds (Scharnberg 1995; Larson et al. 2006). Limited work relating to lentic aquatic macroinvertebrates can be found from tropical volcanoes, concentrating on large caldera lakes (Ball and Glucksman 1978, 1980) or 100+-year-old lentic water bodies on the Krakatau Islands, Indonesia (Yukawa and Yamane 1985; Thornton and New 1988; van Tol 1990), but we are unaware of any work in lentic habitats from temperate volcanoes.

Aquatic macroinvertebrates are ecologically important members of temperate freshwater ecosystems. Their high diversity, abundance, biomass, and varied trophic relationships influence primary productivity, nutrient cycling, predator and prey dynamics, and community structure of aquatic systems. When the regional pool of potential colonists and environmental conditions (e.g., elevation, biome, geology, and climate) are shared, local variations in habitat conditions are expected to play a large role in community assembly. Habitat conditions considered important in determining the distribution of aquatic macroinvertebrates in lakes and ponds include substrate and vegetation, pond size and permanence, water chemistry, and temperature (Larson 1985; Heino 2000; Della Bella et al. 2005; Tarr et al. 2005).

Of these factors, hydroperiod, the duration over the course of a year that a pond basin holds water, is often thought to be

S.M. Claeson (✉)
U.S. Department of Agriculture, Forest Service, Pacific Northwest Research Station, Wenatchee Forestry Sciences Laboratory, 1133 North Western Avenue, Wenatchee, WA 98801, USA
e-mail: sclaeson@fs.fed.us

C.M. Crisafulli
U.S. Department of Agriculture, Forest Service, Pacific Northwest Research Station, Mount St. Helens National Volcanic Monument, 42218 NE Yale Bridge Rd., Amboy, WA 98601, USA
e-mail: ccrisafulli@fs.fed.us

W.J. Gerth
Department of Fisheries and Wildlife, Oregon State University, 104 Nash Hall, Corvallis, OR 97331, USA
e-mail: william.gerth@oregonstate.edu

C.M. Crisafulli, V.H. Dale (eds.), *Ecological Responses at Mount St. Helens: Revisited 35 years after the 1980 Eruption*,
https://doi.org/10.1007/978-1-4939-7451-1_13

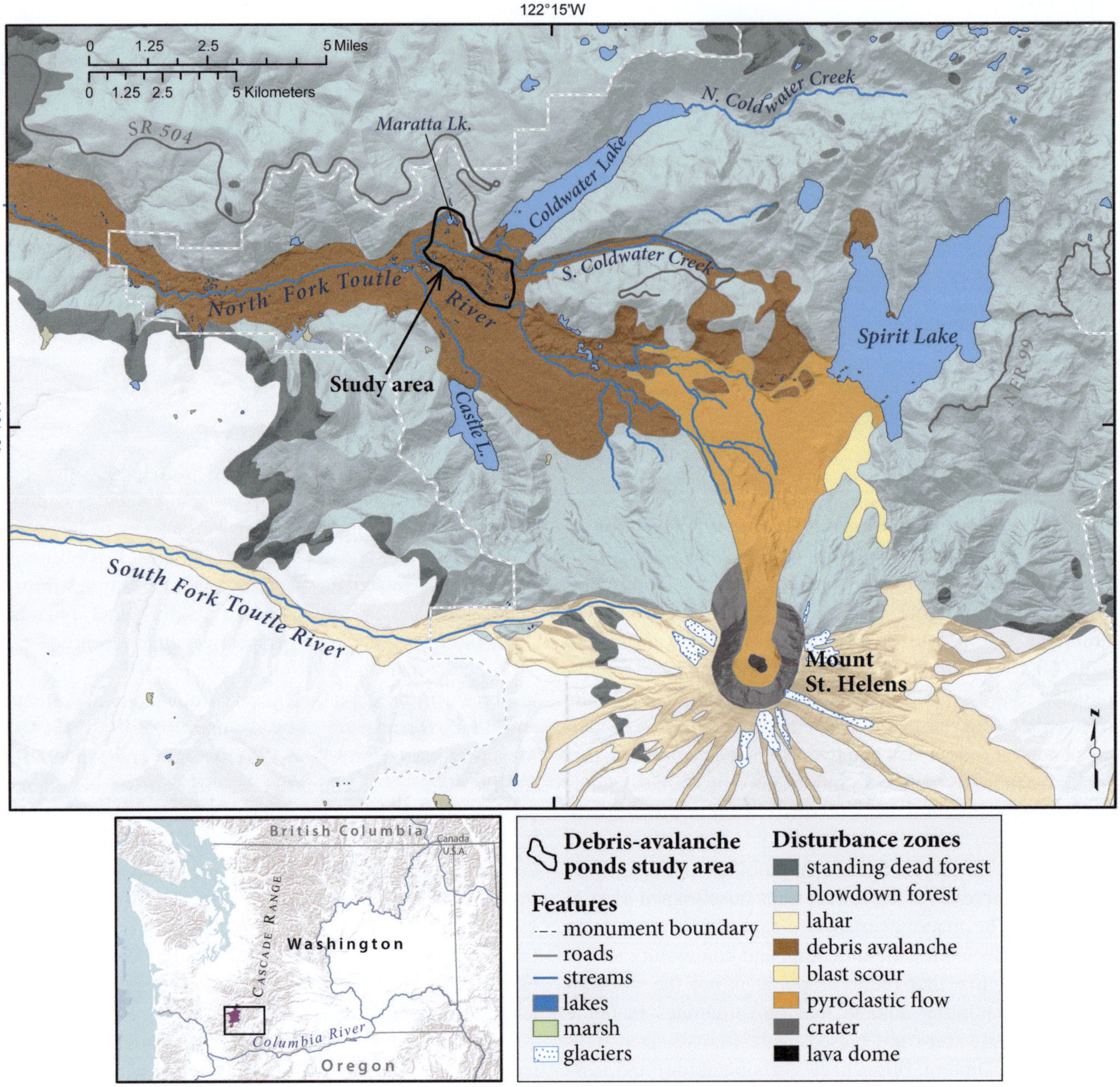

Fig. 13.1 Distribution of primary volcanic deposits and disturbance zones of the 1980 Mount St. Helens eruptions, showing location of study area. (Cartography by Kathryn Ronnenberg, USDA Forest Service, Pacific Northwest Research Station).

the most important and positively related to taxa richness (Della Bella et al. 2005; Williams 2006; Silver et al. 2012). In temporary ponds, desiccation and extreme or unpredictable environmental conditions are expected to exert a strong influence on community composition, supporting species that can colonize quickly, develop rapidly, and exploit temporary food-rich resources. Conversely, permanent pond communities are thought to be primarily structured by biotic interactions, such as competition and predation, because environmental conditions are relatively stable and predictable (Tarr et al. 2005; Verberk et al. 2008; Silver et al. 2012). In fishless ponds and lakes, invertebrate predators are typically more diverse and abundant in permanent waters compared with temporary habitats (Wellborn et al. 1996; Tarr et al. 2005).

Our objective for this study was to document and address how macroinvertebrate assemblages in these recently created, natural ponds are structured. Given the data available, we attempt to answer the following questions: (1) does macroinvertebrate community structure vary among ponds or between years; (2) do different rates of seasonal drying of pond habitats (hydroperiods) affect the number of species present (taxa richness) or community composition; and (3) are identifiable aquatic or riparian habitat features associated with macroinvertebrate community composition? Owing to the nonstandard sampling methods and potential biases in the data set, we limit our analyses to large-bodied, mobile species presence/absence (not detected). Large-bodied and mobile species are typically quite abundant in established ponds and lakes, especially lakes without fish (Schilling et al. 2009); therefore our findings are relevant to newly created ponds in landscapes of natural disturbance, as well as to artificially created ponds.

13.2 Ponds on the Debris-Avalanche Deposit

The debris-avalanche deposit consisted of a mix of sand- to boulder-sized particles, extending 25 km from the volcano, with an average thickness of 45 m and a maximum thickness of 195 m (Voight et al. 1981; Glicken 1996). As the avalanche traveled downslope, numerous hummocks (pointed or domed-shaped mounds up to 75 m tall) and depressions were created. The melting of glacial ice trapped in the deposit and collapse between adjacent avalanche block facies created additional depressions. Overall, more than 130 ponds were formed as they filled with groundwater and precipitation over the first year post eruption (Crisafulli et al. 2005; Swanson and Major 2005).

The Mount St. Helens area has a wet, mild, maritime climate with total annual precipitation around 200 cm, most of it falling as rain or snow in the winter. Average annual air temperature is about 7.0 °C, and average July air temperature is about 15 °C. The ponds are located in the Western Hemlock Zone, where the potential natural vegetation is coniferous forest dominated by the seral tree species Douglas-fir (*Pseudotsuga menziesii*), and at later stages of development by western hemlock (*Tsuga heterophylla*) and western redcedar (*Thuja plicata*) (Franklin and Dyrness 1988). However, in the aftermath of the 1980 eruption, all vegetation was stripped away or buried beneath a thick rocky deposit, leaving a barren landscape devoid of vegetation. At the time of this study (23–25 years post eruption), plant succession was underway, and the vegetation was a complex mosaic of early-seral patches of herbs, willows (*Salix* spp.), and young red alder (*Alnus rubra*) trees (Dale et al. 2005; Dale and Denton, Chap. 8, this volume).

Unlike most lakes of the Cascade Range that were historically fishless and later stocked with non-native trout (Liss et al. 1995), the debris-avalanche ponds are free of fish, but most support amphibians. Northwestern salamanders (*Ambystoma gracile*), in their fully aquatic form (referred to as the neotenic lifeform, retaining larval characteristics into reproductive adulthood), are carnivorous and were present in most of the perennial ponds. Adult and larval rough-skinned newts (*Taricha granulosa*), also carnivorous, occurred in many of the ponds, irrespective of water permanence. Herbivorous tadpoles of Pacific treefrog (*Pseudacris regilla*) and northern red-legged frog (*Rana aurora*) were present at nearly every pond, and western toad (*Anaxyrus boreas*) tadpoles occurred in a third of the ponds (Crisafulli et al. 2005).

13.3 Methods

13.3.1 Study Ponds

Our study focused on 97 ponds distributed over 460 ha on the debris-avalanche deposit, at elevations of 708–781 m. Two groups of ponds are divided by Coldwater Creek (Fig. 13.2). The study ponds range in surface area from 4.9 to 76 465 m^2 and in depth from 0.17 to 8 m, the largest being Maratta Lake. Although not all ponds were sampled every year, over the course of 3 years (2003–2005), we sampled a subset of 97 ponds for amphibians and macroinvertebrates once per summer, for a total of 238 samples: 73 ponds sampled in 2003, 69 ponds in 2004, and 96 ponds in 2005. Sampling occurred in early- to midsummer (late May through July) each year based on prevailing site conditions for that year. The accumulated precipitation for each water year (starting 1 October the previous year, through 30 September) was 167 cm in 2003, 216 cm in 2004, and 169 cm in 2005, as determined from daily SnoTel recordings at nearby Spirit Lake (NRCS 2013).

13.3.2 Macroinvertebrate Collection

We set aquatic funnel (minnow) traps (50 × 24 × 24 cm, mesh size 2.5 mm^2, with 5-cm diameter openings at each end) for 24 ± 2 h in the littoral zone of each pond. Trap locations were distributed at more-or-less-equidistant intervals around the pond perimeters and in water 25–100 cm deep. The number of traps per pond ranged from 2 to 9 traps, depending on pond size. We pooled macroinvertebrates from each trap into a single composite sample for each pond. Aquatic macroinvertebrate specimens were immediately removed from traps and placed in 95% ethanol. We later sorted, counted, and identified individuals to the lowest practical taxon, generally species or genus depending on the life stage (Stonedahl and Lattin 1982, 1986; Larson et al. 2000; Needham et al. 2000; Westfall and May 2006; Merritt et al. 2008). Voucher collections for the taxa described in this analysis currently reside at the Mount St. Helens National Volcanic Monument. We assigned each taxon a functional feeding group (FFG) that generally describes an individual's feeding mechanism (Merritt et al. 2008).

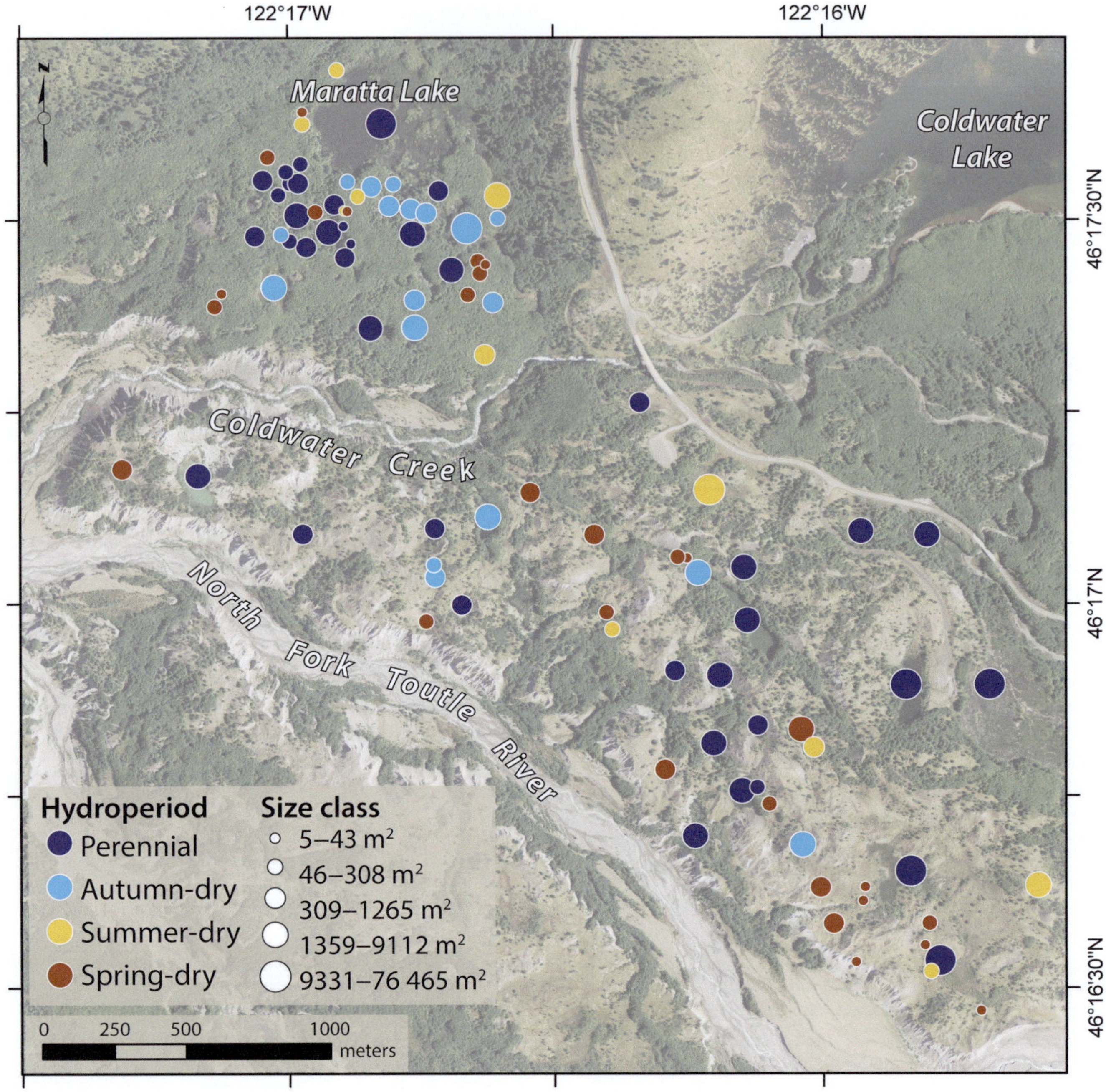

Fig. 13.2 Location of the 97 study ponds on the debris-avalanche deposit. Pond symbols are color-coded by hydroperiod and sized relative to pond surface area, following natural breaks in size groupings (5 *circle* size classes from *small* to *large*: 5–43 m^2; 46–308 m^2; 309–1265 m^2; 1359–9112 m^2; 9331–76 465 m^2). (Photo: USDA National Agricultural Imagery Program, 2013).

13.3.3 Habitat Parameters

During the summer of 2003, we recorded biophysical characteristics of all 97 ponds and riparian areas (Table 13.1). For our analyses, we assumed these habitat parameters to be constant across the three sampling years (2003–2005). We placed primary emphasis on hydrologic condition, substrate, and aquatic and riparian vegetation. We determined the hydro-days (number of days that a basin held water) from a water-level gauge, measured on at least six visits from May through October. For analyses, we divided hydro-days into four group classifications (hydroperiod) based on natural breaks in the data set, using January 1 as day 0: perennial, held water for 365 days; dry in autumn, 279 days or dry by 6 October; dry in summer, 215 days or 3 August; and dry in spring, 147–189 days or 27 May–8 July. Of the 97 ponds, 40 were perennial, 18 autumn-dry, 11 summer-dry, and 28 spring-dry.

Table 13.1 Habitat mean (±1 SE) parameters, by hydroperiod, measured at all 97 ponds during the summer of 2003. Canopy cover over the pond surface area is abbreviated as *Pond CC*.

Habitat	Perennial	Autumn-dry	Summer-dry	Spring-dry
Ponds (#)	40	18	11	28
Area (m^2)	5127 (2006)	1883 (666)	1514 (854)	217 (66)
Depth (cm)	228 (27)	121 (15)	83 (12)	76 (11)
Substrate				
Soil (%)	19 (4)	39 (8)	60 (10)	68 (5)
Sand (%)	55 (4)	51 (8)	29 (7)	21 (4)
Gravel (%)	13 (2)	3 (2)	6 (4)	1 (1)
Cobble (%)	9 (2)	5 (2)	2 (2)	7 (2)
Aquatic vegetation cover				
Emergent (%)	13 (3)	16 (4)	27 (11)	7 (2)
Submergent (%)	22 (5)	12 (4)	5 (2)	0
Floating (%)	14 (4)	9 (5)	1 (1)	0
Riparian vegetation cover				
Willow (%)	12 (3)	11 (3)	31 (10)	36 (6)
Tree (%)	57 (5)	64 (7)	61 (10)	41 (8)
Pond CC (%)	31 (4)	35 (7)	50 (9)	54 (6)

We measured pond surface area at maximum water level with a tape or GPS (Trimble Scientific Instrument Model TSC1). During peak water level, we measured maximum pond depth with a graduated PVC pipe installed at the deepest location in each pond, or in the case of the deepest ponds, by sounding using a graduated, weighted line. We estimated benthic substrate proportions in the littoral zone by visually inspecting or scooping material into a collector at stations equidistant around the pond. We classified substrates as soil (including mineral, organic, or volcanic ash), sand, gravel and cobble, or boulder (Wolman 1954). Water chemistry and temperature are often important factors affecting the distribution of aquatic organisms (Ward 1992); however, we were not able to gather that information from the study ponds over the course of this study.

For each pond, we visually estimated proportions of aquatic and riparian vegetation cover using a modified Daubenmire (1959) cover-class system (eight categories ranging from 0 to 100%). We used the midpoint of each cover class for analyses. We recorded the proportion of the pond covered by rooted emergent, rooted submergent, or free-floating vegetation. We defined the riparian zone as a 3-m-wide band around each pond and recorded the proportion of the riparian area covered by willows or trees (e.g., red alder or cottonwood, *Populus* sp.) separately. We estimated pond canopy cover (CC) with a densiometer at each sampling location around the pond.

13.3.4 Statistical Analyses

We examined taxa-richness patterns at two levels, by sample and by hydroperiod. First, we analyzed taxa richness per sample (each pond separately for each year, n = 238). Because the number of traps set at each pond varied with pond size (two to nine traps/pond at each sampling event, more traps set in larger ponds), we needed to account for the effect of unequal effort on sample taxa richness. An exploratory analysis revealed a linear relationship between numbers of traps set, pond area, and sample taxa richness, so we chose analysis of covariance (ANCOVA). We fit a generalized linear model (GLM, R v.3.3.1) with a Poisson distribution (log link) to sample taxa richness to test for significant effects of number of traps, year, hydroperiod, pond area, depth, and their interactions. We log-10 transformed pond area and depth prior to statistical analyses to normalize their distributions.

Next, we analyzed total richness in the four hydroperiod groups. We used rarefaction to help account for unequal sampling effort within and among groups (different numbers of traps per pond and numbers of ponds per hydroperiod). Rarefaction is a method for estimating species richness through repeated sampling of pooled individuals by group (Gotelli and Colwell 2001). We conducted individual-based rarefaction of abundance data (EstimateS v.9.1) to generate taxa-richness curves with 95% confidence intervals (CIs) for each hydroperiod group and estimated taxa richness with standard error (SE) after standardizing the data to 606 individuals from the spring-dry group, the lowest number of invertebrates collected in any group (Colwell et al. 2012).

To examine assemblage-composition patterns, we used nonmetric multidimensional scaling (NMS, PC-ORD v.6.2) with Sorenson distance to ordinate 238 samples and 110 taxa with presence/absence data (McCune and Grace 2002). We limited multivariate analysis (and the following community analyses) to the presence or absence of taxa observed to reduce abundance biases associated with funnel-trap sampling. The final solution was three dimensional with a stress of 20.85 and instability <0.0001, representing 61% of the total variance: axis 1 26%, axis 2 18%, and axis 3 17%. Joint plots (line vectors) display sample-level taxa richness and habitat parameters associated with each axis (Pearson's correlation coefficient $|R| \geq 0.45$).

We used the multi-response permutation procedure (MRPP, PC-ORD v.6.2), a nonparametric analysis, with Sorenson distance, to test for community differences between samples grouped by hydroperiod (Mielke and Berry 2001). The MRPP provides a p-value and a measure of the effect size, or chance-corrected within-group agreement (A), that is independent of the sample size for each pairwise group comparison. The agreement statistic *A* describes within-group homogeneity compared to random expectation and can range from <0 (less agreement) to 1 (all items are identical).

To highlight large-bodied macroinvertebrate species positively associated with each hydroperiod group (Dufrene and Legendre 1997), we used indicator species analysis (ISA, PC-ORD v.6.2). We calculated indicator values for each taxon based on the relative abundance and frequency of occurrence in each of the four hydroperiods. Indicator values (IV) can range from 0 to 100, with a value of 100 meaning perfect indication (all individuals in a taxon only occur in one group, and all samples in that group contain that taxon). We tested indicator values for statistical significance using Monte Carlo randomizations (4999 permutations). Only species with IV >10 and p-values <0.10 are listed for each hydroperiod.

13.4 Results

13.4.1 Taxonomic Diversity

Macroinvertebrate taxa varied among the 97 study ponds, both spatially and temporally, and were heavily influenced by relatively rare taxa. We collected 10,994 individuals representing 110 taxa from 2 phyla and 4 classes (Arthropoda Insecta, Mollusca Gastropoda), 8 orders, 35 families, 68 subfamilies or genera, and 50 species or groups (Appendix 13.1). Of the individuals collected, the majority were coleopteran, hemipteran, or odonate predators (Table 13.2). All of the scrapers were snails. Collector-gatherers were more diverse, represented by some coleopterans, dipterans, one ephemeropteran, and one trichopteran. Piercer-herbivores were primarily represented by a few coleopterans and hemipterans. The few shredders collected were mostly trichopterans. The most widely distributed species was the predacious diving beetle, *Acilius abbreviatus* (85 ponds). Other frequently observed taxa included another predacious diving beetle (*Agabus* sp., from 79 ponds), ram's horn snails (*Gyraulus* spp. from 64 ponds), water boatmen (*Callicorixa* spp. and *Hesperocorixa* spp.), a backswimmer (*Notonecta kirbyi*), and midge and phantom midge larvae (Chironominae and *Chaoborus* spp.), all from 47 to 57 ponds.

Many taxa were spatially limited in distribution but not in any discernable pattern. Of the taxa observed, 20% (22 out of 110 taxa) were collected from only one pond and 11% (12 taxa) from two ponds. Only 7% (8 taxa) were collected from more than 50% of the ponds (Appendix 13.1). Taxa presence also varied temporally. Over the course of 3 years of sampling, nearly a third of the taxa, 28% (31 taxa), were collected in only 1 year, whereas 48% (53 taxa) were collected in all 3 years. The mean number of taxa per pond in each of the 3 years ranged from 10.2 to 11.4 (0.5–0.6 SE). In total, 76 taxa were collected from 73 ponds in 2003, 73 taxa from 69 ponds in 2004, and 95 taxa from 96 ponds in 2005. Temporal variability in taxa presence was high, but community compositions were ordinated similarly among years (Fig. 13.3a).

Table 13.2 Distribution by hydroperiod of macroinvertebrate taxa richness, catch per unit effort (CPUE, number of individuals/trap), dominance (% abundance by most abundant taxon), and proportional trap densities by functional feeding group and taxonomic order. Sample size (n) is the number of ponds sampled from 2003 to 2005 (total of 238 samples). Most values are the sample average (±1 SE), except for observed total richness, which is the total number of taxa from all samples per hydroperiod. Rarified total richness was standardized to the number of invertebrates collected in the spring-dry group. FFGs that are entirely or primarily (~) represented within each taxonomic order are noted in parentheses.

	Hydroperiod			
Metric	Perennial	Autumn-dry	Summer-dry	Spring-dry
N (no. samples)	120	52	31	35
Observed total richness	96	80	63	45
Rarefied total richness	63 (4.8)	50 (4.2)	54 (2.8)	45 (2.9)
Observed sample richness	12.0 (0.5)	11.4 (0.6)	9.8 (0.7)	6.4 (0.6)
CPUE (individuals/trap)	7.1 (0.5)	10.7 (2.3)	7.1 (0.8)	4.5 (0.6)
Dominance (%)	36 (1.4)	40 (2.9)	38 (3.0)	47 (3.8)
Functional feeding group (%)				
Predator[a]	69 (2.2)	68 (3.1)	73 (4.3)	73 (4.8)
Collector-gatherer[b]	9 (1.3)	9 (1.7)	13 (3.9)	14 (3.4)
Scraper[c]	11 (1.3.)	12 (2.2)	7 (2.4)	4 (2.9)
Piercer-herbivore[d]	10 (1.1)	11 (1.7)	6 (1.1)	7 (1.9)
Shredder[e]	1 (0.4)	1 (0.4)	1 (0.4)	<1 (0.2)
Taxonomic order (%)				
Coleoptera (~PR[a])	33 (2.3)	33 (2.9)	43 (5.2)	60 (5.4)
Hemiptera (~PR[a], ~PH[b])	26 (1.8)	32 (3.5)	22 (3.7)	13 (3.9)
Odonata (PR[a])	18 (1.5)	11 (1.9)	11 (2.2)	2 (0.5)
Diptera (~CG[b],~PR[a])	8 (1.4)	9 (2.3)	15 (4.6)	18 (4.2)
Pulmonata (SC[c])	11 (1.3)	12 (2.2)	7 (2.4)	4 (2.9)
Trichoptera (~CG[b],~SH[e])	3 (0.6)	1 (0.5)	1 (0.5)	<1 (0.2)
Ephemeroptera (CG[b])	1 (0.2)	1 (0.4)	1 (0.5)	2 (1.7)
Megaloptera (~PR[a])	0 (0)	<1 (<0.1)	<1 (<0.1)	0 (0)

Functional feeding groups:
[a]Predator (PR)—attack and engulf or pierce live prey (aquatic insects, tadpoles, or small fish).
[b]Collector-gatherer (CG)—scavenge for decomposing fine particulate organic matter (detritus).
[c]Scraper (SC)—scrape and graze live biofilm (algae, diatoms, bacteria, etc.) from rocks and plants.
[d]Piercer-herbivore (PH)—pierce tissues of live aquatic plants (emergent, submergent, or floating plants).
[e]Shredder (SH)—shred and chew coarse particulate organic matter (leaf litter, wood, etc.).

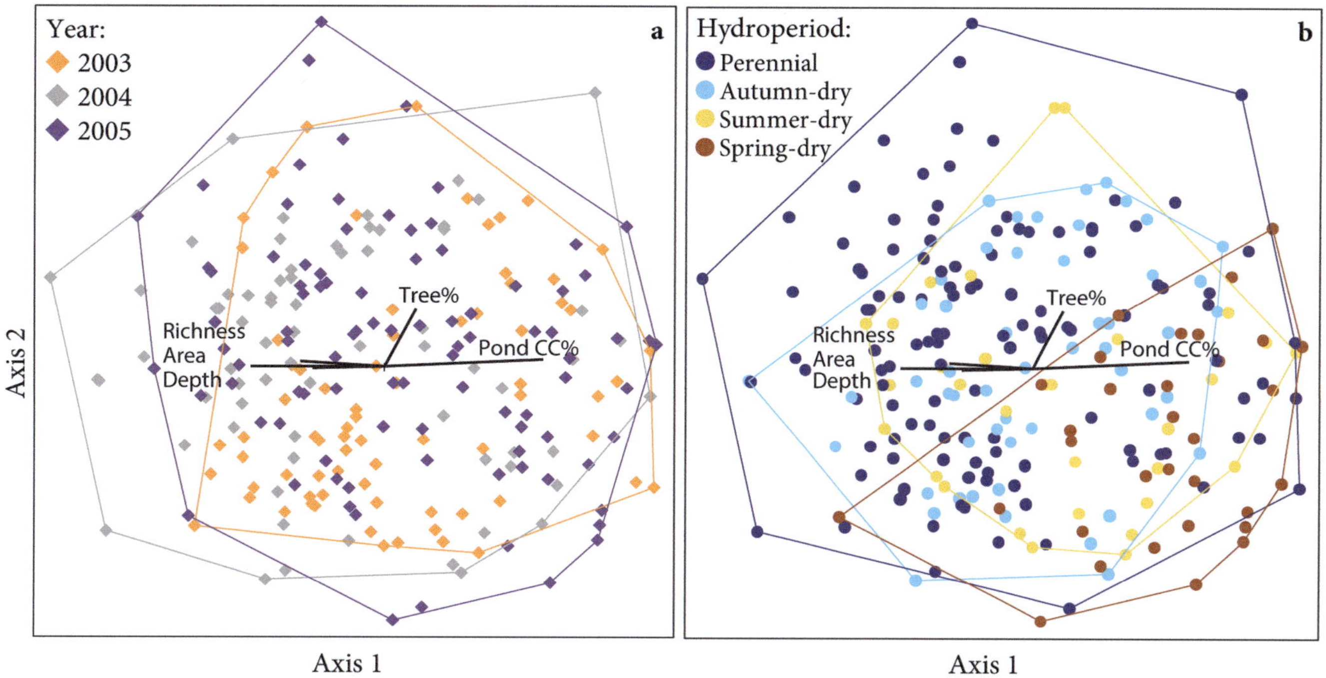

Fig. 13.3 NMS ordination of large-bodied macroinvertebrate communities from 238 samples by 110 taxa of presence/absence data. Each point (sample) represents one pond sampled in 2003, 2004, or 2005. NMS axes 1 (26%) and 2 (18%) show samples coded by (**a**) year and (**b**) hydroperiod. Joint plots (*black line vectors*) show taxa richness and habitat variables correlated with the ordination axes (Pearson's $|R| \geq 0.45$). Axis 3 (17%, not shown) was not correlated with any of the habitat variables measured.

13.4.2 Hydroperiod

Without accounting for differences in sampling effort, mean macroinvertebrate taxa richness was highest in perennial pond samples and lowest in spring-dry pond samples, but % dominance by the most abundant taxon was greatest in spring-dry samples and lowest in perennial pond samples. Measures from summer-dry and autumn-dry samples were intermediate (Table 13.2). At the sample level, taxa richness was linearly related to pond area and to the number of traps set (Fig. 13.4). In Pearson correlation analyses, the number of traps was highly correlated to pond area (0.82), but less so to depth (0.51), whereas area and depth were somewhat correlated to each other (0.58). Thus, the number of traps was the only significant explanatory variable for taxa richness ($p < 0.0001$) when compared against hydroperiod, area, depth, year, and their interactions ($p > 0.05$). Rarefaction curves of taxa richness for each of the four hydroperiods show significantly greater richness in perennial ponds (63 taxa) compared with spring-dry ponds (45 taxa), having non-overlapping 95% CIs (Fig. 13.5), but not enough individuals were collected from spring-dry ponds for the rarefaction curve to reach an asymptote, suggesting that more taxa would likely have been encountered with greater sampling intensity.

The NMS ordination of large-bodied macroinvertebrate communities showed a slight difference among samples grouped by hydroperiod (Fig. 13.3b). However, only spring-dry samples, as a group, were significantly different compared with groups of perennial, autumn-dry, or summer-dry samples (MRPP, $A \geq 0.03$, $p < 0.0001$). The other hydroperiod-group comparisons were either not statistically significant ($p > 0.05$) or had very low within-group agreement ($A < 0.01$). Indicator taxa identified as significant by hydroperiod (higher indicator values than expected by chance) were diverse and represented 15 genera or species from 5 different orders (Table 13.3). However, the indicator taxa all had relatively low indicator values (11.1–30.3), meaning that their association to a hydroperiod, although statistically significant, was not strong.

Predatory macroinvertebrates were by far the most proportionately abundant functional feeding group represented among all hydroperiods (mean 68–73%, Table 13.2). This is not surprising given the use of submerged funnel traps, which primarily capture large-bodied, mobile aquatic fauna such as coleopteran, hemipteran, and odonate predators. On average, proportionately more coleopteran larvae and adults, primarily *Acilius abbreviatus*, were collected from spring-dry ponds as compared with the other hydroperiods. However, a number of other coleopteran species were selected as indicator species of summer-dry, autumn-dry, or perennial ponds (Table 13.3). Hemipteran and odonate individuals were collected from ponds of all hydroperiods but were proportionately most

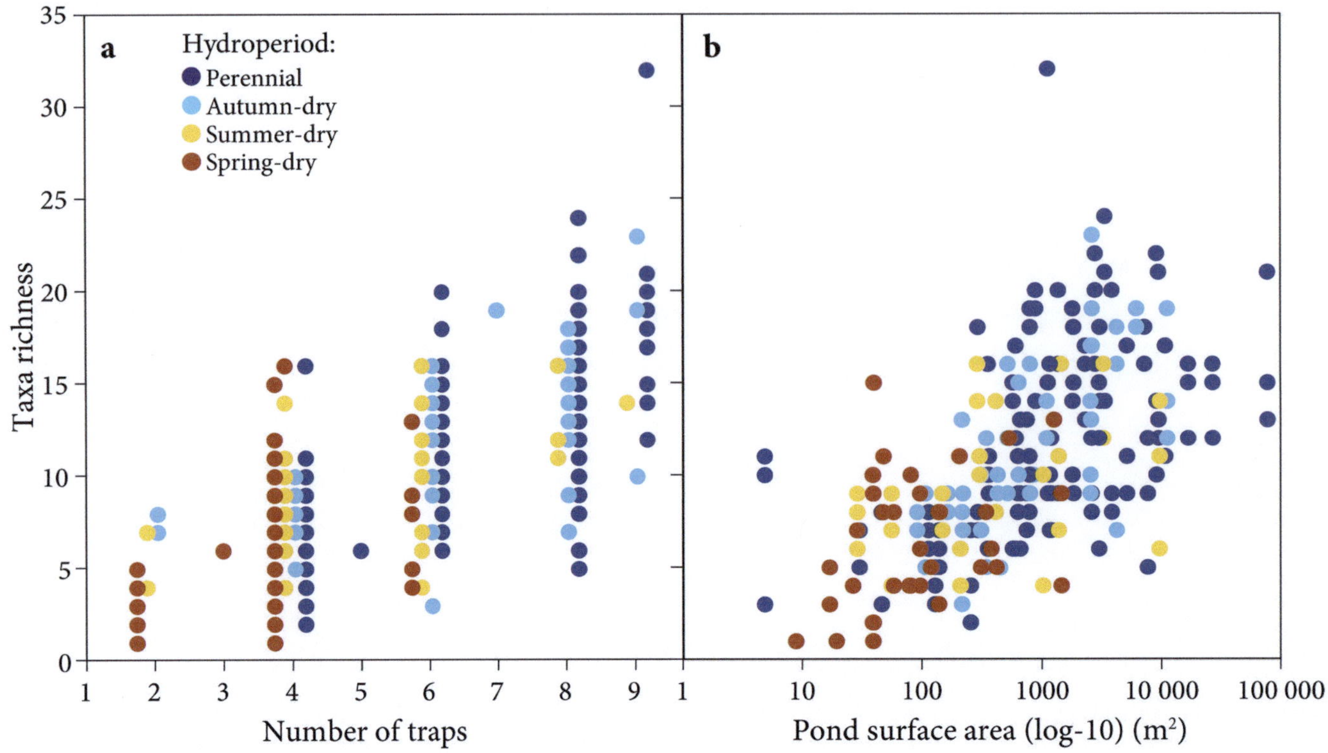

Fig. 13.4 Taxa richness by the (**a**) number of traps set and (**b**) pond surface area. Each point represents one pond sampled in 2003, 2004, or 2005; color coded by hydroperiod ($n = 238$). In **a**, points are jiggered by hydroperiod along the x-axis around each whole number of traps. In **a** and **b**, not all points are visible owing to overlapping x- and y-axis values.

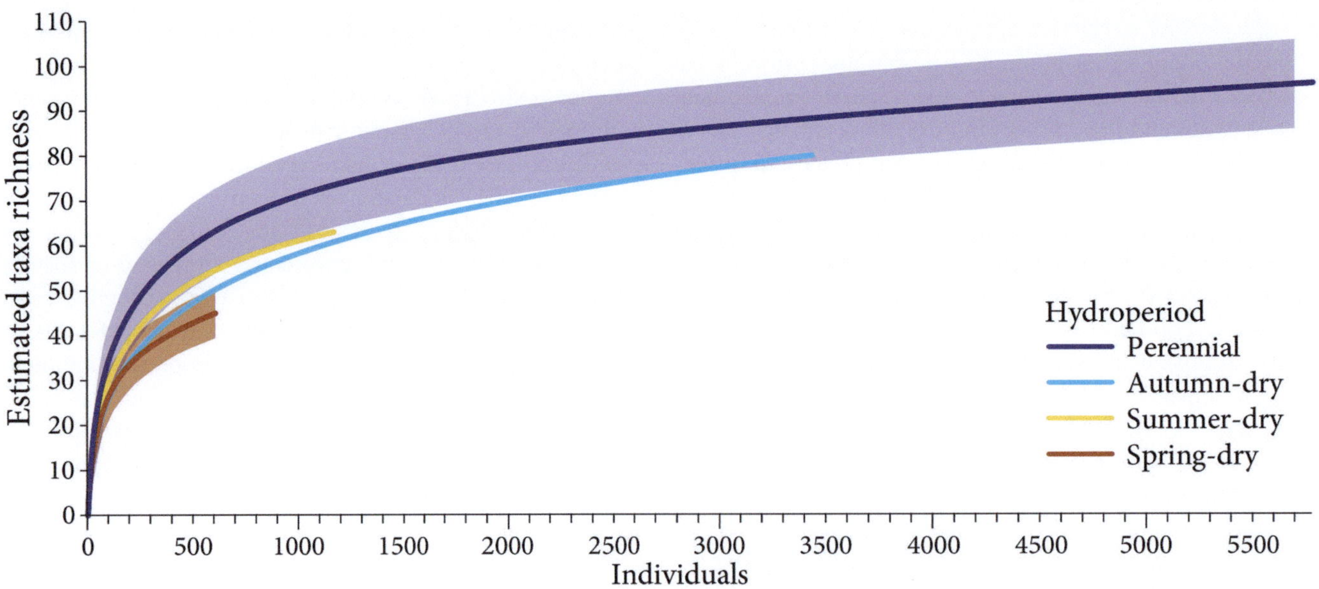

Fig. 13.5 Rarefaction curves (*solid lines*) with 95% CIs (*shaded area*) depicting estimated number of macroinvertebrate taxa by the number of individuals sampled, grouped by hydroperiod. Confidence intervals are not shown for autumn-dry or summer-dry curves, as they always overlapped the other curves.

abundant in perennial or autumn-dry ponds. Water boatman (Corixidae spp.) and damselfly species (*Enallagma*, *Ischnura*, and *Lestes*) were indicators of perennial or autumn-dry ponds. Diptera, representing many different FFGs (Appendix 13.1), were proportionately most abundant at spring-dry ponds, primarily Chironominae and *Chaoborus* species, the latter an indicator species for spring-dry ponds. The snails encountered in this study were all pulmonates (air-breathing),

Table 13.3 Community composition indicator species for each hydroperiod with indicator values (IV) >10 and $\alpha = 0.10$. Functional feeding group abbreviations match those defined in Table 13.2.

Order	Family	Genus/species	Functional feeding group	Indicator value	*p*-value
Perennial					
Odonata	Coenagrionidae	*Ischnura* sp.	PR	30.3	0.0002
Odonata	Coenagrionidae	*Enallagma* sp.	PR	18.1	0.0042
Coleoptera	Hydrophilidae	*Tropisternus* sp.	PR	17.7	0.0844
Hemiptera	Corixidae	*Hesperocorixa* sp.	PH	17.3	0.0886
Pulmonata	Lymnaeidae	*Radix auricularia*	SC	12.5	0.0066
Coleoptera	Dytiscidae	*Neoporus undulatus*	PR	11.7	0.0088
Pulmonata	Physidae	*Physa/Physella* sp.	SC	11.1	0.0264
Autumn-dry					
Hemiptera	Corixidae sp.		PH	20.5	0.0736
Coleoptera	Dytiscidae	*Rhantus consimilis*	PR	18.9	0.0112
Odonata	Lestidae	*Lestes* sp.	PR	18.3	0.0296
Summer-dry					
Coleoptera	Dytiscidae	*Dytiscus* sp.	PR	19.4	0.0088
Odonata	Libellulidae	*Sympetrum* sp.	PR	14.1	0.0128
Spring-dry					
Coleoptera	Dytiscidae	*Acilius abbreviatus*	PR	24.7	0.0790
Diptera	Chaoboridae	*Chaoborus* sp.	PR	18.4	0.0164
Coleoptera	Hydrophilidae	*Hydrobius* sp.	CG	14.4	0.0074
Coleoptera	Dytiscidae	*Rhantus binotatus*	PR	11.3	0.0092

scrapers of algae and detritus, and proportionately most abundant from perennial and autumn-dry ponds. The bladder snails (*Physa* or *Physella* spp.) and European or big-ear radix (*Radix auricularia*, syn. *Lymnaea auricularia*), a non-native pond snail, were indicator species for perennial ponds.

13.4.3 Habitat

Only a few of the habitat parameters measured in this study were correlated with the ordination of large-bodied macroinvertebrate community compositions from 238 samples collected over a 3-year period (Fig. 13.3). Macroinvertebrate communities distributed along axis 1 were correlated with pond area, depth, and % canopy cover (Pearson's $R = -0.67$, -0.49, and $+0.72$, respectively). In general, ponds with small areas and shallow depths were heavily shaded (high % canopy cover) when riparian vegetation was present. In contrast, larger and deeper ponds were more open to solar exposure (low % canopy cover) and tended to have greater macroinvertebrate taxa richness. Cover (%) by both emergent and submergent aquatic plants was weakly correlated with axis 1 ($R = -0.37$ and -0.35, respectively) and tended to be greater in large, open ponds. Taxa negatively correlated with axis 1 and large, open ponds were the water-scavenger beetles (Hydrophilidae, *Tropisternus* sp.), backswimmers (*Notonecta kirbyi*), and damselflies (*Ischnura* and *Lestes*) ($R = -0.49$, -0.48, -0.40, -0.41, respectively). The predacious diving beetle *Colymbetes densus* was positively correlated with axis 1 and small, shaded ponds ($R = 0.43$).

Macroinvertebrate communities distributed along axis 2 were positively correlated with % riparian tree cover ($R = 0.44$) and weakly negatively correlated with % riparian willow cover and soil substrate ($R = -0.41$ and -0.35). *Gyraulus* snails, Chironominae larvae, and *Aeshna* dragonflies were positively correlated with axis 2 ($R = 0.63$, 0.48, 0.46, respectively); and *A. abbreviatus* was negatively correlated ($R = -0.42$). No measured habitat parameters were correlated with axis 3.

13.5 Discussion

The numerous ponds created in the wake of the 1980 eruption of Mount St. Helens provided a unique opportunity to investigate fauna, including macroinvertebrate communities, in young (23- to 25-year-old) ponds following a large natural disturbance. Although the macroinvertebrates collected in this study were generally limited to large-bodied and mobile taxa owing to sampling methods meant for amphibians, the taxa reported here are the first and only documentation of invertebrates from this unique complex of ponds on the Mount St. Helens debris-avalanche deposit. The aquatic macroinvertebrates observed in these young ponds are already more diverse and functionally complex than those found in older volcanic lakes isolated on oceanic islands (e.g., Lake Wisdom and Lake Dakataua, Papua New Guinea) (Ball and Gluckman 1980). This diversity is likely due to the relatively close proximity of the Mount St. Helens ponds to other water bodies and sources of potential colonists, as compared with freshwater-limited islands. Although taxa richness was greater at Mount St. Helens, the general types of aquatic invertebrates observed here were similar to those found at other volcanoes (e.g., snails, dytiscid beetles, dragonflies, damselflies, backswimmers, water striders, chironomid larvae, and some caddisflies).

Invertebrate predators were proportionately the most abundant FFG collected from both seasonal and perennial ponds, and they increased in richness with increasing water permanence. Sampling methods undoubtedly influenced our results; however, in water bodies without fish, a robust invertebrate-predator assemblage is expected (Larson 1990; Wellborn et al. 1996; Tarr et al. 2005). Plus, many of the taxa identified here were also reported from 41 oligotrophic lakes in North Cascades National Park, Washington (Hoffman et al. 1996). Of the insect orders that are predominantly predators, we observed five of six Odonata families (all but Cordulegastridae), four of six Hemiptera families, the one Megaloptera family and genus, and four of five Coleoptera families (all but Chrysomelidae) that Hoffman and colleagues found using sweep nets, benthic cores, and handpicking of invertebrates. We also observed three of four gastropod families (all but Ancylidae), three of five Ephemeroptera families (all but Heptageniidae and Siphlonuridae), five of six Trichoptera families (all but Brachycentridae), and eight of 14 Diptera families found by Hoffman and colleagues. However, we did not collect any stoneflies, whereas Hoffman and colleagues found seven Plecoptera families. Surveys of lakes in Glacier National Park, Montana, in the Northern Rocky Mountains, yielded 5 families and 11 genera of stoneflies (Newell and Baumann 2013). Plecopteran nymphs are primarily invertebrate shredders or predators, and many have body sizes as large as or larger than individuals of other species we collected in the funnel traps. Although typically lotic, some plecopteran species can occur in lentic environments such as windswept shorelines that are similar in many respects to lotic habitats. Plecopterans may be present in Mount St. Helens ponds at low densities or in habitats not sampled. More thorough invertebrate and habitat-sampling efforts would be required to detect whether they are present.

Although other studies have reported decreased species richness with decreasing hydroperiod length (Collinson et al. 1995; Della Bella et al. 2005; Tarr et al. 2005; Silver et al. 2012), in our analysis at the sample level, we were not able to find a clear, significant relationship between taxa richness and hydroperiod when sampling effort was taken into account. However, we did observe lower rarified taxa-richness estimates and different community compositions in the shortest hydroperiod, spring-dry ponds, as compared with perennial ponds. Communities from perennial ponds had odonate, corixid, and snail indicator species, whereas communities from spring-dry ponds had dipterans and the dytiscid *Acilius abbreviatus* as indicator species. These results are similar to others who have observed odonates to be dominant predators in permanent waters and dytiscids, especially *Acilius* spp., to be more abundant in temporary habitats (Larson 1990; Tarr et al. 2005).

Hydroperiod, both in terms of seasonality and predictability, can affect species survival and population dynamics. For some species, like odonates, seasonal ponds could be population sinks, as many species require 1–2 years in water to mature (Merritt et al. 2008). However, we collected both dragonfly (*Aeshna*, *Libellula*) and damselfly (*Enallagma*, *Lestes*) nymphs in a few spring-dry ponds. These individuals may not have survived to metamorphosis, but at least some species of *Lestes* are known to successfully reproduce in temporary ponds (Westfall and May 2006). Aquatic invertebrates have evolved different ways to survive a dry phase to complete their life cycles, including burrowing, desiccation-resistant or diapausing eggs, residing under moist benthic material, and adult emergence before pond desiccation. The majority of the species observed in this study are capable of aerial dispersal as adults and many are considered strong fliers, especially the beetles, hemipterans, and dragonflies (Poff et al. 2006). Although not capable of active overland dispersal, aquatic snails are able to use water birds as passive dispersal vectors through internal (e.g., stomach) or external (e.g., legs, feathers, or bill) transport (van Leeuwen and van der Velde 2012; van Leeuwen et al. 2012). Therefore the distance between ponds was likely not as important to species establishment as the availability and synchronization of wet and dry habitats that coincide with species life-history requirements.

In general, differences in macroinvertebrate communities between the ponds appeared to be influenced more by pond size and canopy cover, rather than by hydroperiod alone. Larger ponds may have greater within-pond habitat diversity and may also be larger targets for discovery and colonization by immigrating species. Pond hydroperiod, size, and canopy cover are all somewhat correlated, as perennial ponds were generally large and deep, with low canopy cover, although this was not always the case. Some ponds were quite large in surface area but shallow and seasonal. Other ponds were small in area, but deep and perennial, and may or may not have been surrounded by riparian vegetation that shaded the pond.

13.6 Conclusions

Large-bodied macroinvertebrate communities from ponds on the Mount St. Helens debris-avalanche deposit were highly variable in spite of their overall close geographical proximity, shared regional species pool, and similar creation/disturbance history. The ponds comprised a broad range of hydroperiods and biophysical characteristics. Much of the variability in taxa presence may be due to seasonal drying, stochastic colonization events, or habitat differences, including factors not measured in this study. The vegetation on the debris-avalanche deposit during our study was at an early-successional stage and ranged from meadow to early deciduous-forest communities. As the plant communities become dominated by coniferous forest, the aquatic communities will likely change in response to environmental conditions. This study was the first to look at aquatic macroinvertebrates inhabiting these ponds; however, it is clear that to better understand their faunal assembly, we need macroinvertebrate community data from unbiased sampling methods and water-quality measurements of temperature, dissolved oxygen, and chemistry.

Acknowledgments We acknowledge the taxonomic skills and efforts provided by the National Aquatic Monitoring Center (Logan, UT) and specialists Dr. Luke Jacobus, Dr. Dennis Paulson, Doug Post, and Rich Van Driesche. We thank Hans Purdom, Eric Lund, Aimee McIntyre, and Juniper Tagliabue for assistance with field work and Rachel Chai for assistance with lab work. The Gifford Pinchot National Forest, Mount St. Helens National Volcanic Monument, provided access to the study area. This work was funded through the USDA Forest Service, Pacific Northwest Research Station.

Appendix

Appendix 13.1 Taxa collected from 97 ponds on the MSH debris-avalanche deposit from 2003 to 2005 via funnel traps in the littoral zone. Species in parentheses were identified from only a few reference individuals; therefore, other species are possibly present. Functional feeding groups (*FFG*) are collector-gatherers (CG), piercer-herbivores (PH), predators (PR), scrapers (SC), shredders (SH), and unknown (UN). The number of ponds each taxon was captured from is listed below. *CPUE* (individuals/trap) are the sum of the number of individuals, by life stage, per pond per year standardized by the number of traps.

				CPUE (indiv./trap)		
Class, order, family	**Subfamily, genus, and/or species**	**FFG**	**Ponds (no.)**	**Larvae**	**Pupae**	**Adults**
Insecta						
Coleoptera						
Dytiscidae	spp.	PR	30	6.4	0	2.3
	Acilius abbreviatus	PR	85	80.7	0	127.0
	Agabus spp. (*austinii, lutosus, seriatus*)	PR	79	15.0	0	76.9
	Boreonectes sp.	PR	21	0.5	0	8.4
	Boreonectes griseostriatus	PR	17	0	0	4.7
	Colymbetes spp.	PR	5	1.9	0	0
	Colymbetes densus	PR	34	0.3	0	15.7
	Colymbetes sculptilis	PR	1	0	0	0.2
	Coptotomus longulus	PR	18	0.9	0	7.7
	Dytiscus spp.	PR	45	30.7	0	0.3
	Dytiscus cordieri	PR	9	0	0	1.6
	Dytiscus hatchi	PR	1	0	0	0.3
	Dytiscus marginicollis	PR	20	0	0	4.5
	Graphoderus spp.	PR	5	1.3	0	0
	Graphoderus liberus	PR	7	0	0	1.5
	Graphoderus perplexus	PR	14	0	0	6.0
	Hygrotus spp.	PR	6	0	0	1.9
	Hygrotus sayi	PR	4	0	0	0.8
	Hydaticus aruspex	PR	3	0	0	0.7
	Hydroporus spp.	PR	8	0	0	1.8
	Ilybius spp.	PR	16	0	0	3.8
	Ilybius fraterculus	PR	1	0	0	0.3
	Ilybius quadrimaculatus	PR	16	0	0	5.4
	Laccophilus maculosus decipiens	PR	31	0.8	0	18.1
	Liodessus obscurellus	PR	1	0	0	0.1
	Neoporus undulatus	PR	14	0.1	0	6.9
	Rhantus spp.	PR	17	5.5	0	0.3
	Rhantus binotatus	PR	13	0	0	5.3
	Rhantus consimilis	PR	42	0	0	30.3

(continued)

Appendix 13.1 (continued)

Class, order, family	Subfamily, genus, and/or species	FFG	Ponds (no.)	CPUE (indiv./trap) Larvae	Pupae	Adults
Gyrinidae	*Gyrinus* spp.	PR	17	0.3	0	3.5
Haliplidae	*Haliplus* spp.	PH	15	0.4	0	2.5
	Peltodytes spp. *(callosus)*	PH	10	3.7	0	1.3
Hydrophilidae	spp.	PR	9	2.8	0	0
	Berosus spp.	PH	39	1.6	0	15.3
	Hydrochus spp.	SH	2	0	0	0.4
	Hydrobius spp.	CG	24	5.4	0	8.3
	Tropisternus spp. *(lateralis, limbalis)*	PR	44	2.0	0	34.5
Diptera						
	spp.	UN	6	0.3	0.9	0
Ceratopogonidae	spp.	PR	2	0.3	0.1	0
	Bezzia spp.	PR	3	0.3	0	0
	Chaoborus spp.	PR	49	42.0	23.9	1.5
Chaoboridae	spp.	CG	18	0.4	3.6	2.3
	Chironominae spp.	CG	52	93.1	0	0
	Orthocladiinae spp.	CG	11	3.0	0	0
	Tanypodinae spp.	PR	14	10.4	0	0
Culicidae	spp.	CG	3	0	6.8	0
	Aedes sp.	CG	2	3.0	0	0
	Anopheles sp.	CG	1	0.3	0	0
	Culiseta sp.	CG	4	17.0	0	0
	Culex sp.	CG	2	0.4	0	0
Dixidae	*Dixella* sp.	CG	6	2.0	0	0
Sciomyzidae	spp.	PR	2	0.3	0	0
Tabanidae	spp.	PR	1	0.1	0	0
Tipulidae	*Tipula* sp.	SH	1	0.1	0	0
Ephemeroptera						
	sp.	UN	1	0.3	0	0
Baetidae	*Callibaetis fluctuans*	CG	33	14.4	0	0.3
Caenidae	*Caenis youngi*	CG	7	3.2	0	0
Leptophlebiidae	*Paraleptophlebia* spp.	CG	2	0.5	0	0
Hemiptera						
Belostomatidae	*Lethocerus americanus*	PR	16	0.5	0	2.7
Corixidae	spp.	PH	66	80.5	0	9.7
	Callicorixa spp. *(vulnerata)*	PR	46	0.7	0	29.7
	Hesperocorixa spp. *(atopodonta, laevigata, michiganensis, vulgaris)*	PH	57	0	0	49.1
Gerridae	*Gerris* spp.	PR	17	3.0	0	1.7
Nepidae	*Ranatra fusca*	PR	11	0	0	2.7
Notonectidae	spp.	PR	43	140.4	0	0
	Buenoa spp.	PR	1	0	0	0.1
	Notonecta spp.	PR	55	63.6	0	16.8
	Notonecta kirbyi	PR	49	0.3	0	172.0
	Notonecta undulata	PR	26	0	0	23.0
Megaloptera						
Sialidae	*Sialis rotunda*	PR	2	0.4	0	0
Odonata						
	spp.	PR	2	0.1	0	0.1
Aeshnidae	spp.	PR	16	3.6	0	0
	Aeshna sp.	PR	46	22.0	0	0

(continued)

Appendix 13.1 (continued)

Class, order, family	Subfamily, genus, and/or species	FFG	Ponds (no.)	CPUE (indiv./trap) Larvae	Pupae	Adults
	Aeshna palmata	PR	5	0.9	0	0.1
	Anax junius	PR	3	0.5	0	0
Coenagrionidae	spp.	PR	29	25.6	0	0.1
	Enallagma spp.	PR	31	22.7	0	0
	Ischnura spp.	PR	35	38.1	0	0
	Ischnura cervula	PR	1	0	0	0.1
	Ischnura perparva	PR	1	0	0	0.1
Corduliidae	*Somatochlora minor*	PR	1	0.2	0	0
Lestidae	*Lestes* spp.	PR	44	65.6	0	0.3
	Lestes dryas	PR	1	0.2	0	0
Libellulidae	spp.	PR	29	13.9	0	0
	Libellula forensis	PR	10	2.1	0	0
	Sympetrum spp.	PR	27	16.8	0	0
Trichoptera						
Hydroptilidae	*Hydroptila* spp.	PH	1	0.1	0	0
Lepidostomatidae	*Lepidostoma* spp.	SH	2	0.5	0	0
	Lepidostoma cinereum	SH	1	0.3	0	0
	Lepidostoma unicolor group	SH	1	0.1	0	0
Leptoceridae	*Oecetis* spp.	PR	5	1.0	0	0
	Triaenodes spp.	SH	2	0.9	0	0
	Ylodes spp.	SH	1	0.1	0	0
Limnephilidae	spp.	SH	4	0.9	0	0.1
	Halesochila taylori	CG	22	9.7	0	0
	Limnephilus spp.	SH	25	8.5	0	0
Phryganeidae	*Banksiola crotchi*	SH	6	1.1	0	0.1
Polycentropodidae	*Polycentropus* spp.	PR	1	0.7	0	0
Gastropoda						
Pulmonata						
Lymnaeidae	spp.	SC	15	0	na	7.6
	Fossaria spp.	SC	4	0	na	3.7
	Stagnicola spp.	SC	6	0	na	3.9
	Radix auricularia	SC	14	0	na	8.3
Physidae	*Physa/Physella* spp.	SC	16	0	na	9.1
Planorbidae	spp.	SC	2	0	na	0.3
	Gyraulus spp.	SC	63	0	na	128.4
	Helisoma spp.	SC	1	0	na	0.3
	Helisoma anceps	SC	1	0	na	0.2
	Planorbella trivolvis	SC	1	0	na	0.2

References

Anderson, N.H. 1992. Influence of disturbance on insect communities in Pacific Northwest streams. *Hydrobiologia* 248: 79–92.

Anderson, N.H., and R.W. Wisseman. 1987. Recovery of the Trichoptera fauna near Mt. St. Helens five years after the 1980 eruption. In *Proceedings of the fifth international symposium on Trichoptera*, ed. M. Bournaud and H. Tachet, 367–373. Dordrecht: Dr W. Junk Publishers.

Ball, E., and J. Glucksman. 1978. Limnological studies of Lake Wisdom, a large New Guinea caldera lake with a simple fauna. *Freshwater Biology* 8: 455–468.

———. 1980. A limnological survey of Lake Dakataua, a large caldera lake on West New Britain, Papua New Guinea, with comparisons to Lake Wisdom, a younger nearby caldera lake. *Freshwater Biology* 10: 73–84.

Collier, K.J. 2002. Effects of flow regulation and sediment flushing on instream habitat and benthic invertebrates in a New Zealand river influenced by a volcanic eruption. *River Research and Applications* 18: 213–226.

Collinson, N.H., J. Biggs, A. Corfield, M.J. Hodson, D. Walker, M. Whitfield, and P.J. Williams. 1995. Temporary and permanent ponds: An assessment of the effects of drying out on the conservation value of aquatic macroinvertebrate communities. *Biological Conservation* 74: 125–133.

Colwell, R.K., A. Chao, N.J. Gotelli, S. Lin, C.X. Mao, R.L. Chazdon, and J.T. Longino. 2012. Models and estimators linking individual-based and sample-based rarefaction, extrapolation and comparison of assemblages. *Journal of Plant Ecology* 5: 3–21.

Crisafulli, C.M., L.S. Trippe, C.P. Hawkins, and J.A. MacMahon. 2005. Amphibian responses to the 1980 eruption of Mount St. Helens. In *Ecological responses to the 1980 eruption of Mount St. Helens*, ed. V.H. Dale, F.J. Swanson, and C.M. Crisafulli, 183–197. New York: Springer.

Dale, V.H., D.R. Campbell, W.M. Adams, C.M. Crisafulli, V.I. Dains, P.M. Frenzen, and R.F. Holland. 2005. Plant succession on the Mount St. Helens debris avalanche deposit. In *Ecological responses to the 1980 eruption of Mount St. Helens*, ed. V.H. Dale, F.J. Swanson, and C.M. Crisafulli, 59–73. New York: Springer.

Daubenmire, R. 1959. A canopy-coverage method of vegetational analysis. *Northwest Science* 33: 43–64.

Della Bella, V., M. Bazzanti, and F. Chiarotti. 2005. Macroinvertebrate diversity and conservation status of Mediterranean ponds in Italy: Water permanence and mesohabitat influence. *Aquatic Conservation: Marine and Freshwater Ecosystems* 15: 583–600.

Dorava, J.M., and A.M. Milner. 1999. Effects of recent volcanic eruptions on aquatic habitat in the Drift River, Alaska, USA: Implications at other Cook Inlet region volcanoes. *Environmental Management* 23: 217–230.

Dufrene, M., and P. Legendre. 1997. Species assemblages and indicator species: The need for a flexible asymmetrical approach. *Ecological Monographs* 67: 345–366.

Franklin, J.F., and C.T. Dyrness. 1988. *Natural vegetation of Oregon and Washington*. Corvallis: Oregon State University Press.

Glicken, H. 1996. *Rockslide-debris avalanche of May 18, 1980, Mount St. Helens volcano, Washington*, Open-File Report 96-677. Vancouver: U.S. Geological Survey.

Gotelli, N.J., and R.K. Colwell. 2001. Quantifying biodiversity: Procedures and pitfalls in the measurement and comparison of species richness. *Ecology Letters* 4: 379–391.

Hawkins, C.P. 1988. Effects of watershed vegetation and disturbance on invertebrate community structure in western Cascade streams: Implications for stream ecosystem theory. *Verhandlungen des Internationalen Vereinigung für Theoretische und Angewandte Limnologie* 23: 1167–1173.

Heino, J. 2000. Lentic macroinvertebrate assemblage structure along gradients in spatial heterogeneity, habitat size and water chemistry. *Hydrobiologia* 418: 229–242.

Hoffman, R.L., W.J. Liss, G.L. Larson, E.K. Deimling, and G.A. Lomnicky. 1996. Distribution of nearshore macroinvertebrates in lakes of the Northern Cascade Mountains, Washington, USA. *Archiv für Hydrobiologie* 136: 363–389.

Karlstrom, E.L. 1986. Amphibian recovery in the North Fork Toutle River debris avalanche area of Mount St. Helens. In *Mount St. Helens: Five years later*, ed. S.A.C. Keller, 334–344. Cheney: Eastern Washington University Press.

Larson, D.J. 1985. Structure in temperate predaceous diving beetle communities (Coleoptera: Dytiscidae). *Holarctic Ecology* 8: 18–32.

———. 1990. Odonate predation as a factor influencing dytiscid beetle distribution and community structure. *Quaestiones Entomologicae* 26: 151–162.

Larson, D.J., Y. Alarie, and R.E. Roughley. 2000. *Predaceous diving beetles (Coleoptera: Dytiscidae) of the Nearctic Region, with emphasis on the fauna of Canada and Alaska*. Ottawa: NRC Research Press.

Larson, D.W., J. Sweet, R.R. Petersen, and C.M. Crisafulli. 2006. Posteruption response of phytoplankton and zooplankton communities in Spirit Lake, Mount St. Helens, Washington. *Lake and Reservoir Management* 22: 273–292.

Liss, W.J., G.L. Larson, E. Deimling, L. Ganio, R. Gresswell, R. Hoffman, M. Kiss, G. Lomnicky, C.D. McIntire, R. Truitt, and T. Tyler. 1995. *Ecological effects of stocked trout in naturally fishless lakes, North Cascades National Park Service Complex, WA, USA*. Technical Report NPS/PNROSU/NRTR-95-03. U.S. Department of the Interior, National Park Service.

McCune, B., and J.B. Grace. 2002. *Analysis of ecological communities*. Gleneden Beach: MJM Software Design.

Merritt, R.W., K.W. Cummins, and M.B. Berg, eds. 2008. *An introduction to the aquatic insects of North America*. 4th ed. Iowa: Kendall Hunt Dubuque.

Meyerhoff, R.D. 1991. Post-eruption recovery and secondary production of grazing insects in two streams near Mt. St. Helens. Ph.D. dissertation. Corvallis: Oregon State University.

Mielke, P.W., Jr., and K.J. Berry. 2001. *Permutation methods: A distance function approach*. New York: Springer Series in Statistics.

Miserendino, M.L., M. Archangelsky, C. Brand, and L.B. Epele. 2012. Environmental changes and macroinvertebrate responses in Patagonian streams (Argentina) to ashfall from Chaiten Volcano (May 2008). *The Science of the Total Environment* 424: 202–212.

Natural Resources Conservation Service [NRCS]. 2013. Washington SNOTEL site Spirit Lake (777). National water and climate center, Natural resources conservation service, USDA. Accessed 5 May 2013. http://www.wcc.nrcs.usda.gov/nwcc/site?sitenum=777&state=wa.

Needham, J.G., M.J. Westfall Jr., and M.L. May. 2000. *Dragonflies of North American*. Gainesville: Scientific Publishers.

Newell, R.L., and R.W. Baumann. 2013. Studies on distribution and diversity of nearshore ephemeroptera and plecoptera in selected lakes of Glacier National Park, Montana. *Western North American Naturalist* 73: 230–236.

Poff, N.L., J.D. Olden, N.K.M. Vieira, D.S. Finn, M.P. Simmons, and B.C. Kondratieff. 2006. Functional trait niches of North American lotic insects: Traits-based ecological applications in light of phylogenetic relationships. *Journal of the North American Benthological Society* 25: 730–755.

Scharnberg, L.D. 1995. Zooplankton community structure in lakes near Mount St. Helens, Washington. Master's thesis. Portland: Portland State University.

Schilling, E.G., C.S. Loftin, and A.D. Huryn. 2009. Macroinvertebrates as indicators of fish absence in naturally fishless lakes. *Freshwater Biology* 54: 181–202.

Silver, C.A., S.M. Vamosi, and S.E. Bayley. 2012. Temporary and permanent wetland macroinvertebrate communities: Phylogenetic structure through time. *Acta Oecologica* 39: 1–10.

Stonedahl, G.M., and J.D. Lattin. 1982. *The Gerridae, or water striders, of Oregon and Washington (Hemiptera: Heteroptera)*, Agriculture Experiment Station, Technical Bulletin 144. Corvallis: Oregon State University Press.

———. 1986. *The Corixidae of Oregon and Washington (Hemiptera: Heteroptera)*, Agriculture Experiment Station, Technical Bulletin 150. Corvallis: Oregon State University Press.

Swanson, F.J., and J.J. Major. 2005. Physical events, environments, and geological-ecological interactions at Mount St. Helens—March 1980 to 2004. In *Ecological responses to the 1980 eruption of Mount St. Helens*, ed. V.H. Dale, F.J. Swanson, and C.M. Crisafulli, 27–44. New York: Springer.

Tarr, T.L., M.J. Baber, and K.J. Babbitt. 2005. Macroinvertebrate community structure across a wetland hydroperiod gradient in southern New Hampshire, USA. *Wetlands Ecology and Management* 13: 321–334.

Thornton, W.B., and T.R. New. 1988. Freshwater communities on the Krakatau Islands. *Philosophical Transactions of the Royal Society of London series B* 322: 487–492.

Turner, A.M., and J.C. Trexler. 1997. Sampling aquatic invertebrates from marshes: Evaluating the options. *Journal of the North American Benthological Society* 16: 694–709.

van Leeuwen, C.H.A., and G. van der Velde. 2012. Prerequisites for flying snails: External transport potential of aquatic snails by waterbirds. *Freshwater Science* 31: 963–972.

van Leeuwen, C.H.A., G. van der Velde, B. van Lith, and M. Klaassen. 2012. Experimental quantification of long distance dispersal potential of aquatic snails in the gut of migratory birds. *PLoS ONE* 7: e32292.

van Tol, J. 1990. Zoological expeditions to the Krakatau Island, 1984 and 1985: Odonata. *Tijdschrift voor Entomologie* 133: 273–279.

Verberk, W.C.E.P., H. Siepel, and H. Esselink. 2008. Applying life-history strategies for freshwater macroinvertebrates to lentic waters. *Freshwater Biology* 53: 1739–1753.

Voight, B., H. Glicken, R.J. Janda, and P.M. Douglas. 1981. Catastrophic rockslide avalanche of May 18. In *The 1980 Eruptions of Mount St. Helens, Washington*, Professional Paper 1250, ed. P.W. Lipman and D.R. Mullineaux, 347–377. Washington, DC: U.S. Geological Survey.

Ward, J.V. 1992. *Aquatic insect ecology: 1. Biology and habitat*. New York: Wiley.

Wellborn, G.A., D.K. Skelly, and E.E. Werner. 1996. Mechanisms creating structure across a freshwater habitat gradient. *Annual Review of Ecology and Systematics* 27: 337–363.

Westfall, M.J., Jr., and M.L. May. 2006. *Damselflies of North America*. Gainesville: Scientific Publishers.

Williams, D.D. 2006. *The biology of temporary waters*. New York: Oxford University Press.

Wolman, M.G. 1954. A method of sampling coarse river-bed material. *Transactions of the American Geophysical Union* 35: 951–956.

Yukawa, J., and S. Yamane. 1985. Odonata and Hemiptera collected from the Krakataus and surrounding islands, Indonesia. *Kontyu Toyo* 53: 690–698.

Characteristics of a New Rainbow Trout Population: Spirit Lake, Mount St. Helens Volcano, 2000–2015

14

Tara E. Blackman, Charles M. Crisafulli, and Shannon M. Claeson

14.1 Introduction

Intense large-scale natural disturbances such as deep-seated landslides, extreme floods, and volcanism have shaped freshwater fish communities in the Pacific Northwest for millennia. Extreme disturbance may result in evolution of aquatic habitats on new landforms or, in other cases, create habitats that retain legacies of the prior landscape. Yet, little is known about the immediate ecological responses of fish to these post-disturbance aquatic habitats. In the Pacific Northwest, volcanism is a pervasive agent of disturbance, and much of the region's physiography and associated biota have been shaped by Cascade Range eruptions. Mount St. Helens, Washington, is the most active stratovolcano in the Cascade chain, with at least 20 eruptions in the last 4000 years, including the most recent large eruption on 18 May 1980. This eruption was highly energetic, grossly transforming more than 570 km^2 of Cascadian forest, meadow, riverine, and lake habitats. Several 1000s km^2 were disturbed to a lesser extent by the deposition of wind-borne pumice and ash (Foxworthy and Hill 1982). The geophysical events during this eruption created multiple disturbance zones (Fig. 14.1) that collectively form a disturbance mosaic or gradient (Swanson and Major 2005). The resulting landscape provides a rare opportunity to observe responses of fish in a post-eruption landscape. Our objective was to observe long-term responses of ***adfluvial*** rainbow trout (*Oncorhynchus mykiss*) following their establishment in the post-eruption Spirit Lake basin, the water body nearest to Mount St. Helens.

Terms in ***bold italic face*** are defined in the glossary at the end of the chapter.

T.E. Blackman (✉) • C.M. Crisafulli
U.S. Department of Agriculture, Forest Service, Pacific Northwest Research Station, Mount St. Helens National Volcanic Monument, 42218 NE Yale Bridge Road, Amboy, WA 98601, USA
e-mail: tarablackman@gmail.com; ccrisafulli@fs.fed.us

S.M. Claeson
U.S. Department of Agriculture, Forest Service, Pacific Northwest Research Station, Wenatchee Forestry Sciences Laboratory, 1133 North Western Avenue, Wenatchee, WA 98801, USA
e-mail: sclaeson@fs.fed.us

14.1.1 Pre-eruption Fish at Mount St. Helens

Streams and rivers draining Mount St. Helens are part of the lower Columbia River system, the largest drainage in the Pacific Northwest. The lower Columbia River network has historically supported 37 of the 52 freshwater fishes found throughout the entire Columbia River basin, 12 of which exist only in that part of the network (McPhail and Lindsey 1986). Fish fauna in the three major drainage networks at Mount St. Helens, the Lewis River, Kalama River, and Cowlitz River, are represented by 8 families and 26 species (Bisson et al. 2005). This includes 11 species of salmonids, making this region among the most productive for ***anadromous*** fish in southern Washington (McPhail and Lindsey 1986). The distributions of these fish communities are determined primarily by natural barriers (e.g., waterfalls) that limited dispersal (Swanson et al. 2005) and gene flow (McPhail and Lindsey 1986).

Natural geologic barriers also influence fish access to lakes in the region. Prior to 1980, there were 37 lakes in the Mount St. Helens vicinity, most of which were probably devoid of fish prior to the arrival of European settlers in the late nineteenth century (Crawford 1986). The notable exception was Spirit Lake (Fig. 14.1), owing to its connectivity with the Toutle River and the lack of physical barriers to fish dispersal. As a result, Spirit Lake was accessed by sea-run coastal cutthroat (*O. clarkii clarkii*), winter steelhead (sea-run *O. mykiss*), and coho salmon (*O. kisutch*) (Bisson et al. 2005). Numerous other species that are present throughout the region may also have inhabited Spirit Lake, though no comprehensive survey records exist for the pre-1980 fish community. Additionally, Spirit Lake was inhabited by fish

C.M. Crisafulli, V.H. Dale (eds.), *Ecological Responses at Mount St. Helens: Revisited 35 years after the 1980 Eruption*,
https://doi.org/10.1007/978-1-4939-7451-1_14

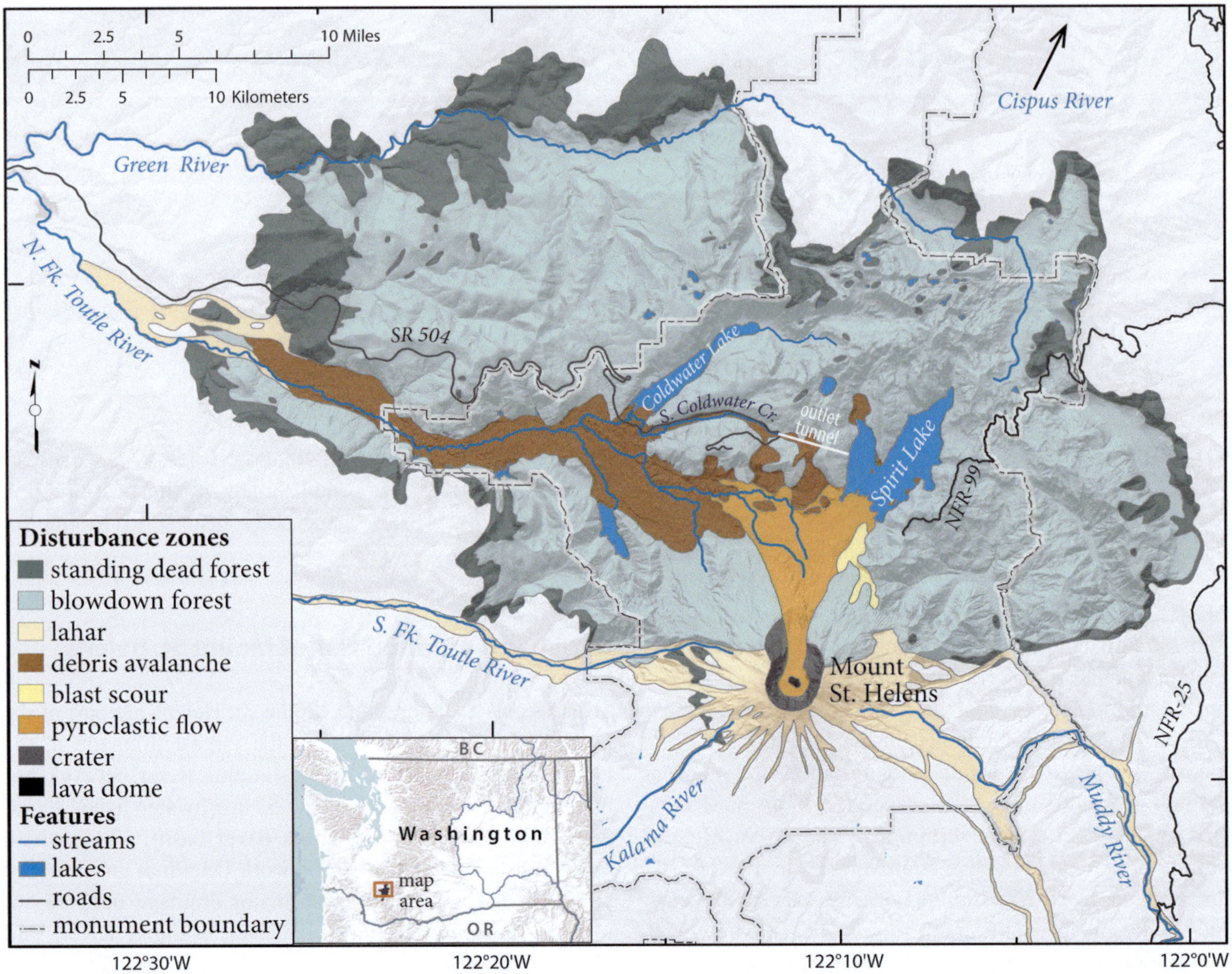

Fig. 14.1 Disturbance zones from the 1980 eruption of Mount St. Helens (Cartography by Kathryn Ronnenberg, USDA Forest Service, Pacific Northwest Research Station).

that arrived via stocking activities, a common practice among Mount St. Helens lakes in the early to mid-1900s. Rainbow trout (resident form of *O. mykiss*) were the first documented species to be stocked in Spirit Lake, beginning circa 1913 and continuing through 1979 (Lucas and Weinheimer 2003). Limited records from the Washington Department of Fish and Wildlife, Vancouver, document that summer-run steelhead, brook trout (*Salvelinus fontinalis*), and cutthroat trout stocking activities also occurred prior to the eruption.

14.1.2 Post-eruption Aquatic Environments

How aquatic environments were affected by the 1980 eruption depended on the type and intensity of disturbance they received. Several major tributaries of the lower Columbia River, including the North Fork Toutle and Cowlitz Rivers, were affected by ***lahars*** and subsequent flooding related to the debris-avalanche deposit (Swanson and Major 2005) that was created by the catastrophic collapse of the volcano's northern flank. The South Fork Toutle River also received a large, intense lahar spawned from melting glaciers on the volcano's western flank. The most severely affected was the Toutle River drainage, a tributary to the Cowlitz River, in which the entire main stem and the lower reaches of numerous tributaries were obliterated (Major et al., Chap. 2, this volume). The immediate outcome was likely a loss of all fishes throughout the main stem Toutle River and a reduction of fishes throughout some tributaries of the watershed (Bisson et al. 1988, 2005). Smaller lahars affected the Muddy River and Smith Creek, tributaries of the Lewis River, when pyroclastic flows melted snow and ice on the north and east flank of the volcano (Swanson and Major 2005; Lisle et al., Chap. 3, this volume). The Green River and the upper reaches of Clearwater Creek received lithics, pumice, and ash from the ***blast pyroclastic density current***, commonly referred to as the lateral blast and hereafter called the ***blast PDC***, which altered channel substrate and morphology (Bisson et al. 2005). Drainages in the tephrafall zone such as the Cispus River and Clear Creek received varying thicknesses of pumice and ash.

Lakes were disturbed at varying intensities ranging from a light dusting of tephra in distant locations to the total transformation of basin morphometry in the Spirit Lake basin. The primary disturbance to Spirit Lake was displacement of the lake's water and filling of the former basin by a gigantic debris avalanche generated by the collapse of the volcano's northern flank. The debris avalanche was succeeded by a blast PDC. When the avalanche entered the lake, it generated an enormous wave that sloshed water up the basin's steep walls, stripping off the forest and carrying back down into the elevated basin where it created a massive floating log mat on the lake's surface (Fig. 14.2a). In the aftermath, the optical, chemical, and biological properties of the water had been grossly transformed; this included diminished light transmission and water clarity, elevated water temperatures, and increased levels of nutrients and chemically reduced metals. The substantial input of organic materials caused proliferation of heterotrophic bacteria leading to anoxic conditions, resulting in a loss of all biota requiring oxygen (Dahm et al. 2005). Moreover, the gross nutrient enrichment increased productivity and transformed the lake from an ***ultra-oligotrophic*** system (Bortleson et al. 1976; Wissmar et al. 1982) typical of high lakes in the Cascades to a ***mesotrophic*** system. The 1980 eruption created conditions such that in-lake survival of fish was not possible and the lake was no longer accessible to fish from source populations downstream, as all water connectivity was blocked by the debris-avalanche deposit.

Studies of fish in the post-eruption Mount St. Helens aquatic landscape focused primarily on the responses of game fish (steelhead, coho salmon, and cutthroat trout) in the Toutle River (Stober et al. 1981; Martin et al. 1982; Redding and Schreck 1982; Lucas and Crawford 1986; Lucas and Pointer 1987; Bisson et al. 1988; Leider 1989; Olds 2002), probably because it was the most intensely disturbed watershed and represented an important fishery in the lower Columbia River Basin. Other work assessed the influence of in-stream habitat on survival and recolonization. Between 1981 and 1982, studies of nine streams in the debris-avalanche deposit found that high winter mortality of coho salmon was associated with unstable channel morphology,

Fig. 14.2 Photoplate of (**a**) the Spirit Lake basin and (**b**–**d**) tributaries in three disturbance zones. In Spirit Lake photo (**a**), note floating logs forming a large mat in the western portion of the lake. Photos: (**a**) Douglas Larson, (**b**–**d**) Tara Blackman.

high suspended-sediment loads, and insufficient cover as compared with undisturbed reference streams (Martin et al. 1986). In the blowdown-forest and tephrafall zones (see Fig. 14.1), survival of resident fish was attributed to ***refugia*** provided by headwater lakes covered by snow and ice during the 1980 eruption (Hawkins and Sedell 1990; Crisafulli and Hawkins 1998) and to tributaries with connectivity to the less disturbed tephrafall zone (Bisson et al. 2005).

Trout surveys in lakes within the blast area were also conducted in the post-eruption landscape. Between July 1980 and August 1985, 12 of 18 previously stocked lakes in the blast area contained surviving fish populations, attributed to protective snow and ice covering the lakes during the eruption (Crawford 1986). Char (*S. fontinalis* and *S. namaycush*) subsequently persisted in lakes for decades in the absence of stocking (Lucas and Weinheimer 2003), presumably because of their lentic-spawning characteristics. In contrast, stream-spawning rainbow trout and cutthroat trout perished as a result of initially lethal water quality and the absence of suitable spawning and rearing habitat (Bisson et al. 2005). In Spirit Lake, gillnetting in the years immediately following the eruption yielded no fish (Crawford 1986; Lucas and Weinheimer 2003), which was no surprise given the post-eruption conditions. Unexpectedly, rainbow trout were observed in Spirit Lake in 1993, likely the result of clandestine stocking, and have since maintained a self-sustaining population (Bisson et al. 2005). No other fish species are known to live in the lake as of 2015. Some of Spirit Lake's tributaries were opportunistically surveyed (visually and by electrofishing) between 1983 and 2005; fish were not observed (Lucas and Weinheimer 2003), although the extent and rigor of these surveys is unclear. In 2011, rainbow trout were visually observed by the authors in several streams on the southwest shoreline that drain the pyroclastic-flow zone. By 2012, we detected fish in six of the nine flowing tributaries. These observations coincided with major geomorphic and ecological changes to the terrestrial landscape in the basin (Fig. 14.2b–d), during which time drainage networks began to stabilize and incipient riparian vegetation had markedly increased. These observations, in part, provide the foundation for this study because lake-dwelling rainbow trout use streams for reproduction (Wydoski and Whitney 2003) and the availability of stream habitat may strongly influence the recruitment of new individuals and, by extension, may directly affect trout population characteristics of the lake.

In this study, we observed changes in the Spirit Lake rainbow trout population between 2000 and 2015 using hook-and-line and gillnet sampling to assess how individual length, mass, size-at-age, and population size have changed. We expected that population size would increase over time but individual lengths and mass would decline as a result of increased intraspecific competition. We were also interested in the diet of rainbow trout in Spirit Lake because prey type and relative availability provide additional information on shifts in population dynamics. Lastly, we inventoried stream habitat available to trout in the Spirit Lake basin and surveyed trout presence and abundance to provide a snapshot of stream use. We anticipated that presence and abundance of trout would be greater in reaches with more slow-water habitat, a characteristic of stable channel morphology.

14.2 Materials and Methods

14.2.1 Study Site

The Spirit Lake basin (0565687 E 5122921 N, zone 10T, datum WGS84) is located on the north side of Mount St. Helens. The morphometry of the lake was severely altered by the debris avalanche, which displaced the surface elevation roughly 60 m above its pre-eruption level of 975 m (masl), with a current surface area almost double that of the pre-1980 lake. Because Spirit Lake no longer had an outlet into the North Fork Toutle River drainage basin after the eruption, the Army Corps of Engineers started construction on a tunnel in 1985 to begin an artificial drawdown to manage the rising volume of the lake. The steep exit point into South Coldwater Creek (Fig. 14.1) creates an impassable fish barrier, as does the ~10-m vertical waterfall on Coldwater Creek just upstream from the confluence with North Fork Toutle River. Therefore, fish cannot enter Spirit Lake from any downstream locations.

The lake currently has a volume of 0.235 km^3, surface area of 10.33 km^2, and a maximum depth of 36 m. Much of the extensive coniferous forest surrounding the lake that was leveled during the eruption has for the past 36 years floated as a giant log mat covering roughly 25–35% of the lake's surface. Biogeochemical conditions changed rapidly over the first few post-eruption years, leading to improved water quality and establishment of species-rich phytoplankton and zooplankton communities by 1986 (Larson 1993; Dahm et al. 2005; Larson et al. 2006; Gawel et al., Chap. 4, this volume).

We sampled for fish primarily in Duck Bay, a long, shallow bay that extends southeast from the main lake body. Duck Bay is characterized by an extensive, shallow littoral zone dominated by dense macrophytes including water crowfoot (*Ranunculus aquatilis*), hornwort (*Ceratophyllum demersum*), and several pondweeds (*Potamogeton* spp.). Other sampling areas included Donnybrook and Bear Cove (Fig. 14.3); however, we sampled these areas much less frequently.

We surveyed all streams draining into Spirit Lake, encompassing three disturbance zones: pyroclastic flow, debris-avalanche deposit, and blowdown forest (Fig. 14.3). Inlet streams draining the pyroclastic-flow zone include Willow Springs, Clear Creek, Geothermal Creek, and Camp Creek. The pyroclastic-flow deposit is roughly 40 m thick and is composed of pumiceous gravels and cobbles that are highly

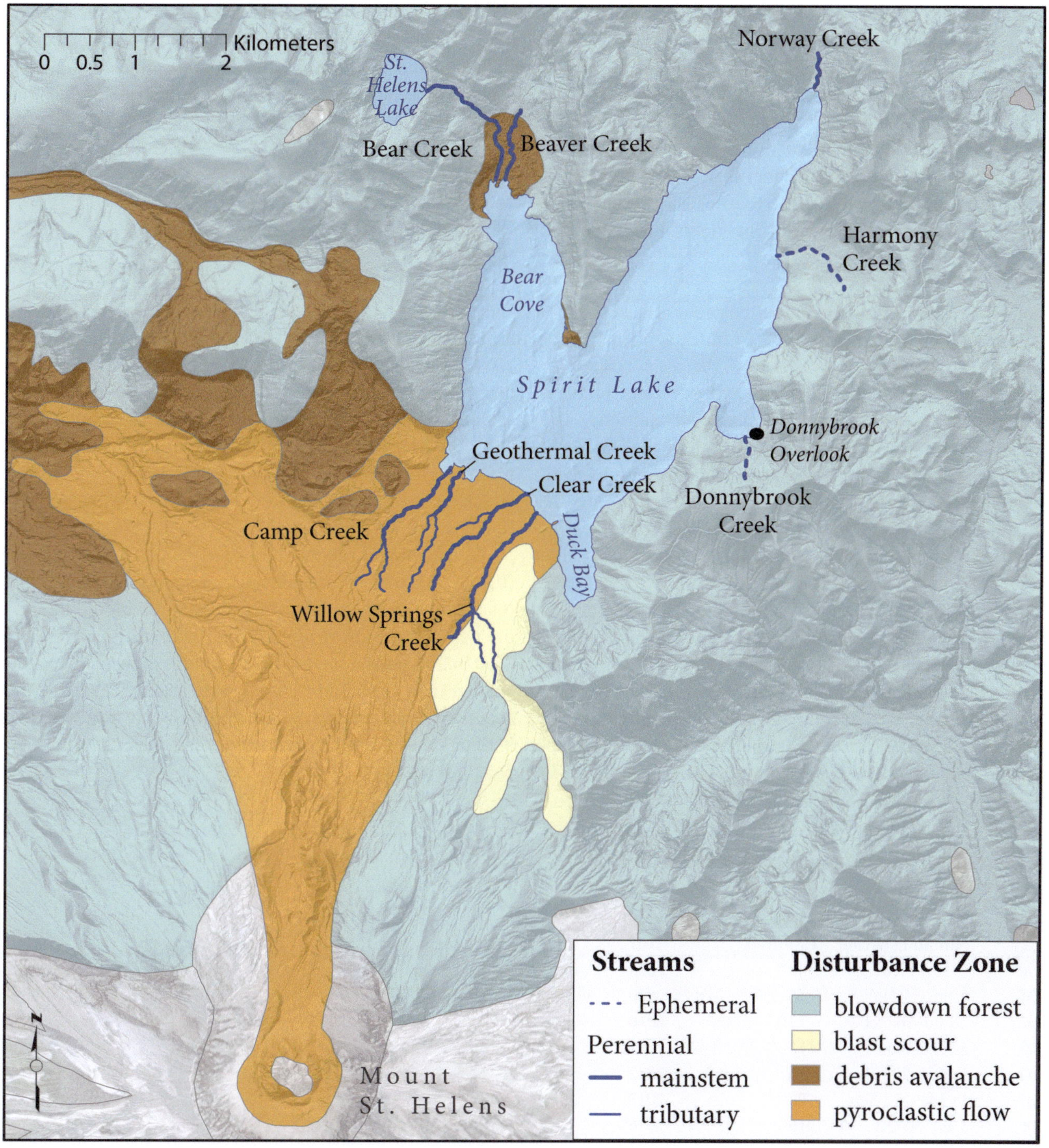

Fig. 14.3 Study sites in the Spirit Lake basin (Cartography by Kelly Christiansen, USDA Forest Service, Pacific Northwest Research Station).

susceptible to erosion. This material overlays very deep layers (several 10s m) of the volcano's former summit that were deposited as part of the debris avalanche, along with lithics from the blast PDC. Streams flowing through this substrate are characterized by chronic channel instability (e.g., high sediment transport, channel avulsions, seasonal dewatering); even low-intensity rain events can result in flows that entrain and transport sediment. Thirty-five years post eruption, the broader landscape on the pyroclastic flow is still in early stages of plant primary succession dominated by graminoids and forbs (del Moral and Titus, Chap. 7, this volume), although some stream sections have developed riparian communities of deciduous shrubs including Sitka willow (*Salix sitchensis*) and green alder (*Alnus viridis*).

Catchments in the debris-avalanche deposit include Beaver Creek and Bear Creek, which are characterized by steep hummocks composed of landslide material displaced from the former north flank and summit of Mount St. Helens. Adjacent tributaries flowing over the portion of the debris-avalanche deposit that traveled down the Toutle drainage were subjected to significant channel shifts, scour, and deposition from 1980 through 2000 (Swanson and Major 2005) and had annual suspended-sediment yields that exceeded 500 times the background level of typical Western Cascades rivers (Major et al. 2000). The streams in this study likely underwent similar initial post-eruption processes as channel development proceeded but stabilized much sooner. In contrast to streams on the pyroclastic flow, the upper reaches of debris-avalanche streams in the blowdown zone retained physical legacies of pre-eruption channel conditions. The riparian community is in the shrub/sapling successional stage, dominated by large willows and alder shrubs with some red alder trees (*A. rubra*). Additionally, there is significant American beaver (*Castor canadensis*) activity, particularly near the outlets into Spirit Lake, creating a large complex of interconnected pools.

The blowdown zone was created by the blast PDC, which toppled forest over 370 km^2 (Swanson and Major 2005). Norway Creek was the only perennial stream flowing into Spirit Lake in the blowdown zone during this study. This drainage is characterized by large woody debris in the stream channel and a well-developed riparian habitat in a shrub/sapling successional phase dominated by willow and alder. In contrast to both the debris-avalanche and pyroclastic-flow streams, which underwent significant channel development following the eruption, Norway Creek retained its original channel morphology and has embedded and stable substrate.

14.2.2 Trout Sampling

14.2.2.1 Spirit Lake Gillnet Sampling

We sampled trout with gillnets, generally between May and October (Table 14.1). We measured fish fork length (mm) and mass (g) and collected scale samples. Additionally, we took sagittal otoliths from 163 individuals during the 2012, 2014, and 2015 gillnet surveys. We selected 24 trout for determining both otolith-based ages and scale-based ages to evaluate accuracy of the two aging methods, with 96% concordance between them. We explored relationships between trout length and mass through simple linear regression, in which the estimated mean length (or mass) was modeled as a function of time (years).

We calculated relative biomass and catch per unit effort (CPUE) of trout captured in gillnet surveys by dividing the total mass (kg) or number of fish captured by the total time the net was deployed (hours) and the net's area (m^2). Nets used in 2015 had a section of 2.5-cm mesh, which was much smaller than mesh on nets used in previous years; the 2.5-cm section did not catch any fish. For these reasons, we did not include the 2.5-cm portion of the net in the CPUE calculations.

Table 14.1 Summary of sampling captures and dates for (a) angling and (b) gillnet methods.

(a) Angling

Year	Sampling period	Number captured	Year	Sampling period	Number captured
2000	7 Oct–3 Nov	48	2008	18 Jun–8 Oct	188
2001	4 Jun–17 Nov	16	2009	1 Jun–15 Sep	207
2002	28 Jul–26 Oct	73	2010	17 May–2 Nov	300
2003	5 Jul–8 Nov	115	2011	12 Jun–25 Oct	236
2004	22 Jun–22 Nov	46	2012	10 May–7 Sep	110
2005	11 May–12 Sep	88	2013	30 Jul–7 Sep	58
2006	1 Jul–30 Oct	216	2014	4 Jul–29 Jul	55
2007	7 May–6 Nov	242	2015	28 Jul–9 Sep	42

(b) Gillnet

Year	Number captured	Sampling dates	Duration (hours)	Net location (no. of nets)	Net dimensions (m)	Mesh size (cm)
2000	15	25–26 Jul	18.42	Duck Bay (2)	15.24 × 2.0	5, 10, 12.5
2006	26	25–26 Jul	20.5, 21.5	Duck Bay (2)	15.24 × 2.0	5, 10, 12.5
2012	27	18–19 Sep	22.83, 27.0	Duck Bay (1), Donnybrook (1)	15.24 × 2.0	5, 10, 12.5
2014	80	3–8 Sep*	18.0, 117.5[a]	Duck Bay (1), Bear Cove (1)	66.0 × 2.2	5.5, 10, 12.5
2015	59	9–10 Sep	23.72, 23.4	Duck Bay (2)	49.5 × 2.2	2.5[b], 5, 7.5, 10

[a]Net in Bear Cove was deployed from 3 to 8 September and the net in Duck Bay from 3 to 4 September.
[b]The 2.5-cm portion of net was excluded from analysis.

14.2.2.2 Spirit Lake Angling Surveys

We used barbless-hook lures in angling surveys for mark-recapture sampling (Table 14.1). We marked individual fish with uniquely numbered spaghetti tags, collected scale samples, and recorded fork length (mm) and mass (g) before release. Opportunistically, we took sagittal otoliths from 123 individuals that perished.

We determined population estimates for Spirit Lake using a closed-population, multiple capture-recapture calculation. We included any marked fish in the model for 3 consecutive years after their initial capture and not thereafter, as fish were never recaptured more than 3 years after initial capture. Although multiple sampling sessions took place in a year, we treated each year as an independent sampling period. Owing to restricted lake access, we used only captures from Duck Bay in the calculation, because a disproportionate number of the captures (>80%) occurred in this location, limiting the scope of inference to Duck Bay.

14.2.2.3 Stream Habitat and Trout Sampling

In order to quantify the amount and type of flowing water available to stream-dwelling fish, we surveyed stream habitat accessible to rainbow trout between 9 July and 9 August 2013. We began a survey at the mouth of each stream where it entered the lake and progressed upstream until we reached a significant fish barrier (i.e., subterranean flow or a steep waterfall). We identified all geomorphically and hydrologically distinct habitat units 1 m^2 or larger as one of three habitat classifications modified from Hawkins et al. (1993): "slow," which encompassed all nonturbulent flow (pools and runs); "intermediate," defined as having a gradient <4% (low-gradient riffles); and "fast," defined as gradients ≥4% (high-gradient riffles and cascades). We recorded the linear distance (m) and wetted channel width (m) for each habitat unit using a calibrated hip-chain and measuring tape. We measured specific conductivity (uS cm^{-1}) once in all streams using a multiparameter water quality instrument (YSI, model 650 MDS) at a single, mid-channel location.

We took reach-scale habitat measurements in selected 10-m-long reaches in each study stream. We subdivided the entire length of each stream into 10-m-long reaches that were characterized by their dominant habitat type based on the classification described above. In total, there were ~1500 reaches, with the majority in the pyroclastic-flow zone. We selected a stratified random sample of 63 reaches so that the frequency of sample reach types was roughly proportional to the occurrence of those reach types within each disturbance zone (Blackman 2014). In a subsample of reaches, we measured the wetted width (m) of the channel perpendicular to stream flow at a single cross section of the channel. We measured depth (m) in the ***thalweg*** at a single representative point in each reach. Using modified Daubenmire (1959) classes (0, 1–5, 6–25, 26–50, 50–75, >75), we estimated percent of the channel covered by riparian vegetation. We measured wood in the active channel in three dimensions, length, width, and height/diameter, and calculated volume (m^3) using formulas for paraboloid, cylindrical, and neiloid shapes:

$$V_{\text{neiloid}} = (h/4)\left(\pi \cdot r_1^2\right) + \left(r_1^2 + r_2^{1/3} + \left(r_1 \cdot r_2^2\right)^{1/3} + r_2\right)$$

$$V_{\text{paraboloid}} = (\pi \cdot h/2)\left(r_1^2 + r_2^2\right)$$

$$V_{\text{cylinder}} = h \cdot \left(\pi \cdot r_1^2\right)$$

where h = *height* and r = *radius*.

We captured rainbow trout from the selected stream reaches (described above) using three-pass electrofishing. To confine fish, we placed block nets simultaneously on the downstream and upstream ends of reaches. We electrofished each reach with a backpack unit (Smith-Root LR-24), beginning at the downstream end of the reach and ending upstream. We euthanized all captured fish using MS-222 (Argent Labs *Finquel*), weighed them to the nearest 0.1 g, and measured their fork length to the nearest 1.0 mm. We collected sagittal otoliths to determine the precise age for the young-of-year fish from each stream by counting daily growth rings between the emergence check and the date of capture, a common benchmark used for precise age determination (Campana and Neilson 1985).

From the three-pass electrofishing surveys, we estimated the number of fish within each sample reach using the depletions/removal estimates vignette (Ogle 2013). This method assumes a closed population with no mortality, recruitment, immigration, or emigration. We estimated fish densities in a stream by dividing summed fish estimates by total sampled area. For stream reaches where fish were present, we fitted the relationship between slow-water habitat and the estimated number of fish to a negative binomial model that included predictor variables for disturbance zone, proportion of slow-water habitat, and an interaction term.

14.2.2.4 Trout Diet

Between May and November from 2000 through 2006, we collected 111 rainbow trout between ages 2+ and 4+ (Table 14.2) with considerable variation in length (195–665 mm, Fig. 14.4) and mass (108–2800 g) from Spirit Lake. Upon reaching shore, we euthanized these fish and removed and preserved their gastrointestinal tracts in 90% ethanol. In the laboratory, we opened fish stomachs and flushed all contents into petri dishes to be identified to the lowest taxonomic level possible. We could identify amphibians, zooplankton, and numerous insects to species level, but for several insect and snail taxa identifications, we were limited to genus or family level owing to partial decomposition and could identify

Table 14.2 Number of fish sampled for dietary analyses by month for years 2000 through 2006, except 2001.

	Year					
Month	**2000**	**2002**	**2003**	**2004**	**2005**	**2006**
May			5			
June					1	
July			6	2	5	38[a]
August		2	2	3	5	
September			3	1	2	3
October	17	4				5
November	3		3			

[a]26 fish were captured by gillnetting in July 2006.

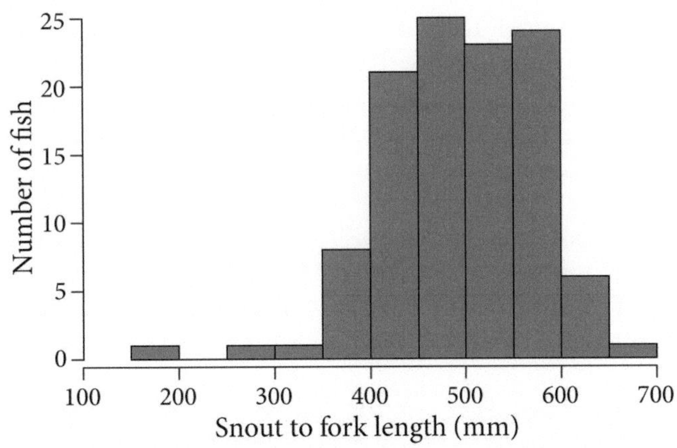

Fig. 14.4 Length distribution of fish used in dietary analysis.

most noninsect invertebrates only to class (e.g., Oligochaeta). All prey items were counted, dried, and either weighed the post-digestion dry mass (snails and salamanders) or estimated the predigestion dry mass (invertebrates) using length-mass regressions of the form $M = aL^b$ where M is dry mass (mg), L is length (mm), and a and b are fitted parameters. Based on the dominance of prey items taken by fish in Spirit Lake, we created a subset of prey categories by lumping similar prey types into groups that included aquatic invertebrates, aquatic snails, and terrestrial invertebrates. We converted the estimated pre- or post-digestion dry mass for each prey category to a percentage of the individual's overall diet and summarized the mean percentage for all fish by year and month.

14.3 Results

14.3.1 Individual Length, Mass, and Size-at-Age

Both length and biomass of angled lake fish declined sharply between 2000 and 2015, especially between 2001 and 2006 (Fig. 14.5a, c). We estimate that mean individual trout biomass declined 77.6 g per year (95% CI from 72.5 to 82.8 g) and the total percent change in mean biomass was 56% over the 16-year period of record. We also estimate that mean individual fish length declined 8.6 mm per year (95% CI from 7.7 to 9.4 mm), a 22.3% change over the period of record. Gillnet-captured trout revealed similar declining trends in length and biomass (Fig. 14.5b, d). In both angling and gillnet surveys, we detected fish less than 300 mm in length only after 2005.

Size-at-age showed declining trends for individuals age 1–4 years from 2000 through 2014 (Fig. 14.6). The age structure of fish caught using both angling and gillnets was composed primarily of fish age 1–4 years, with over 50% age 3+, and no fish older than age 5+ were observed.

14.3.2 Population Structure

Angling mark-recapture data (Table 14.3) reveal a declining trend in trout population size beginning in 2011. Estimates generated from angling surveys contradict the overall increasing trend in catch per unit effort (CPUE) observed in gillnet surveys between 2006 and 2015 (Fig. 14.7a). Gillnet data show a slight decline in CPUE between 2006 and 2012 and almost a twofold increase between 2012 and 2014 surveys. There also appears to be a slight size bias between sampling methods—smaller trout were proportionally more abundant in gillnets than in angling surveys (see Fig. 14.5c, d).

The relative trout biomass in gillnet samples varied over the 16-year study period (Fig. 14.7b). Biomass remained relatively static between 2000 and 2006 (ranging from 17.57 to 20.77 g h^{-1} m^{-2}), greatly fluctuated between 2012 and 2014, and returned to values ranging from 18.28 to 19.00 g h^{-1} m^{-2} in 2015.

The 2014 gillnet in Bear Cove was set for 4 consecutive days because the drifting log mat prevented us from accessing the net. Trout captured in this net were omitted from abundance and biomass calculations because mass was compromised by decomposition and catch per unit effort was confounded by changes in the capture efficacy after trout became crowded and putrid.

14.3.3 Diet

A total of 65 prey taxa were identified that included 54 insect taxa (77% aquatic and 23% terrestrial), plus Ostracoda (seed shrimp), Amphipoda (*Hyalella*), Oligochaeta (segmented worms), Nematoda (roundworms), Acari (mites), Arachnida (spiders), Cladocera (*Daphnia pulicaria*), Gastropoda

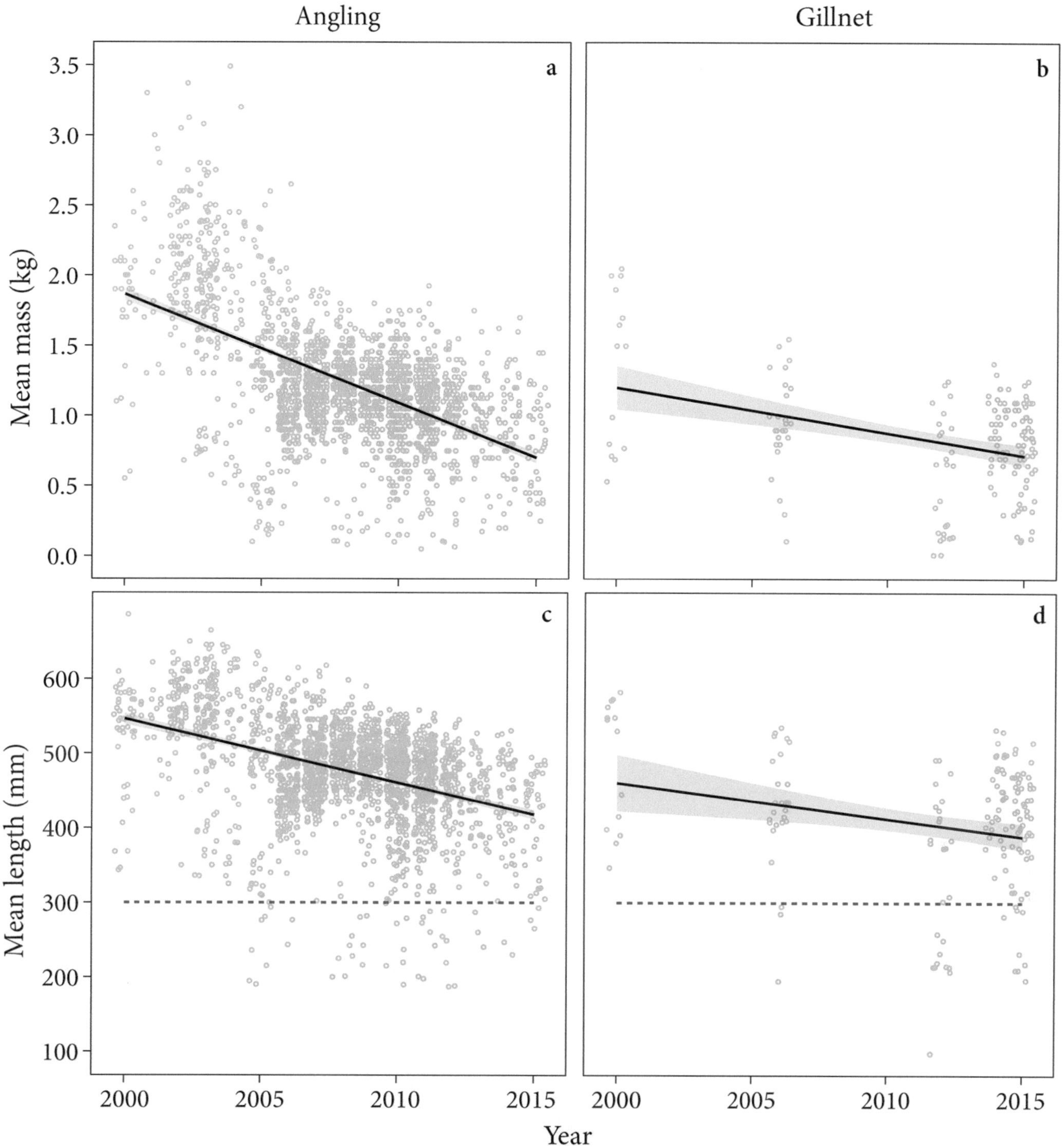

Fig. 14.5 Estimated change in rainbow trout (**a–b**) mass (kg) and (**c–d**) length (mm) captured in angling (**a**, **c**) and gillnet (**b**, **d**) surveys from 2000 to 2015 with 95% confidence band (*gray shading*). *Gray dots* are individual observations. *Dashed line* shows fish <300 mm.

(snails; *Gyraulus deflectus*, *Lymnaea humilis*, *L. auricularia*), and the northwestern salamander (*Ambystoma gracile*). Based on mean percentage dry mass of prey items in May through November samples from 2000 through 2006, aquatic insects and aquatic snails made up the vast majority of the rainbow trout diet each year (Fig. 14.8a). More specifically, snails, chironomid midges, and mayflies (*Callibaetis ferrugineus*) alone accounted for between 70% and 96% of the mean dry mass of prey items. Our analyses of monthly mean percentage of prey types show a decline in the consumption of aquatic insects from May through November and a corresponding increase in the consumption

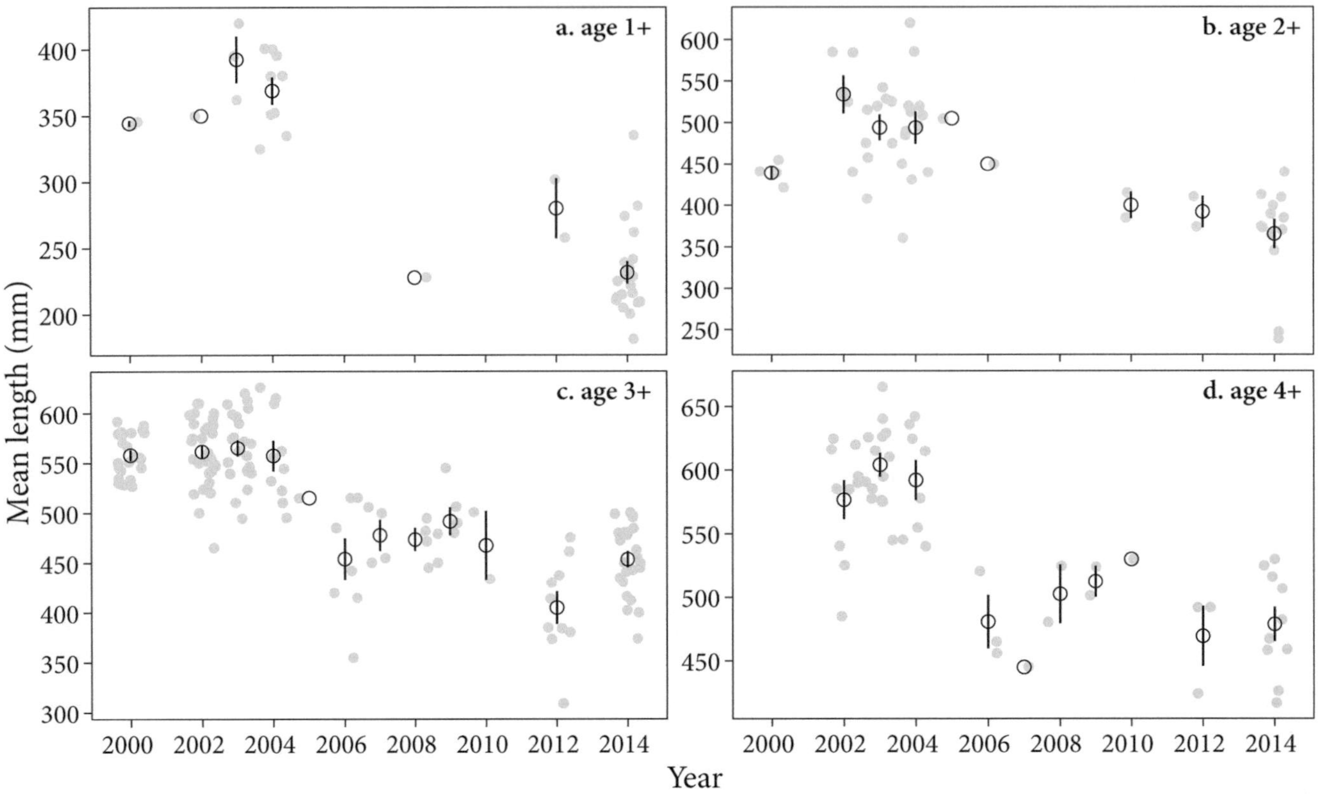

Fig. 14.6 Annual mean size-at-age of rainbow trout (ages 1–4 are shown in graphs **a**–**d**, respectively). *Open circles* are annual mean; *gray dots* are individual observations. *Bars* are standard error on the mean.

Table 14.3 Population estimates from angling surveys using the Schnabel mark-recapture method. n^* = number of marked fish for 3-year period.

Year	Population estimate	95% confidence interval Lower	Upper	n^*
2009	14,963	8682	27,661	506
2010	14,193	9255	22,820	590
2011	12,061	8151	18,579	658
2012	11,024	7327	17,324	721
2013	8262	5331	13,454	631
2014	5356	3108	9902	395

of aquatic snails during that period (Fig. 14.8b). There was also roughly uniform consumption of terrestrial insects from spring to late summer, with an increase in September (primarily flying ants) followed by a sharp decline in October and November.

14.3.4 Stream Habitat and Use

Young-of-year and age 1 fish were present in all study streams except Willow Springs, where no fish were detected. Age 2+ fish were rare among all streams, and no fish older than age 2+ were observed. Bear Creek, Clear Creek, and Norway Creek had the highest densities of young-of-year, and Bear Creek had the highest density of age 1+ fish (Table 14.4).

When fish were present in streams, the relationship between estimated age 1+ fish abundance and the proportion of slow-water habitat differed among zones (drop in deviance F-test for different slopes, $p = 0.02$). The predicted relationship between the mean estimated number of age 1+ fish and the proportion of slow-water habitat had a positive relationship in stream reaches flowing through the debris-avalanche deposit, whereas reaches in blowdown and pyroclastic-flow streams show no clear relationship (Fig. 14.9). There was no evidence that the relationship between mean abundance of young-of-year fish and the proportion of slow-water habitat varied between streams located in different disturbance zones (drop in deviance F-test for different slopes, $p = 0.24$).

Stream habitat was most abundant on the pyroclastic-flow zone, which accounts for more than 75% of the stream habitat available to fish in the entire Spirit Lake basin. Further, habitat structure was much more uniform in pyroclastic-flow streams compared with debris-avalanche and blowdown streams (Fig. 14.10). Pyroclastic-flow streams had long unbroken reaches of low-gradient riffle habitat,

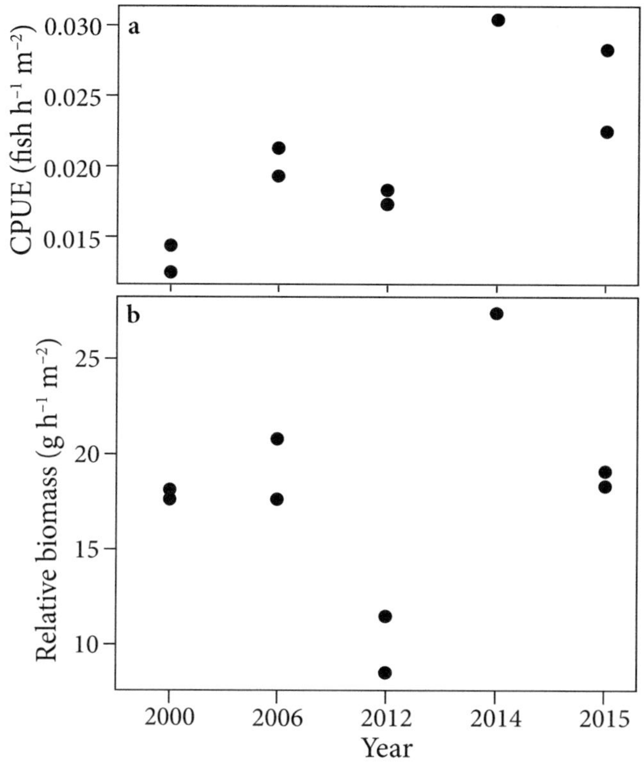

Fig. 14.7 Gillnet (**a**) catch per unit effort (CPUE) and (**b**) relative biomass of trout captured from 2000 to 2015.

those on the debris avalanche were dominated by pool habitat, and the single blowdown stream was a relatively uniform sequence of high-gradient riffles interrupted by pools. Several other biophysical habitat features were similar across debris-avalanche and blowdown streams but were strikingly different from pyroclastic-flow streams (Table 14.4). Riparian plant cover and volume of large woody debris were generally lowest in pyroclastic-flow streams and greatest in debris-avalanche and blowdown streams. The distance between pools was greatest in pyroclastic-flow streams compared with other zones. Specific conductance values were very high (88–520 μS cm^{-1}) in pyroclastic-flow streams and similar (28–36 μS cm^{-1}) among all other streams (Table 14.4). Average stream depths were shallow (8–17 cm) in all but Bear Creek, which was approximately two to four times deeper. Overall, streams on the pyroclastic flow had less riparian cover and less in-stream wood compared with debris-avalanche and blowdown zone streams.

14.4 Discussion

14.4.1 Population Dynamics

Rainbow trout in Spirit Lake appear to be increasing in abundance and experiencing a decline in size-at-age. We consider the overall increase in gillnet CPUE to be a more accurate measure of population growth than angling surveys, which were limited by the very low recapture rates (<10%), variation in marking effort, and potential size bias that excluded smaller fish. The shift in population structure from fewer large individuals to numerous smaller individuals does not appear to have changed the overall standing crop of fish between 2000 and 2015, although it does fluctuate from year to year. This phenomenon suggests that the population is constrained by the availability of resources. Changes in abundance may reflect the increased suitability of stream habitat for both spawning and rearing as channels became more stable. The result could be both an increase in the number of young produced and the amount of time those individuals spend in the stream. Stream-dwelling residents grow much slower than adfluvial rainbow trout (Wydoski and Whitney 2003); prolonged stream rearing would therefore likely shift the size-at-age of fish entering the lake. Thus, the capacity of streams to support fish may mediate declines in size-at-age both indirectly by increasing competition in Spirit Lake and directly through age at emigration to the lake.

14.4.2 Stream Habitats

Densities of fish were highly variable both among and within disturbance zones; however, there was no clear relationship between slow-water habitat and fish abundance in the pyroclastic-flow and blowdown streams. The spatial distributions of habitat types in stream reaches were highly variable among the zones, ranging from completely uniform riffles on the pyroclastic flow to repeating sequences of pools and riffles in the blowdown zone, which may have hindered our ability to relate fish abundance to discrete habitat types. For example, in the pyroclastic-flow zone, slow-water habitat was so scarce that fish likely occupy all available space. Conversely, pool habitat in the blowdown stream occurred with such uniform frequency that sample reaches almost always included or were immediately adjacent to pool habitat, which likely explains the uniform distribution of fish throughout the stream. The positive relationship between age 1+ fish abundance and slow-water habitat in streams on the debris-avalanche deposit may be explained simply by the much larger amount of slow-water habitat relative to other zones, providing a more adequate sample of discrete slow-water reaches.

Overall, streams in the pyroclastic-flow zone had less complex habitat and a high degree of variability in fish demographics. Among these streams, Clear Creek had the highest abundance of young-of-year trout. Clear Creek also had more in-stream wood, resulting from the artificial drawdown of Spirit Lake in the mid-1990s; these large old-growth logs were observed to increase pool frequency in lower stream reaches. More large woody debris is associated with higher overwinter survival of juvenile rainbow trout (Bisson

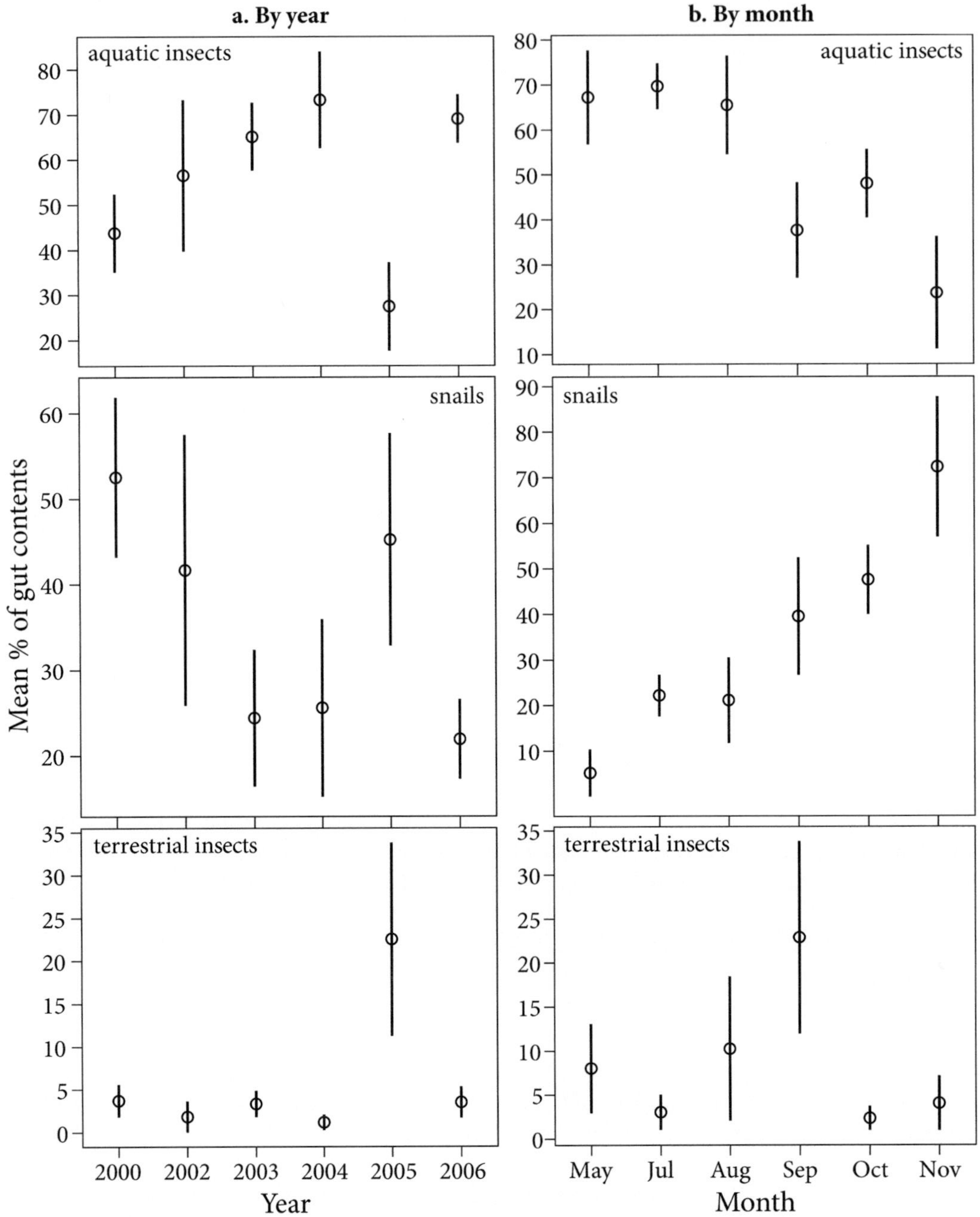

Fig. 14.8 Gut content of rainbow trout shown as the mean percentage of aquatic invertebrates, aquatic snails, and terrestrial invertebrates by (**a**) year and (**b**) month. Bars show standard error of the mean with June omitted owing to sample size: $n = 1$.

et al. 1987; Roni and Quinn 2001) and favorably influences habitat structure and food availability. Milner et al. (2000) also found that juvenile fish abundance and diversity in recently de-glaciated streams in Alaska at various stages of development were positively associated with in-stream wood and abundant pool habitat, which is consistent with our results.

Initial disturbance types have played a large role in the successional trajectory of the Spirit Lake basin landscape and its streams. We believe that the hydrogeologic features

Table 14.4 Habitat parameter and fish estimates for streams in the Spirit Lake basin.

Streams by disturbance zone	Water source	Reach depth (m) ± SE[a]	Riparian cover[a]	Nearest pool (m) ± SE[a]	Volume LWD (m^3) ± SE[a]	Specific conductivity μS cm^{-1}	No. of sampled reaches	Estimated young-of-year/m^2	Estimated age 1+/m^2
Pyroclastic flow									
Clear Creek	Spring	0.09 ± 0.01	1–5%	108.3 ± 39.2	1.94 ± 0.81	333	19	0.73	0.08
Camp Creek	Glacial melt	0.16 ± 0.05	0%	131.8 ± 48.5	0.40 ± 0.31	239	8	0.12	0.01
Geothermal Creek	Glacial melt	0.12 ± 0.01	1–5%	34.2 ± 7.0	0.00	520	17	0.28	0.07
Willow Springs	Spring/glacial melt	0.08 ± 0.01	0%	NA[b]	0.00	88	8	0	0
Debris avalanche									
Bear Creek	Lake	0.36 ± 0.08	50–75%	1.0 ± 1.0	3.67 ± 3.47	30	5	0.76	0.17
Bear Ponds[c]	Lake	–	–	–	–	–	8	–	–
Beaver Creek	Headwall seeps	0.14 ± 0.02	6–25%	0.8 ± 0.8	4.07 ± 4.07	28	4	0.13	0.03
Blowdown forest									
Norway Creek	Headwall seeps	0.17 ± 0.02	6–25%	1.8 ± 0.55	3.50 ± 1.62	36	10	0.61	0.09

[a]Averages calculated from all sampled reaches in stream ± standard error of the mean.
[b]No pools present.
[c]Ponds in Bear Creek were surveyed using different methods than streams.

of individual streams may also have contributed to the variation we observed in fish abundance. For example, water sources were variable among streams, and those having either lake or spring-fed headwater sources did not experience significant channel shifting over the course of the study, with the exception of Willow Springs, which had mixed headwater sources. Streams fed solely or partially by glacial melting (Geothermal Creek, Camp Creek, and Willow Springs Creek) seemed to be associated with less riparian cover and lower estimated trout abundance. This result is consistent with the findings of Milner et al. (2000, 2008) that streams on early-successional landscapes in Alaska with lakes as headwater sources possessed greater channel stability and subsequent riparian development and pool habitat than streams fed by snowmelt, rain, and glacial melt. We believe that streams fed by glacial melt have increased susceptibility to flooding and dewatering events, creating channel instability that may influence fish abundance. Moreover, instability may be compounded by channels flowing through unconsolidated pyroclastic-flow deposits from which sediment is easily mobilized during heavy precipitation events. Reproductive success and juvenile fish survival in unstable stream channels may be lower during years of high flow, which can physically scour redds (aggregated fish eggs deposited by adults in streambed substrates), remove smaller fish, and potentially abrade breathing structures (gills) when suspended-sediment loads are high. High flow can also physiologically increase the energy required to maintain position in the water column.

In previous studies conducted in portions of the debris-avalanche and blowdown zones that are a considerable distance from the volcano's vent, fish responses to the 1980 eruption of Mount St. Helens demonstrated a trend of fish recovery within 3 years, attributed to stream stabilization and development of riparian habitat (Bisson et al. 2005). In contrast to those studies, we believe the pace of recovery in stream habitat in the Spirit Lake basin has progressed at a much slower pace because of the high intensity of the initial disturbance. No studies were done on stream development in the pyroclastic-flow zone in the decades following initial disturbance, but we believe all four pyroclastic-flow streams in this study remain unstable because of the unconsolidated nature of their substrates. During our stream survey, three of the four pyroclastic-flow streams shifted portions of their channel, dewatered in upper reaches, and were scoured by heavy sediment loads. We observed no changes related to secondary disturbance in streams on the debris-avalanche deposit or blowdown zones during our relatively short-term survey.

As an observational case-study, the scope of inference for our stream surveys is limited to the Spirit Lake basin for the period of record. Because we sampled for only a single season (summer 2013), we cannot discount the possible contribution of streams where we observed low or no fish abundance to overall recruitment in the population. Stream

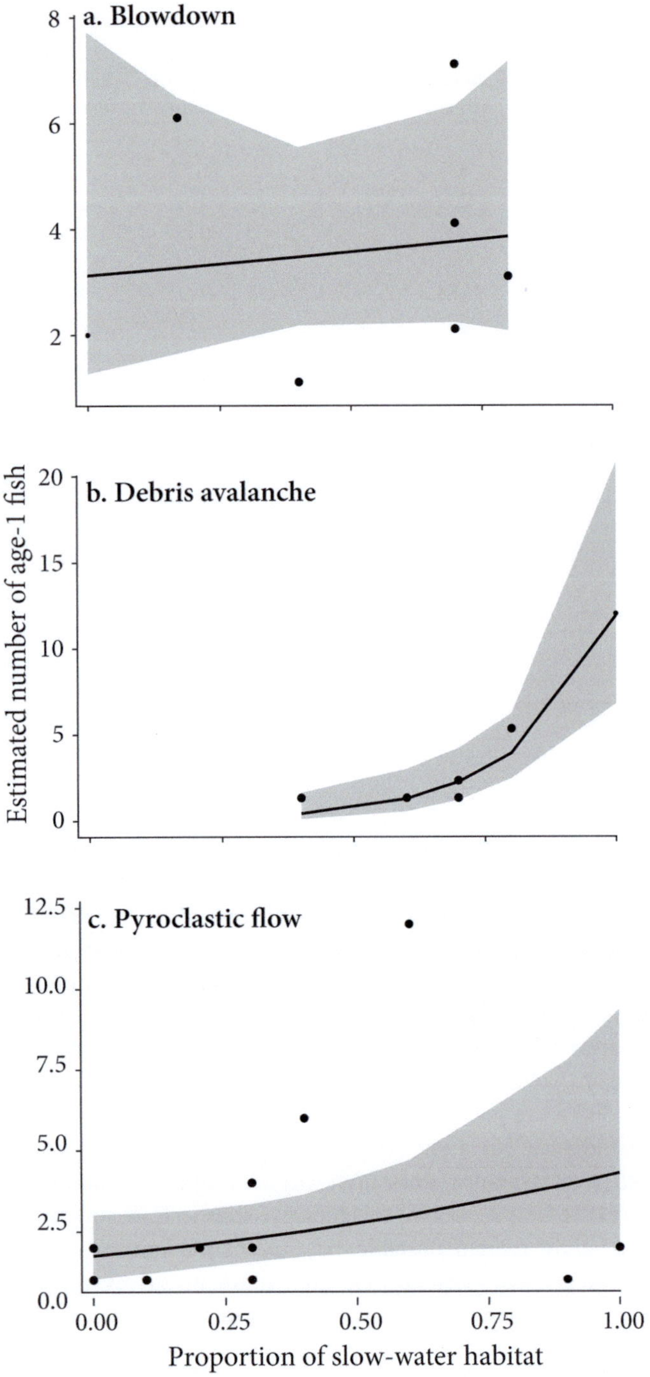

Fig. 14.9 The estimated relationship between the predicted number of age 1 fish and proportion of slow-water habitat in each disturbance zone. Lines represent the predicted relationship from a negative binomial regression model, with a 95% confidence interval (*gray shading*). *Black dots* are observed values for sampled reaches.

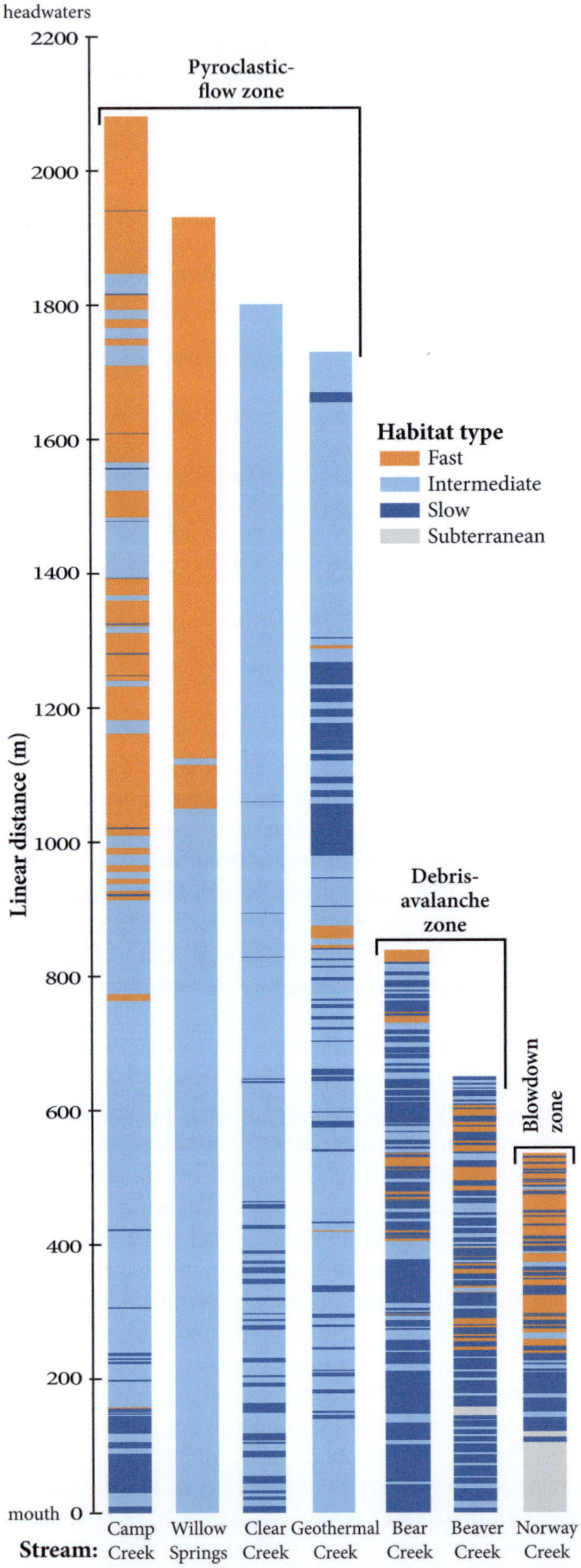

Fig. 14.10 Structure of habitat units along the linear distance of streams grouped by disturbance zone.

habitats can sometimes promote egg development but preclude or limit rearing (Quinn 2005). Streams on the pyroclastic flow were dominated by low-gradient riffle habitat, providing more than three times the spawning habitat of other zones yet having the lowest fish abundances. Conversely, despite less spawning habitat, fish densities were high in debris-avalanche and blowdown streams where pool habitat was more frequent and abundant. Shifts in habitat use between good spawning and good rearing habitat were demonstrated by Pavlov et al. (2008); in their study, a smaller stream with less complex habitat prompted migration (anadromy), and a larger stream with more complex habitat allowed for residency. Thus, stream conditions may in fact be conducive to different life-history strategies (e.g., adfluvial versus resident and different age-at-emigration), although it is beyond the scope of this study to distinguish life-history variants in the population.

14.4.3 Diet

Rainbow trout diet in Spirit Lake consisted of numerous aquatic and terrestrial prey items but was overwhelmingly dominated by just a few taxa. Chironomid midge pupa, *Callibaetis ferrugineus* mayfly nymphs, and aquatic snails contributed 70–96% of the diet. Other aquatic and terrestrial prey were presumably taken when they became locally or ephemerally abundant. For example, in late summer, winged ants (Formicidae) and terrestrial beetles (Coleoptera) contributed up to 20% of trout diet in some years. Rainbow trout are known to be generalist consumers and consume a wide range of prey items that varies by location throughout Washington state (Wydoski and Whitney 2003). At some sites, zooplankton (e.g., *Daphnia* spp.) may constitute a large percentage of prey items, whereas at others prey fish are important. Interestingly, zooplankton were rarely found in our diet samples, and fish were never observed. Overall, the size of prey taken by Spirit Lake trout was small, with the dominant snail and midge pupa measuring a few to 10 mm. Given the large size of most fish in our study, tremendous numbers of these small prey items must be consumed to support such growth in the absence of piscivory. Our data show that rainbow trout are gathering their prey from several different substrates within the nearshore environment that we sampled. For example, dozens to hundreds of snails, often admixed with large masses of aquatic vegetation (e.g., *Ranunculus aquatilis*), were found in the gut contents of fish, evidence that the trout were presumably swallowing vegetation to gather small, low-mobility snails. In contrast, fish also contained benthic detritus (e.g., wood, pumice, and muck) and numerous chironomid pupae, suggesting that they were dredging the lake bottom for food. Fish that foraged in this manner were observed to have abrasions on the underside of their lower mandibles. Based on the presence of aquatic beetles (*Dytiscidae*) and emerging mayflies in their stomachs, fish also forage from the water column. In addition, adult terrestrial and aquatic insects were taken from the water's surface. In the future, year-round sampling could offer a more comprehensive view of seasonal prey items, and the use of stable isotopes might provide insight regarding the relative importance of different prey items to the diet of this trout population.

14.5 Conclusions

Our findings provide a snapshot of the rainbow trout population in Spirit Lake during the late spring through autumn seasons but over a relatively long period (2000–2015). During that time, adult trout in Spirit Lake appeared to be increasing in abundance and decreasing in size. Fish with diets primarily composed of small prey items (insects and snails) likely spend considerable time and effort foraging. Should the diet switch to piscivory, the size and age structure of the population in Spirit Lake could change dramatically. The relatively recent (2011) observation of juvenile fish occupying streams draining into Spirit Lake demonstrates the increased suitability of stream habitat for both spawning and rearing, which may account for the changing population dynamics and annual variation. The landscape surrounding the Spirit Lake basin is in early stages of succession and, in the absence of future eruptions, will develop more complex vegetation that is eventually dominated by forest. Such changes will lead to more stable channel conditions and concomitant shifts in light and temperature regimes as well as food resources in streams. These conditions will undoubtedly influence instream fish production and residency. It is our hope that work on this rainbow trout population will continue well into the future and provide additional insights on post-eruption ecological processes in this remarkable living laboratory.

Acknowledgments This work was supported by funding from the USDA Forest Service, Pacific Northwest Research Station, and the National Science Foundation (LTREB Program, DEB-0614538) to CMC. Our project benefitted from early collaborations and discussions with Bob Lucas, John Weinheimer, and Chris Wageman of the Washington Department of Fish and Wildlife. Juvenile fish aging was made possible through the support of Gordon Reeves's lab. We thank the numerous field and lab assistants whose help made this project possible. We thank Kathryn Ronnenberg and Kelly Christiansen for producing final figures and tables.

Glossary

Volcanic Processes

Blast pyroclastic density current (blast PDC) In the case of the 1980 Mount St. Helens eruption, failure of the volcano's north flank unroofed pressurized magma and superheated water. Rapid exsolution of magmatic gases and conversion of superheated water to steam produced a laterally directed blast, which formed a density current that flowed across rugged topography. The current contained fragmented rock debris as well as shattered forest material.

Lahar An Indonesian term for a rapid granular flow of a fully saturated mixture of volcanic rock particles (± ice), water, and commonly woody debris. A lahar that has ≥50% solids by volume is termed a *debris flow*; one that has roughly 10–50% solids by volume is termed a *hyperconcentrated flow*. Flow type can evolve with time and distance along a flow path as sediment is entrained or deposited.

Pyroclastic flow Rapid flow of a dry mixture of hot (commonly >700 °C) solid particles, gases, and air that has a ground-hugging flow that is often directed by topography. Flows are generally gravity driven but may be accelerated initially by impulsive lateral forces of directed volcanic explosions. Flows typically move at high velocity (up to several hundred km hr^{-1}).

Tephrafall A rain of volcanic particles to the ground following ejection into the atmosphere by an explosive eruption. Tephra is a collective term for particles of any size, shape, or composition ejected in an explosive eruption.

Other Terms

Adfluvial (adj) Describes life-history strategy in which adult fish spawn and juveniles subsequently rear in streams but migrate to lakes for feeding as subadults and adults.

Anadromous (adj) Describes fish that ascend rivers or streams to spawn. Subdivision of diadromous.

Mesotrophic (adj) Describes a body of water having a moderate amount of dissolved nutrients as compared to high levels (eutrophic) or low levels (oligotrophic).

Refugia (n, pl) Locations that support isolated populations of a once more widespread species. Isolation at Mount St. Helens was caused by intense and high severity disturbance from volcanic forces during the eruption that reduced or eliminated organisms from areas surrounding the refugia.

Thalweg (n) The trajectory of the connection of lowest points of the channel bed along the length of a stream channel. The thalweg marks the natural direction of a watercourse.

Ultra-oligotrophic (adj) Having low primary productivity; pertaining to waters having low levels of the mineral nutrients required by green plants. Used of a lake which the hypolimnion does not become depleted of oxygen during the summer.

References

Bisson, P.A., R.E. Bilby, M.D. Bryant, C.A. Dolloff, G.B. Grette, R.A. House, M.L. Murphy, K.V. Koski, and J.R. Sedell. 1987. Large woody debris in forested streams in the Pacific Northwest: Past, present, and future. In *Streamside management: Forestry and fishery interactions, Proceedings of the symposium* Contribution No. 57, ed. E.O. Salo and T.W. Cundy, 143–190. Seattle: College of Forest Resources, University of Washington.

Bisson, P.A., J.L. Nielsen, and J.W. Ward. 1988. Summer production of coho salmon stocked in Mount St. Helens streams 3–6 years after the 1980 eruption. *Transactions of the American Fisheries Society* 117: 322–335.

Bisson, P.A., C.M. Crisafulli, B.R. Fransen, R.E. Lucas, and C.P. Hawkins. 2005. Responses of fish to the 1980 eruption of Mount St. Helens. In *Ecological responses to the 1980 eruption of Mount St. Helens*, ed. V.H. Dale, F.J. Swanson, and C.M. Crisafulli, 163–181. New York: Springer.

Blackman, T.E. 2014. Life-history strategies of rainbow trout (*Oncorhynchus mykiss*) cross a volcanic disturbance gradient at Mount St. Helens, Washington. Masters thesis. Corvallis: Oregon State University.

Bortleson, G.C., N.P. Dion, J.B. McConnell, and L.M. Nelson. 1976. Reconnaissance data on lakes in Washington. Volume 2, King and Snohomish Counties. State of Washington, Department of Ecology Water-Supply Bulletin No. 43.

Campana, S.E., and J.D. Neilson. 1985. Microstructure of fish otoliths. *Canadian Journal of Fisheries and Aquatic Sciences* 42: 1014–1032.

Crawford, B.A. 1986. *Recovery of game fish populations impacted by the May 18, 1980 eruption of Mount St. Helens. Part II. Recovery of surviving fish populations within the lakes in the Mount St. Helens National Volcanic Monument and Adjacent Areas*. Fishery Management Report 85-9B. Vancouver: Washington Department of Fish and Game.

Crisafulli, C.M., and C.P. Hawkins. 1998. Ecosystem recovery following a catastrophic disturbance: Lessons learned from Mount St. Helens. In *Status and trends of the nation's biological resources*, Vol. 2. ed. M.J. Mac, P.A. Opler, C.E. Puckett Haecker, and P.D. Doran, 253–310. Reston: U.S. Geological Survey. https://pubs.er.usgs.gov/publication/70103848.

Dahm, C.N., D.W. Larson, R.R. Petersen, and R.C. Wissmar. 2005. Response and recovery of lakes. In *Ecological responses to the 1980 Eruption of Mount St. Helens*, ed. V.H. Dale, F.J. Swanson, and C.M. Crisafulli, 255–274. New York: Springer.

Daubenmire, R. 1959. A canopy-coverage method of vegetational analysis. *Northwest Science* 33: 43–64.

Foxworthy, B.L., and M. Hill. 1982. *Volcanic eruptions of 1980 at Mount St. Helens: The first 100 days*, Professional Paper 1249. Washington, DC: U.S. Geological Survey.

Hawkins, C.P., and J.R. Sedell. 1990. The role of refugia in the recolonization of streams devastated by the 1980 eruption of Mount St. Helens. *Northwest Science* 64: 271–274.

Hawkins, C.P., J.L. Kershner, P.A. Bisson, M.D. Bryant, L.M. Decker, S.V. Gregory, D.A. McCullough, C.K. Overton, G.H. Reeves, R.J. Steedman, and M.K. Young. 1993. A hierarchical approach to classifying stream habitat features. *Fisheries* 18 (6): 3–12.

Larson, D.W. 1993. The recovery of Spirit Lake. *American Scientist* 81: 166–177.

Larson, D.W., J.W. Sweet, R.R. Peterson, and C.M. Crisafulli. 2006. Posteruption response of phytoplankton and zooplankton communities in Spirit Lake, Mount St. Helens, Washington. *Lake and Reservoir Management* 22: 273–292.

Leider, S.A. 1989. Increased straying by adult steelhead trout, *Salmo gairdneri*, following the 1980 eruption of Mount St. Helens. *Environmental Biology of Fishes* 24: 219–229.

Lucas, R.E., and B.A. Crawford. 1986. *Recovery of game fish populations impacted by the May 18, 1980, eruption of Mount St. Helens. Part I. Recovery of winter-run steelhead in the Toutle River watershed.* Fishery Management Report 85-9A. Olympia: Washington Department of Game, Fisheries Management Division.

Lucas, R.E., and K. Pointer. 1987. *Wild steelhead spawning escapement estimates for Southwest Washington streams—1987*, Fishery Management Report. 87–6. Olympia: Washington Department of Wildlife, Fisheries Management Division.

Lucas, R.E., and J. Weinheimer. 2003. *Recovery of fish populations affected by the May 18th, 1980 eruption of Mount St. Helens.* Olympia: Washington Department of Fish and Wildlife. http://wdfw.wa.gov/publications/00942/wdfw00942.pdf.

Major, J.J., T.C. Pierson, R.L. Dinehart, and J.E. Costa. 2000. Sediment yield following severe volcanic disturbance—A two-decade perspective from Mount St. Helens. *Geology* 28: 819–822. https://doi.org/10.1130/0091-7613(2000)28<819:SYFSVD>2.0.CO;2.

Martin, D.J., L.J. Wasserman, R.P. Jones, and E.O. Salo. 1982. Effects of the eruption of Mount St. Helens on salmon populations and habitat of the Toutle River. In *Proceedings from the conference, Mount St. Helens: Effects on water resources*, Report No. 41, 235–254. Pullman: State of Washington Water Research Center, Washington State University.

Martin, D.J., L.J. Wasserman, and V.H. Dale. 1986. Influence of riparian vegetation on posteruption survival of coho salmon fingerlings on the west-side streams of Mount St. Helens, Washington. *North American Journal of Fisheries Management* 6: 1–8.

McPhail, J.D., and C.C. Lindsey. 1986. Zoogeography of the freshwater fishes of Cascadia (the Columbia system and rivers north to the Stikine). In *Zoogeography of North American freshwater fishes*, ed. C.H. Hocutt and E.O. Wiley, 615–637. New York: Wiley.

Milner, A.M., E.E. Knudsen, C. Soiseth, A.L. Robertson, D. Schell, I.T. Phillips, and K. Magnusson. 2000. Colonization and development of stream communities across a 200-year gradient in Glacier Bay National Park, Alaska, U.S.A. *Canadian Journal of Fisheries and Aquatic Sciences* 57: 2319–2335. https://doi.org/10.1139/f00-212.

Milner, A.M., A.L. Robertson, K.A. Monaghan, A.J. Veal, and E.A. Flory. 2008. Colonization and development of an Alaskan stream community over 28 years. *Frontiers in Ecology and the Environment* 6: 413–419.

Ogle, D.H. 2013. FSA: Fisheries stock analysis. R package version 0.4.3. https://fishr.wordpress.com/.

Olds, C. 2002. *Fisheries studies at the sediment retention structure on the North Fork Toutle River 1993, 2001, and 2002.* Prepared for the U.S. Army Corps of Engineers, Portland District, Contract 36001648. Olympia: Washington Department of Fish and Wildlife.

Pavlov, D.S., K.A. Savvaitova, K.V. Kuzishchin, M.A. Gruzdeva, A.Yu. Mal'tsev, and J.A. Stanford. 2008. Diversity of life strategies and population structure of Kamchatka mykiss *Parasalmo mykiss* in the ecosystems of small salmon rivers of various types. *Journal of Ichthyology* 48: 37–44.

Quinn, T.P. 2005. *The behavior and ecology of Pacific salmon and trout.* American Fisheries Society, Bethesda, and University of Washington Press, Seattle.

Redding, J.M., and C.B. Schreck. 1982. Mount St. Helens ash causes sublethal stress responses in steelhead trout. In *Proceedings from the conference, Mount St. Helens: Effects on water resources*, Report No. 41, 300–307. Pullman: State of Washington Water Research Center, Washington State University.

Roni, P., and T.P. Quinn. 2001. Density and size of juvenile salmonids in response to placement of large woody debris in western Oregon and Washington streams. *Canadian Journal of Fisheries and Aquatic Sciences* 58: 282–292.

Stober, Q.J., B.D. Ross, C.L. Melby, P.A. Dinnel, T.H. Jagielo, and E.O. Salo. 1981. *Effects of suspended volcanic sediment on coho and Chinook salmon in the Toutle and Cowlitz rivers.* Technical Completion Report, Contract No. 14-34-0001-1417. Washington Department of Fisheries; Office of Water Research and Technology, U.S. Department of the Interior; State of Washington Water Research Center. https://digital.lib.washington.edu/researchworks/handle/1773/3985.

Swanson, F.J., and J.J. Major. 2005. Physical events, environments, and geological–ecological interactions at Mount St. Helens: March 1980–2004. In *Ecological responses to the 1980 eruption of Mount St. Helens*, ed. V.H. Dale, F.J. Swanson, and C.M. Crisafulli, 27–44. New York: Springer.

Swanson, F.J., C.M. Crisafulli, and D.K. Yamaguchi. 2005. Geological and ecological settings of Mount St. Helens before May 18, 1980. In *Ecological responses to the 1980 eruption of Mount St. Helens*, ed. V.H. Dale, F.J. Swanson, and C.M. Crisafulli, 13–26. New York: Springer.

Wissmar, R.C., A.H. Devol, A.E. Nevissi, and J.R. Sedell. 1982. Chemical changes of lakes within the Mount St. Helens blast zone. *Science* 216: 175–178. https://doi.org/10.1126/science.216.4542.175.

Wydoski, R.S., and R.R. Whitney. 2003. *Inland fishes of Washington.* Bethesda/Seattle: American Fisheries Society/University of Washington Press.

Mammal Community Assembly During Primary Succession on the Pumice Plain at the Mount St. Helens Volcano (1983–2015)

15

Charles M. Crisafulli, Robert R. Parmenter, Tara E. Blackman, and James A. MacMahon

15.1 Introduction

Ecologists have a long-standing interest in disturbance events and the successional processes that follow them (Thoreau 1887; Cowles 1899; Clements 1916; Gleason 1926). The 18 May 1980, eruption of the Mount St. Helens volcano (hereafter MSH) created an outstanding opportunity to study ecosystem responses to a large, intense forest disturbance (Franklin et al. 1985; Dale et al. 2005). This energetic eruption involved a complex ensemble of geophysical forces that over a 12-h period grossly transformed >570 km^2 of Cascadian ecosystems and disturbed to a much lesser extent an area of several 1000 km^2 with wind-deposited pumiceous lapilli and ash (tephra). Briefly, the eruption included a massive ***debris avalanche***[1], a powerful, horizontally directed pyroclastic density current (commonly referred to as a lateral blast, but which we will call the ***blast PDC***), volcanic mudflows known as ***lahars***, searing-hot ***pyroclastic flows***, ***tephrafalls***, and creation of a lava dome confined to the crater (see Lipman and Mullineaux 1981 for detailed accounts of the eruption). In the aftermath, a mosaic of disturbance zones was created that collectively represent a disturbance gradient, reflecting the different types and intensities of volcanic processes that occurred during the eruption (Fig. 15.1). The severity of disturbance to the biota along this gradient ranged from the obliteration of all vestiges of the pre-eruption old-growth forest immediately north of the crater in the pyroclastic-flow zone to areas of high survival beyond the reach of the blast PDC in the tephrafall zone, where <30 cm of cool pumice and ash was deposited.

Ecologists quickly recognized the significance of this event for documenting initial and longer-term responses of organisms and biological communities to a contemporary eruption, and scores of investigations ensued that encompassed a diverse array of taxa, system types, and ecological processes (Dale et al. 2005). Since the earliest post-eruption days, mammals, both large and small, have captured the interest of investigators at MSH, who recognized the ecological importance of these animals because of their high diversity, their large population sizes, their positions at multiple levels in food webs, and their roles as bioengineers and keystone species. As a result, mammal research has been conducted in each of the disturbance zones created during the 1980 eruption, and some of these studies have been ongoing from 1980 to the present (2015).

In this chapter, we expand our understanding of mammal responses to the 1980 eruption by providing a focused look at small-mammal community assembly from 1983 through 2015 on the Pumice Plain, an area undergoing primary succession, located immediately north of the crater in the pyroclastic-flow disturbance zone. As far as we know, there are no other detailed published accounts that chronicle the assembly of small-mammal communities in a primary-successional context over the course of decades following a large, intense natural disturbance. Hence, our story should provide novel insights of long-standing interest of ecologists, and serve as a case study for comparisons with community assembly occurring at other locations and for other taxa in the future. Other works that may offer interesting comparisons to our study include mammal responses to volcanism at volcanoes other than MSH (Burt 1961; Pearson 1994; Saba and de Lamo 1994; Irie and Tsuyuzaki 2007), to

[1]Terms in ***bold italic face*** are defined in the glossary at end of the chapter.

C.M. Crisafulli (✉) • T.E. Blackman
U.S. Department of Agriculture, Forest Service, Pacific Northwest Research Station, Mount St. Helens National Volcanic Monument, 42218 NE Yale Bridge Road, Amboy, Washington 98601, USA
e-mail: ccrisafulli@fs.fed.us; tarablackman@gmail.com

R.R. Parmenter
Valles Caldera National Preserve, National Park Service, P.O. Box 359, 090 Villa Louis Martin, Jemez Springs, NM 87025, USA
e-mail: Robert_Parmenter@nps.gov

J.A. MacMahon
Department of Biology and the Ecology Center, Utah State University, 1688 Willow Way, Logan, UT 84341, USA
e-mail: jim.macmahon@usu.edu

C.M. Crisafulli, V.H. Dale (eds.), *Ecological Responses at Mount St. Helens: Revisited 35 years after the 1980 Eruption*, https://doi.org/10.1007/978-1-4939-7451-1_15

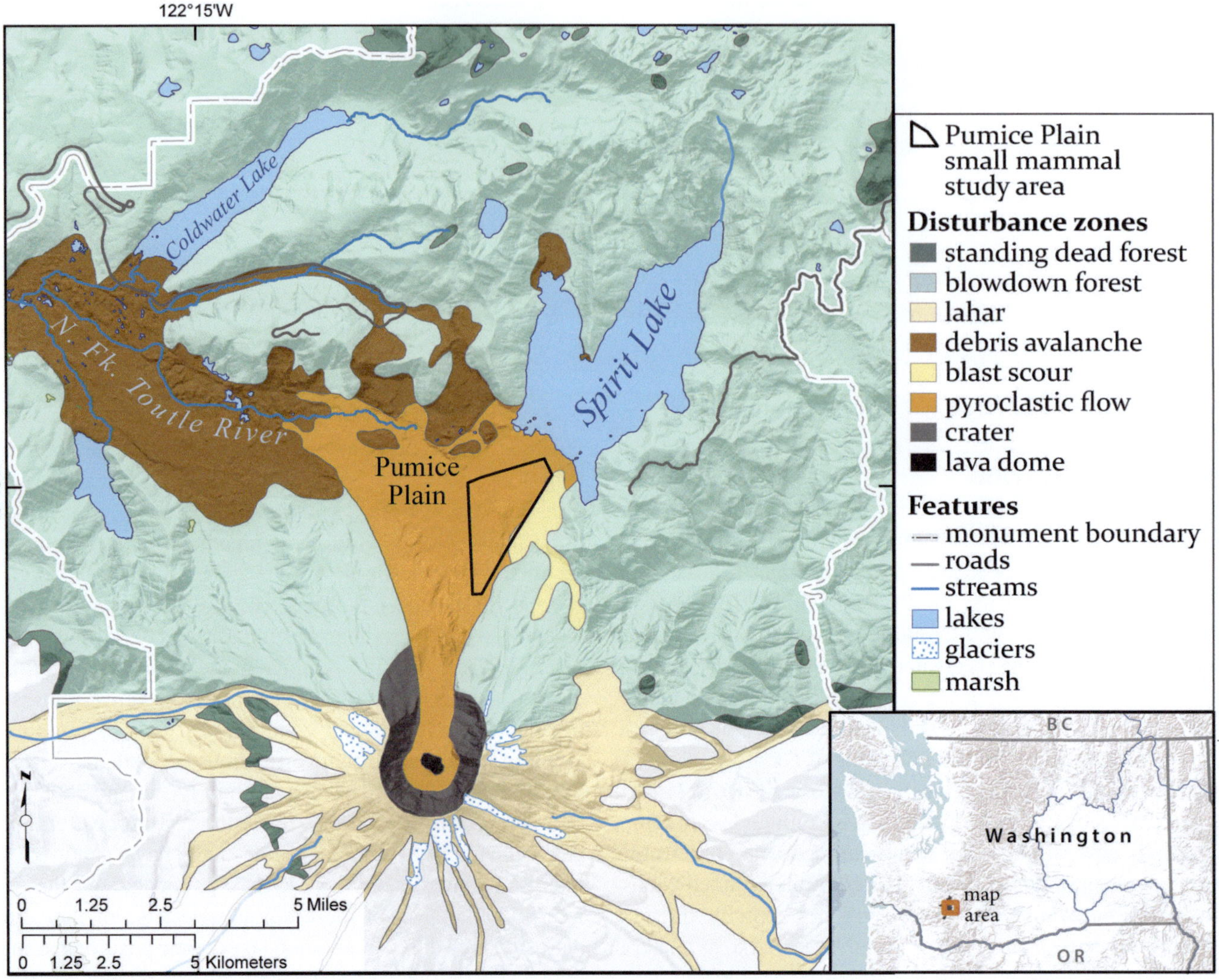

Fig. 15.1 Disturbance zones created during the 1980 eruption of Mount St. Helens. The 16 study sites reported in this chapter are located in areas undergoing primary succession, in or immediately adjacent to the pyroclastic-flow zone. See Fig. 15.4 for a more detailed depiction of these sites (Cartography by Kathryn Ronnenberg, USDA Forest Service, Pacific Northwest Research Station).

mining (Parmenter et al. 1985; Larkin et al. 2008; Craig et al. 2012), to wildfire (Fox 1982; Monamy and Fox 2000; Fisher and Wilkinson 2005; Zwolak 2009; Banks et al. 2011; Mendonca et al. 2015), and to forest management (Kirkland 1990; Fisher and Wilkinson 2005; Bogdziewicz and Zwolak 2014). However, most of these other studies focused on less-severe disturbances in which mammals and components of their biophysical habitat survived, and ecological response proceeded under secondary-successional scenarios. Crisafulli et al. (2005) compared MSH mammal responses to other disturbances; in this chapter, we provide updates to our previous observations as the small-mammal community has continued to develop over the subsequent 15 years.

We first present a review of published small-mammal studies at MSH, providing a context for our work (see Sect. 15.2). Next, we describe a conceptual model of community assembly during primary succession that we use as a framework in which to place our small-mammal results from the Pumice Plain (detailed in Sect. 15.3 and Fig. 15.2). Our model considers three primary processes, ***dispersal***, ***establishment***, and ***community development***, and each has a number of key factors (e.g., dispersal distance from less-disturbed habitats) that influence these processes. Previous studies at MSH (Andersen and MacMahon 1985a; MacMahon et al. 1989; Crisafulli et al. 2005) documented a regional species pool of mammals that could serve as source populations to the Pumice Plain. Further, we predicted that variation in species-specific dispersal characteristics and life history would lead to differential rates of ***arrival*** on the Pumice Plain. For example, common, highly mobile, and generalist taxa would arrive first, fossorial and habitat specialists next, and taxa associated with mature forest conditions would be last to arrive, but overall dispersal distance per se would not limit mammal community development. This led us to predict that, ultimately, the most important factors driving mammal community assembly would be the development of

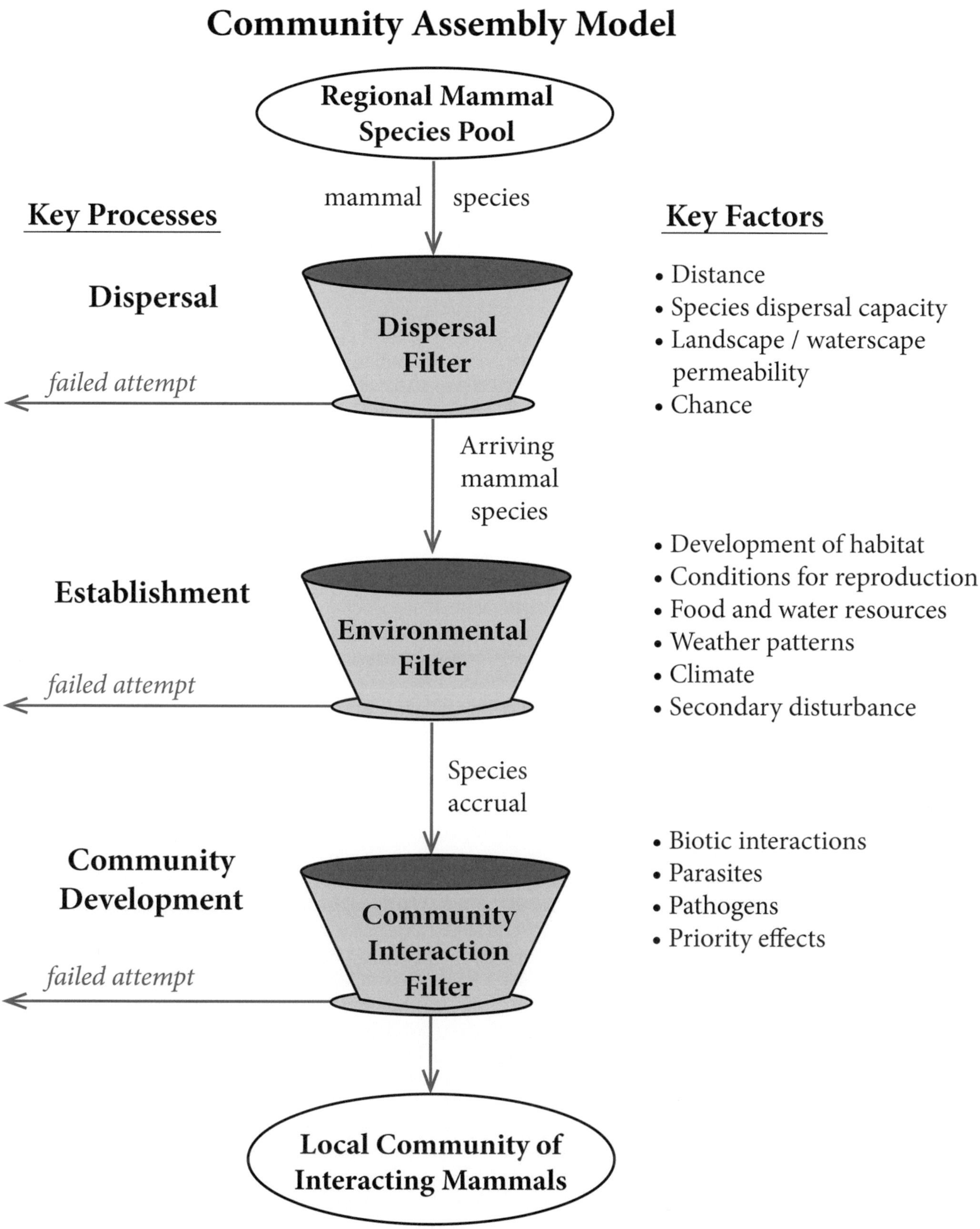

Fig. 15.2 Conceptual model of mammal community assembly during primary succession at Mount St. Helens. A regional species pool, three key processes, and several factors that are either deterministic or stochastic form the bases of this simple but instructive model.

soils and plant communities that provide requisite habitat resources for mammals, such as food, reproduction sites, and cover from weather and predators. Erosional processes—stream and swale formation, and redistribution of gravels, sands, and silts—would parse the microhabitats for development of meadows, grasslands, shrublands, wetlands, and willow-/alder-dominated riparian zones. We anticipated that mammal communities in barrens would have low species richness and abundances and be dominated by habitat generalists. In contrast, habitats with complex vegetation structure would support comparatively species-rich assemblages, higher abundances, greater trophic diversity, and more habitat specialists.

In 1989, we first noticed the presence of the non-native willow stem-boring beetle (Coleoptera, Curculionidae, *Cryptorhynchus lapathi*) on the Pumice Plain, where it was using Sitka willow *(Salix sitchensis)* as a host and causing stem mortality through its feeding behavior. The beetle has persisted and spread since its initial establishment, inflicting a heavy toll on willow that retards growth and spread of willow in upland habitats (Che-Castaldo 2014). Therefore, given the effects of this beetle on willow stand structure, density, and spatial extent, we predicted that there would be a negative influence of increasing beetle damage on mammal species richness and abundance. Thus, both plant succession and herbivory, such as by the willow stem-boring beetle, were expected to influence mammal community assembly.

In this chapter, we address the following questions about mammal community assembly on the Pumice Plain:

1. What is the pattern and rate of mammal species arrival and establishment relative to the regional species pool? (Herein, we include all mammal species, both large and small, with the exception of bats.)
2. What were the patterns of species richness, abundance, guild structure, and sex ratios during community assembly of the small-mammal fauna?
3. How has vegetation structure, as influenced by successional processes and herbivory, influenced small-mammal community assembly?

15.2 Background: A Review of MSH Small-Mammal Literature

Previous small-mammal research at MSH has focused on (1) patterns of survivorship, (2) individual species and community responses to the eruption, and (3) biotic interactions and ecosystem processes. Andersen (1982) documented patterns of survivorship of the northern pocket gopher (*Thomomys talpoides*) at several locations within the severely disturbed blast area and tephrafall zone, based on surveys in 1980 and 1981. He concluded that the overlying soil and, at higher elevations, the additional protection created by snowpack facilitated the survival of this fossorial rodent and that ample belowground plant parts (e.g., roots, bulbs, rhizomes) provided food resources during the first post-eruption year. Pyke (1984) investigated potential health effects of distal (~400 km NE of the volcano) fine-grained (0.11–116 μm diameter) ash deposits on the deer mouse (*Peromyscus maniculatus*) and the montane vole (*Microtus montanus*). He conducted both field and laboratory experiments and documented hair loss in the vicinity of the eye and swelling of the eyelids in both species exposed to ash, but the deer mouse response was more prevalent and severe. Histopathological assays of lung and gastrointestinal tissue and eyes did not demonstrate ash-associated pathology. He concluded that hair loss around the eye and swelling of the eyelid were likely caused by scratching as a behavioral response to ash irritation, rather than as a pathology. Andersen and MacMahon (1985a) documented the distribution and abundance patterns of residual small mammals across several disturbance zones in 1981–1983. They reported an inverse relationship between mammal survival and disturbance intensity and concluded that in the tephrafall, blowdown, and blast-PDC zones there were 11, 7–8, and 3 species, respectively, that survived and persisted during their 40-month post-eruption study. Adams et al. (1986) reported on small-mammal surveys conducted in 1984 on the debris-avalanche deposit and in the adjacent blowdown zone, including areas of toppled old-growth forest and areas that had been clearcut logged within 10 years of the eruption. They captured deer mice across all sites, with the fewest captures on the structurally simple debris-avalanche deposit, where the species did not survive but had immigrated after the eruption. Four additional species were captured only in the structurally complex toppled old-growth forest, where they were thought to have possibly survived the eruption.

MacMahon et al. (1989) studied small-mammal responses from 1980 through 1987 across four broadly defined habitats (forests, clearcuts, subalpine meadows, and riparian ravines) arranged along a volcanic disturbance gradient ranging from areas of low to high disturbance severity. They documented the establishment of 6 species in the defaunated pyroclastic-flow zone, 15 species in the tree blowdown zone, and 22 species in the tephrafall zone and suggested that establishment of mammals during these early post-eruption years was influenced by (1) the presence of residual mammals throughout much of the disturbed landscape, (2) post-eruption erosional processes that promoted the development of vegetation, and (3) highly fluctuating weather conditions from 1983 to 1986. Bevers (1998) studied the colonization and genetic structure of American pika (*Ochotona princeps*) to the north of the volcano from 1992 through 1994. He reported the establishment of eight new colonies in the blast area that were founded sometime between 1987 and 1992. Six of these sites were in typical talus/scree habitat, and the remaining two sites were atypical and included log piles that mounded up by the Spirit Lake surge wave during the 1980 eruption. He suggested that these new colonies originated from individuals dispersing from source populations 3 to 6 km to the south. Protein electrophoresis supported a founder effect with associated rapid genetic divergence. Crisafulli et al. (2005) described small-mammal assemblages and species' relative abundance patterns from 40 sites distributed across tephrafall, scorch, blowdown, tree-removal, and pyroclastic-flow zones from 1980 through 2002. They reported a surprisingly rapid assembly of small-mammal communities along the disturbance gradient, with nearly all members of the regional species pool having colonized much of the landscape during the first 22 years post eruption, the notable exception being four species closely associated with forest structure. The authors suggest that although large (570 km^2), the size of the blast area did not preclude colonization by most small-mammal species, that individuals of these species readily dispersed from source populations, and if they encountered suitable habitat they established, often resulting in an unusual mix of species. This observation largely supports Gleason's individualistic hypothesis of succession (Gleason 1926).

The role of small mammals in ecosystem processes, including plant succession, symbioses, and nutrient cycling, is also part of the MSH small-mammal research portfolio. Andersen and MacMahon (1985b) investigated the role of the northern pocket gopher in facilitating plant establishment and growth in areas that received up to 25 cm of tephra. They reported that a greater number of plant species and individuals survived burial on gopher mounds than in adjacent off-mound sites and that there was greater seedling diversity on mounds. They concluded that pocket gophers can exert a strong influence on plant populations and thus influence successional processes of volcanically disturbed landscapes.

Mammals also influenced MSH ecosystem succession by facilitating interactions between plants and mycorrhizal fungi. Allen et al. (1984, 1992, 2005, Chap. 11, this volume), Allen (1987), and Allen and MacMahon (1988) report on the formation of symbioses between mycorrhizal fungi and vascular plants mediated by mammals. This body of literature demonstrates the transport of mycorrhizal fungal spores by the northern pocket gopher from forest soils buried beneath ≤50 cm of tephra to the surface, the establishment of mycorrhizae (a symbiosis between a fungus and a vascular plant, located on or in plant roots), and the benefits of this association to plants, such as increased photosynthetic rate, growth, and seed production. They also discuss the movement of mycorrhizal spores by several other small-mammal species, as well as elk (*Cervus elaphus*) that are drawn to patches of vegetation across the landscape and during these visits deposit spores in their fecal material. Parmenter (2005) documented differences in decomposition and nutrient-cycling rates of small-mammal carcasses across the volcanic disturbance gradient in 1986, with the slowest rates occurring in carcasses on the relatively hot, dry Pumice Plain and the fastest rates in cool, moist, undisturbed old-growth forests; rodent carcass decomposition and nutrient-cycling rates were also significantly faster belowground than on the ground surface. Yurkewycz et al. (2014) evaluated nutrient dynamics, soil physical properties, and plant communities on and adjacent to northern pocket gopher mounds in the pyroclastic-flow zone in 2009. They found that gophers occupied a small area (~2.5%) of the landscape and that gopher mounds had lower plant species density compared with adjacent areas not influenced by gophers. Plant abundance was greater on mound margins, plant burial beneath mounds led to increased soil carbon and nitrogen, and crusts that formed on mounds reduced infiltration of water.

Collectively, this literature provides the most thorough body of knowledge in the world on ground-dwelling small-mammal responses to volcanism.

15.3 Conceptual Model of Small-Mammal Community Assembly on the Pumice Plain

We use a simple conceptual model to describe key processes and factors involved in small-mammal community assembly during primary succession on the Pumice Plain at MSH (Fig. 15.2). In the aftermath of the 1980 eruption, all components of the pre-eruption old-growth forest ecosystem were removed, and the assembly process began with no mammals present. In order for community assembly to proceed, there must be a regional species pool of small mammals that serves as a source population for the defaunated landscape. At MSH, areas beyond the limit of the Pumice Plain and the adjacent debris-avalanche deposit experienced less-intense volcanic disturbance, and some mammals survived (Andersen and MacMahon 1985a; MacMahon et al. 1989; Crisafulli et al. 2005). These less-disturbed areas, as well as expansive tracts of late-seral and younger forests, undoubtedly supported populations of all members of the regional species pool. Given the assumed availability of mammals in this pool, the first key process of community assembly is dispersal of mammals from the regional species pool to the Pumice Plain. We assumed that there were several factors that acted as dispersal "filters" that limited or prevented successful arrival on the Pumice Plain, including (1) species-specific dispersal capacity and within-species variation based on sex and age; (2) distance between source populations and the Pumice Plain; (3) the ability of an individual mammal to traverse the landscape or to navigate waterways between source populations and the Pumice Plain based on biophysical conditions, referred to as landscape/waterscape permeability; and (4) the "condition" (density, demographics) of the source population at the time of dispersal. Given the population cycles exhibited by small mammals, if the eruption occurred during the low-density phase of the surrounding populations' cycle, then dispersal would likely be delayed, because individuals would have plenty of resources in undisturbed habitats and would not be particularly motivated to leave. However, if the population was at the "high" point, with lots of social/resource pressures on individuals to disperse, then we would expect to see more dispersal attempts and ultimately "arrivals"—assuming that the cycles occur every 4–6 years, this would add to the observed variability of arrival times and to the community-assembly rate (Kirkland 1990). Chance also plays a role in dispersal, but is not specifically addressed in our study. Collectively, these factors should result in variation in the time at which different species arrive in our study area.

Arriving on the Pumice Plain is no guarantee of establishment, the second key process of community assembly. Several factors may be important in determining whether individuals of a species establish on the Pumice Plain. These establishment "filters" include but are not limited to (1) development of appropriate biophysical habitat conditions that provide food, water, cover from physical elements and predators, and sites for reproduction, and (2) availability of suitable mates. Stochastic factors, such as unusual weather and secondary disturbances (e.g., floods, erosion, burial, subsequent eruptions), may influence establishment. Acting singly or in combination, these factors may lead to failed establishment. On the Pumice Plain, we assume that patterns and rates of plant succession have played an important role in mammal establishment because vegetation provides food for mammals directly (plant parts) and indirectly by providing food for invertebrates that, in turn, serve as prey for mammals. Vegetation is also important because it allows small mammals to avoid, hide, and escape from predators and also provides protection from thermal extremes. Arthropods arrived on the Pumice Plain in great abundance prior to plant community development (Edwards 1986; Edwards and Sugg 1993; Parmenter et al., Chap. 12, this volume), and these organisms likely served as an important food resource for early mammal immigrants. In addition, vegetation establishment accelerates pedogenesis (soil formation), increasing soil organic matter, changing soil texture, and facilitating microbial and fungal communities—all important components for soil-dwelling arthropod larvae, which, in turn, are important prey for several small-mammal species in the MSH regional species pool. Mammals that arrive on the Pumice Plain but are unable to establish may perish onsite, and their bodies decompose, enriching the site with nutrients that promote the development of carrion communities, soils, and vegetation (Parmenter 2005).

The third and final key process is community development, which is defined by a suite of biotic interactions among species that may serve as "filters" influencing the success of a species in the community-assembly process. These interactions come in many forms and include examples such as predation, competition for space or resources, agonistic behavior, and adverse effects related to parasites and pathogens. Indirect effects of species interactions may also influence persistence of mammal species and community assembly. For example, insect herbivory may alter plant community structure, resulting in changes in habitat having either positive or negative effects on mammal community assembly, and we provide some evidence for this based on willow-boring beetle activities. The order in which species arrive, often referred to as priority effects, may influence community development based on niche breadth and overlap. We present our model in a way that might suggest a clean-cut, stepwise progression of community assembly, but certainly the process is much more complicated, and most or all processes clearly occur concomitantly and are spatially uneven across different microhabitats. Thus, our model is admittedly oversimplified but nonetheless has heuristic value by providing a framework to view mammal community assembly on the Pumice Plain.

15.4 Study Area

Prior to the 1980 eruption, our study area, located between Mount St. Helens and Spirit Lake to the north, supported old-growth coniferous forest dominated by Pacific silver fir (*Abies amabilis*) and mountain hemlock (*Tsuga mertensiana*), with a handful of other less-common conifer species (Fig. 15.3a). The understory was composed of ericaceous shrubs, herbs, and mosses associated with the Pacific silver fir zone (Franklin and Dyrness 1973). The area has a moist maritime climate, with relatively mild, wet winters and warm, dry summers. Long-term (1987–2015) mean annual precipitation data for the Spirit Lake basin is 2255 mm, most of which falls as snow (NRCS SNOTEL Data, Spirit Lake Station 777, data extracted 28 March 2016). Mean growing season (June through September) precipitation is 251 mm or 11% of mean annual precipitation. Mean annual temperature is 6.4 °C, mean July temperature is 14.1 °C (range 7.7 to 19.5 °C), and mean January temperature is 0.69 °C (range −1.95 to 3.7 °C). After the eruption, the absence of forest cover and presence of high-porosity volcanic substrates led to a "functionally" different climate from the pre-eruption conditions, having increased insolation at ground-level increased wind velocity, lower water-holding capacity in substrates, and the near absence of a winter snowpack throughout this 33-year study. This situation is in stark contrast to pre-eruption conditions, when snowpack in

Fig. 15.3 (**a**) Pre-eruption (1979) view from the summit of Mount St. Helens looking north, with the Spirit Lake basin in the foreground and Mount Rainier in the background. Clearly visible is the expansive presence of mature and old-growth coniferous forest. (**b**) Post-eruption photo (1980) with the same view as above, taken from a helicopter, showing the nudation and filling of the landscape with several tens of meters of volcanic deposits during a rapid series of volcanic processes. All vestiges of the pre-eruption forest were scoured from the area of the study sites, located between the volcano and the south shoreline of Spirit Lake. Note the floating log raft covering ~40% of the lake's surface. (Photos courtesy of the USDA Forest Service).

the basin would commonly accumulate up to 3 m and persist into July. Thus, from the biota's perspective during the growing season, it was a warmer, drier, and windier environment.

In the aftermath of the eruption, a vast, undulating, steaming-hot, barren landscape devoid of all life had replaced the former old-growth forest (Fig. 15.3b), setting the stage for primary succession and the assembly of biological communities. In the 35 years since the eruption, the Pumice Plain has undergone substantial change in substrate characteristics (Halvorson et al. Chap. 5, this volume) and vegetation structure (del Moral et al. 2012). Cobble to large-gravel-sized dacite pumice weathered to finer gravel and sand. The once-barren plain was colonized slowly by a diverse assemblage of herbaceous plants and a handful of shrub and tree species (del Moral et al. 2012). Plant cover in the upland habitat expanded from epicenters of initial establishment, with small patches eventually coalescing into larger patches as succession proceeded. Shrubs (Sitka willow and green alder, *Alnus viridis*) established sparsely throughout the upland habitat beginning about 10 years post eruption and, since that time, have created thickets in more mesic locations but otherwise remain sparse. Several species of conifers established at low densities, and by 30 years post eruption, some were producing a local seed source. Plant cover in the upland matrix habitat 30 years post eruption ranged from 21.6 to 68.6% (del Moral et al. 2012).

Sitka willow colonized seeps and stream courses beginning in the mid-1980s and then quickly expanded throughout the watercourses, creating dense and sharply defined plant communities in about a decade or so. Our "Willow Spring" site (WILSPR, Fig. 15.4) was the first riparian plant community to develop on the Pumice Plain, with an initial community present in 1986. Other riparian communities did not develop on the Pumice Plain until a number of years later. Successional processes have been spatially variable across the Pumice Plain, leading to habitat patches in numerous successional states at any given time.

15.5 Field Methods

15.5.1 Mammal Surveys

We investigated small mammals on the Pumice Plain from 1983 to 2015. Our mammal data came from two sources. The first focused on small mammals and employed mark-recapture sampling using Sherman live traps that were installed across a network of permanent study sites (Fig. 15.4). Sherman traps are widely used for monitoring small-mammal populations, but are known to be more effective for some species than others, introducing some bias (Crisafulli et al. 2005). Our targeted species for live-trapping were mammals with a mass of ~4 to 300 g, including shrews, shrew-moles, voles, mice, chipmunks, squirrels, and weasels.

Most (15 of 16) sites were sampled using permanent 120-m line transects with trap stations every 5 m (25 traps total); the remaining site used a web-trapping design (Anderson et al. 1983) with 148 traps. Traps were open continuously for four to five consecutive days and checked each morning. Traps were baited with rolled oats and peanut butter, and cotton batting was placed in each trap as bedding material and to prevent hypothermia. We identified captured individuals to species

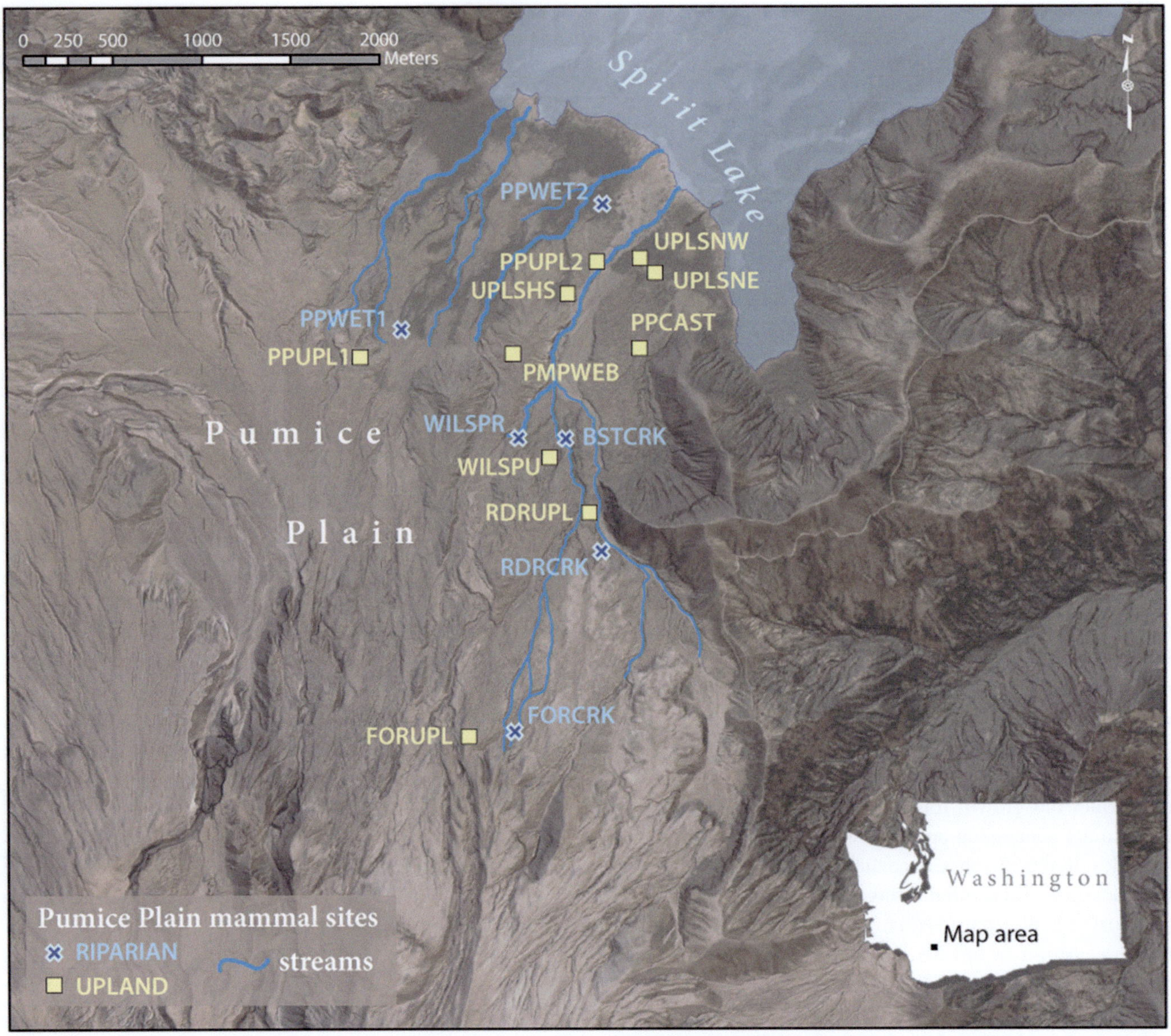

Fig. 15.4 Small-mammal study sites on the Pumice Plain at Mount St. Helens depicted on imagery (Imagery source, World Imagery service from ESRI). See Appendix 15.1 for physical information and sampling history of these sites. (Cartography by Kelly Christiansen, USDA Forest Service, Pacific Northwest Research Station).

and recorded sex, body mass, and reproductive condition, followed by marking with either ear tags, toe clips, or indelible ink. Additional details of our live-trapping methods can be found in Crisafulli et al. (2005).

Trapping was conducted at sites representing three general habitat types: upland (seven sites), riparian/wetland (six sites), and upland shrub (three sites). These sites occur within a 3-km^2 area in the eastern portion of the 11-km^2 Pumice Plain (Figs. 15.1 and 15.4). The number of sites sampled increased from just a few sites in the early 1980s, when the volcano was still erupting intermittently and access was greatly limited, to 16 sites that were sampled on a nearly annual basis since establishment. Appendix 15.1 lists the sampling period, total years sampled, number of traps, and physical information for our sites. Our goal was to document the rate and pattern of small-mammal arrival and establishment on the Pumice Plain. Thus, as new habitats developed, we expanded our sampling to include these sites. We also increased representation of certain habitats in our sampling design over the survey years by adding additional sites; we refer to these as replicate sites. In most years, each site was trapped during a single midsummer (July–August session), but in some years, sites were trapped during two sessions.

When traps were "sprung" and did not have a mammal inside, they were assigned a value of 0.5 trap night, following Nelson and Clark (1973), to account for the trap not being available to mammals for some unknown amount of time each trap night. To characterize our trapping effort, we summed the number of trap nights (number of traps operated times the number of nights trapped) and tallied the number of captured unique individuals (marked mammals).

We calculated recapture rates for dominant species both within session and between years. For within-session recaptures, percent recapture rates were calculated for a subset of species that were numerically dominant and marked with ear tags. We summed the number of individuals captured >1 time during our midsummer trapping session, divided this value by the total number of individuals captured at the end of the session (recaptured animals plus animals captured only once), and multiplied this quotient by 100. These calculations were performed on data spanning 2000 through 2015 and included only individuals that did not die during the first capture. Our approach considered captures separately for upland and riparian habitats, pooling replicate sites and years, and presented as mean and SE recapture rates for a session. For between-year recaptures, we first generated recapture rates for each species at each site for each of the sample years 2000 through 2015. Recapture rate was calculated as the percentage of marked (ear-tagged) individuals of each species in a given sample year that appeared in the sample of the following year. This value revealed consistently low annual recapture rates through time, and based on this consistency, we pooled the annual recapture rates for all sites (within a habitat type) and for all years and provide means and SE for upland and riparian sites.

The second data source focused on mid-to-large mammals and was based on observations made by the authors and field crew members who were trained in mammal identification. These data were visual observations of mammals, often accompanied by photo documentation. The majority of these species were conspicuous in the denuded post-eruption landscape and include species such ungulates (e.g., elk), large rodents (e.g., hoary marmot, *Marmota caligata*), and carnivores (e.g., coyote, *Canis latrans*). Each observation was entered in a field notebook with details of date, time, specific location, and diagnostic features leading to species identification. Each entry was then scrutinized by the first author, and any questionable observations were removed. Other corroborative observations, such as tracks, scat, and foraging activities, were added to the visual observations database.

Our research was approved and permitted by Washington Department of Wildlife, and permits are on file with the first author. Mammal taxonomy follows the Integrated Taxonomic Information System (http://www.itis.gov/, extracted February 2016). Mammal voucher specimens were deposited at the Burke Museum, University of Washington, and the Museum of Southwestern Biology, University of New Mexico, and include representative specimens of all species trapped.

15.5.2 Vegetation Structure Surveys

To address relationships between vegetation structure and small-mammal community structure, we measured vegetation at our study sites in 2009 and again in 2013 using a modified "foliage height diversity" (FHD) method (MacArthur and MacArthur 1961). This method provides information on plant physiognomy by measuring the vertical distribution of vegetation by life-form. At each of our study sites (except PMPWEB and UPLSHS), nine transects were sampled at 15-m intervals along and perpendicular to the mammal-trapping transect. The FHD transect length was fixed at 30 m at upland sites and varied at the riparian/wetland and upland shrub sites based on the width of those habitats (range 12–38 m). That is, riparian transects began and ended based on the presence of shrubs. The FHD measurement stations (N = 66–144) were placed every 2 m along each transect, and at each station, a 3-m graduated PVC pole was positioned vertically and assessed for the presence (contact with pole) of graminoids, forbs, and shrubs at each of the following height classes: 0–10 cm, 11–25 cm, 26–50 cm, 51–100 cm, 101–150 cm, 151–200 cm, 200–300 cm, and >300 cm. Thus, across sites there were between 400 and 1152 points in space where vegetation presence was assessed.

Next, we multiplied the number of FHD stations on each transect by the eight height classes to derive the total number of points sampled for vegetation. Then, we summed the number of intercepts with vegetation present (positive hits). We divided the number of positive hits along each transect by the total number of possible points along that transect to calculate the frequency of vegetation present. This value provided an index of vegetation complexity that we used to explore correlations between mammal species richness and abundance with vegetation structure.

15.5.3 Herbivory Effects on Small-Mammal Assembly

To assess the effects of herbivory on small-mammal community assembly, we measured the attack rate of willow stem-boring beetles on upland willow shrubs at the plant and stem level for 3 years: 2010 (n = 583), 2011 (n = 577), and 2015 (n = 463). We defined plant-level attack as any stem or the caudex (root crown) attacked on a willow shrub. We defined stem-level attack as the presence of beetle attack anywhere along the stem axis from where it emerged from the caudex upward toward the distal end. We uniquely tagged stems on each living willow and recorded their ground-level diameter (mm) and length (cm). We then scored stems for the presence of beetle herbivory and recorded their vigor (dead or alive). Through successive annual sampling, these data provided information on herbivory prevalence and severity, as well as plant and stem mortality. We use these data to make a correlative case for herbivory retarding the development of shrub habitat and, by doing so, influencing mammal community structure.

15.5.4 Temperature and Relative Humidity

We recorded temperature and relative humidity hourly during the growing season (June through early October) from 2003 through 2015, except for 2004 and 2005, using Onset Computer data logger models: HOBO Pro Series model H08-032-08 and HOBO Pro v2. We attached data loggers to wooden solar shields and fastened them to metal t-posts at a height of 1 m above the ground surface at each mammal-sampling site. We partitioned data by upland and riparian habitat and trimmed them to include only July and August values, which coincided with our mammal trapping. We reported these data in two formats: mean temperature and mean relative humidity, with standard deviation, for 2003–2015, and mean maximum and mean minimum temperature with standard deviation for each upland and riparian habitat for all years. This approach provides a general (average) understanding of the micrometeorology as well as the extremes between upland and shrub habitats.

15.6 Data Compilation, Manipulation, and Analyses

15.6.1 Arrival and Establishment

What is the pattern and rate of arrival and establishment of mammal species relative to the regional species pool?

We compiled a list of mammals (excluding bats, Chiroptera) that represents the regional species pool for the southwest Washington Cascade Range (Wilson and Ruff 1999) and used this list to identify the pool of species potentially available to the Pumice Plain. Then, using all available trapping (all sites, sessions, and years) and observation (1983–2015) data, we recorded the year that each mammal species arrived (first detection) and/or established (assumed breeding population) on the Pumice Plain. This simple analysis relates to dispersal and establishment processes of the model. We used these data to describe mammal arrival/establishment chronology in four time periods: 1983–1989, 1990–2000, 2001–2010, and 2011–2015.

In a similar analysis, but only for small mammals, we used our complete trapping data set (all sites, sessions, and years, 1983–2015) to examine the relationship between annual species accrual rate and time since the eruption using simple linear regression. This model assumed equal variance and sample independence:

$$Y_i = \beta o + \beta 1(x_i)$$

where y is the estimated number of species established on the Pumice Plain for the ith year post eruption, x_i the year post eruption (ype), βo the y-intercept, and $\beta 1$ the estimated change in species for every ype. We compared the observed species-accrual pattern to our predictions based on assumed dispersal capacity, life-history characteristics, and habitat associations.

We quantified annual rates of mammal species gain, loss, and turnover using the rate metrics provided by Anderson (2007). These metrics were species gain rate (proportion of new species between years), species loss rate (proportion of species lost between years), and species turnover rate (proportional changes between years). Annual species gain rate (G_p) was computed as:

$$G_p = G / \left[\left(St_1 + St_2\right)/2\right]$$

where G is the number of new species gained during the time period, and St_1 and St_2 are the number of species present at the beginning (t_1) and end (t_2) of the sample time period (usually 1 year). In calculating species gain rate in any given year, we included not only newly arrived species but also species that had previously occupied the site but had temporarily disappeared and then reappeared in subsequent years.

We calculated annual species loss rate (L_p) as:

$$L_p = L / \left[\left(St_1 + St_2\right)/2\right]$$

where L is the number of species lost during the year.

We calculated annual species turnover rate (T_p) as:

$$T_p = \left(G + L\right) / \left(St_1 + St_2\right)$$

We determined annual mammal species gain rate, loss rate, and turnover rate for each year-to-year sample period during which a comparable sampling effort was conducted; this criterion eliminated the years 1984, 1985, 1986, 1989, 2001, and 2010. For years following eliminated years, we divided rate values by the number of years between samples. We assumed that species richness on the Pumice Plain was zero immediately after the 1980 eruption. We then graphed annual rate values during 1980–2015, and linear or nonlinear regressions (exponential, polynomial, logarithmic, power function) were fitted to the data. Selected regression models had the best fit indicated by maximum coefficient of determination.

15.6.2 Species Richness, Abundance, Foraging-Guild Structure, and Sex Ratios

What were the patterns of species richness, abundance, guild structure, and sex ratios during community assembly of the small-mammal fauna?

We addressed this question using two data sets. The first includes data for one riparian site (WILSPR) and one upland site (PMPWEB) that were trapped for 28 and 32 years, respectively. These two sites were trapped the longest of all our sites and thus provide the best temporal record of early post-eruption community assembly.

The second data set was derived from six pairs of riparian and upland sites that were established in 2000 and sampled through 2015. These data have a shorter temporal span than the first data set but have larger spatial extent. Data for each of the two data sets were trimmed to four continuous days across all sites and years to provide standardized data (i.e., equal sampling effort) for analyses. We report species richness (mean number of species captured during a trapping session), capture rates (number of unique individuals captured/100 trap nights), species biomass by foraging guild (percent contribution for each species relative to the total biomass for all taxa in that guild), and guild structure (species composition and number of captures for each species within each guild). We calculated guild biomass by summing the mass (g) of all unique individuals of each species captured during a sampling session each year. Foraging guilds included omnivore, herbivore, and insectivore.

We plotted long-term dynamics of small-mammal abundance based on unique captures for each species present during the period of record (1984–2015) at the PMPWEB site and for 2003–2015 at the paired upland and riparian sites. We summed the number of unique individuals of each species captured during a sampling session each year and classified them into abundance categories, which were plotted as "bubbles" of varying size, equating to capture numbers. The number of categories and the range of capture values were consistent across species and years for each habitat type but varied across upland and riparian habitats based on the range of values in our data. For example, the number of unique captures ranged from 10 to 50 in riparian sites and from 25 to 125 in upland sites. Further, we color-coded each species by foraging guild to reveal information on its relative importance within the temporal variation of guild structure.

To explore small-mammal sex ratios using our 2000–2015 data, we counted captured male and female individuals of 11 species and compared the ratios for riparian and upland sites. We pooled sites by riparian and upland habitat and summed the number of male and female unique individuals for each sample year. To assess sex ratios within habitat types, we used proportion of female captures (%), with 95% confidence intervals for a one-sample test of equality of proportions (i.e., sex ratios different from 1:1).

Using mark-recapture studies, we opportunistically assessed movement patterns of small mammals by compiling a list of individuals that were initially captured (and ear-tagged) at one site and were later captured at other sites. These observations provided insights on spatial ecology of the small-mammal assembly. We restricted our analysis to five species with sufficient data records and considered all 15 transect sites for the years 2000–2015. For each species, we tracked the number of individuals that moved between each site and the direction of movement. We also calculated the average and maximum distances moved and provided a few examples of movement rate (time between captures at different sites).

15.6.3 Vegetation Structure

What influence did vegetation structure and herbivory have on small-mammal community assembly?

We expected that vegetation structure would strongly influence mammal community assembly, such that closely juxtaposed patches representing different plant communities would support different mammal assemblages. In this regard, vegetation characteristics function as filters that sort mammals into different community compositions. We performed a Welch two-sample *t*-test to examine differences between mean percent vegetation structures of riparian and upland habitats for 2009 and 2013. We defined mammal species richness as the number of species trapped at each site for 2009 and 2013. Species lacking evidence of establishment and those associated with water (water shrew, *Sorex palustris*; marsh shrew, *S. bendirii*; and water vole, *Microtus richardsoni*) were removed from the analysis in an attempt to focus on the differences between mammal species richness and vegetation architecture and not water as a habitat feature per se. The vegetation-structure parameter we used was the percent of three-dimensional space occupied by all life-forms of vegetation (as described in Sect. 15.5.2). We also used this analysis to indirectly evaluate the influence of willow stem-boring beetles on mammal species richness, comparing riparian with upland sites. We examined relationships between mammal species richness and vegetation structure using a paired t-test for 2009 and 2013 data.

15.7 Results

15.7.1 Arrival and Establishment: Comparison of the Pumice Plain Mammal Assemblage to the Regional Species Pool

The regional mammalian species pool of the southern Washington Cascade Range totals 45 species, excluding Chiroptera (bats) (Wilson and Ruff 1999). This diverse assemblage includes 6 orders, 19 families, and 34 genera (Table 15.1), and shows tremendous variation in body size and trophic status, ranging from tiny insectivorous shrews weighing 5 g to massive herbivorous elk weighing up to about 300 kg. At least 34 (76%) of the 45 species included in the regional species pool arrived on the Pumice Plain between 1983 and 2105, including members of all orders, 17 of 19 families, and 25 of 34 genera.

Arrival of mammals from the regional species pool, grouped by body size, included 22 (85%) of 26 small mammals (Fig. 15.5a inset), 8 (57%) of 14 medium-sized mammals, and 4 (80%) of 5 large mammals. There was no pattern of arrival associated with species' taxonomic affinity and trophic status, but midsized mammals were disproportionately underrepresented. That is, species of different body size, taxonomic affinity, and position in food webs (trophic status) arrived throughout most of the 33 years of study. Undetected species include those associated with forest structure (8 species) and those that are water dependent (2 species) or fossorial (1 species). We have evidence, based on the presence of pregnant or abundant individuals and successive years of capture, that 25 of 34 species that arrived on the Pumice Plain established populations there, whereas 5 species lacked support of establishment, and establishment status was uncertain for an additional four species (Table 15.1).

Small-mammal arrival sequence is shown in Fig. 15.5a. The first small mammals to arrive on the Pumice Plain were the deer mouse and the long-tailed vole (*Microtus longicaudus*) (both recorded in 1983), and it wasn't until 1988 (5 years later) that an additional species, another rodent, the water vole, was captured. Nocturnal rodents were the only mammals captured from 1983 to 1988; diurnal rodents would not arrive until several years later (1995 and 1998), which coincided with the development of shrub thickets in riparian habitats. The temporal pattern of arrival was variable, with 12 rodent species arriving over a 31-year period (1983–2013) and 7 insectivore species arriving during a 22-year period (1990–2011), whereas carnivores (Mustelidae) were not detected until 23 and 27 years after the eruption.

Eighteen (82%) of the 22 species arriving on the Pumice Plain established; however, 4 species, the cinereus shrew (*Sorex cinereus*), Douglas's squirrel (*Tamiasciurus douglasii*), bushy-tailed wood rat (*Neotoma cinerea*), and American pika, appear to have failed to establish breeding populations. Our first year of trapping (1983) revealed that two species had colonized since the 18 May 1980 eruption, when no mammals were present on-site. After that time, the rate of increase was approximately linear over the next 32 years. Our linear regression analysis based on successful establishment of 18 species over a 33-year period extending from 1983 to 2015 produces an annual accrual rate of 0.57 species per year ($r^2 = 0.98$, $p < 0.01$, Fig. 15.5b). In most of the years in which we trapped mammals (19 of 30), new species were observed in our sample, and when there was a hiatus in arrival, it was generally 1–2 years; the maximum duration was 3 years (1984–1987). The 4 small-mammal species that consistently went undetected in our samples are forest obligates, including Townsend's chipmunk (*Tamias townsendii*), northern flying squirrel (*Glaucomys sabrinus*), southern red-backed vole (*Myodes* [= *Clethrionomys*] *gapperi*), and the terrestrial insectivore, coast mole (*Scapanus orarius*).

Analysis of annual rates of species gain, loss, and turnover on the Pumice Plain (based on the metrics of Anderson 2007) revealed the long-term dynamics of continuous gains and losses of colonizing mammal species (Fig. 15.6). Annual species gain (G_p) was proportionately high during 1980–1983 as species arrived on the Pumice Plain, but the gain rate dropped to zero through 1987 before increasing again in the late 1980s (Fig. 15.6a). The decade of the 1990s showed considerable fluctuation in species gain rates, eventually dropping to significantly lower levels during 2000–2015 as an increasing proportion of the regional species pool arrived on the Pumice Plain (Fig. 15.6a).

The annual species loss rate (L_p) ranged from 0% during 1980–1983 as mammal species began to accumulate on the Pumice Plain to 53% in 1995 during the decade of the 1990s (Fig. 15.6b). Although no significant trend in species loss rates was observed over the entire observation period (1980–2015, Fig. 15.6b), we did find a significant negative trend from 1987 to 2015 ($L_p = 0.0003Yr^2 – 1.063Yr + 1071$, $r = 0.453$, $df = 24$, $P < 0.05$). Overall, from 1987 through 2015, annual species loss rates averaged 16.4% per year, and annual species gain rates averaged 22.2% per year.

Annual turnover rate (T_p) exhibited a significant stabilization trend over the entire observational record (1980–2015), with turnover steadily declining from a period of fluctuating turnover in the 1980s, through a decade of relatively high turnover rates in the 1990s, to the era of stabilization during 2000–2015 (Fig. 15.6c).

15.7.2 Trapping Results

From 1983 through 2015, we sampled small mammals for 35,978 trap nights, yielding 9623 total captures and 3651 animals, representing 21 species. Deer mice comprised 55% of all unique individuals, and this species, plus two additional rodents, the yellow-pine chipmunk (*Tamias amoenus*) and the Pacific jumping mouse (*Zapus trinotatus*), accounted for 75% of all individuals captured. Rodents, insectivores, and carnivores represented 90.0%, 9.0%, and 1.0% of our samples, respectively.

Table 15.1 Mammal species of the southern Washington Cascade Range. *Light gray shading* indicates species that have arrived on the Pumice Plain; *medium gray shading* denotes species that have established. Species with no shading have not been observed as arriving. Habitat association and guild assignment based on information from *Mammalian Species* (https://academic.oup.com/mspecies) and from Wilson and Ruff (1999). *Obs* observed, *Estab* established, *First obs* year first observed, *Y* yes, *N* no, *U* unknown, *N/A* not applicable.

Order/Family	Scientific name	Common name	obs.	Estab.	First obs.	Literature habitat association	MSH habitat association	Guild
Didelphimorphia								
Didelphidae	*Didelphis virginiana*	Virginia opossum	Y	U	2013	deciduous forest	Lake shoreline, riparian and upland shrub thicket	omnivore
Soricomorpha								
Soricidae	*Sorex bendirii*	marsh shrew	Y	Y	2009	marsh and moist riparian	riparian shrub thicket	insectivore
	S. cinereus	cinereus shrew	Y	U	2011	forest, shrub thicket	riparian shrub thicket	insectivore
	S. monticolus	montane shrew	Y	Y	1990	moist forest and shrub thicket	riparian shrub thicket	insectivore
	S. palustris	water shrew	Y	Y	1994	wet riparian shrub thicket	riparian shrub thicket	insectivore
	S. trowbridgii	Trowbridge's shrew	Y	Y	2004	forest	riparian shrub thicket	insectivore
	S. vagrans	vagrant shrew	Y	Y	2002	mesic forest and shrub thicket	riparian shrub thicket	insectivore
Talpidae	*Neurotrichus gibbsii*	American shrew mole	Y	Y	2010	forest	riparian shrub thicket	insectivore
	Scapanus orarius	coast mole	N	N	N/A	open forest and meadow	not present	insectivore
Carnivora								
Canidae	*Canis latrans*	coyote	Y	Y	1990	generalist	upland early-seral forb and riparian shrub thicket	omnivore
	Vulpes vulpes cascadensis	Cascade red fox	N	N	N/A	forest, subalpine meadow	not present	omnivore
Ursidae	*Ursus americanus*	American black bear	N	N	N/A	forest, meadow	not present	omnivore
Mustelidae	*Martes americana*	American marten	N	N	N/A	mature forest	not present	omnivore
	Mustela erminea	ermine	Y	Y	2003	generalist	riparian shrub thicket	carnivore
	M. frenata	long-tailed weasel	Y	Y	2007	generalist	riparian shrub thicket	carnivore
	Neovison vison	American mink	N	N	N/A	riparian zones of forest and shrub thicket	not present	carnivore
	Lontra canadensis	northern river otter	Y	Y	2012	river, lake, pond, marsh	lake	carnivore
Procyonidae	*Procyon lotor*	northern raccoon	Y	Y	2014	generalist	lake shoreline, riparian shrub thicket	omnivore
Felidae	*Puma concolor*	puma	Y	Y	2008	generalist	upland early-seral forb and riparian shrub thicket	carnivore
	Lynx rufus	bobcat	N	N	N/A	generalist	not present	carnivore
Artiodactyla								
Cervidae	*Cervus elaphus*	elk	Y	Y	1983	forest and meadow	upland early-seral forb and riparian and upland shrub thicket	herbivore
	Odocoileus hemionus columbianus	black-tailed deer	Y	N	1998	forest	upland shrub thicket and lake shorelines	herbivore
Bovidae	*Oreamnos americanus*	mountain goat	Y	Y	2000	alpine meadow, cliffs, scree	upland early-seral forb	herbivore
Rodentia								
Aplodontiidae	*Aplodontia rufa*	mountain beaver	N	N	N/A	moist forest	not present	herbivore
Sciuridae	*Tamias amoenus*	yellow-pine chipmunk	Y	Y	1998	open forest, shrub thicket	riparian and upland shrub thicket	omnivore
	T. townsendii	Townsend's chipmunk	N	N	N/A	forest	not present	omnivore

(continued)

Table 15.1 (continued)

Order/Family	Scientific name	Common name	obs.	Estab.	First obs.	Literature habitat association	MSH habitat association	Guild
	Marmota caligata	hoary marmot	Y	N	2010	montane meadow, talus	lava flows and associated scree	herbivore
	Callospermophilus saturatus	Cascade golden-mantled ground squirrel	Y	Y	1996	open forest, shrub thicket, meadow	riparian and upland shrub thicket	omnivore
	Tamiasciurus douglasii	Douglas's squirrel	Y	N	1992	forest	riparian shrub thicket	granivore
	Glaucomys sabrinus	northern flying squirrel	N	N	N/A	forest	not present	omnivore
Geomyidae	*Thomomys talpoides*	northern pocket gopher	Y	Y	1992	Meadow, open forest	upland early-seral forb	herbivore
Castoridae	*Castor canadensis*	American beaver	Y	Y	2002	Lake, pond, stream	lake	herbivore
Cricetidae	*Peromyscus keeni*	Keen's mouse	Y	Y	1997	Forest, shrub thicket	riparian and upland shrub thicket	omnivore
	P. maniculatus	deer mouse	Y	Y	1983	generalist	upland early-seral forb and riparian and upland shrub thicket	omnivore
	Neotoma cinerea	bushy-tailed woodrat	Y	U	2012	caves, cliffs, rock crevices, scree	lava flows and associated scree	herbivore
	Myodes gapperi	southern red-backed vole	N	N	N/A	forest	not present	omnivores
	Phenacomys intermedius	western heather vole	Y	Y	2013	moist meadow, open forest, riparian	upland early-seral forb	herbivore
	Microtus longicaudus	long-tailed vole	Y	Y	1983	generalist	riparian shrub thicket	herbivore
	M. oregoni	creeping vole	Y	Y	1994	meadow, forest	riparian shrub thicket	herbivore
	M. richardsoni	water vole	Y	Y	1988	wet meadow, riparian	riparian shrub thicket	herbivore
	Ondatra zibethicus	muskrat	N	N	N/A	Marsh, pond, lake, stream	not present	omnivore
Dipodidae	*Zapus trinotatus*	Pacific jumping mouse	Y	Y	1991	riparian, moist meadow, shrub thicket	riparian shrub thicket	omnivore
Erethizontidae	*Erethizon dorsatum*	North American porcupine	Y	N	2007	forest	riparian shrub thicket	herbivore
Lagomorpha								
Ochotonidae	*Ochotona princeps*	American pika	Y	U	1994	talus, scree	lava flows and associated scree	herbivore
Leporidae	*Lepus americanus*	snowshoe hare	Y	N	1983	forest	riparian and upland shrub thicket	herbivore
Totals: **6**/19	45		34	25				

Based on mark-recapture live-trapping for the dominant species, within-year recapture values were very high in most cases, and varied across species, and between riparian and upland habitats (Table 15.2). All species except the deer mouse had higher recapture values in riparian versus upland habitats, and this was most pronounced for three species: Cascade golden-mantled ground squirrel (*Callospermophilus saturatus*), yellow-pine chipmunk, and northwestern mouse (*Peromyscus keeni*).

Mean between-year recapture rates were generally low, had high variance, and showed marked differences among species and across upland and riparian habitats (Table 15.2). Not surprisingly, long-lived species (e.g., yellow-pine chipmunk and Cascade golden-mantled ground squirrel) generally had more recaptures between years as compared with shorter-lived species, but there were some exceptions to this pattern, for example, the Pacific jumping mouse in upland habitats and northwestern mouse in riparian habitats.

15.7.3 Mammals in Upland and Riparian Habitats

Species richness at the PMPWEB upland site changed very little over a 32-year sampling period (1984–2015), ranging from 1 to 4 species, and during most years, only 1 to 2 species

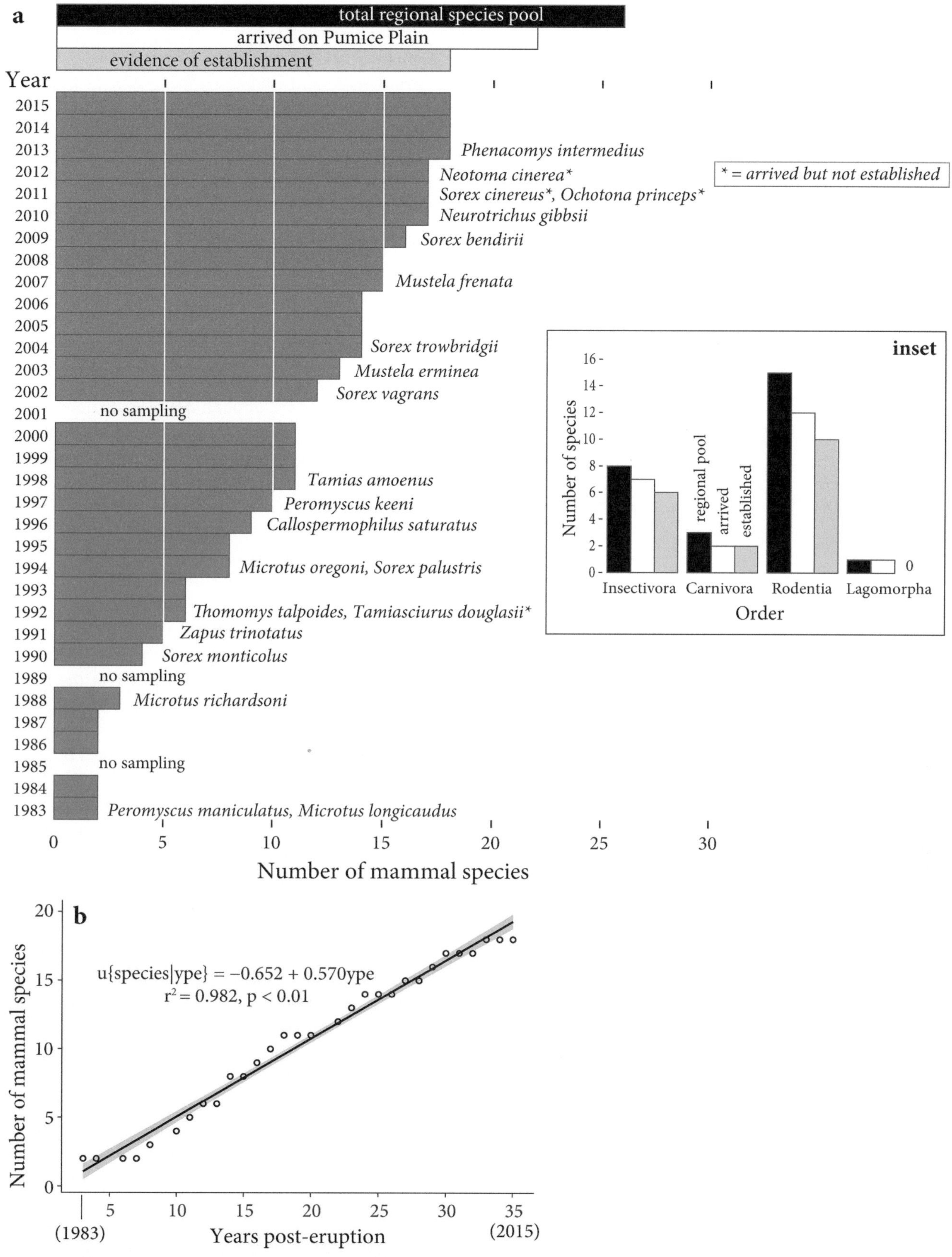

Fig. 15.5 (**a**) Small-mammal species accrual rate from 1983 to 2015, based on live-trapping at 16 sites on the Pumice Plain at Mount St. Helens; includes only taxa that could be captured in Sherman trap. Latin binomials are listed for the year that a species was first trapped. The graph is cumulative; newly arriving species are added to all previously recorded species, but this does not imply that all species were present in any given year. See text for additional information. The *inset* graph shows the number of small-mammal species, listed by taxonomic order, in the regional species pool, arriving on the Pumice Plain, and having established. (**b**) Simple linear regression plot of small-mammal species accrual on the Pumice Plain at Mount St. Helens (1983–2015). *Gray band* represents 95% confidence limit on the mean estimated number of species.

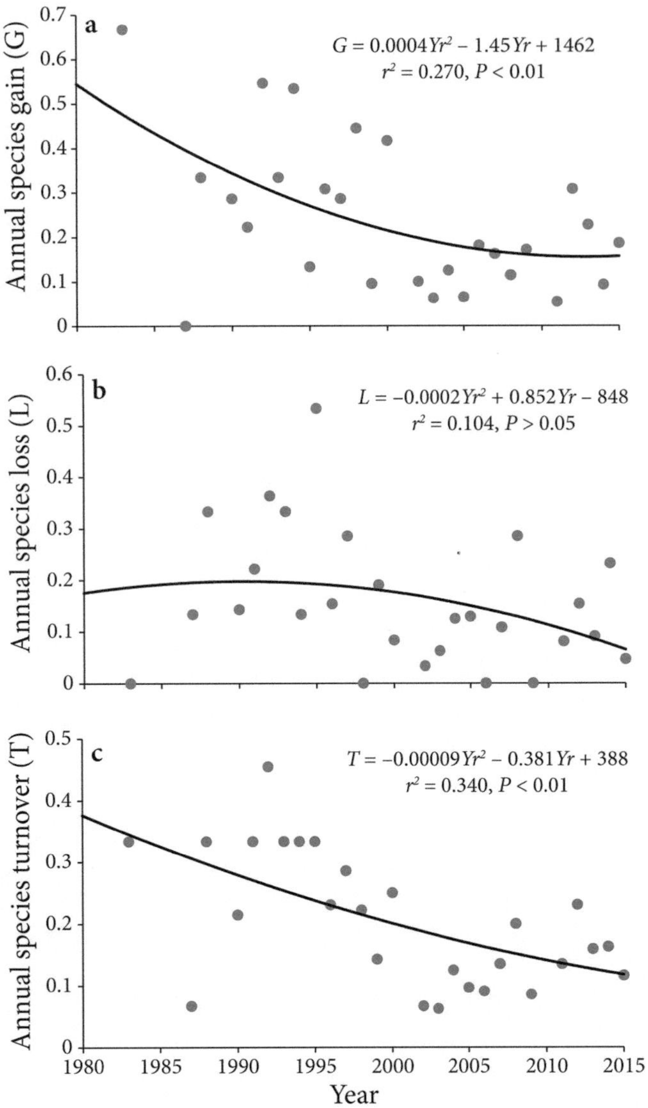

Fig. 15.6 Successional patterns of mammal species (**a**) gain rates, (**b**) loss rates, and (**c**) turnover rates on the Pumice Plain at Mount St. Helens.

were represented (Fig. 15.7a). In contrast, species richness in the WILSPR riparian site increased over the 28-year sampling period (1988–2015), with a maximum value of 10 species in 2008 (Fig. 15.7a).

Small-mammal capture rates at PMPWEB upland site were very consistent during the period of record, with <5 captures per 100 trap nights for all but one of the 32 years (Fig. 15.7b). The WILSPR riparian site had highly variable capture rates, with six clear swings from low to high during the 28-year period of record that included values ranging from 10 to 45 captures per 100 trap nights (Fig. 15.7b). The riparian site commonly had capture rates 4–12 times greater than those at the upland site.

Foraging-guild structure at the PMPWEB upland site included two guilds and, based on both biomass and number of captures, was dominated by terrestrial omnivores (Fig. 15.8). Within this three-member guild, the deer mouse consistently contributed the greatest biomass to the mammal assemblage (Fig. 15.8b) and also had the highest capture rate (Fig. 15.8a). The remaining four species in our sample represented a small percentage of the fauna and were temporally variable in their presence. The yellow-pine chipmunk, first detected in 2002 and present on and off through 2015, contributed 10–25% of the small-mammal biomass during this period. The long-tailed vole, Pacific jumping mouse, and water shrew were rarely observed and likely represented dispersing individuals rather than residents.

The WILSPR riparian site exhibited a complex foraging-guild structure that included six guilds, each with diverse membership. From 1988 through 2015, terrestrial omnivores consistently contributed the greatest biomass to the assemblage, but the contribution of this biomass shifted dramatically within this guild from deer mouse dominance from 1988 through 2000 to dominance by the diurnal rodents Cascade golden-mantled ground squirrel and yellow-pine chipmunk from 2002 through 2015 (Fig. 15.9a; data shown for 1988–1999 are for WILSPR only; see caption for explanation). Pacific jumping mouse biomass contribution was temporally variable and ranged from 0% to ~30%, whereas northwestern mouse was the most variable species, frequently absent and never contributing >5% of the biomass. The terrestrial herbivore foraging guild at WILSPR riparian site had three-member species, all ***microtine rodents***, of which a subset of these species was represented in 18 of the 26 years sampled (Fig. 15.9b; data shown for 1988–1999 are for WILSPR only; see caption for explanation). The biomass allocation was primarily contributed by the water vole, which in many years (11 of 26) was the only member of this guild present. The long-tailed vole was present in 5 years and contributed all the biomass for this guild in 3 of 26 years, and creeping vole was only present during 2 of the 26 years and made a minor contribution to the guild biomass. The terrestrial herbivore guild was absent from the site in 8 of 26 sampling years from 1988 through 2015. The insectivore guild first appeared at this site in 1990 (10 years post eruption) and was present most years (18 of 26) through 2015 (Fig. 15.9c; data shown for 1988–1999 are for WILSPR only; see caption for explanation). This guild had six members and included all but two members found in the regional species pool. Member species increased through time and reached a high of six species in 2012, after which time there was a decline that ranged from two to five species each year. Biomass has largely been dominated by the semiaquatic water shrew, followed by the terrestrial montane shrew. The remaining four species have been temporally variable in their presence and abundance and thus biomass contribution.

For our comparison between paired riparian and upland sites, mean species richness values were consistently higher

Table 15.2 Mean recapture rates within trapping sessions and between years.

	Mean annual recapture rate within sessions						Mean annual recapture rate between sessions					
	Upland			Riparian			Upland			Riparian		
Species	Mean [±SE]	No of years	Total captures	Mean [±SE]	No. of years	Total captures	Mean [±SE]	No. of years	Total captures	Mean [±SE]	No. of years	Total captures
Cascade golden-mantled ground squirrel	9.26% [6.28]	9	17	41.80% [9.36]	13	97	0 [0]	9	18	5.92 [3.18]	11	84
Long-tailed vole	NA		NA	64.53% [15.38]	7	31	0 [0]	2	2	0 [0]	5	23
Water vole	NA		NA	59.88% [10.04]	10	68	NA		0	0 [0]	8	44
Northwestern mouse	39.00% [11.69]	10	31	74.76% [8.92]	13	63	0 [0]	9	27	5.29 [3.84]	9	53
Deer mouse	68.89% [1.93]	15	1006	60.06% [4.16]	15	439	1.07 [0.41]	13	783	1.66 [1.07]	13	354
Yellow-pine chipmunk	22.49% [6.31]	13	67	41.68% [5.07]	14	203	13.42 [6.28]	11	52	6.44 [2.43]	13	176
Pacific jumping mouse	7.14% [7.14][a]	7	12	12.79% [3.92]	15	164	8.33 [8.33][a]	6	11	0.95 [0.67]	13	185

NA not applicable.
[a]These values reflect only a single recapture across all years.

(2–6×) in the riparian sites compared with the upland sites. Mean species richness in upland sites increased slightly from 2000 to 2015 but remained low (1–3 species) (Fig. 15.10a). There was low within-year variance across upland sites. Mean species richness in the riparian sites was relatively high (~6 species) in 2000 and then fluctuated up or down over the ensuing 15 years (Fig. 15.10a).

Mean capture rates in the upland sites were low, fluctuating from about 6 to 15 captures per 100 trap nights, and showed multi-year periods of both highs and lows (Fig. 15.10b). Mean capture rates in the riparian sites were consistently higher than the upland sites (~20 to 35 captures/100 trap nights), yet the two habitats showed a strikingly similar temporal pattern of annual variation during the 15 years of measurements (Fig. 15.10b).

Terrestrial omnivore was the dominant foraging guild in terms of both biomass and capture rates in the upland sites from 2000 through 2015 (Fig. 15.11). During this period, the number of omnivore species increased from one species in 2000 to four to six species from 2012 through 2015. Deer mouse contribution to the overall guild biomass decreased steadily throughout the period of record, whereas that of two diurnal rodents, Cascade golden-mantled ground squirrel and yellow-pine chipmunk, increased. The remaining three species made minor contributions to the guild biomass.

Five foraging guilds were recorded at our riparian sites from 2000 through 2015, and the terrestrial omnivore guild contributed the greatest biomass (Fig. 15.9a, d; for Fig. 15.9a, data shown for 1988–1999 are for WILSPR only). This guild had five member species, and each of their biomass contributions was relatively stable across most years, except in 2000 when deer mice were dominant. In most years, the deer mouse, yellow-pine chipmunk, and Pacific jumping mouse had the highest capture rate (Fig. 15.9d), and the yellow-pine chipmunk and Cascade golden-mantled ground squirrel contributed the greatest biomass (Fig. 15.9a). The herbivore guild was present in the assemblage every sampling year and included three members, all voles. The presence of these species varied tremendously through time, with the water vole generally contributing the most biomass from 2000 through 2009, followed by a shift to the creeping vole and long-tailed vole (Fig. 15.9b). The insectivore guild was present through most of the period of record (1990–2015) and represented by six species. The number of member species present each year ranged from 2 to 6, and the overall diversity increased from 2000 through 2015 (Fig. 15.9c; data shown for 1988–1999 are for WILSPR only). The semiaquatic water shrew had the highest capture rate (Fig. 15.9d) and contributed the greatest biomass to this guild (Fig. 15.9c). The semi-fossorial American shrew mole (*Neurotrichus gibbsii*) first appeared in 2010 and has been present periodically since. Carnivores, ermine (*Mustela erminea*) and long-tailed weasels (*M. frenata*), appeared sporadically in our sample (Fig. 15.9d).

15.7.4 Mammal Sex Ratios

Seven of the 11 species assessed for sex ratios in the riparian habitat were not significantly different from 1:1, and, of the remaining four species, there were two species with male or female biases (Table 15.3). In the upland habitat, where only

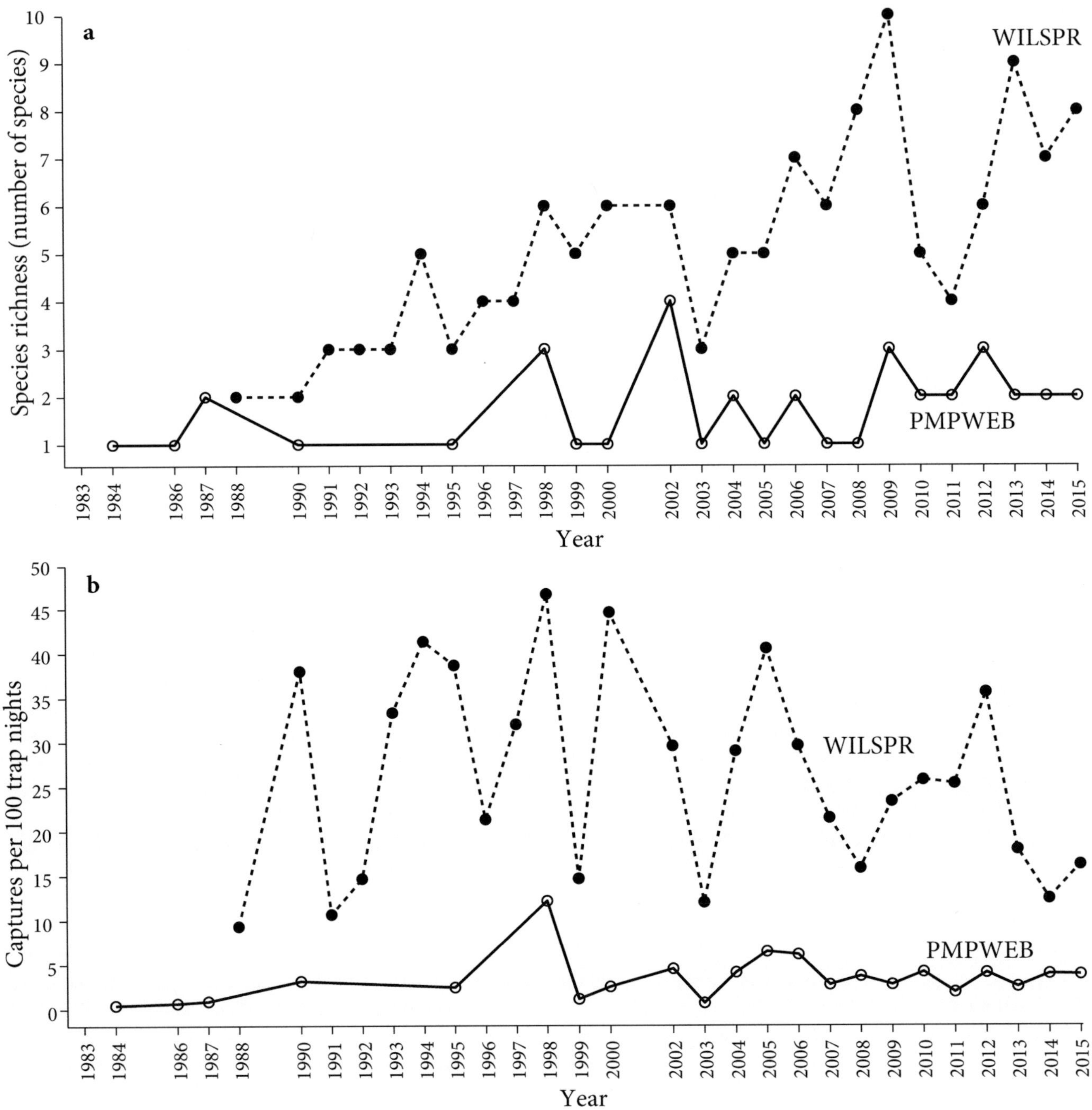

Fig. 15.7 Small-mammal (**a**) species richness and (**b**) capture rate (number captures/100 trap nights) live-trapped at two long-term study sites representing upland (PMPWEB) and riparian (WILSPR) habitats on the Pumice Plain at Mount St. Helens. Capture rate data were standardized for equal sampling effort (4 nights/year).

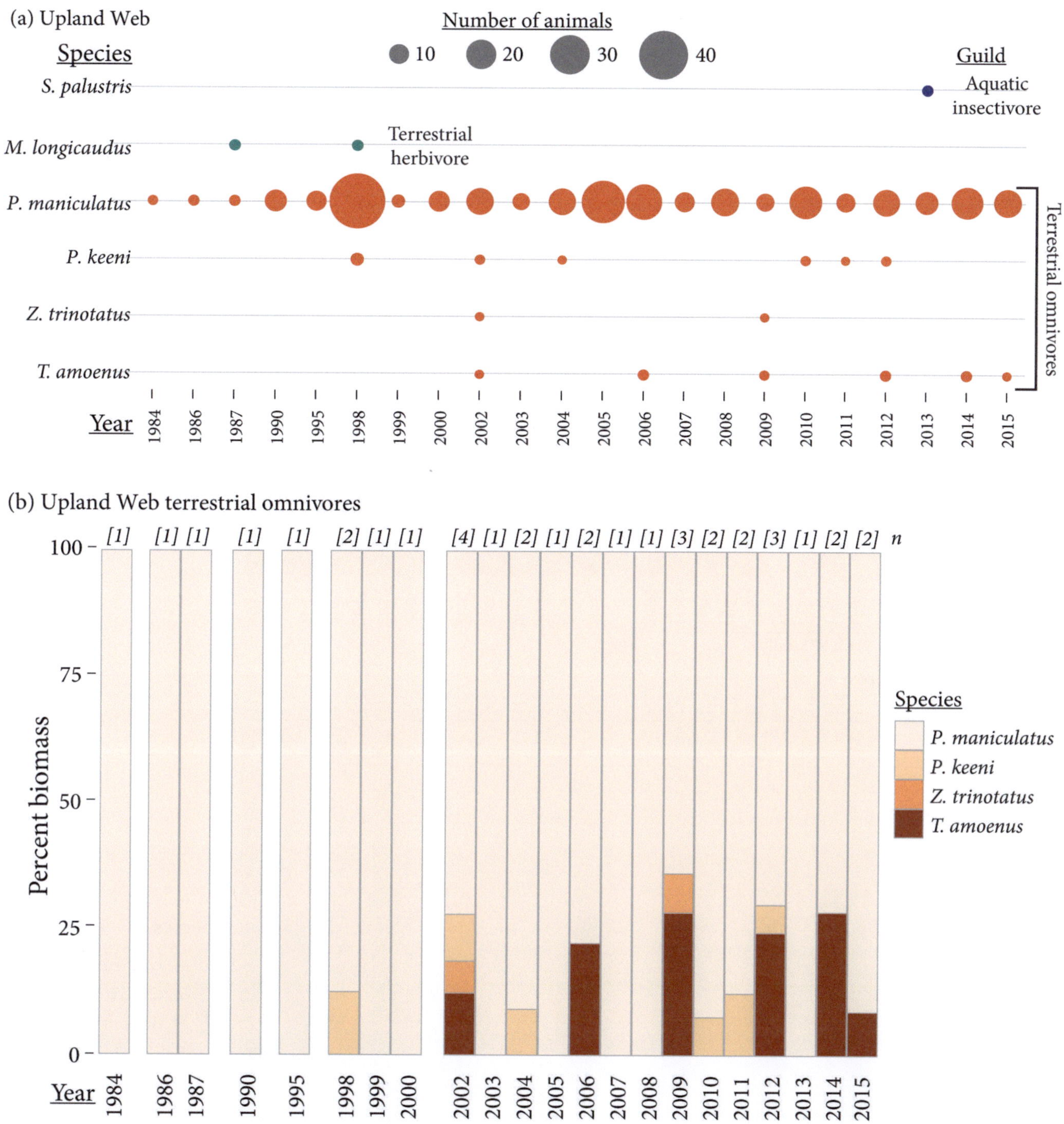

Fig. 15.8 (**a**) Frequency of occurrence and relative abundance of small-mammal species according to foraging guild at our long-term upland trapping site (PMPWEB) on the Pumice Plain at Mount St. Helens. (**b**) Guild structure expressed as percent biomass attributed to the terrestrial omnivore foraging guild at the upland site (1984–2015).

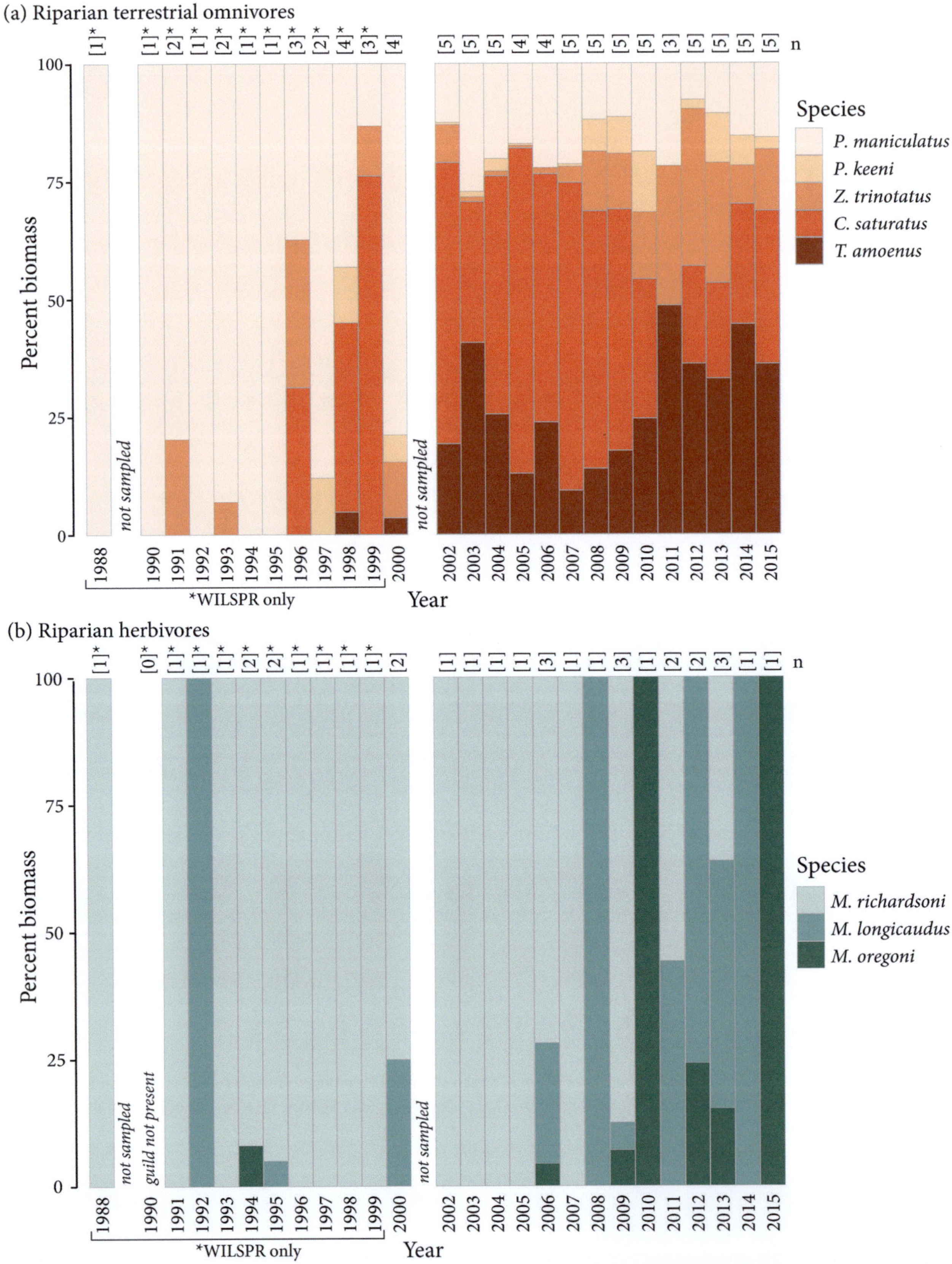

Fig. 15.9 Foraging-guild structure expressed as percent biomass for (**a**) terrestrial omnivore, (**b**) herbivore, and (**c**) insectivore guilds at riparian sites on the Pumice Plain at Mount St. Helens from 1988 through 2015. Data for 1988 through 1999 are only for WILSPR site, whereas 2000–2015 are for replicate sites. (**d**) Small-mammal species frequency of occurrence and relative abundance according to foraging guild.

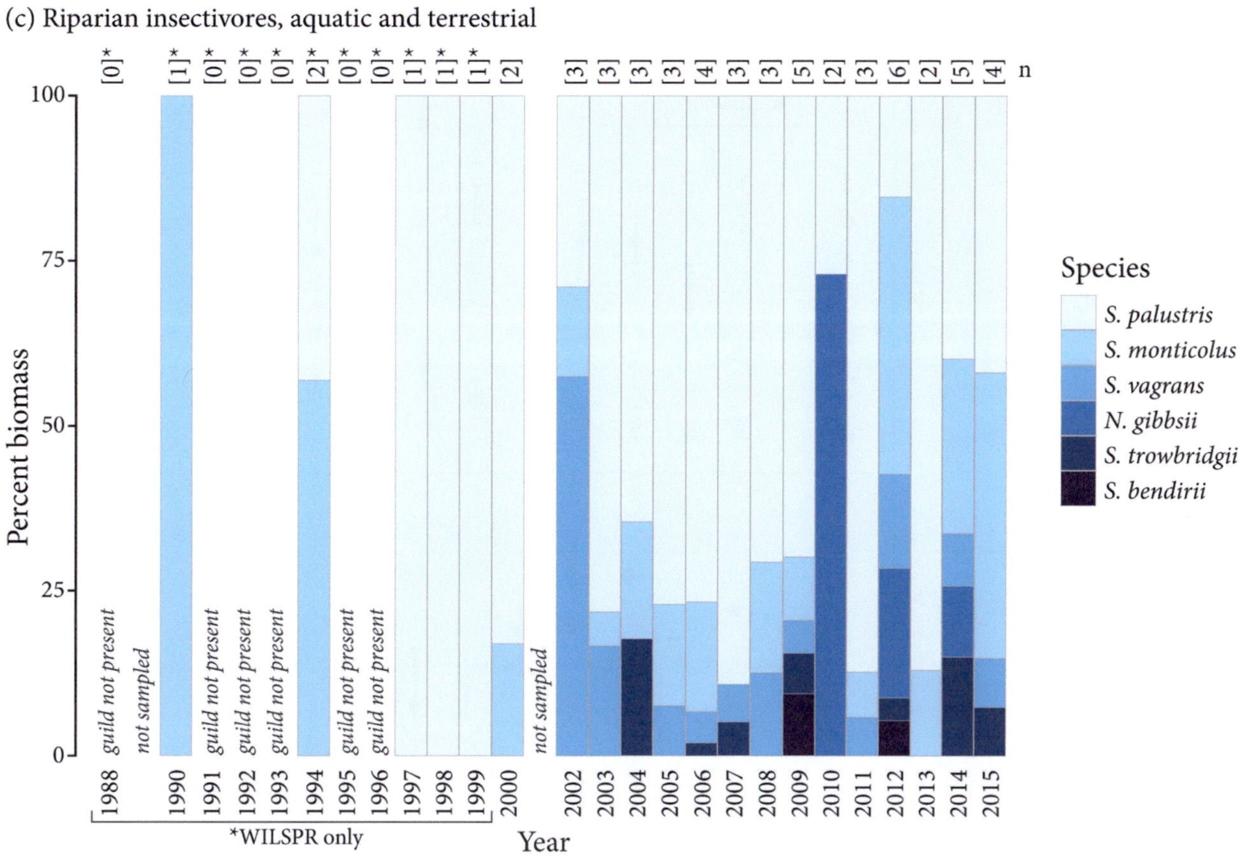

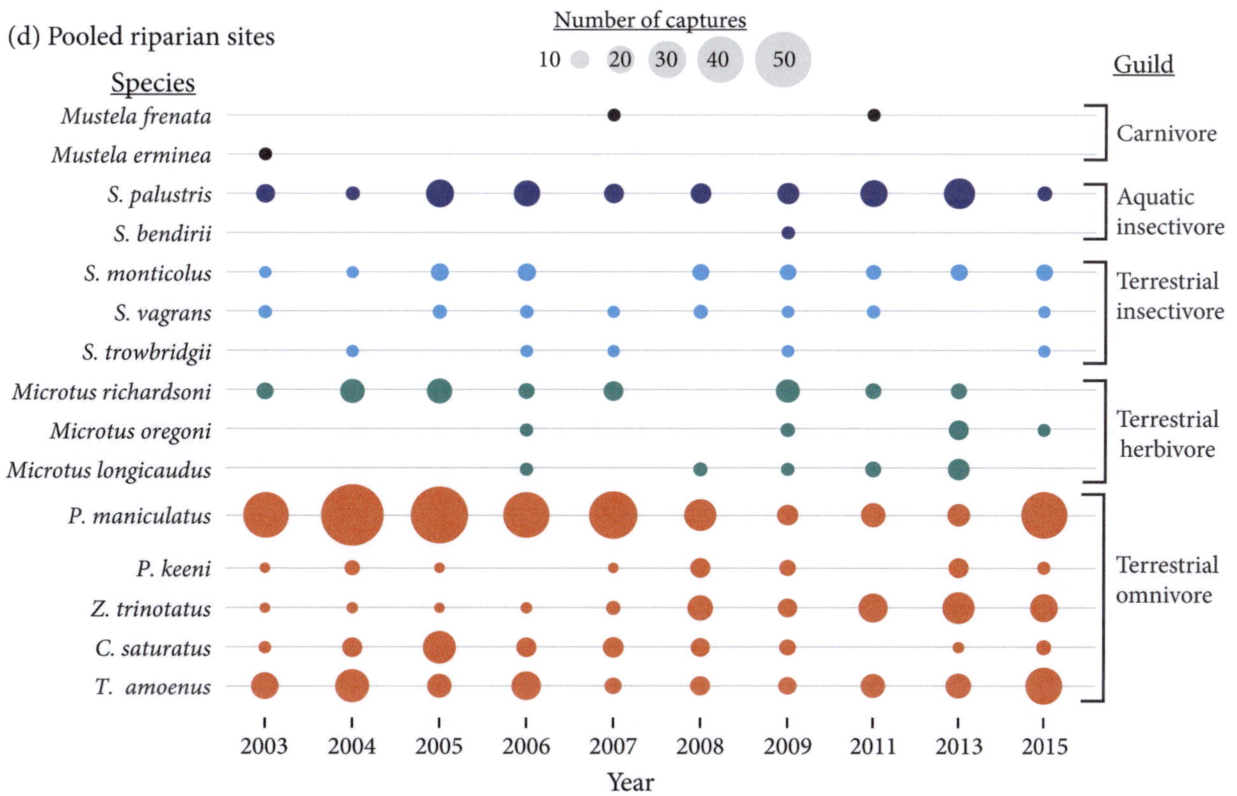

Fig. 15.9 (continued)

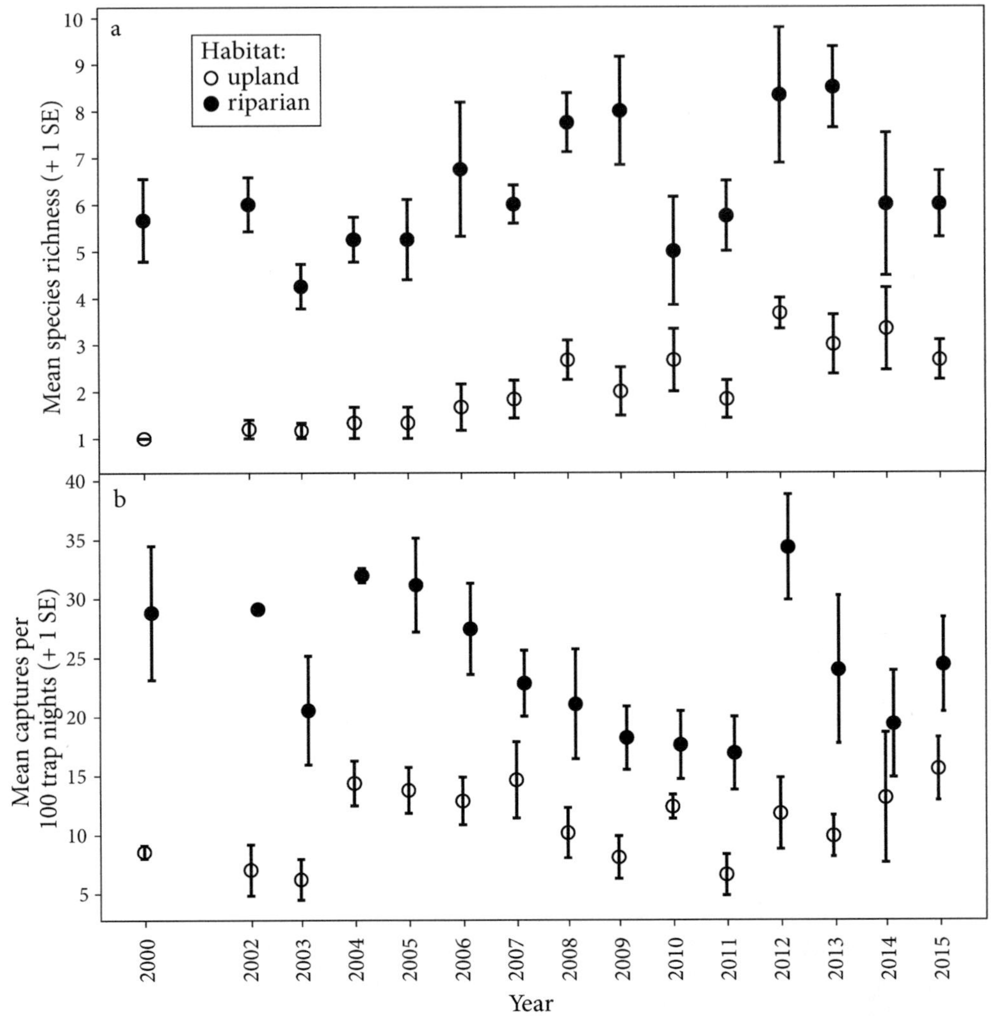

Fig. 15.10 (**a**) Number of small-mammal species and (**b**) their capture rates (number captures/100 trap nights) live-trapped at replicate study sites on the Pumice Plain at Mount St. Helens from 2000 through 2015. Data are presented as means ±1 SE.

four species were assessed for sex ratios, two species had significant male bias, and the remaining two species were not different from 1:1 (Table 15.3).

15.7.5 Vegetation Structure and Mammal Species Richness

Welch two-sample t-test showed a significant difference ($t = 17$, df = 11.6, $p < 0.001$) in vegetation structure between riparian and upland habitats. Mean percent vegetation structure was estimated to be 55.98% (95% CI, 38.4–49.5%) greater in riparian than upland habitats. Vegetation structure was somewhat bimodal, with upland matrix sites having low complexity and riparian shrub having comparatively high complexity. Vegetation structure increased in the riparian shrub sites from 2009 to 2013 as shrubs grew and spread in the mesic habitats. This pattern was not observed in the upland matrix sites, where the vegetation structure was similar between sample years (Fig. 15.12).

As expected, small-mammal species richness was positively related to vegetation complexity for both years assessed but most evident for 2013. The paired t-test showed that mammal species richness was significantly different ($t = 5.8$, df = 7, $p < 0.001$) between riparian and upland habitats. The mean species richness was 3.6 species greater in riparian habitat (95% CI, 2.1–5.1 species).

The plant-level attack rates of willow stem-boring beetles on upland willow shrubs for years 2010, 2011, and 2015 were 83.5, 67.9, and 79.2, respectively. The mean attack rates on willow stems were 38.7, 33.4, and 45.7%. The mean mortality rate of attacked stems on upland shrubs from 2009 to 2010 was 67.0% (95% CI, 64.0–71.0) (Chris Che-Castaldo, unpublished data). Thus, most upland willow shrubs were attacked; on those attacked shrubs, more than one-third of the stems were attacked; and most (67%) of attacked stems died. Plants generate new stems annually, but the effect incurred from this herbivory prevents increases in shrub volume owing to stem turnover.

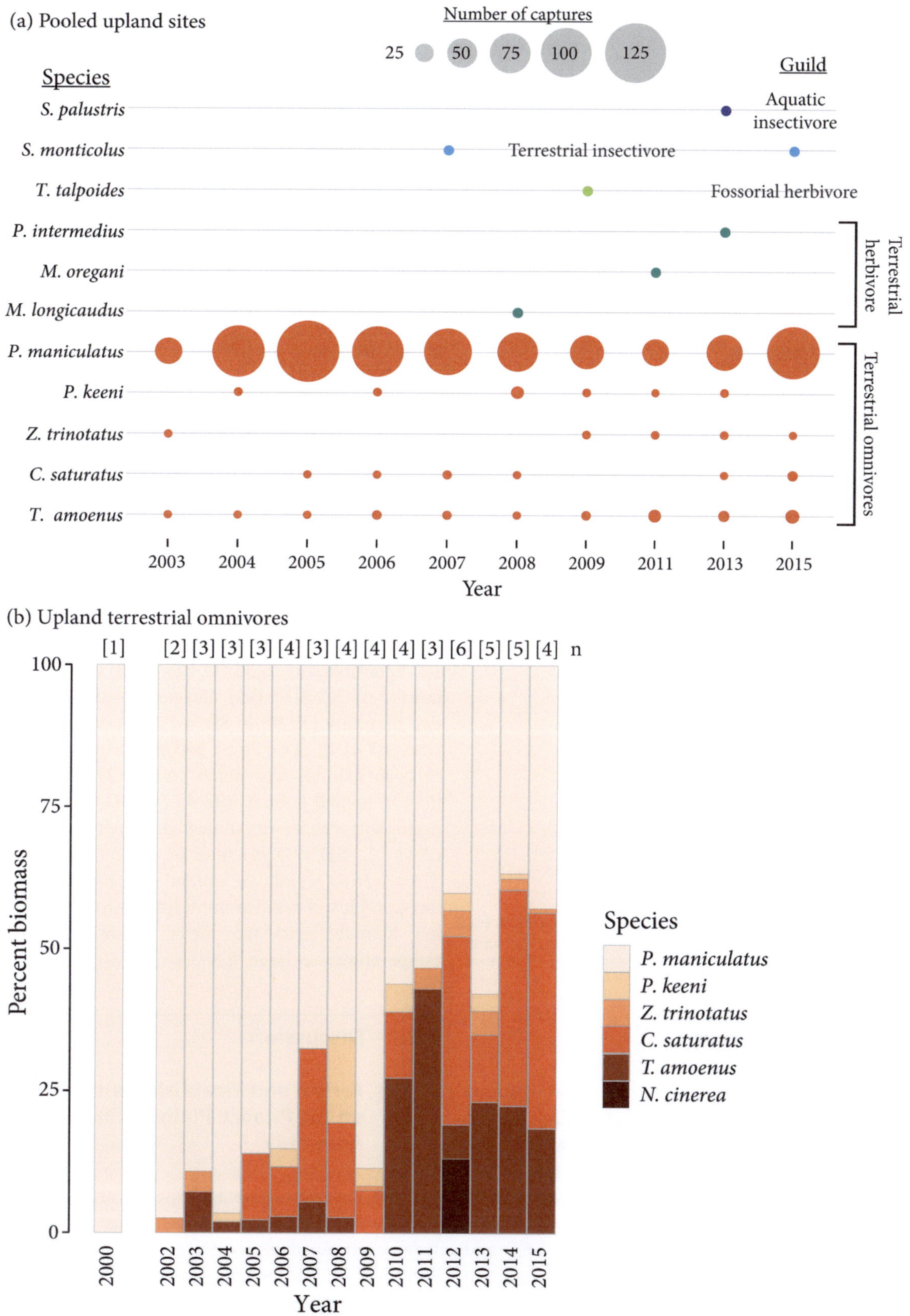

Fig. 15.11 (**a**) Frequency of occurrence and relative abundance of small-mammal species according to foraging guild, and guild structure expressed as percent biomass for the terrestrial omnivore guild (**b**) at replicate upland sites on the Pumice Plain at Mount St. Helens from 2000 through 2015.

Table 15.3 Proportion of female captures (%), with 95% confidence intervals for a one-sample test of equal proportions.

	Riparian/wetland				Upland			
Species	*n*	% female	ChiSq	95% CI	*n*	% female	ChiSq	95% CI
Montane shrew	61	59	1.64	[46–71]				
Water shrew	**70**	**66**	**6.3**	**[53–76]**				
Vagrant shrew	34	47	0.029	[30–65]				
Yellow-pine chipmunk	**265**	**43**	**5.45**	**[37–49]**	**72**	**35**	**6.13**	**[24–47]**
Cascade golden-mantled ground squirrel	107	49	0.037	[39–58]	20	60	0.45	[36–80]
Northwestern mouse	67	49	0	[37–62]	32	47	0.031	[30–65]
Deer mouse	**565**	**41**	**19.14**	**[36–45]**	**1043**	**44**	**12.46**	**[41–48]**
Long-tailed vole	38	47	0.026	[31–64]				
Creeping vole	42	52	0.024	[37–68]				
Water vole	80	55	0.061	[44–66]				
Pacific jumping mouse	**228**	**68**	**27.37**	**[61–73]**	14	64	0.64	[36–86]

Statistically significant values ($\alpha = 0.05$) are shown in **bold** face.

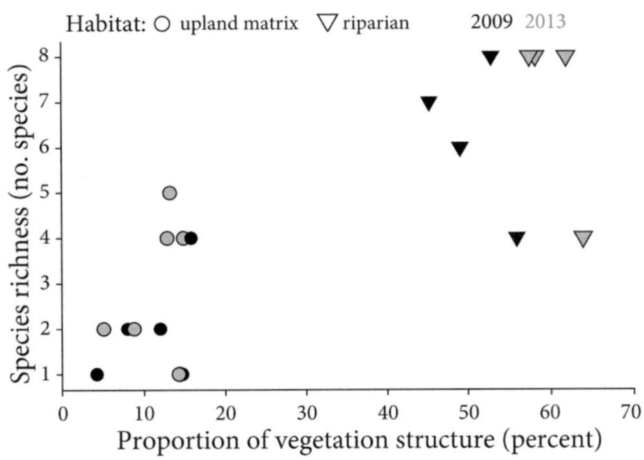

Fig. 15.12 Relationship between small-mammal species richness and three-dimensional vegetation structure on the Pumice Plain at Mount St. Helens in 2009 (*black*) and 2013 (*gray*).

15.7.6 Mammal Movement

We used our long-term (2000–2015) records to examine movement of tagged mammals among all study sites. In total, we had 126 movement observations that included 24 site combinations and 6 species. The deer mouse and yellow-pine chipmunk accounted for 81% of the observed movements. The northwestern mouse and Pacific jumping mouse moved the greatest distances, 1584 and 1021 m, respectively. However, most movements were in the 200- to 400-m range across the six species. Figure 15.13 gives a graphic representation of our mammal movement data.

15.7.7 Temperature and Relative Humidity: Comparing Upland and Riparian Habitats

Average July and August temperatures were cooler by ~2 °C in riparian (13.38 °C, SD = 5.2) than in upland habitats (15.58 °C, SD = 6.3), and riparian sites had higher relative humidity (79.96%, SD = 19.6) compared with upland sites (67.44%, SD = 26.5). Mean maximum temperature of riparian habitats was much cooler (27.62 °C, SD = 2.0) than in upland habitats (33.43 °C, SD = 2.6). Mean minimum temperature was somewhat cooler in riparian habitats (2.73 °C, SD = 0.96) than in upland habitats (3.42 °C, SD = 1.36). Overall, the upland sites had warmer temperatures and lower relative humidity compared with the riparian habitats, and the mean maximum difference in temperature was large, 5.8 °C.

15.8 Discussion

15.8.1 Arrival and Establishment of Mammals on the Pumice Plain During Community Assembly

Overall, the rate of mammal species arrival on the Pumice Plain was surprisingly fast and was undoubtedly much faster than our data demonstrate, as our detection of species' arrival must be interpreted as a "best approximation." Notwithstanding, at least 34 of the 45 mammal species in the regional species pool had arrived on the Pumice Plain from 1983 to 2015, and at least 25 of these had established

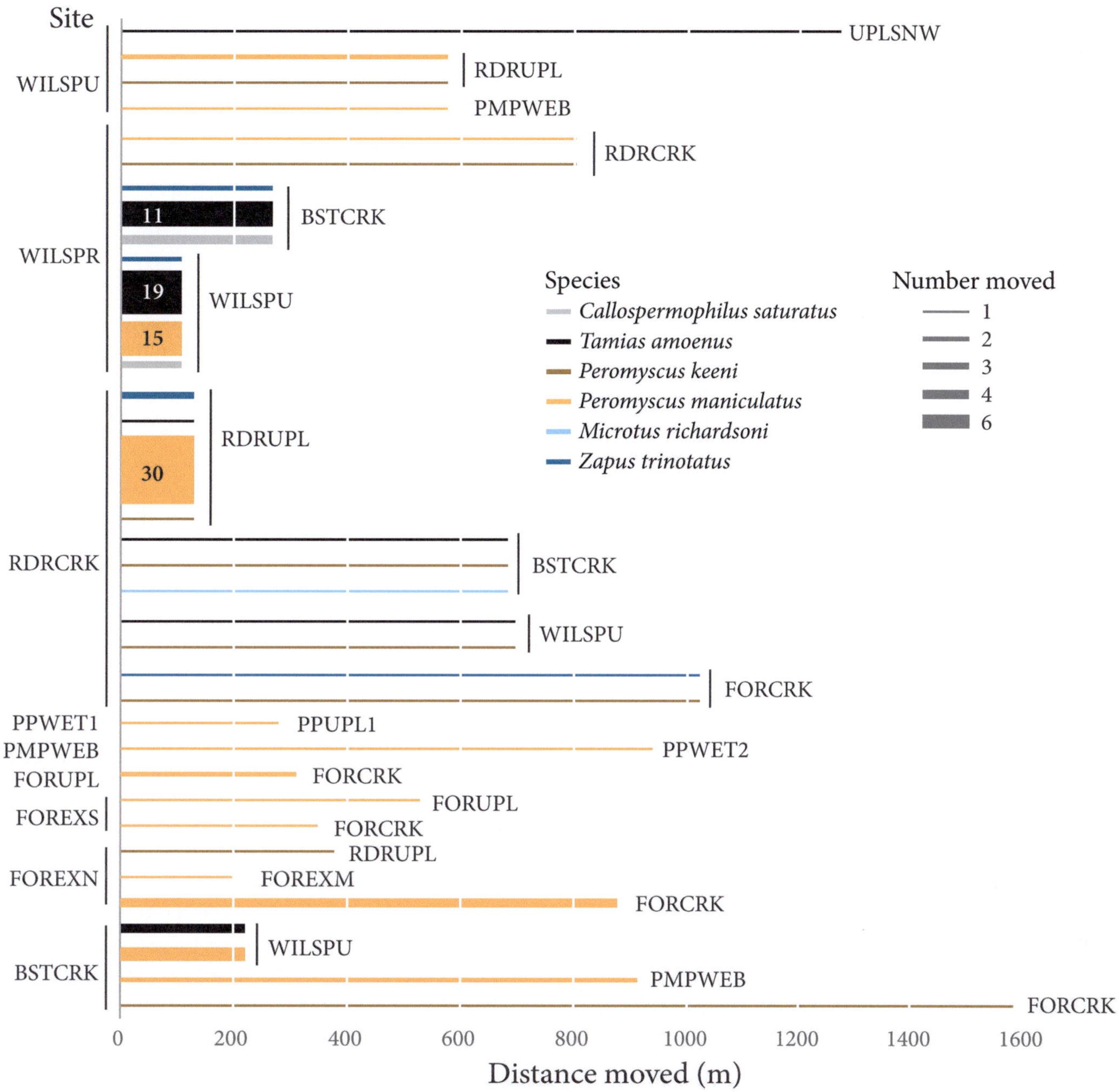

Fig. 15.13 Movement of small-mammal species between study sites on the Pumice Plain based on mark-recapture live-trapping, 2000–2015. Line color denotes species, and line thickness represents the number of animal movement observations. For additional information about these study sites, see Table 15.1 and Fig. 15.4.

breeding populations. Relating this to our community-assembly conceptual model (Fig. 15.2) provides clear evidence that source populations for the majority of species belonging to the regional species pool were available in the surrounding landscape. The presence of source populations was supported by early work in less-disturbed areas adjacent to the Pumice Plain (Andersen and MacMahon 1985a; Adams et al. 1986; Crisafulli et al. 2005), and it also was not surprising, given the relatively intact condition of forested areas beyond the reach of the blast PDC. The availability of source populations, and hence community assembly, would have been very different if the eruption had occurred in the summer months, when mammals did not benefit from a protective layer of snow, or at night (coinciding with the nocturnal activity of most mammals), potentially leading to much lower survivorship in the adjacent disturbance zones and a more protracted process of arrival on the Pumice Plain.

Consistent with our prediction, dispersal, the first process in the conceptual model, was not an obstacle to most species arriving in our study area during the first few post-eruption decades. Key factors influencing dispersal success, such as distance, species-specific dispersal capacity, landscape/waterscape permeability, and chance, undoubtedly played a role, but the extent to which each of these factors operated is unknown. Dispersal distance was generally (but not always) inversely related to body size, for all large mammals (e.g., elk, mountain goats [*Oreamnos americanus*], black-tailed deer [*Odocoileus hemionus columbianus*], black bear [*Ursus americanus*], and puma [*Puma concolor*]) were killed within the 570-km^2 blast area, creating a large (several to tens of km depending on species) distance between source populations in the regional species pool and the Pumice Plain. This is in stark contrast to most small mammals, among which at least some individuals of many species survived in the adjacent disturbance zones, resulting in much shorter dispersal distances (few to several km) for these taxa. A number of these species (e.g., deer mouse and long-tailed vole) were the first mammals to arrive on the Pumice Plain. On the other hand, large mammals, such as elk, obviously have a much greater capacity to disperse because of their size and energy reserves as compared to smaller mammals, yet they arrived the same year as the rodents. Additional information on dispersal distance, species dispersal capacity, and permeability are discussed below.

Successful dispersal and arrival of mammals on the Pumice Plain was no guarantee of establishment, the second process of community assembly. We predicted that the key factor governing establishment would be development of plant communities that could provide requisite habitat resources such as food, reproduction sites, and cover from weather and predators and contribute to the development of organic matter and soil. Further, we predicted that generalist species associated with early-seral conditions would be among the first species to establish, with fossorial and habitat specialists next. Those species associated with mature forest conditions would be the last to establish and not be likely to appear until forest develops. For the 18 species of small mammals that established between 1983 and 2015, our results generally support our predictions but with some exceptions. Species associated with early-successional herb/grass communities or habitat generalists were the first species to establish, but species with these characteristics continued to establish through 2013. One habitat specialist established early on, and three additional species later. Three of these species are riparian-meadow or shrub-thicket specialists, and the fourth is associated with high organic content on the ground surface. The number of shrub-thicket-associated mammal species increased through time and coincided with establishment and spread of willow and alder shrubs. In all cases, the establishment of mammal species was related to vegetation and litter conditions, corresponding to the habitat development factor of our model under the establishment process (Fig. 15.2).

A general overview of arrival and establishment of mammals on the Pumice Plain includes seven general patterns. First, three species arrived early in our study and were consistently present throughout or during most years. These included a habitat generalist (the deer mouse in 1983), an early-seral meadow specialist (the long-tailed vole in 1983), and a riparian specialist (the water vole in 1988).

Second, most species arriving on the Pumice Plain appeared sporadically in our samples for several years prior to maintaining a consistent or nearly consistent presence. The Cascade golden-mantled ground squirrel and Pacific jumping mouse are two examples of the numerous taxa following this pattern. Presumably, this pattern emerged either because individuals arriving on the Pumice Plain were few and the probability of capture was low, and/or habitat and important resources were in short supply or temporally variable and species failed to gain a foothold and establish, essentially the "environmental filter" of our conceptual model.

Third, by 2000–2002 most of the species, such as the northwestern mouse and montane shrew that were temporally variable in their presence up to this time, began to persist from that point forward. This pattern marked a major transition in the assembly process, which likely coincided with increased arrival from populations that had recovered more quickly in adjacent zones that were less disturbed by the eruption. But also, in the 20 years that had elapsed since the eruption, key characteristics of the habitat had developed (plant cover, organic matter, soil genesis) on the Pumice Plain, providing food (and other) resources for herbivores and indirectly supporting arthropods consumed by omnivores and insectivores. Production of surplus individuals in adjacent zones could by itself increase the "propagule pressure" toward the Pumice Plain, leading to apparent persistence even if there was high turnover of individuals because of poor habitat quality. If this phenomenon were to prove correct, the Pumice Plain could be viewed as a sink habitat (*sensu lato* Pulliam 1988). Our low between-year recapture rates for most taxa hint that this situation is possible (Table 15.2). We know that plant cover increased (del Moral et al. 2012) and substrate conditions were ameliorated through time (Halvorson et al., Chap. 5, this volume), but the extent to which increased immigration versus increased habitat suitability influenced persistence is unknown. Regardless, as plant succession advanced in areas adjacent to our study area, permeability likely increased from the early post-eruption days to two decades later, when species persistence increased.

Fourth, a suite of additional mammal species not limited by dispersal arrived on the Pumice Plain but lacked evidence of establishment, probably because appropriate habitat did not exist. This included talus specialists such as the bushy-tailed wood rat, American pika, and hoary marmot and forest obligates such as the Douglas's squirrel, snowshoe hare (*Lepus americanus*), and North American porcupine (*Erethizon dorsatum*). Up to 2015, the lack of appropriate resources created a sink habitat that apparently precluded these species from establishing. In the case of the forest-associated species, over the next few decades, it is likely that the development of coniferous forest will facilitate their establishment.

Fifth, one group of mammal species that appears to be rare in the regional species pool arrived late in the assembly process (i.e., 29–33 years post eruption), was represented by few captures, and was temporally variable in their presence. This includes the marsh shrew, cinereus shrew, and heather vole (*Phenacomys intermedius*), all of which have been absent or rare in other long-term studies at MSH (MacMahon et al. 1989; Crisafulli et al. 2005) and in the region (West 1991; Carey and Johnson 1995; Carey and Harrington 2001; Gitzen and West 2002). The marsh shrew is likely limited by the lack of suitable habitat (e.g., American skunk cabbage [*Lysichiton americanus*], red alder wetlands), whereas the underlying reasons for the observed pattern for the heather vole and cinereous shrew remain unclear.

Sixth, two small carnivores, the ermine (first detected in 2003) and the long-tailed weasel (first detected in 2007), arrived late in the assembly process and have been sampled infrequently and at low capture rates. Their low and infrequent captures were likely influenced by our sampling methods. Had we used Tomahawk traps baited with fish or chicken, capture rates might have been much higher. Regardless, the timing of their initial detection and their subsequent capture rate coincides with the persistence of small mammals, their primary prey.

Seventh, there are several other species in the regional species pool that have gone undetected in our 1983–2015 sampling effort. These species include forest obligates—Townsend's chipmunk, southern red-backed vole, and northern flying squirrel—as well as a fossorial insectivore (coast mole) and semi-fossorial herbivore (mountain beaver, *Aplodontia rufa*). It is likely that the forest-obligate species are limited by landscape permeability issues in terms of arrival and the lack of forest habitats in terms of establishment, rather than dispersal distance per se, whereas the coast mole and mountain beaver are likely limited by distance between source populations and the Pumice Plain.

15.8.2 1983 Through 1989

Here, we take a more detailed look at the pattern of mammal arrival and establishment during four timeframes: 1983–1989, 1990–2000, 2001–2010, and 2011–2015.

In the first year of the study (1983), which was the third post-eruption growing season, we documented the deer mouse, long-tailed vole, snowshoe hare, and elk on the Pumice Plain. The former two species were assumed to be residents and likely came from source populations in the adjacent blowdown zone where they survived the eruption in belowground refugia (Andersen and MacMahon 1985a; Adams et al. 1986), requiring a dispersal distance of perhaps a few to several km. In contrast, elk and hares were killed by the eruption within the 570-km^2 blast area and must have arrived from more distant source populations. Elk were simply passing through the area, as there was no available forage or hiding cover available to them. The single hare that we observed was found freshly dead (i.e., just hours deceased); its carcass later devoured by ravens (*Corvus corax*), with only bones remaining by postmortem day three. This individual hare had likely dispersed several km from the tephrafall zone but perished in the absence of food or cover from the hot August sun. The next species to arrive was the water vole in 1988, which was live-trapped at the WILSPR riparian site, where willows had established 2 years earlier, marking the arrival of the first habitat specialist.

15.8.3 1990 Through 2000

During the following decade, an additional 13 mammal species arrived on the Pumice Plain. These included seven rodents, two shrews, two ungulates, pika, and coyote. The rodents included a trophically diverse assemblage, including the Pacific jumping mouse, a granivorous riparian specialist; the yellow-pine chipmunk and Cascade golden-mantled ground squirrel, both diurnal omnivores; the northwestern mouse, a nocturnal omnivore; the creeping vole, an herbivore; the northern pocket gopher, a fossorial herbivore; and the Douglas's squirrel, a tree-dwelling granivore. Source populations for the Pacific jumping mouse, creeping vole, northwestern deer mouse, and yellow-pine chipmunk were likely progeny of surviving individuals in the blowdown zone, whereas the Cascade golden-mantled ground squirrel likely arrived from source populations located in clearcuts at least several km distant from the Pumice Plain. Pocket gophers took 12 years to colonize the Pumice Plain, presumably from source populations immediately north of Spirit Lake identified as residuals in August 1981 by Andersen (1982). Given the fossorial life history of this species, it is not surprising that it took 12 years for gophers to travel ~3 km from the assumed source population, although the extent of surface versus burrow movement is not known. The Douglas's squirrel, which arrived on the Pumice Plain in 1992, is a forest species like the snowshoe hare (arrived 1983), and both of these species failed to establish populations in the absence of requisite habitat. Individuals of each of these species have appeared sporadically on the Pumice Plain in subsequent years but as of 2015 have failed to establish. Nonetheless, each arrived at the site, demonstrating that dispersal was not limiting their establishment, but instead that habitat genesis was the most likely constraining factor. The remaining six rodents that arrived on the Pumice Plain between 1990 and 2000, the jumping mouse, pocket gopher, yellow-pine chipmunk, Cascade golden-mantled ground squirrel, creeping vole, and northwestern mouse, all appeared to have established breeding populations, based on numerous individuals captured or varied age classes and pregnant females included in our sample. In contrast, the ground squirrel and creeping vole were represented by just a few adults and may not have established breeding populations. These two species would eventually establish breeding populations in 2002 and 2008, respectively.

Of the two insectivores that were detected on the Pumice Plain from 1990 to 2000, the montane shrew had been captured in the blowdown zone by Andersen and MacMahon (1985a). Montane shrews were first trapped in 1990 (10 years post eruption) but were only sporadically present on the Pumice Plain until 2000, after which time they were recorded annually through 2015. Following a somewhat similar pattern, the water shrew was first trapped in 1994 (14 years post eruption) and was consistently present from 1997 through 2015.

Pika were first observed in 1994 and again in 2007, and in each case, a single individual was occupying scree at the base of the same basalt flow. Bevers (1998) reported source populations of pika in areas surrounding intensely disturbed sites at MSH; the closest source population he identified was ~5.2 km from the Pumice Plain. As of 2015, pika have not established breeding populations on the Pumice Plain despite abundant forage and small areas of scree habitat. We first observed coyotes in 1990 and have seen individuals as well as their scat and tracks through 2015. These animals have large home ranges, undoubtedly integrate the Pumice Plain as well as other adjacent disturbance zones into their territories, and were likely to have arrived prior to our first observation. Source populations were at least several km from the Pumice Plain, which is not an insurmountable distance for such a mobile species.

Black-tailed deer (a coastal subspecies of the mule deer) were first observed on the Pumice Plain in the vicinity of Spirit Lake in 1998 and have since occasionally been seen along the margins of the lake. These animals prefer dense vegetation and appear to spend relatively little time on the Pumice Plain.

Mountain goats were first observed on the volcano's flanks in 1987, and individuals were occasionally sighted throughout the 1990s. Beginning in 2003, we consistently observed small groups of goats, including adults with kids, on the north flank of the volcano and on the adjacent Pumice Plain. From 2010 to 2015, their populations appeared to have grown steadily, and it was common to see several groups scattered across the area, collectively numbering 30–50 individuals. We speculate that by 2015 the entire herd on and adjacent to the volcano could exceed 150 individuals. Since the early 2000s, a dense growth of seashore bent grass (*Agrostis pallens*) has developed, and numerous upland sedge plants (*Carex spectabilis* and *C. mertensii*) colonized this area (del Moral et al. 2012). We observed heavy grazing of these graminoids by goats. Although the source population for the newly established goat herd on Mount St. Helens is unknown, the most likely source is either Mount Adams or the Goat Rocks Wilderness area, located 54 and 67 km to the east and northeast. Supporting evidence for this assertion includes the dozens of sightings we have made from 2005 through 2015 of goats, including nannies with kids, moving toward Mount St. Helens through the blowdown zone in each of three major drainages east of the volcano.

15.8.4 2001 Through 2010

During the third post-eruption decade, we documented the arrival of nine additional mammal species on the Pumice Plain, and seven of these showed evidence of establishing populations. These species included three each of large rodents, insectivores, and carnivores. The rodents were the American beaver (*Castor canadensis*), a semiaquatic herbivore; the North American porcupine, a forest-dwelling terrestrial herbivore; and the hoary marmot, an alpine-montane meadow herbivore. Beavers colonized Spirit Lake in 2002 and have persisted through 2015. Individuals forage on the adjacent Pumice Plain for willow and alder shrubs from autumn through spring, and in summer add various herbs to their diet.

During summer 2007, an adult porcupine took up temporary residence in our Red Rock Creek riparian study site (RDRCRK, Fig. 15.4), a dense thicket of willow and alder shrubs embedded in a matrix of barren upland habitat, where the animal routinely tripped most of our Sherman traps each night. Eventually, this individual moved on, and since that time, no other porcupines have been observed on the Pumice Plain. Similarly, a single hoary marmot was observed in a small talus area in 2010 (the same site that a pika occupied as described in Sect. 15.8.3). This individual remained in the area for 3 days and was never seen again. Known potential source populations include a site to the north at the edge of the blast area and also a site on the south side of the volcano, 11.75 km and 8.8 km from the observation site on the Pumice Plain. Both the porcupine and marmot were presumably dispersing individuals that encountered the Pumice Plain in their travels and found a mismatch between their habitat requirements and the available resources. Eventually, once coniferous forest returns to the Pumice Plain, a process that is well underway but will take decades longer, porcupines will likely establish. However, it is unlikely that appropriate habitat conditions will develop for the marmot in the future, and this species is not expected to be a member of the mammalian assemblage of the Pumice Plain.

Insectivores arriving on the Pumice Plain were the Trowbridge's shrew, marsh shrew, and American shrew mole. Trowbridge's shrew and the shrew mole are both forest-associated, semi-fossorial species and were observed only in riparian habitats where litter and organic soil matter had developed over the preceding decades (Halvorson et al., Chap. 5, this volume). Source populations for these species were likely from the blowdown zone, where the latter species was captured by Andersen and MacMahon (1985a) and reported to be residual. Trowbridge's shrew was numerically dominant in several studies of forest small-mammal communities (West 1991; Wilson and Carey 2000; Carey and Harrington 2001) and likely survived in the blowdown zone but was not detected in earlier trapping efforts (MacMahon et al. 1989; Crisafulli et al. 2005). The marsh shrew creates a more challenging scenario, because it is relatively uncommon and generally thought to be tightly associated with lower-elevation marshes and wet riparian forests; hence, its detection was surprising.

Two small carnivores, the ermine and long-tailed weasel, first appeared in our samples in 2003 and 2007 and then occurred sporadically thereafter. These animals are highly vagile, and it was surprising that they were not detected before the third post-eruption decade, particularly given the early colonization and high densities of their preferred small-mammal prey. Low capture rates may be related to our use of Sherman traps with rolled oats and peanut butter bait targeting rodents and insectivores; Tomahawk traps with fish or chicken might have yielded more captures. The third carnivore we detected was the puma, an observation based on tracks and scat in 2008. Given the relatively high frequency that we see these big cats in the adjacent blowdown zone, and the high densities of prey (elk), it is likely that these elusive animals used the Pumice Plain long before we confirmed their presence.

15.8.5 2011 Through 2015

Our study covered the first half of the fourth post-eruption decade, and during these 5 years, we documented the arrival of six additional mammal species on the Pumice Plain, three of which established populations. The cinereus shrew was trapped only in 2011 and with just a single individual recorded lacks evidence of having established. This generalist shrew has the broadest geographical distribution of any shrew in our study area, but Mount St. Helens is located at the edge of the species' range, an area in which it appears to be uncommon. A single adult female bushy-tailed wood rat was captured at an upland matrix study site (RDRUPL, Fig. 15.4) in 2012. This species is an obligate rock-dweller living among boulders, talus, cracks, and crevices in cliffs and caves, and this single animal should be considered as a dispersing individual. There are a few small patches of suitable habitat on the Pumice Plain that could potentially support a small population of wood rats in the future. The heather vole appeared only in our 2013 sample but was represented by adults of each sex and by subadults. Toward the end of our study (2012–2015), much of the Pumice Plain supported meadow-like and riparian habitat that appears suitable for this semi-fossorial herbivore, and it may become more abundant and widespread in the future. The northern river otter (*Lontra canadensis*) was first observed during 2012 along the shores of Spirit Lake. From 2012 through 2015, adults and juveniles have been observed, providing evidence of an established population. It is likely that this species dispersed up the Toutle River from source populations located lower in the drainage. In addition, adults have been observed in the blowdown zone ~12 km to the north since 1996, and also on the debris-avalanche deposit ~10 km to the west since 2005, so the Spirit Lake population may have originated from one of these sites. The northern raccoon (*Procyon lotor*) was first observed in the northern portion of the Pumice Plain, primarily along the south shoreline of Spirit Lake, in 2014 and again in 2015. Tracks of both adults and young suggest that a population has established in this area. The Virginia opossum (*Didelphis virginiana*) was also observed in the same area as the raccoons, but its establishment status remains unknown. Opossums were recorded in the blowdown zone in 1998 (Crisafulli et al. 2005), and animals in this zone may have served as the source population.

15.8.6 Species Gains, Losses, and Turnover

The successional patterns of mammal species gain, loss, and turnover rates were summarized with metrics of species presence/absence through time (Anderson 2007, Fig. 15.6). The early years on the post-eruption Pumice Plain were characterized by high annual species gain rates and low species loss rates, followed by the decade of the 1990s with high fluctuations in both gain and loss rates as various species "winked in and out" during their initial establishment attempts; this decade was followed by a 15-year period of stabilization illustrated by lower annual rates of species gain and loss (Fig. 15.6). When rates of species gain and loss are combined to produce a species turnover rate, the results revealed a significant trend of decreasing turnover through time. These patterns are consistent with (1) habitat development (plant succession and soil formation) on the Pumice Plain, in that greater numbers of mammal species accrue through time as suitable habitat patches develop and expand, and (2) the site supports an increasingly larger proportion of the available regional species pool, reducing the remaining number of potential colonist species. We expect to observe continued stabilization in gain, loss, and turnover rates as more mammal species establish permanent populations across the various Pumice Plain habitats, with a potential future pulse (increase) in gain/loss/turnover rates as forest habitat develops and allows forest-specialist species to colonize, while meadow-dwelling species decline and vanish.

15.8.7 Species Richness and Guild Structure in Upland and Riparian Habitats

Species richness increased through time in both upland and riparian habitats. We clearly (and unsurprisingly) demonstrate that riparian habitats, with their complex plant physiognomy, support greater mammal species richness (Fig. 15.12) and more complex guild structure (Figs. 15.9d and 15.11a) compared with the structurally simple upland sites. From 1984 through 2000, there was only one species in the upland sites; however, from 2002 through 2015, there was a general increasing trend in the number of species until mean richness reached three species. During this same period, cumulative richness steadily increased, reaching five to six species after 2012. The deer mouse maintained numerical dominance throughout the study, and although overall diversity increased, further increases appear to be limited by habitat complexity. Foraging-guild structure was simple in the upland sites and overwhelmingly dominated by omnivores. The omnivore guild had a single member from 1984 to 2000, and then steadily increased through the years, reaching four to five species during the most recent years of study. There was also a major shift in the contribution of species biomass during this same period, from the generalist deer mouse (1984–2000) to the Cascade golden-mantled ground squirrel and yellow-pine chipmunk. This increase in guild membership and associated shift in biomass among species within the guild coincided with increases in plant cover on the Pumice Plain (del Moral et al. 2012). The authors report an increase in mean plant cover from about 5–15% in 1989 to 21.6–68.6% in 2000, depending on community type, followed by relatively static values through 2010. Both the Cascade golden-mantled ground squirrel and the yellow-pine chipmunk, as well as the Pacific jumping mouse and northwestern mouse, have strong associations with shrub architecture, and their establishment coincides with increases in Sitka willow from 1988 through 2000 (del Moral et al. 2012). Although upland willow cover increased dramatically during this period, it was still sparse compared with riparian sites (<1% versus >80%). Regardless, willow at relatively low cover values appears to offer habitat conditions conducive to ground squirrels and chipmunks, but under such conditions, their abundancies are low, as inferred from capture rates (Fig. 15.11a). An additional factor likely influencing increases in the number of guilds in upland sites is the sharp increase in herb cover reported by del Moral et al. (2012), which provides flowers and seeds used as forage by these rodents.

The temporally variable but increasing use of upland habitats by three voles, Pacific jumping mice, and montane shrews, from 2008 through 2015, is likely coupled to the increases in herbaceous vegetation. Indeed, many of the mammals sampled may be using upland sites primarily for foraging, returning to adjacent shrub thickets for reproduction, resting, and hibernation. Regardless, the data support the concept of increased use of upland habitat by these mammals (Fig. 15.11). The extent to which these species are integrating riparian shrub and upland habitats into their home ranges warrants further investigation. Still, colonization, growth, and spread of vegetation from 1984 through 2015 appear to have facilitated increases in mammal species richness and the number of guilds in upland habitats.

Like the upland sites, there was a trend of increasing mean richness values in the riparian sites from 1988 (first year of sampling) to 2013, when richness reached eight species, after which time it dropped to six species. Mean species richness was typically five to eight species after 2000, and a maximum total richness of ten species was reached in 2009. Similarly, the number of guilds and within-guild membership increased during our study, and generally most of this increase occurred rapidly, beginning 16–20 years post eruption. These increases appear to be primarily associated with habitat development and associated resource availability, but arrival time on the Pumice Plain certainly played a role as well. In the case of omnivores, guild structure underwent dramatic change in 1996 as new members were added and then largely stabilized by 2002 with little change over the next 13 years. We infer that rapid development and spread of willow and alder thickets drove this response as the generalist deer mouse was joined by four species closely associated with shrub habitat. These species differ from the deer mouse in that most rely on a plant diet (seeds, flowers, fruits), three hibernate, two are diurnal, and two are considerably larger (2–30×). The herbivore guild shifted dominance from the large and semiaquatic water vole to the much smaller and more terrestrial long-tailed and Oregon voles (*Microtus oregoni*). The underlying reason for the decrease and eventual absence of the water vole is unclear, whereas increases in the other two voles are likely related to increases in seed and foliage production at the margins of the riparian area that are more strongly influenced by light, compared with the interior habitat where the water vole primarily resides.

Insectivores were represented by a diverse assemblage with increasing membership through time. It is somewhat astonishing that these small (generally 4–6 g) animals with extraordinary metabolic rates successfully arrived on the Pumice Plain. Further, seven of eight species in the regional species pool had arrived, and six had established populations by 2010. All are semi-fossorial species that forage in litter for a variety of arthropods and to a much smaller extent consume seeds and other plant parts from the surface of the ground. These animals (except for an occasional montane shrew captured in upland habitat) are confined to habitats with dense thickets of willow and alder shrubs. In these sites, annual leaf senescence creates a litter layer (average thickness, 33.6 mm; SD, 11.51 mm) that presumably provides requisite microhabitat features and associated prey. We are currently comparing arthropod faunas between riparian and upland sites, and that study may provide additional insights into food resource availability for insectivores in these disparate habitats. By 20 years post eruption, riparian thickets had developed along most stream courses and continued to expand through the end of our study. However, for the past 15 years or more, our riparian study sites have not undergone any substantial changes in physiognomy, but there has been a slow change in community structure as alder replaces willow. This relatively static vegetation condition has provided conditions allowing the establishment of the full complement of mammals from the regional species pool that have shrub-habitat associations. We do not anticipate a major shift in mammal community composition until a more structurally complex vegetation develops that includes trees.

15.8.8 The Influence of Herbivory on Community Assembly

The role of herbivory during primary succession has received relatively little attention by ecologists, and its importance has been historically underappreciated (Connell and Slatyer 1977). However, recently examples have emerged, showing the strong influences that vertebrate (Bryant and Chapin 1986; Walker and Chapin 1986; Walker et al. 1986; Van Cleve et al. 1993; Kielland and Bryant 1998) and invertebrate (Bach 1990, 1994, 2001; Fagan and Bishop 2000; Bishop 2002; Bishop et al. 2005) herbivores can have on successional processes. Our study on the Pumice Plain showed large differences in small-mammal community structure, comparing upland habitat with forb-dominated vegetation to riparian habitat with dense shrub thickets. These differences were greatest during the first two post-eruption decades, but even after 35 years, these communities were markedly dissimilar. Sitka willow colonized upland habitats during the late 1980s, but individual shrubs failed to attain typical stature (~75% <1.0 m), even when they persisted for years. In contrast, conspecifics in riparian habitats attained heights of 3–5 m. Close inspection of the upland willows revealed that a large proportion (76.8%) of the shrubs showed herbivory effects caused by the non-native willow stem-boring beetle, which attacked ~40% of the stems, with a mortality rate of 67%. The effect of this chronic annual herbivory has been essentially "no net growth" for infested shrubs, because a large percentage of stems are killed; once new stems grow, they are attacked and many of them perish. Plants appear to put much of their energy into growing new shoots, and our long-term data revealed a near-static condition for willow volume.

We suggest that in the absence of beetle herbivory, willows would grow to several times their current stature, and in many cases, canopies of shrubs would coalesce to a point of altering light regimes and microclimates. This, in turn, would alter the extant forb-dominated plant communities and create conditions that could facilitate the colonization of most of the mammal species that are found in the riparian willow thickets, with the notable exception of the water-associated species (water shrew, water vole). In this sense, we propose that a non-native beetle has influenced the pace and pattern of plant succession on the Pumice Plain, and more specifically the shift from forb- to shrub-dominated vegetation, and in doing so has shaped mammal community assembly during the first few post-eruption decades. Interestingly, since 2010, green alder has begun colonizing many of our upland sites. This shrub attains a growth form very similar to Sitka willow but is not a preferred host of the willow stem borer. If alder continues to grow and spread, it may well result in the transition of upland habitats from forb- to shrub-dominated plant communities and facilitate the establishment of small mammals.

15.9 Summary and Outlook for the Future

The 1980 eruption of Mount St. Helens created an outstanding opportunity to investigate mammal community assembly during primary succession on the Pumice Plain. From 1983 to 2015, we documented the arrival of 34 of the 45 mammals in the regional species pool and the successful establishment of at least 25 species. The majority of small mammals that established on the Pumice Plain were likely derived from source populations that survived in isolated refugia in adjacent areas that were less disturbed during the eruption, requiring dispersal distances of a few to several kilometers. In contrast, large mammals likely arrived from more distant source populations, perhaps up to tens of km away. Most species appeared intermittently for a few to several years prior to maintaining a consistent presence. This pattern of occurrence is likely related either to missed detections owing to low abundances or to failed establishment related to lack of suitable habitat and associated resources. Species richness, guild structure, and abundances were all positively associated with increased vegetation abundance and complexity. Herbivory by the willow stem-boring beetle has limited development of shrub cover in upland habitats and, by doing so, has indirectly influenced the pace and pattern of small-mammal community assembly across most of the Pumice Plain. We anticipate that in the next decade green alder will continue to spread, altering plant physiognomy of upland habitats and facilitating establishment of several additional mammal species. Plant communities on the Pumice Plain will likely require at least a century to develop a coniferous forest canopy, and several seral stages will form, persist for various lengths of time, and then transition. Our study covers the early stages of this sere, beginning with barrens, transitioning to herb/graminoids, and, toward the end of our study, some sites becoming dominated by shrubs and others by scattered conifer saplings. However, successional processes have been spatially variable across the Pumice Plain, leading to habitat patches in numerous successional states at any given time. The next important transition in mammal community assembly will likely occur in three or four decades, as coniferous tree density increases as a result of production of local seed and concomitant growth of extant conifer trees leads to development of open-forest conditions that provide habitat for forest-associated species not yet established on the Pumice Plain. This process of forest succession will likely lead to a reduction in the numbers of several early mammal colonists and perhaps even to their local extirpation.

Acknowledgments Long-term studies such as ours require the contributions by numerous individuals. We are particularly thankful to the cadre of technicians and graduate students who assisted with the fieldwork. Leslie Carraway at Oregon State University and Jeff Bradley at the Burke Museum aided in identification of voles and shrews. We thank the Burke Museum and Museum of Southwestern Biology for accession of our voucher collections. Kathryn Ronnenberg produced figures and made editorial improvements to this manuscript. Kelly Christiansen assisted with GIS-related figures. Chris Che-Castaldo generously provided mortality rate data for willow stems. This manuscript benefitted from comments provided by Virginia Dale and three anonymous reviewers. We thank the USDA Forest Service and Mount St. Helens National Volcanic Monument for providing access to our study sites. Funding for this research was provided by grants from the National Science Foundation (DEB81-16914, BSR 84-07213 to J.A.M., and LTREB Program DEB-0614538 to C.M.C.), and from the USDA Forest Service Pacific Northwest Research Station.

Glossary

Arrival A mammal is detected at a site either through capture in a trap or by visual observation.

Blast PDC In the case of the 1980 Mount St. Helens eruption, failure of the volcano's north flank unroofed pressurized magma and superheated water. Rapid exsolution of magmatic gases and conversion of superheated water to steam produced a laterally directed blast, which formed a density current that flowed across rugged topography. The current contained fragmented rock debris as well as shattered forest material.

Community development Process of any number of mammals belonging to a number of different species that co-occur in the same habitat or area and interact through trophic and spatial relationships.

Debris avalanche A rapid granular flow of an unsaturated or partly saturated mixture of volcanic rock particles (± ice) and water, initiated by the gravitational collapse and disintegration of part of a volcanic edifice. Debris avalanches differ from debris flows in that they are not water saturated. Although debris avalanches commonly occur in association with eruptions, they can also occur during periods when a volcano is dormant.

Dispersal Movement of a mammal from its point of origin or home site to another.

Establishment/established A species that is assumed to have a breeding population at a site based on the presence of one or more of the following three criteria: (a) $\geq$1 adult male and female detected at the same site during the same sampling session, (b) $\geq$1 pregnant or lactating females detected at a site, or (c) several juveniles of a species detected at a site during a single trapping session.

Lahar An Indonesian term for a rapid granular flow of a fully saturated mixture of volcanic rock particles (± ice), water, and commonly woody debris. A lahar that has $\geq$50% solids by volume is termed a *debris flow*; one that has roughly 10–50% solids by volume is termed a *hyperconcentrated flow*. Flow type can evolve with time and distance along a flow path as sediment is entrained or deposited.

Microtine rodents A subfamily of rodents (Arvicolinae) that includes the voles, lemmings, and muskrats. At Mount St. Helens, they include three genera of herbivorous voles (*Microtus*, *Myodes*, and *Phenacomys*).

Pyroclastic flow Rapid flow of a dry mixture of hot (commonly >700 °C) solid particles, gases, and air that has a ground-hugging flow often directed by topography. Flows are generally gravity driven but may be accelerated initially by impulsive lateral forces of directed volcanic explosions. Flows typically move at high velocity (up to several hundreds of km h^{-1}).

Tephra Fragmental rock material ejected from a volcano during an eruption and deposited by airfall. It is typically composed of ash (less than 4 mm in diameter), lapilli (4- to 32-mm particles), and blocks (angular stones larger than 32 mm).

Appendix 15.1 Study site characteristics. All sites are located in the pyroclastic-flow disturbance zone and were trapped during the summer season (June through August) for each sampling year. UTM coordinates are zone 10T, datum WGS84.

Site name	Code	Years active	No. of years trapped	Pattern	Traps	Habitat	UTM East (m)	UTM North (m)	Elevation (m)	Aspect
Basalt Creek	BSTCRK	2000, 2002–2009, 2011–2015	14	Transect	25	Riparian	564727.2	5121206.9	1025	NW
Forsyth Creek	FORCRK	2003–2011, 2013, 2015	11	Transect	25	Riparian	564456.1	5119646.1	1293	N
Forsyth Creek Upland	FORUPL	2003–2011, 2013, 2015	11	Transect	25	Upland	564171.4	5119523.3	1318	N
Pumice Plain Castilleja	PPCAST	2000, 2002–2009, 2011–2015	14	Transect	25	Upland	564953.8	5121344.8	1040	NW
Pumice Plain Upland #1	PPUPL1	2002–2009, 2011, 2013, 2015	11	Transect	25	Upland	563570.3	5121675.1	1104	N
Pumice Plain Upland #2	PPUPL2	2002–2009, 2011, 2013, 2015	11	Transect	25	Upland	564908.6	5122209.6	1119	N
Pumice Plain Web	PMPWEB	1984, 1987, 1990, 1995, 1998, 1999, 2000, 2002–2015	21	Web	148	Upland	564429.5	5121689	1160	0
Pumice Plain Wetland #1	PPWET1	2000, 2002–2009, 2011, 2013, 2015	12	Transect	25	Riparian	563803.6	5121827	1020	N
Pumice Plain Wetland #2	PPWET2	2000, 2002–2009, 2011, 2013, 2015	12	Transect	25	Riparian	564966	5122459	1000	N
Red Rock Creek	RDRCRK	2000, 2002–2015	15	Transect	25	Riparian	564925.9	5120553.7	1030	N
Red Rock Creek Upland	RDRUPL	2000, 2002–2015	15	Transect	25	Upland	564889.9	5120677.7	1030	N
Upland Shrub Northeast	UPLSNE	2013, 2015	2	Transect	25	Shrub	565141	5122081	1060	NE
Upland Shrub Northwest	UPLSNW	2013, 2015	2	Transect	25	Shrub	565152	5122227	1060	NE
Upland Shrub South	UPLSHS	2013, 2015	2	Transect	25	Shrub	564742	5122033	1080	N
West Lupine Plot	WLP	1983, 1988	2	Transect	2	Upland	564097.9	5121550.5	1155	0
Willow Springs	WILSPR	1988, 1990–2000, 2002–2015	26	Transect	25	Riparian	564460	5121207.4	1020	N
Willow Springs Upland	WILSPU	2002–2015	14	Transect	25	Upland	564523.4	5121121.4		

References

Adams, A.B., K.E. Hinckley, C. Hinzman, and S.R. Leffler. 1986. Recovery of small mammals in three habitats in the northwest sector of the Mount St. Helens National Volcanic Monument. In *Mount St. Helens: Five years later*, ed. S.A.C. Keller, 345–358. Cheney: Eastern Washington University Press.

Allen, M.F. 1987. Re-establishment of mycorrhizas on Mount St. Helens: Migration vectors. *Transactions of the British Mycological Society* 88: 314–417.

Allen, M.F., and J.A. MacMahon. 1988. Direct VA mycorrhizal inoculation of colonizing plants by pocket gophers (*Thomomys talpoides*) on Mount St. Helens. *Mycologia* 80: 413–417.

Allen, M.F., J.A. MacMahon, and D.C. Andersen. 1984. Reestablishment of endogonaceae on Mount St. Helens: Survival of residuals. *Mycologia* 76: 1031–1038.

Allen, M.F., C.M. Crisafulli, C.F. Friese, and S.L. Jeakins. 1992. Re-formation of mycorrhizal symbioses on Mount St. Helens, 1980–1990: Interactions of rodents and mycorrhizal fungi. *Mycological Research* 96: 447–453.

Allen, M.F., C.M. Crisafulli, S.J. Morris, L.M. Egerton-Warburton, J.A. MacMahon, and J.M. Trappe. 2005. Mycorrhizae and Mount St. Helens: Story of a symbiosis. In *Ecological responses to the 1980 eruption of Mount St. Helens*, ed. V.H. Dale, F.J. Swanson, and C.M. Crisafulli, 221–231. New York: Springer.

Andersen, D.C. 1982. Observations on *Thomomys talpoides* in the region affected by the eruption of Mount St. Helens. *Journal of Mammology* 63: 652–655.

Andersen, D.C., and J.A. MacMahon. 1985a. The effects of catastrophic ecosystem disturbance: The residual mammals at Mount St. Helens. *Journal of Mammology* 66: 581–589.

———. 1985b. Plant succession following the Mount St. Helens volcanic eruption: Facilitation by a burrowing rodent, *Thomomys talpoides*. *American Midland Naturalist* 114: 62–69.

Anderson, D.R., K.P. Burnham, G.C. White, and D.L. Otis. 1983. Density estimation of small-mammal populations using a trapping web and distance sampling methods. *Ecology* 64: 674–680.

Anderson, K.J. 2007. Temporal patterns in rates of community change during succession. *American Naturalist* 169: 780–783.

Bach, C.E. 1990. Plant successional stage and insect herbivory: Flea beetles on sand-dune willow. *Ecology* 71: 598–609.

———. 1994. Effects of a specialist herbivore (*Altica subplicata*) on *Salix cordata* and sand dune succession. *Ecological Monographs* 64: 423–445.

———. 2001. Long-term effects of insect herbivory and sand accretion on plant succession on sand dunes. *Ecology* 82: 1401–1416.

Banks, S.C., M. Dujardin, L. McBurney, D. Blair, M. Barker, and D.B. Lindenmayer. 2011. Starting points for small mammal population recovery after wildfire: Recolonization or residual populations? *Oikos* 120: 26–37.

Bevers, E.J. 1998. *Colonization and population biology of pikas (Ochotona princeps) on Mount St. Helens*. Doctoral dissertation, New Mexico State University, New Mexico, USA.

Bishop, J.G. 2002. Early primary succession on Mount St. Helens: The impact of insect herbivores on colonizing lupines. *Ecology* 83: 191–202.

Bishop, J.G., W.F. Fagan, J.D. Schade, and C.M. Crisafulli. 2005. Causes and consequences of herbivory on prairie lupine (*Lupinus lepidus*) in early primary succession. In *Ecological responses to the 1980 eruption of Mount St. Helens*, ed. V.H. Dale, F.J. Swanson, and C.M. Crisafulli, 151–161. New York: Springer.

Bogdziewicz, M., and R. Zwolak. 2014. Responses of small mammals to clear-cutting in temperate and boreal forests of Europe: A meta-analysis and review. *European Journal of Forest Resources* 133: 1–11.

Bryant, J.P., and F.S. Chapin III. 1986. Browsing–woody plant interactions during boreal forest plant succession. In *Forest ecosystems in the Alaskan Taiga*, ed. K. Van Cleve, F.S. Chapin III, L.A. Viereck, and C.T. Dyrness, 213–225. New York: Springer.

Burt, W.H. 1961. *Some effects of Volcan Paricutin on vertebrates*, Occasional Paper No. 620. Ann Arbor: Museum of Zoology, University of Michigan.

Carey, A.B., and C.A. Harrington. 2001. Small mammals in young forests: Implications for management for sustainability. *Forest Ecology and Management* 154: 289–309.

Carey, A.B., and M.L. Johnson. 1995. Small mammals in managed, naturally young, and old-growth forests. *Ecological Applications* 5: 336–352.

Che-Castaldo, C. 2014. *The attack dynamics and ecosystem consequences of stem borer herbivory on Sitka willow at Mount St. Helens*. Doctoral dissertation. University of Maryland, College Park, MD, USA.

Clements, F.E. 1916. *Plant succession: An analysis of the development of vegetation*, Publication No. 242. Washington, DC: Carnegie Institution.

Connell, J.H., and R.O. Slatyer. 1977. Mechanisms of succession in natural communities and their role in community stability and organizations. *American Naturalist* 111: 1119–1144.

Cowles, H.C. 1899. The ecological relations of vegetation on the sand dunes of Lake Michigan. *Botanical Gazette* 27: 95–117. 167–202, 281–308, 361–391.

Craig, M.D., E. Giles, St.J. Hardy, J.B. Fontaine, M.J. Garkakalis, A.H. Grigg, C.D. Grant, P.A. Fleming, and R.J. Hobbs. 2012. Identifying unidirectional and dynamic habitat filters to faunal recolonization in restored mine pits. *Journal of Applied Ecology* 49: 919–928.

Crisafulli, C.M., J.A. MacMahon, and R.R. Parmenter. 2005. Small mammal survival and colonization on the Mount St. Helens Volcano: 1980–2002. In *Ecological responses to the 1980 eruption of Mount St. Helens*, ed. V.H. Dale, F.J. Swanson, and C.M. Crisafulli, 199–218. New York: Springer.

Dale, V.H., F.J. Swanson, and C.M. Crisafulli, eds. 2005. *Ecological responses to the 1980 eruption of Mount St. Helens*. New York: Springer.

del Moral, R., L.A. Thomason, A.C. Wenke, N. Lazanoff, and M.D. Abata. 2012. Primary succession trajectories on pumice at Mount St. Helens, Washington. *Journal of Vegetation Science* 23: 73–85.

Edwards, J.S. 1986. Arthropods as pioneers: Recolonization of the blast zone on Mount St. Helens. *Northwest Environmental Journal* 2: 263–273.

Edwards, J.S., and P. Sugg. 1993. Arthropod fallout as a resource in the recolonization of Mount St. Helens. *Ecology* 74: 954–958.

Fagan, W.F., and J.G. Bishop. 2000. Trophic interaction during primary succession: Herbivores slow a plant reinvasion at Mount St. Helens. *American Naturalist* 155: 238–251.

Fisher, J.T., and L. Wilkinson. 2005. The response of mammals to forest fire and timber harvest in the North American boreal forest. *Mammal Review* 35: 51–81.

Fox, B.J. 1982. Fire and mammalian secondary succession in an Australian coastal heath. *Ecology* 63: 1332–1341.

Franklin, J.F., and C.T. Dyrness. 1973. *Natural vegetation of Oregon and Washington*, General Technical Report PNW-8. Portland: U.S. Department of Agriculture, Forest Service, Pacific Northwest Forest and Range Experiment Station.

Franklin, J.F., J.A. MacMahon, F.J. Swanson, and J.R. Sedell. 1985. Ecosystem responses to the eruption of Mount St. Helens. *National Geographic Research* 1: 198–216.

Gitzen, R.A., and S.D. West. 2002. Small mammal response to experimental canopy gaps in the southern Washington Cascades. *Forest Ecology and Management* 168: 187–199.

Gleason, H.A. 1926. The individualistic concept of the plant association. *Bulletin of the Torrey Botanical Club* 53: 1–20.

Irie, K., and S. Tsuyuzaki. 2007. Faunal make-up and abundance of rodents 17 years after volcanic eruptions. *Northwest Science* 81: 333–336.

Kielland, K., and J.P. Bryant. 1998. Moose herbivory in taiga: Effects on biogeochemistry and vegetation dynamics in primary succession. *Oikos* 82: 377–383.

Kirkland, G.L., Jr. 1990. Patterns of initial small mammal community change after clearcutting of temperate North American forests. *Okios* 59: 313–330.

Larkin, J.L., D.S. Maehr, J.J. Krupa, J.J. Cox, K. Alexy, D.E. Unger, and C. Barton. 2008. Small mammal response to vegetation and spoil conditions on a reclaimed surface mine in eastern Kentucky. *Southeastern Naturalist* 7: 401–412.

Lipman, P.W., and D.R. Mullineaux. 1981. *The 1980 eruptions of Mount St. Helens, Washington*, Professional Paper 1250. Washington, DC: U.S. Geological Survey.

MacArthur, R.H., and J.W. MacArthur. 1961. On bird species diversity. *Ecology* 42: 594–598.

MacMahon, J.A., R.R. Parmenter, K.A. Johnson, and C.M. Crisafulli. 1989. Small mammal recolonization on the Mount St. Helens volcano: 1980–1987. *American Midland Naturalist* 122: 365–387.

Mendonca, A.F., T. Armond, C.L. Camargo, N.F. Carmago, J.F. Ribeiro, P.L. Zangrandi, and E.M. Vieira. 2015. Effects of an extensive fire on arboreal small mammal populations in a neotropical savanna woodland. *Journal of Mammalogy* 96: 368–379.

Monamy, V., and B.J. Fox. 2000. Small mammal succession is determined by vegetation density rather than time elapsed since disturbance. *Austral Ecology* 25: 580–587.

Nelson, L., Jr., and F.W. Clark. 1973. Correction for sprung traps in catch/effort calculations of trapping results. *Journal of Mammalogy* 54: 295–298.

Parmenter, R.R. 2005. Patterns of decomposition and nutrient cycling across a volcanic disturbance gradient: A case study using rodent carcasses. In *Ecological responses to the 1980 eruption of Mount St. Helens*, ed. V.H. Dale, F.J. Swanson, and C.M. Crisafulli, 233–242. New York: Springer.

Parmenter, R.R., J.A. MacMahon, M.E. Waaland, M.M. Stuebe, P. Landres, and C.M. Crisafulli. 1985. Reclamation of surface coal mines in western Wyoming for wildlife habitat: A preliminary analysis. *Reclamation and Revegetation Research* 4: 93–115.

Pearson, O.P. 1994. The impact of an eruption of Volcan Hudson on small mammals in Argentine Patagonia. *Mastozoologia Neotropical* 1: 103–112.

Pulliam, H.R. 1988. Source, sinks and population regulation. *American Naturalist* 132: 652–661.

Pyke, D.A. 1984. Initial effects of volcanic ash from Mount St. Helens on *Peromyscus maniculatus* and *Microtus montanus*. *Journal of Mammalogy* 65: 678–680.

Saba, S.L., and D.A. de Lamo. 1994. Dynamic responses of mammals to the eruption of Volcan Hudson. *Mastozoologia Neotropical* 1: 113–122.

Thoreau, H.D. 1887. *The succession of forest trees*. Cambridge, MA: The Riverside Press, Houghton, Mifflin and Co.

Van Cleve, K., L.A. Viereck, and G.M. Marion. 1993. Introduction and overview of a study dealing with the role of salt-affected soils in primary succession on the Tanana River floodplain, interior Alaska. *Canadian Journal of Forest Research* 23: 879–888.

Walker, L.R., and F.S. Chapin III. 1986. Physiological controls over seedling growth in primary succession on an Alaskan floodplain. *Ecology* 67: 1508–1523.

Walker, L.R., J.C. Zasada, and F.S. Chapin III. 1986. The role of life history processes in primary succession on an Alaskan floodplain. *Ecology* 67: 1243–1253.

West, S.D. 1991. Small mammal communities in the southern Washington Cascade Range. In *Wildlife and vegetation of unmanaged Douglas-fir forests*, General Technical Report PNW-GTR-285, ed. L.F. Ruggiero, K.B. Aubry, A.B. Carey, and M.H. Huff, 268–283. Portland: U.S. Department of Agriculture, Forest Service, Pacific Northwest Research Station.

Wilson, D.E., and S. Ruff, eds. 1999. *The Smithsonian book of North American mammals*. Washington, DC: Smithsonian Institution Press.

Wilson, S.M., and A.B. Carey. 2000. Legacy retention versus thinning: Influences on small mammals. *Northwest Science* 74: 131–145.

Yurkewycz, R.P., J.G. Bishop, C.M. Crisafulli, J.A. Harrison, and R.A. Gill. 2014. Gopher mounds decrease nutrient cycling rates and increase adjacent vegetation in a volcanic primary succession. *Oecologia*. https://doi.org/10.1007/s0042-014-3075-7.

Zwolak, R. 2009. A meta-analysis of the effects of wildfire, clearcutting, and partial harvest on the abundance of North American small mammals. *Forest Ecology and Management* 258: 539–545.

Volcano Ecology: State of the Field and Contributions of Mount St. Helens Research

16

Frederick J. Swanson and Charles M. Crisafulli

16.1 Introduction

The effects of volcanic eruptions on ecosystems have long interested ecologists. Pioneering studies at Katmai (Griggs 1918, 1922), Krakatau (Dammerman 1922, 1948; Docters van Leeuwen 1936), several volcanoes in the western United States and Mexico (Eggler 1941, 1948; Brattstrom 1963), Hawaiian volcanoes (Eggler 1971; Smathers and Mueller-Dombois 1974), and Surtsey (Fridrikssen 1975) identify fascinating ecological responses to a variety of volcanic disturbance processes. Many important contributions to this field of science have emerged since these early works, among them studies triggered by the 1980 eruption of Mount St. Helens. This eruption and the emergence of disturbance ecology as an important subdiscipline of ecology have helped propel development of the field of volcano ecology—the study of immediate effects of volcanic processes on ecosystems and the response of those ecosystems through the course of ***succession***[1]. Volcano ecology investigations have led to syntheses of studies focused on individual sites (e.g., Griggs 1922; Fridriksson 1975; Dammerman 1948; Thornton 1996; Dale et al. 2005a; DeGange et al. 2010) and comparing multiple volcanoes (e.g., Fridriksen and Magnusson 1992; del Moral and Grishin 1999; Thornton 2000, 2007; Dale et al. 2005a; Edwards 2005; del Moral and Magnusson 2014; Ruggiero and Kitzberger 2014; Crisafulli et al. 2015; Veblen et al. 2016).

The current state of global knowledge regarding volcano ecology provides a basis for evaluating the contributions of ecological research conducted at Mount St. Helens following the 1980 eruption. The work and findings to date range from case studies of ecological responses to an individual eruption to synthesis publications, including journal articles, books, book chapters, and reports. This literature reveals the broad scope of volcano ecology research by several aspects of its contexts, including (1) the relation of volcano ecology to allied fields of science, such as volcanology and disturbance ecology, (2) the global distribution of volcanoes and types of volcanism, (3) the diversity of biogeographic and climatic settings, and (4) societal aspects of volcanoes in terms of opportunities for ecological study, science outreach to the public, and implications for land management. This overview reveals common themes of analysis among the hundreds of studies, such as mechanisms of survival, importance of ***biological legacies*** from the pre-disturbance ecosystem, physiological sensitivity of organisms, site amelioration, processes of community assembly and succession, and biotic interactions operating in ecosystems influenced by contingencies and stochastic factors. In this chapter, we mention many of these phenomena but do not attempt a critical analysis and synthesis.

This chapter presents salient features of the developing field of volcano ecology based largely on assessment of the published literature and the global distribution of volcanoes that have erupted since 1883, when the eruption of Krakatau marked the beginning of ecological studies of recent eruptions. We first characterize volcano ecology in terms of the types of scientific inquiry involved: the volcanic, biogeographic, and climatic systems where volcanoes have been studied; and social settings of the volcanoes. Next we review the record of volcano ecology studies, noting their distribution among types of volcanic and biogeographic systems, and how the pace of this research, as measured by publication

[1]Terms in ***bold italic face*** are defined in the glossary at the end of the chapter.

F.J. Swanson (✉)
U.S. Department of Agriculture, Forest Service, Pacific Northwest Research Station, Corvallis Forestry Sciences Laboratory, 3200 Jefferson Way, Corvallis, OR 97331, USA
e-mail: fred.swanson@oregonstate.edu

C.M. Crisafulli
U.S. Department of Agriculture, Forest Service, Pacific Northwest Research Station, Mount St. Helens National Volcanic Monument, 42218 NE Yale Bridge Rd., Amboy, WA 98601, USA
e-mail: ccrisafulli@fs.fed.us

C.M. Crisafulli, V.H. Dale (eds.), *Ecological Responses at Mount St. Helens: Revisited 35 years after the 1980 Eruption*,
https://doi.org/10.1007/978-1-4939-7451-1_16

records, has varied since 1883. A brief review of findings from work at many volcanoes reveals common lessons, despite the great variety of eruption types and affected ecosystems. Comments on the state of the science provide a basis for assessing how the work at Mount St. Helens has contributed to the emergence of volcano ecology. We conclude with suggestions for capitalizing on learning opportunities presented by ongoing research as well as future eruptions.

16.2 Settings of Volcano Ecology: Science, Geology/Volcanic Systems, Biogeography/Climate, and Society

16.2.1 Science Setting

To understand its scope, it is helpful to view volcano ecology in the context of the wide variety of disciplines, topics, and time scales involved. Ecological research on ecosystems influenced by volcanism encompasses numerous taxa, system types, and ecological processes across nearly all of the major biomes on earth (Crisafulli et al. 2015). Terrestrial and freshwater ecologists have tended to study initial effects of disturbances on ecosystems within a few days to a few years of an eruption and to track succession over decades to a few centuries. Soil scientists and geomorphologists are concerned about initial effects of volcanism, but also stretch their time scales of interest over millennia and beyond, sometimes using approaches based on the study of chronosequences and a wide array of dating techniques. Oceanographers examine "extremophiles" associated with the lightless environments of deep-sea vents to address a broad range of questions, such as the potential to find life on other planets and hypotheses about the origin of life. Paleontologists, geophysicists, and geochemists consider the likelihood of flood-basalt eruptions versus bolide impacts in triggering global extinction events over the past half billion years. In this chapter, we address only the interactions of terrestrial and, to a lesser extent, freshwater ecosystems at the time scales relevant to documenting immediate ecosystem response and to ecological succession spanning years to a few centuries following disturbance.

The field of volcano ecology has prospered during recent decades with the recent surge of interest in disturbance ecology, which addresses the roles of diverse disturbance processes in ecological systems (Pickett and White 1985; Turner 1987; Turner et al. 1997; Turner and Dale 1998; Peters et al. 2011). Ecological science has long involved the study of succession following disturbance, leading to some general findings. Spatial variability in intensity and severity of large disturbance events often leads to heterogeneous patterns of surviving organisms (Turner et al. 1998). When disturbance intensity is high, (a) initial density of organisms is low; (b) recovering patches may serve as foci for additional colonization and expand spatially; (c) competition is less important relative to chance arrival in determining community composition; (d) community composition is not initially predictable; and (e) the rate of recovery of community composition is slow (Turner et al. 1998). In many types of disturbance processes that have been examined in the ecological literature, such as flooding, wildfire, windstorms, and insect defoliation, single disturbance types are involved. However, explosive volcanism adds complexities of multiple types and mechanisms of disturbance, and our review of several hundred publications on volcano ecology reveals a generally poor characterization of the volcanic disturbance types and mechanisms that altered the ecosystem, information critical to robust interpretation, especially for comparative studies.

Ecosystem interactions with volcanism make some distinctive contributions to disturbance ecology, partially because a critical feature of explosive volcanism is the involvement of several geophysical processes with varied properties as disturbance agents. The resulting mix of disturbance types in a single eruption event is unlike more frequent, single-process, nonvolcanic types of disturbances, such as wildfire, ice storms, and insect defoliation events. This complexity of volcanic disturbances in a single eruption fosters general thinking about multiple disturbance types and the importance of recognizing the mechanisms of disturbance within each disturbance type (Table 16.1). Our comparison of a sample of volcanic and nonvolcanic disturbance processes reveals that it is actually common for a single disturbance type to involve more than one mechanism and that volcanic disturbance types are more likely to include three or four mechanisms (Adams and Dale 1987; Turner et al. 1997).

The most immediate scientific approach to volcano ecology is undertaken during and immediately following an eruption. This can best be accomplished where a record of pre-eruption ecological conditions exists from earlier monitoring or studies, but such a record is rarely available. Rapid post-eruption assessments can provide much critical information that would be difficult to ascertain at a later time, such as identity of initial survivors and roles of transient ***refugia*** like snowbanks, factors that may have important long-term ecological consequences for disturbed systems. One-time sampling years to decades after an eruption can give a snapshot of then-current biological conditions, but critical information on ecosystem development is revealed only by assessments shortly after an eruption. It may be possible to gain a sense of the pace of ecological change—the development of the ecological complexity and successional trajectory—using a chronosequence approach in which biotic conditions are observed on substrates of different age, commonly referred to as space-for-time substitutions (Pickett 1989). This approach requires the assumption that the sites were similar before the eruption, were disturbed in the same way, and experienced similar ecological processes since the

Table 16.1 Natural disturbance process types and physical disturbance mechanisms for a sampling of volcanic and nonvolcanic events. Capitalized and bold **X** denotes the possibility of especially high-intensity disturbance.

Process type	Impact force	Abrasion	Heat	Erosion	Deposition	Canopy loading[a]
Volcanic						
Tephrafall	x	x			x	x
Lava flow	**X**		**X**		**X**	
Blast PDC	**X**	x	x		x	
Debris avalanche	**X**	x			**X**	
Lahar	x	x		x	x	
Pyroclastic flow	**X**	x	**X**		**X**	
Nonvolcanic						
Ice, wet snow						**X**
Wildfire			x			
Wind	x	x				
Landslide	**X**	x		**X**	**X**	
Flood	x	x		x	x	

[a]"Canopy loading" refers to accumulation of tephra and precipitation in tree canopies, leading to canopy collapse under the weight of the load.

time of disturbance—assumptions that may not be valid. Another approach taken at a handful of volcanoes has been to establish a network of plots shortly after an eruption (days to months) and follow-up measurements at annual or some other increments over decades. Any of these approaches to timing of sampling can be taken in studies that are either narrowly focused or broadly interdisciplinary. Remote places, such as some oceanic islands, favor expedition-style visits (such as Surtsey, Kasatochi), which may foster interdisciplinary work. On the other hand, readily accessible places make possible near-continuous observations by teams with varying degrees of disciplinary diversity and collaboration. Some sites have been subjected to a blend of some repeated, annual sampling plus single or repeated, multidisciplinary pulses of effort—for example, Mount St. Helens (Dale et al. 2005b), Chaiten (Pallister et al. 2010), and Kasatochi (DeGange et al. 2010).

16.2.2 Geologic Setting of Volcanic Systems

Geologic settings of volcanoes vary greatly. They may be volcanic arcs associated with subduction zones in either marine (Aleutian archipelago) or continental margin (Andes, Cascade Range) settings, linear chains of volcanoes associated with hot spots in marine (Hawaii) or continental (Snake River Plain–Yellowstone) settings, linear vent systems associated with spreading centers (East Pacific Rise), or complex archipelagos related to hot spots associated with spreading centers (Galapagos, Iceland) (Fig. 16.1). The tectonic setting of volcanic systems influences the dominant style (explosive versus effusive), products, and frequency of eruptions (Siebert et al. 2015). There are two general types of eruption styles, explosive and effusive. Explosive eruptions occur when magma and solid rock material are violently fragmented and expelled from a vent by the tremendous force of expanding gases, which is common in volcanic chains associated with subducting tectonic plates of oceanic origin. Effusive eruptions, involving a steady flow of lava from a vent, are characteristic of oceanic ridge systems, as in the cases of mid-ocean ridges. Hot-spot volcanism has been responsible for the formation of island chains; the Hawaiian archipelago is a type example dominated by effusive eruptions. Intracontinental volcanic systems occur in conjunction with both hot spots (Yellowstone) and rift systems (East African Rift) where thick, silica-rich, continental crust contributes to highly explosive eruptions. Flood basalts are extremely voluminous and infrequent (nine events in the past 300 million years) and have profoundly altered the world's biota through periodic mass extinction events (Wignall 2005).

Water plays a critical role in many volcanic processes and is therefore an important factor in interpreting volcanic events and disturbance to ecosystems. Phreatic eruptions, debris avalanches triggered by sector collapse, and lahars are all critically influenced by the presence of groundwater and/or surface water. The water content of mobilized material influences runout and, consequently, the distance that volcanic disturbance processes can extend outward from the source. Large volcanic edifices tend to be water-rich because the orographic effect of the mountain entrains moisture from the atmosphere, and that moisture may be retained in the form of ice and snow at high elevation where eruptions commonly occur. Explosive eruptions can deposit fragmental ejecta over vast areas, potentially altering water runoff, sediment transport, river networks, and even climate (Pierson and Major 2014). Sediment deposition in channels and across floodplain and terrace surfaces disturbs in-stream and riparian ecosystems. This volcanic sediment

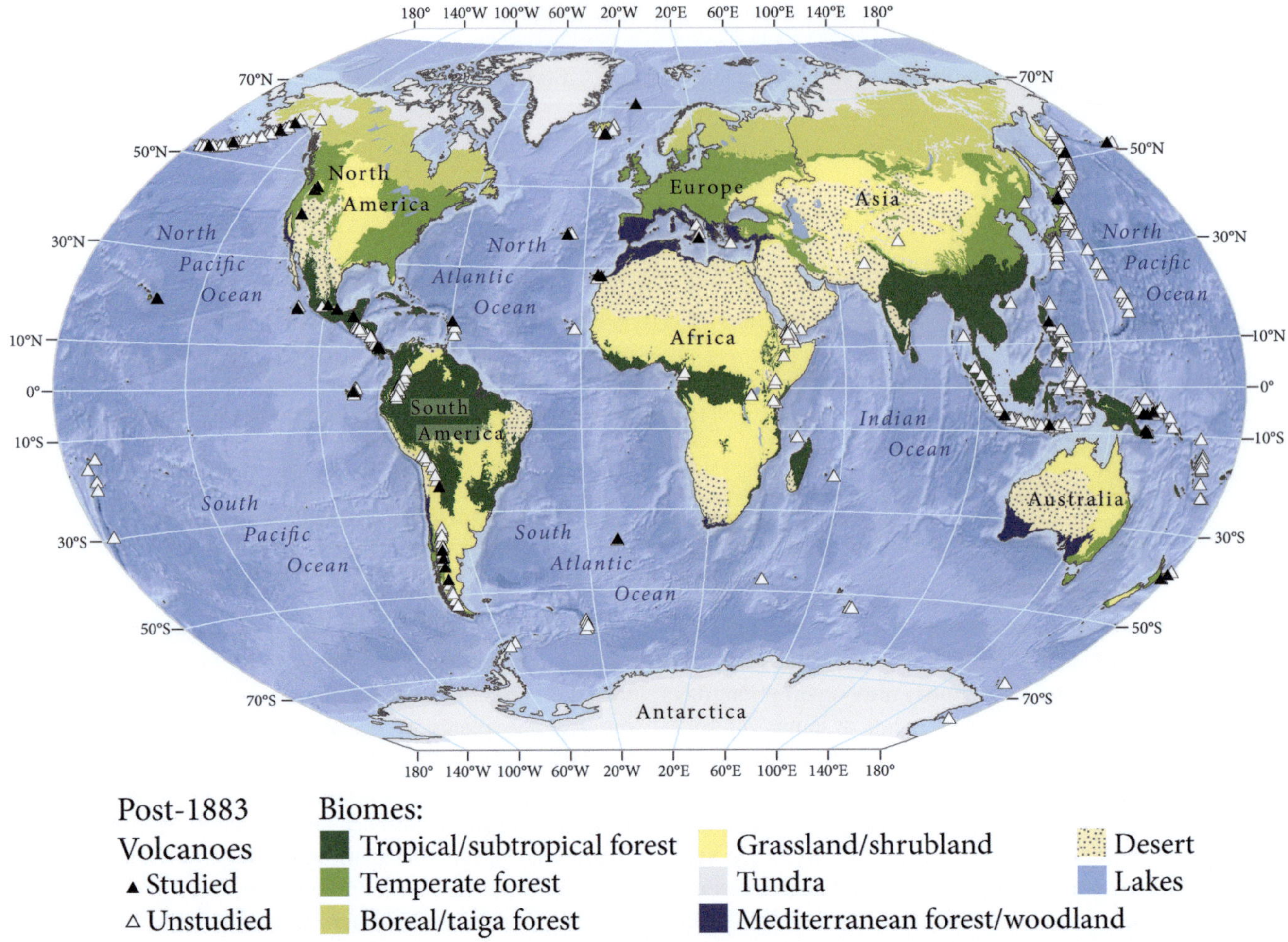

Fig. 16.1 Volcanoes listed in the Smithsonian Institution/US Geological Survey Global Volcanism Programs catalog that were reported to have erupted subaerially since 1883 CE, plotted on the World Wildlife Fund's map of terrestrial biomes of the globe. Volcanoes that have been the subject of published ecological study are symbolized by *black triangles*, those with no published studies by *white triangles*. (From Crisafulli et al. 2015).

may be stored and later remobilized, creating chronic or episodic disturbance for downstream ecosystems and human communities that may persist for decades.

The extensive reach of tephrafall, lahars, and other consequences of explosive volcanism can result in diverse ecosystems—such as desert, grassland, steppe, meadow, forest, lake, river, and marine systems—being affected by a single eruption. Single explosive eruptions often involve a half dozen processes; for example, the 1980 eruption of Mount St. Helens involved a major ***debris avalanche***, pyroclastic density current (here referred to as the ***blast PDC***), ***lahars***, ***pyroclastic flows***, ***tephrafall***, and lava-dome growth (Lipman and Mullineaux 1981). A common feature of most of these processes is deposition of fragmental volcanic rock debris across the surrounding landscape. The thickness, texture, and temperature of materials and the force with which they are emplaced vary greatly among processes. Ecological effects of explosive events can range from benign to total obliteration; they may occur in quick succession; and their effects may overlap in extent, compounding effects on ecosystems. In contrast, effusive eruptions result in relatively slow flow of lava across the land surface, and the heat and resulting solid-rock deposit can eliminate preexisting biota and slow establishment of invading organisms.

Initial interactions among volcanic processes and biota are best interpreted by considering properties of individual disturbance types in terms of the mechanisms of each geophysical process that act as ecosystem disturbance agents (Table 16.1). Principal mechanisms involved in volcanic processes include deposition of volcanic material on vegetation and on the ground (burial); erosion of soil; heat; impact force (e.g., leading to toppling of trees); abrasion by flying, flowing, or falling rock debris and organic matter; and chemical pollutants (in rainwater, as aerosols, or in gaseous form). Individual volcanic processes may damage, kill, or utterly obliterate biota by combinations of these mechanisms.

Across eruptions the frequency, extent, and severity of ecological disturbance vary greatly among volcanic processes. The intensity of each disturbance type commonly decreases along the flow path and laterally away from the axis of the flow path of the process. Tephrafall is by far the most extensive (sometimes circum-global) and common volcanic disturbance process. Deposit thickness may exceed several 10s of m close to the vent, but over most of the depositional area deposit thickness is less than 1 mm. Lava flows are also very common and range widely in spatial scale, though they are commonly constrained in area (a few km^2) and strongly influenced by topography, but they consistently cause severe damage to ecosystems and the resulting raw, stony landscape is slow to revegetate. Lahars are very common and can lead to severe disturbance of rivers and riparian zones, potentially extending 10s of km from volcanoes. Debris avalanches, on the other hand, are rather rare, and proximal zones of impact generally measure 10s of km^2 in extent and 10s of m in thickness, completely resetting terrestrial ecosystems and also creating new aquatic environments. PDCs include several types of high-intensity flow processes that consist of highly turbulent mixtures of fragmented rock and hot gas. These events are commonly associated with plinian and subplinian eruptions and are generated by a variety of mechanisms, such as dome collapse, column collapse, dome explosion, and lateral phreatic blasts. Generally, the zones of substantial ecological disturbance by these processes cover a few 10s of km^2, but the PDC associated with the Mount St. Helens 1980 eruption extended over several 100s of km^2. PDCs typically leave deposits only a few m thick or less, thus creating a landscape rich in biological legacies. Pyroclastic flows, a specific form of PDC, create high-intensity, high-severity disturbance with high temperature (300 to >800 °C), high velocity (up to 700 km h^{-1}), and highly variable deposit thicknesses (<1 m to 10s of m), and consequently few or no biological legacies survive.

16.2.3 Biogeographic/Climatic Setting

The geography of volcanoes and their environmental context strongly influence the types and pace of ecosystem response to disturbance. Historically active, terrestrial/subaerial volcanoes occur from 77.5° south (Mt. Erebus, Antarctica) to 71.1° north (Beerenberg volcano, Jan Mayen Island, Norway), spanning a wide range of biogeographic and climatic settings. Moist temperate and tropical regions are likely to support more rapid biomass accumulation, species assembly, and soil development in response to volcanic disturbance than are more arid and colder regions. The preponderance of volcanoes known to have been historically active is located in temperate and tropical settings, including volcanic chains associated with subduction zones tracing nearly the entire length of the Americas, and the Southeast and East Asian volcanic chains (Fig. 16.1). The stature of these volcanic mountains orographically entrains precipitation, potentially adding to the wetness of their local and regional environments. For example, the continental–interior location of the East African Rift places it in an arid setting, but moisture from summer monsoons entrained by the topographic relief supports savannah and some forest cover. High-latitude volcanic landscapes such as Iceland, the Aleutians, and Kamchatka support tundra. A general point is that biomes and environmental settings differ in their regimes of native disturbance by nonvolcanic processes, which may add uncertainty to predictions of ecosystem response to volcanic disturbance (Paine et al. 1998).

The biogeographic context of an eruption-affected landscape also influences the geographic reach of nutrient fluxes and the dispersal of organisms and propagules, both locally and over vast expanses of the globe. Terrestrial and oceanic systems provide a striking contrast in limitations and opportunities for species and nutrient accrual. Newly emergent oceanic volcanoes (e.g., Surtsey) or severely disturbed islands (e.g., Krakatau) may experience very limited immigration of most terrestrial species that have to traverse great distances of open ocean (Fridriksson 1975; Thornton 1996). On the other hand, highly vagile organisms, such as many spider species, are an exception and quickly arrive at such sites in great numbers (Edwards and Thornton 2001). New islands can experience intense, locally derived nutrient subsidies from the marine system via nesting seabirds and congregations of large marine mammals on land (Sigurdsson and Magnusson 2010). Disturbed terrestrial sites and newly formed lakes and ponds, on the other hand, may be populated from adjacent areas but also benefit from long-distance dispersal via wind or from biotic vectors such as migratory birds and anadromous fish.

16.2.4 Societal Setting

The geographic distribution of erupting volcanoes, their eruptive products, and nearby human settlements profoundly affects hazards for human lives, property, and infrastructure, and also prospects for the pursuit of volcano ecology research and public education. A wide array of societal perspectives is relevant to volcano ecology: public education, assessment of economic impacts of eruptions, tourism, archaeology, art, literature, and film (Sigurdsson et al. 2015). Ecologists benefit where physical infrastructure (e.g., roads and trails), landowners, and administrators provide easy access to volcanic study sites. Eruption sites that include highly human-modified landscapes obviously limit options for studying relatively pristine systems and may also reduce options for research because of limited access to privately owned property and increased potential for vandalism to study infrastructure and equipment. Even on public lands, such as parks and reserves, research opportunities may be compromised as the land and waterways are modified for hazard mitigation, planned and unplanned development, recreation, and ecosystem restoration. Intentional or inadvertent introduction of non-native species may also be an issue.

The capacity to support long-term volcano ecology research and interpretive programs for the public are key intersections of society and science, which vary widely from country to country. By communicating key lessons from ecological research at eruption sites, societies gain an appreciation of the geological forces that shape their wild and human-manipulated environments, as well as the stunning resilience of nature. Inspired by these powerful geological and ecological forces, the arts and humanities have helped shape public perception (Goodrich et al. 2008; Sigurdsson 2015; Sigurdsson and Lopes 2015). Such lessons also have pragmatic value for both individuals and societies in understanding the consequences of eruptions for important resources that humans rely upon, such as potable water, livestock, food crops, and forest and range products. A critical feature of research capacity is embodied in academic institutions, national research organizations, and science-funding agencies in a societal context that values natural resources, the environment, and science. Inventory and monitoring of environmental conditions, such as vegetation surveys and long-term gaging of streamflow and measurement of sediment yield, provide extremely valuable background information for interpreting effects of new landscape disturbances, including volcanic processes. In an extreme case of social factors affecting opportunity for volcano ecology research, areas of civil unrest and military exclusion zones (e.g., Kamchatka during the Cold War) may preclude access to interesting research opportunities.

16.3 Characteristics of Volcano Ecology Studies: Volcanoes, Affected Biomes, Research Themes, and Findings

16.3.1 Historically Active Volcanoes

To place the progress and biogeographic coverage of volcano ecology research in context, we examine records of the volcanoes that have experienced recent eruptions, the affected proximal terrestrial biomes, and the geographic and temporal distributions of published volcano ecology studies, drawing on the compilation and initial analyses of published literature referenced in Crisafulli et al. (2015) (Figs. 16.1, 16.2, and Table 16.2). We consider only volcanoes with documented eruptions beginning in or after 1883, when the major eruption of Krakatau launched the first such studies; we refer to these volcanoes as "historically active." We consider only subaerial eruptions, because the locations and eruptive history of submarine volcanoes are poorly known and they have not been a focus for most ecologists. Historic eruptions have occurred at 404 (26%) of the 1551 volcanoes in the Smithsonian–US Geological Survey catalog of volcanoes of the world (http://www.volcano.si.edu/search_volcano.cfm, accessed 20 March 2014).

Only 44 (11%) of these 404 historically active volcanoes have been studied by ecologists, who report that work in a total of 423 papers (Crisafulli et al. 2015). Although this record of published volcano ecology studies is incomplete, particularly for vegetation, it is sufficient to draw some useful conclusions. The limited number of recent eruptions that have been studied reflects several types of decisions by ecologists. For example, the most intensively studied eruptions appear to be high-magnitude eruptions that attract global attention, stimulating commitment of individuals, institutions, and science communities to undertake research. The Volcano Explosivity Index (VEI), based on the volume of ejected material and eruption column height, is a useful index of eruption magnitude (Newhall and Self 1982). One unit of the 0–8 VEI ranking scheme represents approximately a tenfold difference in eruption frequency, so an event with a VEI of 3 is 100 times more frequent than one with a VEI of 5. Ecologists have given disproportionate attention to major, explosive eruptions: 12 of the 44 studied events have the relatively infrequent VEI ranks of 5 or 6, whereas 22 eruptions of VEI of 1–3 have been the subject of volcano ecology publications (VEI values for each volcano are from the Smithsonian Institution Global Volcanism Program webpage [Venzke 2013]).

Despite the nonrandom, nonsystematic sampling of volcanoes by ecologists, the studied eruptions have occurred at volcanoes that represent a great variety of geophysical settings. Magma types range from low-silica basalt, characterized by nonexplosive eruptions (lava flows), through andesite and dacite rock types of increasing silica content, to the most silica-rich volcanic rock type, rhyolite. The latter three rock types commonly produce explosive eruptions. The volcanic processes that acted as disturbance agents in the studied ecosystems are broadly representative of the suite of important processes, including both violent events (explosions, pyroclastic flows, debris avalanches, lahars) and quiescent ones (lava flows, tephrafall). The representation of processes in the volcano ecology literature roughly follows their occurrence and extent, with tephrafall and PDCs being most common, lava flows and lahars intermediate, and debris avalanches and chemical toxicity least common. Tectonic settings of the 44 studied sites include volcanic arcs associated with subduction

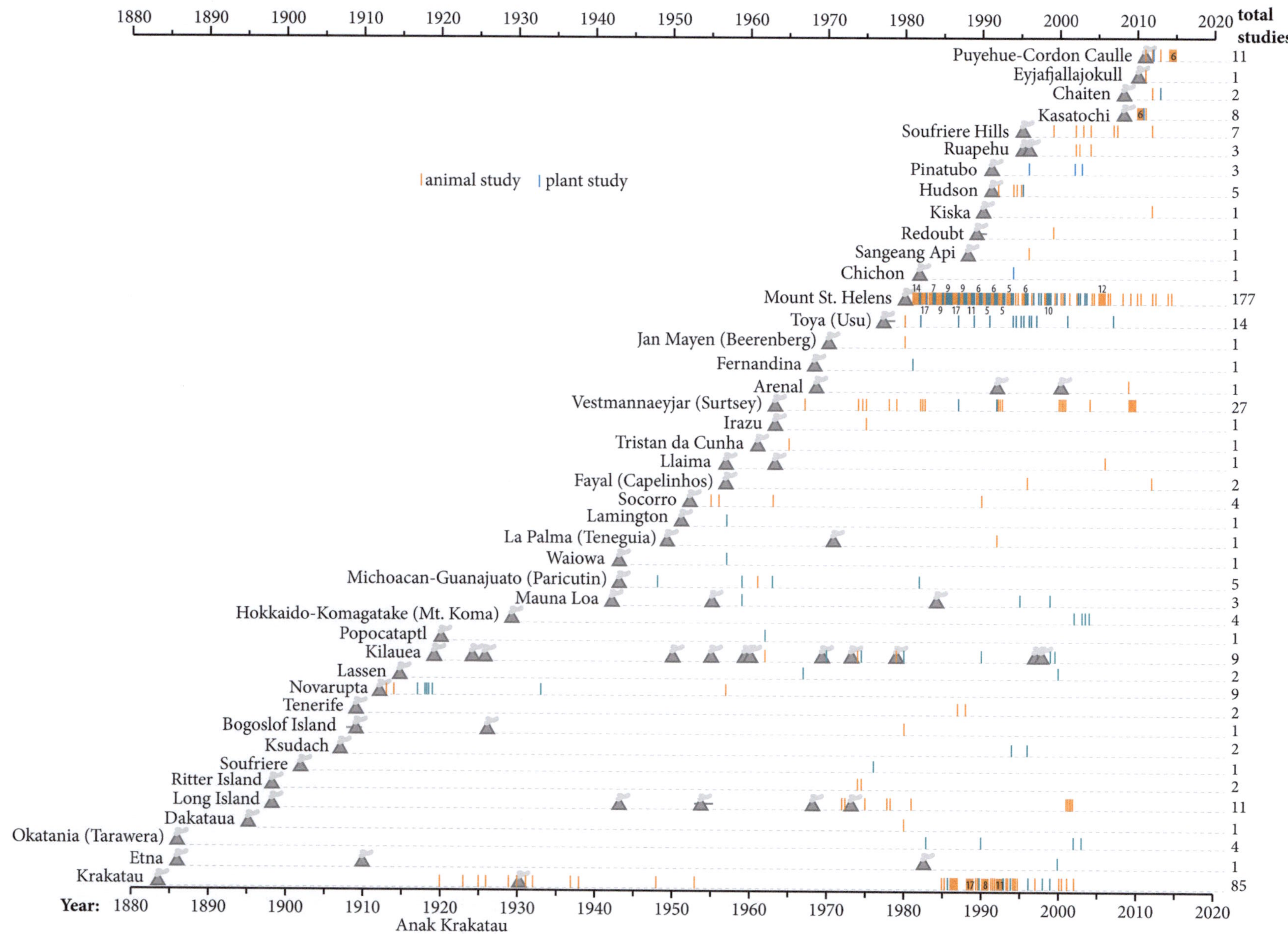

Fig. 16.2 Timeline of volcanic eruptions, beginning with the 1883 Krakatau event, for which we have compiled and analyzed published studies of plant and animal response to eruptions. Subsequent eruptions are shown with volcano symbols, and dates of publications are marked by *vertical lines*. *Numbers* mark years with multiple publications.

Table 16.2 Distribution and number of volcanoes by biome type, with biome area, and associated number of volcano ecology studies with three ranks of study intensity: *H* (high-intensity), continuous, intensive, multiple topics/taxa; *M* (medium-intensity), several samplings over time, multiple topics/taxa; *L* (low-intensity), single sampling, limited subject matter.

Biome	No. active volcanoes	Area (10^6 km^2)	Volcanoes/biome area (no./10^6 km^2)	Number of volcanoes with ecology studies, by study-intensity category			Volcanoes studied (%)
				H	M	L	
Tropical forest	175	24.08	7.27	1	5	9	8.6
Temperate forest	63	16.90	3.73	1	3	8	19.0
Boreal forest	49	15.13	3.24	1	0	2	6.1
Tundra	39	11.65	3.35	0	1	5	15.4
Grassland/shrubland	34	35.99	0.94	0	0	4	11.8
Desert	23	27.89	0.82	0	0	1	4.3
Mediterranean	10	3.22	3.11	0	0	3	30.0

The volcanoes tallied are reported in the Smithsonian Institution/US Geological Survey Global Volcanism Program catalog reported to have erupted since 1883, excluding those noted as "submarine." Locations of these volcanoes were plotted onto the World Wildlife Fund's (WWF) map of terrestrial biomes of the globe. We combined some of the 14 biome types in the WWF system (e.g., three types of tropical and subtropical forests are shown as one, and four types of grassland/shrubland are combined), and several involving very small numbers (<10) of volcanoes and/or small area are not reported (flooded grasslands, mangrove, lake, and rock/ice). The Mediterranean type refers to Mediterranean forest, woodland, and shrubland. For volcanoes designated as "ocean" biome, but not "submarine," we assigned a terrestrial biome type based on published ecological literature from the volcano or surrounding area, web-based searches on vegetation of the target area, and visual inspection of Google Earth imagery (From Table 73.2 in Crisafulli et al. 2015).

zones (33), oceanic rift zones (e.g., Surtsey) (7), and intraplate volcanoes in oceanic contexts (e.g., Hawaii) (4). Most of these volcanoes (26) are in continental settings, and the others are islands of single volcanoes that existed before the studied eruptions (10), part of islands composed of more than two volcanoes (5), and only three are on islands newly emerged from the sea (2) or a lake (1). Despite their rarity, these three cases of entirely new subaerial volcanic substrates have received a great deal of study (Anak Krakatau, 41 published papers; Surtsey, 27; and Long Island, 11), in part to explore the theory of island biogeography, which makes propositions about species diversity in relation to island size and isolation (MacArthur and Wilson 1967).

16.3.2 Affected Biomes

To provide a biogeographic perspective on the global distribution of volcanoes that have erupted since 1883 and also those that have received study by ecologists, we examine these volcanoes in the context of major terrestrial biomes of the world, as defined by the World Wildlife Fund (http://www.worldwildlife.org/publications/terrestrial-ecoregions-of-the-world, accessed 2008). These biomes vary in extent by more than an order of magnitude, and a significant sample (at least 10) of the historically active volcanoes occurs in each biome (Fig. 16.1, Table 16.2). Both the abundance and density of volcano ecology studies vary among terrestrial biomes (Table 16.2) (Crisafulli et al. 2015). Biomes with the highest number of studied volcanoes are tropical (15) and temperate (12) forest biomes, and biomes with the greatest coverage (percent of recently erupted volcanoes studied) are Mediterranean (3 of 10, 30% of recently erupted volcanoes studied) and temperate forest (12 of 63, 19%) biomes. Two of the three most intensively studied volcanoes occur in forest biomes—Krakatau in tropical forest, Mount St. Helens in temperate forest. Although based on its latitude Surtsey, the third, is in the boreal forest biome, its potential natural vegetation is mesic grassland with herbs, presumably owing to its oceanic setting (Borgthor Magnusson, personal communication). On an area basis, the tropical forest biome is more than twice as densely populated with active volcanoes as any other biome, whereas temperate and boreal forest, Mediterranean, and tundra biomes have quite comparable densities of active volcanoes.

The types of volcanic processes involved in an eruption strongly influence the distance that disturbance effects can reach, hence types of biomes that can be affected. Eruptions interact with proximal ecosystems via flowage processes, such as lahars, lava flows, and PDCs, which can extend 10s of km from the source vent. Effects of tephrafall and aerosol deposition can be acute proximally but can also extend to much greater distances, even around the globe, affecting ecosystems dissimilar to the proximal biomes. For example, mid-latitude, explosive eruptions of volcanoes in the Americas deliver tephra to distant grassland/shrubland and desert biomes, arid ecosystems lying in the rain shadows of volcanic chains located in forest biomes. Many of the volcanoes in East Asia and the Aleutians deliver tephra to proximal marine environments; some tephrafall from eruptions elsewhere commonly reaches more distant marine environments. More generally, Ayris and Delmelle (2012, p. 1927) note that, "tephra fallout across the globe may primarily fall into the ocean and mostly into coastal waters."

16.3.3 Volcano Ecology Studies: Topics and Timing

The publication record varies greatly in frequency, continuity, and thematic breadth of ecological study at the 44 volcanoes. Following Crisafulli et al. (2015), we rank study intensity for each volcano in three categories: (1) high—sustained (although it may be episodic), intensive, multiple topics/taxa; (2) moderate—several samplings over time, multiple topics/taxa; (3) low—single sampling, limited subject matter. (Table 16.2). As Crisafulli et al. (2015) report, only three volcanoes have the highest study-intensity ranking: Krakatau (1883), Surtsey (1963), and Mount St. Helens (1980). Dates noted are the initial eruptions in the study period. Each of these volcanoes erupted in subsequent years, but not all eruptions received ecological study. Several sites of current, substantial study may shift from the moderate to the high-intensity category if studies are sustained beyond the initial flush of excitement, for example, Chaiten (2008), Kasatochi (2008), and Cordon Caulle (2011). Most of the studied, recently active volcanoes (32 of the 44 volcanoes), have been the subject of only low-intensity ecological investigation.

Volcano ecology studies have substantial similarity in terms of topics and taxa but also some instructive variation (Crisafulli et al. 2015). Among topics of inquiry, the assembly of biological communities and the factors influencing both the pace and compositional pattern of this process have been dominant themes of study. Volcanoes with published studies of animal response (30 of the 44 recent eruptions) exceed those with studies of plant response; only 24 of the studies and 11 of the volcanoes incorporated both plant and animal response. Of the 302 publications on animal response to eruptions, 47% concern invertebrates and 42% vertebrates, and only 10% of papers concern both vertebrates and invertebrates (Crisafulli et al. 2015). Terrestrial arthropods, birds, and mammals have been common post-eruption faunal taxa studied, perhaps because these organisms are broadly distributed across biomes, play numerous roles in ecosystem functioning, are generally well known, and serve as good barometers of ecological response to disturbance. Arachnids and aquatic arthropods are next-most-common subjects of study. Several other taxonomic groups, such as porifera, cnidaria, and zooplankton, have been subjects of only a handful of studies. Plant survival and subsequent successional processes have been common topics of study, perhaps in part because vegetation is at the base of most food webs, plants are rooted in place and thus are very tractable, and vegetation is well documented to exhibit succession, a process of great interest to many ecologists.

We examine the timing of eruptions of studied volcanoes and related ecology publications by creating a timeline beginning with Krakatau in 1883 and ending with Cordon Caulle in 2011 (Fig. 16.2). The pace of eruptions studied increased from one per 2.5 years in the period 1883–1920, to one per 2 years in 1942–1970, to about one per year for the periods 1988–1995 and 2008–2011. It is not clear to us why there appears to be a brief hiatus in 1996–2007, when no eruptions of volcanoes that had not previously erupted since 1883 triggered publication of volcano ecology studies. The case of periodic eruptions from a single volcano attracting volcano ecology research in the study period has been rather rare, although Mauna Loa and Kilauea in Hawaii, Arenal, and Long Island are good examples where this occurred. The timing of publications after eruptions includes several types: the quick response followed by persistent study, exemplified by the Surtsey and Mount St. Helens cases; single or small numbers of studies that lag the eruption by short or long periods of time, such as Irazu and Fernandina; or a pulse of activity that lags the eruption by many decades, as in the case of Krakatau. Overall, the number of published studies picks up greatly since 1980, with 112 (SE 22.8) publications per decade in 1980–2010 (only 30.3, SE 6.5, per decade excluding Mount St. Helens and Krakatau), compared with 8 (SE 2.0) publications per decade for 1910–1980. Note that all the publications in the period 1883–1950 came from Krakatau, Novarupta, and Paricutin alone.

Human intervention in post-eruption landscapes and waterways may affect ecosystem responses and, therefore, volcano ecology research; however, literature on this topic is scarce (but see Franklin et al. 1988; del Moral and Grishin 1999; Dale et al. 1998, 2005c; Titus and Householder 2007; del Moral and Magnusson 2014; Dale and Denton, Chap. 8, this volume; Karr et al., Chap. 9, this volume). Land management objectives of volcanically disturbed areas are wide ranging and may call for actions such as erosion control on hillslopes and along rivers; removal of damaged trees, with subsequent replanting and fertilization to promote plant growth; fish stocking; and cessation of hunting. Lakes and ponds created by eruption processes are sometimes modified to prevent catastrophic failure of natural dams formed by landslides or deposits of other origins that could result in loss of life and property. Such situations have warranted engineered outlets that can compromise ecological research opportunities. Similarly, rivers flowing through volcanic deposits may carry exceedingly high sediment loads, prompting construction of sediment-retention structures to reduce downstream flood hazards or sediment deposition that interferes with river navigation and associated commerce. These structures and others designed to divert lahars from developed areas downstream may limit post-eruption ecological processes and research opportunities. Arguments about these practices are commonly rooted in competing world views—one holding that nature exists to serve humans, the other that sees the natural world as having intrinsic value independent of humans, in which it is often better to let natural processes take their course. Site manipulation for recreation, public education, and even research itself also alter the landscape, possibly precluding or inhibiting the progress of natural processes.

16.3.4 Volcano Ecology Research

Below we briefly describe key findings from work at volcanically disturbed sites. The topics are ordered to roughly follow the sequence of phenomena over time and their relative frequency in the literature. The first stage of inquiry concerns initial interactions of the primary disturbance processes and biota and between geological and ecological research approaches and objectives—for example, how geologists use damage to biota to interpret geophysical processes and how ecologists use understanding of geophysical processes to interpret biotic responses at site and landscape scales. Ecologists have focused on numerous, often interrelated, themes addressing longer-term plant succession, animal community assembly, nutrient accrual, and the much more protracted processes related to soil genesis.

16.3.4.1 Research Themes at the Geology–Ecology Interface: Immediate Interactions

Generally, ecologists have poorly described the specific types, intensity, and spatial extent of the volcanic disturbances that have affected the sites they studied. In some cases, geologists provide detailed descriptions of geophysical processes and deposits in terms that set a framework for subsequent ecological investigations (at Surtsey and Mount St. Helens, among others). The common disconnect between geologists and ecologists may be related in part to the complexity of volcanic disturbance processes and the arrival of ecologists at eruption sites after some time has elapsed. This limits communication between geologists and ecologists and also allows time for changes in the volcanic deposits and the biota, possibly obscuring some of the interactions.

Table 16.3 Observations of damage to organic matter used to interpret properties of geophysical processes that imposed one or more disturbance mechanisms on terrestrial ecosystems. All references concern the 18 May 1980 eruption of Mount St. Helens and were published in Lipman and Mullineaux (1981).

Process/zone	Observation of organic matter	Mechanism	Reference
Scorch/ standing dead forest	Melting of cuticular waxes on conifer needles	Heat	Winner and Casadevall
PDC/ blowdown forest	Color, smell, flexibility	Heat	Banks and Hoblitt, Moore and Sisson
PDC/ blowdown forest	Tree-fall arrangement	Impact	Kieffer
PDC	Fragmentation, bark/wood removal	Impact, abrasion	Hoblitt et al., Waitt
Debris avalanche	Fragmentation, tree-fall arrangement	Impact, abrasion	Voight et al.
Lahar	Bark/wood removal, tree-fall arrangement	Impact, abrasion	Janda et al.

PDC: pyroclastic density current.

Volcanic and associated hydrologic processes damage vegetation, leaving evidence that geologists can read to help interpret properties of geophysical processes. Examples are widely scattered in the literature, and the 1980 eruption of Mount St. Helens offers a particularly important case where dozens of volcanologists conducted detailed studies, often drawing on evidence recorded by vegetation damage observed within days or weeks of the eruption. The US Geological Survey's initial synthesis of the eruption (Lipman and Mullineaux 1981) provides many examples that capitalize on the rich diversity of geophysical processes and the characteristics of vegetation on the land at the time of the eruption (Table 16.3). Observations of disturbance by mechanisms of abrasion, heat (e.g., charring), and impact force recorded in tree-fall arrangement informed analysis of the PDC (lateral blast), lahar, and debris-avalanche processes. Even before the 1980 eruption, vegetation was used in unraveling the earlier eruptive history detailed in part through radiocarbon dating of organic matter killed at the time of emplacement of deposits and observations of charred wood used to distinguish hot from cold flowage deposits (Mullineaux and Crandell 1981).

Interpretations of initial disturbance caused by eruptions need to be founded on understanding of interactions of properties of the affected biota with properties of the geophysical processes and eruptive products. Nuances in the characteristics of a disturbance type can yield surprisingly different degrees of severity to biota. The 2008 Chaiten and 2011 Cordon Caulle eruptions, for example, both deposited tephra

Fig. 16.3 (**a**) Canopy abrasion by lithic gravel tephra at Chaiten volcano; (**b**) canopy litter (foliage, twigs, ephiphytes) deposited in the basal layer of lithic gravel tephra deposit at Chaiten volcano—samples of tephra deposits on white sheets (left to right) are from 0–5, 5–10, and 10–15 cm above pre-eruption ground surface and dark material is organic matter; (**c**) lack of canopy abrasion or limbfall by 50 cm of pumiceous tephra deposition at Cordon Caulle volcano; (**d**) presence of the understory plants *Drimys andina* in growth position buried in 35 cm of pumiceous tephra at Cordon Caulle volcano; (**e**) high concentration of fallen limbs interpreted as resulting from deposition of fine tephra in the forest canopy, possibly soaked by rain, leading to sufficient weight to cause limbfall at Chaiten volcano; (**f**) bigleaf maple (*Acer macrophyllum*) with foliage amid killed conifers in the standing dead forest zone of Mount St. Helens in 1980 (USDA Forest Service) (Photos: **a** J. Jones; **b–f** F.J. Swanson).

in Chile's Valdivian rainforests, but with different effects depending on particle size and density as well as deposit thickness. A distinctive, single rain of lithic gravel tephra (bulk density >2.0 g cm^{-3}) at Chaiten (the beta layer of Alfano et al. 2011) severely abraded the tree canopy (Fig. 16.3a), depositing foliage and canopy epiphytes in the basal 5–8 cm of tephra deposits (Fig. 16.3b) (Swanson et al. 2013). In contrast, several 10s of cm of low-bulk-density, pumiceous lapilli tephrafall at Cordon Caulle did not abrade the canopy (Fig. 16.3c), failed to produce a concentration of organic matter in the lower part of the tephra profile, and gently buried perennial, understory plants with green foliage in growth position (Fig. 16.3d) (Swanson et al. 2016). At Chaiten, >15 cm of fine tephra (ash) deposition, possibly combined with interception of rainfall, triggered canopy damage when excessive weight in the canopy caused extensive breakage and delivery of limbs to the forest floor (Fig. 16.3e) (Swanson et al. 2013). Both tree phenology in relation to the season when an eruption occurred and the capacity of damaged trees to sprout new limbs and foliage can strongly influence survival after scorching disturbance. In the 110-km^2 standing dead forest zone at Mount St. Helens, for example, all conifers were killed, but a bigleaf maple (*Acer macrophyllum*) had not yet burst bud by the 18 May eruption, so it survived (Fig. 16.3f). Several common tree species in the Valdivian rainforest, in contrast, have the capacity to sprout after disturbance by scorching, but sprouting occurs mainly from major limbs and the bole where the bark is thick enough to protect buds from heat damage.

16.3.4.2 Role of Biophysical Legacies

Ecologists have long known that components of the predisturbance ecosystem that remain after a disturbance can greatly influence the pace, pattern, and direction of ecological response—the course of secondary succession. Because explosive volcanism involves many geophysical processes, and these processes occur as intensity gradients along flow and depositional paths, the type, amount, and spatial distribution of biophysical legacies are commonly present and readily identifiable in post-eruption landscapes. The 1980 eruption of Mount St. Helens galvanized the study of biophysical legacies, as findings from early research there formalized many ideas that were more loosely defined in previous decades. Chief among these ideas is the widespread importance of surviving organisms, propagules, and other organic matter, even at sites that at first appear to have complete mortality. Legacies occur in many circumstances, such as under snow and ice, and within soil, sediment, and decomposed logs. In some cases it may be years before processes such as erosion of new deposits liberate surviving organisms and make pre-eruption soil available to support new life. Biological legacies can profoundly influence the pattern and pace of succession, because they can provide local sources of propagules rather than relying solely on dispersal from distant source populations. Legacies may also provide suites of resources for recolonizing and newly colonizing species, such as food and cover for animals, and safe sites, higher moisture retention, shade, and nutrients that facilitate establishment under harsh environmental conditions (e.g., nutrient-impoverished substrates). However, in many cases organisms may survive the primary disturbance but then fail to persist on the site because of limiting resources, inhospitable environmental conditions, predation, or secondary disturbances, which commonly increase after severe site modification.

16.3.4.3 Plant Succession

A primary focus of ecologists across numerous volcanic settings has been tracking plant species accrual and changes in community structure (composition, richness, and abundance) through time at eruption sites. This work has included classic vegetation science as well as novel experimental work aimed at teasing out the underlying mechanisms causing the observed patterns. Most volcano ecology succession work has adopted a plot-based approach at relatively fine spatial scales (0.25–500 m^2), and these plots are revisited through time to depict the sequence of ecological change. Some of these studies stand as excellent examples of plant responses to disturbance and are widely cited throughout the broader successional literature. Successional development on disturbed sites requires dispersal of seeds and spores from source populations and the amelioration of environmental conditions by many mechanisms, including weathering of inorganic substrates, accumulation of organic matter, and development of biotic structures that serve as habitat for animals. Under primary-successional scenarios, plant community development may be slow and strongly influenced by stochastic processes and contingencies, such as arrival sequence. In contrast, when surviving vegetation is abundant and widespread, vegetation development may proceed quickly and be more deterministic. Del Moral and Magnusson (2014), for example, found that plant species richness in cases of primary succession at Mount St. Helens plateaued in the second decade post-disturbance but cover continued to gradually increase at many sites over longer periods. The more remote, insular, and higher-latitude Surtsey volcano had rather stable richness by the fourth decade at many plots.

16.3.4.4 Animal Community Assembly

Animal ecologists have a long and rich history of describing arrival and accrual of animals, both vertebrates and invertebrates, at eruption sites. In some cases, such as at Krakatau, this work has been accomplished by periodic expeditions to sites where animal taxa are cataloged and compared to assemblages from earlier post-eruption trips. Such investigations appear to have rarely used plot- or transect-based sampling design. In contrast, at a few eruption sites, most notably Mount St. Helens and Surtsey, ecologists arrived shortly after the site was created (Surtsey) or disturbed (Mount St. Helens) and established networks of plots and transects where a diverse array of animals has received detailed and repeated measurements (annually) for decades. Community assembly proceeds at a pace set by extent of survivors and proximity of source populations. In a review of findings from eruptions at seven volcanoes, Edwards (2005) notes rapid accrual of the "aerial plankton" of arthropods adrift on the winds, birds, and other highly mobile organisms. He comments on "pioneers gaining early entry, transients establishing an early beachhead, and permanent colonization proceeding from these sites" (Edwards 2005, p. 268). Across a volcanic disturbance gradient at Mount St. Helens, the pace of small-mammal community assembly was strongly influenced by residual mammals and the amount of living and dead vegetation present in the post-eruption landscape (Crisafulli et al. 2005). For highly mobile taxa, like birds and bats, development of suitable habitat appears to be the single most important factor determining both the rate and pattern of species establishment. Number of bird species climbed through three decades post eruption at Mount St. Helens and at a higher rate than at Surtsey, which appears to have plateaued in the fifth decade (Crisafulli et al. 2015). These two cases also reveal complexities of trajectories of species assembly, such as (1) the role of canopy closure in abruptly reducing plant species diversity after 15 years in some environments at Mount St. Helens; (2) the abrupt increase in bird species at Mount St. Helens in response to habitat structure provided by shrubs and tree saplings; and (3) after 20 years at Surtsey, the effects of nesting seabird colonies in increasing nutrient supply and, thereby, plant accrual and spread (Sigurdsson and Magnusson 2010).

16.3.4.5 Biotic Interactions

Because biological populations are typically reduced or more rarely extirpated during eruptions, in the first years after a volcanic event, ecosystems are rendered relatively simple. Biotic interactions are likely to be more apparent, be unusually strong, and become important in allowing affected systems to initiate biotic assembly, as by mutualism. These mutual interactions are important because of the poor nutrient status and water-holding capacity of volcanic deposits and are therefore of particular interest to ecologists. Examples of mutualistic relationships include those between nitrogen-fixing bacteria and plants, and certain fungi associated with the root systems of plants, called mycorrhizae (see Allen et al., Chap. 11, this volume) in which the plant provides carbon to the fungus, and the fungus supplies the plant with otherwise-limited nutrients, such as phosphorus, along with water. Plant and animal interactions, both positive and negative from a plant perspective, have been reported to be important in volcanic landscapes (Andersen and MacMahon 1985; Che-Castaldo 2014). Herbivory by both ungulates and insects has strongly influenced plant populations and community structure, and seed and spore dispersal by animals has been important for the arrival at eruption sites of plants with poor dispersal mechanisms. Animals influence plant communities by trampling and burrowing (bioturbation), as well as depositing nutrients in the form of urine and feces.

16.3.4.6 Nutrient Inputs and Soil Genesis

Ecologists have been quick to note that volcanic deposits are nutrient impoverished, typically devoid of plant-available nitrogen as well as carbon. The source, accumulation rates, and concentrations of these essential building blocks for ecosystem development have been studied in the context of both oceanic volcanic islands (e.g., Surtsey, Krakatau, San Bernardino) and in continental settings, such as Mount St. Helens. The vectors leading to nutrient enrichment of volcanic deposits can be as varied as aerial fallout of arthropods, plants and their nitrogen-fixing bacterial symbionts, and development of seabird colonies on islands with associated marine-derived guano. In areas with high mortality of trees, remaining boles and branch systems provide nest and perch sites that draw birds and mammals that enrich the site through uric acid/urine and fecal inputs. These animals also deposit seeds and spores that initiate new plant growth. Eventually dead trees decompose, providing a "time-release" source of carbon and increasing the water-holding capacity of the volcanic deposit. Surviving vegetation provides nutrient inputs through litterfall and by snaring wind-blown particles. Longer-term soil genesis processes have been investigated on the Hawaiian Archipelago, among other locations.

16.4 Discussion and Conclusions

16.4.1 Status of Volcano Ecology: A Global Perspective

The pace of volcano ecology research has accelerated over recent decades as the growing body of work calls attention to its value and provides concepts to be tested with study of new eruptions. The pace of study of new volcanic eruptions has increased from one published study per 2.5 years in 1883–1920 to one per year in recent decades. A wide array of biomes, taxa, and ecological processes has been studied around the globe. The studied volcanoes represent a wide range of tectonic settings, biogeographic contexts, volcanic processes, magma composition, and properties of deposits.

The field of volcano ecology is still in an early stage of development. Much of the work has been opportunistic, incomplete in coverage of eruptions, spotty in terms of taxonomic coverage, and lacking in standardized protocols for research approaches. Sustained commitment to study at individual locations has been rare, and, where it has occurred, it is a mix of personal and institutional commitments. However, there have been some notable exceptions, such as the use of common sampling protocols at Mount St. Helens and several Chilean eruption sites; the use of systematic, multidisciplinary planning of campaigns at Krakatau and Kasatochi; and the sustained commitment to sampling at Surtsey and Mount St. Helens. The high incidence of studies in Mediterranean and temperate forest biomes suggests that research potential is influenced by proximity of academic and governmental researchers to recently active volcanoes and, thereby, proximity for governmental investment in research and infrastructure. On the other hand, researchers have conducted studies at very remote volcanoes, but this work has been largely implemented and financed by wealthy nations (Iceland, Japan, the Netherlands, the United Kingdom, and the USA). The tendency for studies to occur at sites of major (high-VEI) eruptions reflects the interesting questions that can be addressed after large, complex eruptions, such as those producing conditions for primary succession, and also the greater public interest and economic and social effects, which may make it easier to secure funding for studies.

Efforts to provide a spatial or geographic framing for volcano ecology studies have taken several forms across the published record of research. Some studies and whole research programs have adopted useful "disturbance-gradient" designs using at least two approaches: (1) study sites located along a gradient of disturbance *severity* (proportion of organisms killed by the primary disturbance event), regardless of the disturbance type; and (2) study sites on a gradient of disturbance *intensity* (physical force exerted on an ecosystem by a single disturbance type or mechanism, such as thickness of tephrafall deposits). A set of core studies at Mount St. Helens, for example, has been arrayed along a gradient of primary disturbance severity ranging from thick, initially sterile, pumiceous pyroclastic-flow deposits to thin tephrafall deposits (several chapters in Dale et al. 2005b). Examples of studies along a gradient of disturbance for a single disturbance type include effects of tephrafall depth on forest understory vegetation (Antos and Zobel 2005), tephra effects on boreal forests (Grishin et al. 1996), and variation in abrasion of vegetation and depth of deposits along lahar flow paths (Frenzen et al. 2005). However, many of the volcano ecology studies we have reviewed do not clearly describe the volcanic disturbance processes that affected study sites and do not place the sites in the larger pattern of the disturbances, which limits interpretation of findings.

We note, however, a near absence of landscape-scale studies that move beyond plot systems and gradients to address large-scale vegetation patterns and the interplay with secondary disturbances (e.g., erosion and deposition during major floods). This work would occur at a scale embracing multiple disturbance types and a sampling of ecosystem types. Such an approach is necessary to test hypotheses about the relative importance of movement into a disturbed area from the edge of the disturbance zone or outward from hot spots of biological legacies within the disturbance zone. Relevant observations (Adams et al. 1987; Lawrence 2005) offer some starting points for this work at Mount St. Helens.

Unlike the field of volcanology, which has established dozens of volcano observatories in many countries since the 1980 eruption of Mount St. Helens, no counterpart ecological observatories focus on volcanoes. The major motivation for volcano observatories is the great risk to public safety posed by eruptions, which justifies the expense. However, volcano ecology studies also serve society by providing information that can be used to anticipate effects of eruptions on the landscape around them, near and far, and what the prognoses are once an eruption occurs, including the potential consequences for a suite of factors of societal importance such as natural resources, agriculture, water, and infrastructure (e.g., transportation systems). Furthermore, volcanoes provide one endpoint in the continuum of large, infrequent disturbances from natural to wholly anthropogenic, and thus can inform ways to formulate and address management goals in the face of disturbances (Dale et al. 1998; Turner and Dale 1998).

In some respects Mount St. Helens and Surtsey have become volcano ecology observatories, with sustained, dedicated staff, an ecological monitoring program, complex portfolios of research activities, and in the case of Mount St. Helens, intensive public outreach.

16.4.2 Mount St. Helens in the Global Context of Volcano Ecology Research

Mount St. Helens has been an important contributor to the field of volcano ecology, as a result of a special combination of circumstances. The major eruption involved diverse processes, which offer a wide array of study opportunities. Precursor studies provided in-depth understanding and knowledge of volcanic and ecological historical settings, giving a strong frame of reference for assessing disturbance effects on ecosystems. Several months of precursor volcanic activity drew the attention of a large, interdisciplinary, regional and national science community that stood ready to take on the opportunity provided by the eruption. Immediate, intensive geological investigations led by the US Geological Survey gave the ecological science community a thorough interpretation of the complex eruption, which formed the template for ecological investigations that followed. Access to the disturbed area was constrained at first because of the destruction of roads, and relied heavily on helicopters, but within a year many roads were repaired and vehicle access was possible. Most importantly, the US Congress created the (44 515-ha) Mount St. Helens National Volcanic Monument, with the goal of allowing geological forces and ecological processes to "proceed substantially unimpeded" and to foster science, education, interpretation, and recreation to the public's benefit. This legislation ensured that research would be a primary focus of the post-eruption landscape and that adequate protection would be in place to safeguard research infrastructure from loss or damage. The federal agency responsible for the new Monument, the US Forest Service, has made a concerted effort to protect research sites and opportunities. Strong institutional support for science has come from the US Forest Service and National Science Foundation; funding for quick response was especially important. Vegetation and aquatic studies completed before the eruption were extremely valuable in assessing system response. Eager, attentive public and media organizations have sustained interest in the area and thereby reinforced agency support.

The contribution of Mount St. Helens to advancement of volcano ecology has taken many forms. The science program has contributed 177 (42%) of 423 publications in the Crisafulli et al. (2015) database. Mount St. Helens scientists have coauthored a large share of the multi-volcano synthesis publications, which has been a rather recent phenomenon (e.g., del Moral and Grishin 1999; Dale et al. 2005a; Edwards 2005; del Moral and Magnusson 2014; Crisafulli et al. 2015). Scientists in the large Mount St. Helens community have frequently consulted with foreign government agencies and academics concerning research opportunities and ecosystem effects associated with new eruptions, co-convened international symposia and field tours on volcano ecology, and been integral players in the emerging global network of scientists working in the field of volcano ecology. Another dimension of advancing volcano ecology has been partnering with colleagues to initiate studies following eruptions elsewhere, using protocols in common with those used at Mount St. Helens: research underway, for example, with Chilean colleagues working at Chaiten, Cordon Caulle, and Calbuco volcanoes. The vibrancy of the Mount St. Helens research program, community of scientists, and outreach specialists signals that these roles are likely to continue.

The 1980 eruption and the setting of Mount St. Helens make it a representative case for study of eruption–ecosystem interactions in some respects and an unusual case in others. The tectonic setting of Mount St. Helens in a volcanic chain associated with a subduction zone is very common, and the highly energetic, explosive 1980 eruption (VEI = 5) in a mountainous landscape with moist, maritime climate involved a wide variety of volcanic and associated hydrologic processes. However, the suite of volcanic processes in the 1980 eruption of Mount St. Helens does not provide examples of several important types of eruptions and settings of volcano ecology studies, such as basaltic lava flows. The pyroclastic flows of the 1980 eruption did not have opportunity to interact with pristine terrestrial or several types of aquatic ecosystems, because those existing just before the eruption had been obliterated by the huge debris avalanche and then by a powerful blast PDC (lateral blast) before the pyroclastic flows occurred. In terms of its ecological setting, Mount St. Helens resides in the temperate forest biome, which is intermediate in extent among the seven types of terrestrial biomes considered (Table 16.2).

Public communications about volcanoes and volcano ecology have been a hallmark of the Mount St. Helens program and a model for other locations around the world. Outreach infrastructure and networking include several visitor centers, dozens of Ranger Naturalists who provide presentations to hundreds of thousands of people annually, sustained relationships with national and international media, a nonprofit organization focused on science education and stewardship at the volcano (the Mount St. Helens Institute), and museums that curate tens of thousands of Mount St. Helens' biological specimens and make them available to researchers around the world. A distinctive feature has been use of the volcanic venue to explore the intersection of arts, humanities, and science, which is a growing phenomenon nationally at sites of long-term ecological research (Swanson 2015). Since 2000, creative writers and scientists have gathered at Mount St. Helens on 5-year eruption anniversaries to share their learning with one another and with the public in performances and written works (Snyder 2004; Goodrich et al. 2008; Buntin 2010; terrain.org 2013).

Highly collaborative, interdisciplinary research and broader inquiry requires a strong, interactive community. Volcano ecology research at Mount St. Helens is unusual in terms of the diversity of disciplines and intensity and duration of studies, which make it one of only three sites in our highest class of study intensity (Table 16.2) and the only one in a continental setting; the other two (Krakatau and Surtsey) are in marine settings. This history highlights the importance of an open, diverse scientific community for dealing with the social and ecological complexity of major volcanic disturbance events. At Mount St. Helens, the Forest Service has made concerted efforts to sustain and enlarge the research community through mechanisms such as week-long field gatherings every 5 years, termed "pulses," which attract more than 100 members of the science community and actively recruit new, early-to-mid-career scientists to maintain long-term studies and to initiate new research. The pulses also host groups of creative writers and involve participation by high-school students and undergraduates. The overall effect is to encourage and sustain interpersonal and interinstitutional collaborations, facilitate group publications, and assist in the intergenerational handoff of research projects.

16.4.3 Future Volcano Ecology Research

This review of the state of global volcano ecology research and the experience of volcano ecology at Mount St. Helens prompts several recommendations for further study:

1. Ongoing research at eruption sites should continue, and at those sites where there has been a lapse in measurements, work should be reinitiated to document long-term patterns of ecosystem responses to volcanic disturbance. This sustained effort is particularly important for ecosystems with long seres following disturbance, such as forests or cold regions that may require centuries to develop following high-intensity eruptions.
2. At sites with multidecadal research, investigators or other responsible parties (e.g., agencies) need to recruit new members into the research community as aging scientists approach retirement, thus allowing for intergenerational transfer of long-term data and plot systems, and stewardship of studies well into the future.
3. In the case of future eruptions, it is important to initiate work as soon as access is possible and conditions are safe,

because early interpretation of the physical disturbance effects across the landscape and initial responses of biota is a key to understanding longer-term processes of biotic assembly. Evidence of ephemeral phenomena (e.g., influence of snow) may quickly vanish yet may be important to explaining organism survival and other phenomena.

4. An integrated multidisciplinary science approach at eruption sites should investigate numerous taxa, ecological interactions, and ecosystem processes, as well as human responses to eruptions at scales of individuals, communities, and nations.
5. Volcano ecology studies should be founded on investigation at the geology–ecology interface to place the work in a clear geographic context by sampling along disturbance gradients and other aspects of landscape context.
6. It is essential to thoroughly document research efforts and data, so they are available to future researchers. Such efforts should include plots that are georeferenced, publically accessible data sets, monumented photopoints, and physical voucher-specimen collections that are deposited in safe, curated repositories.
7. Despite the wealth and diversity of volcano ecology research efforts to date, several areas of inquiry remain unexplored or only sparsely explored. Understudied subject areas that deserve attention include
   (a) Evolutionary processes in sites with limited population size and concomitant isolation in post-eruption landscapes may lead to shifts in genetic structure through genetic drift and bottlenecks. Similarly, immigrants to eruption sites may be subject to founder effects associated with small population size, absence of or low gene flow, and suites of novel selective pressures in the post-eruption landscape. Although tantalizing, such evolutionary processes have only occasionally been assessed at eruption sites, and the results have been variable across taxa.
   (b) Landscape ecology perspectives addressing broad-scale (100–1000 km^2) vegetation change and associated animal community assembly to assess influences on landscape pattern development and consequences for ecological processes. Such approaches would provide a template for follow-up studies of patch dynamics and metapopulations.
   (c) Studies of biogeochemical cycling and soil genesis.
   (d) Development and change in food-web structure during the course of succession.
   (e) Aquatic–land interactions in post-eruption landscapes, such as the flows of energy, matter, and organisms.
   (f) Comparisons of the patterns and rates of change in key ecosystem parameters (e.g., net primary production) and processes among different system types, such as terrestrial, lake, and stream environments that were disturbed by the same eruption.
   (g) Attention to emerging phenomena, such as climate change and invasive non-native species that may alter the course of ecosystem response.

Acknowledgments We greatly appreciate the assistance of E. Schyling in assembly of the bibliographic database, K. Christiansen in creation of the Fig. 16.1, and K. Ronnenberg for creation of timeline, photo plate, and additional editorial assistance. Reviews of the manuscript by V. Dale, J. Franklin, C. Millar, and R. Parmenter were especially helpful. Funding for our research activities at Mount St. Helens and abroad has been provided by the USDA Forest Service, Pacific Northwest Research Station and the National Science Foundation (LTREB Program DEB-0614538). Collaborations with colleagues at Mount St. Helens and in Alaska, Chile, Argentina, China, and Iceland have strengthened our volcano ecology perspectives. We acknowledge and thank the ecologists who since the 1883 eruption of Krakatau have provided important foundational work in the field of volcano ecology.

Glossary

Biological legacy Live and dead organisms and organic matter that survive an ecological disturbance and may affect the pace and pattern of post-disturbance ecosystem development.

Blast PDC (blast pyroclastic density current) A form of pyroclastic density current initiated by rapid decompression of lava domes or cryptodomes (magma bodies cooled high within a volcanic edifice) owing to sudden collapse. Rapid decompression results in a directed explosion that initially impels the current laterally before it becomes a gravity-driven flow. [Sources: a generalized definition based on definitions of PDCs provided in Pierson and Major (2014) and Sigurdsson et al. (2015)]. In the case of the Mount St. Helens 1980 eruption, failure of the volcano's north flank unroofed pressurized magma and superheated groundwater. Rapid exsolution of magmatic gases and conversion of superheated groundwater to steam produced a laterally directed blast, which formed a density current that flowed across rugged topography. The current contained fragmented rock debris as well as shattered forest material (Lipman and Mullineaux 1981).

Debris avalanche A rapid granular flow of an unsaturated or partly saturated mixture of volcanic rock particles (± ice) and water, initiated by the gravitational collapse and disintegration of part of a volcanic edifice. Debris avalanches differ from debris flows in that they are not water-saturated. Although debris avalanches commonly occur in association with eruptions, they can also occur during periods when a volcano is dormant. (Sources: Pierson and Major 2014; Sigurdsson et al. 2015).

Lahar An Indonesian term for a rapid granular flow of a fully saturated mixture of volcanic rock particles (± ice), water, and commonly woody debris. A lahar that has ≥50% solids by volume is termed a *debris flow*; one that has roughly

10–50% solids by volume is termed a *hyperconcentrated flow*. Flow type can evolve with time and distance along a flow path as sediment is entrained or deposited. (Sources: Pierson and Major 2014; Sigurdsson et al. 2015).

Pyroclastic flow Rapid flow of a dry mixture of hot (commonly >700 °C) solid particles, gases, and air, with a ground-hugging flow that is often directed by topography. Flows are generally gravity driven but may be accelerated initially by impulsive lateral forces of directed volcanic explosions. Flows typically move at high velocity (up to several hundred km h^{-1}).

Refuge (refugia) Localized sites where organisms survive a disturbance event at a level greater than the surrounding, disturbance-affected area.

Succession Development of an ecosystem following disturbance, including processes such as species assembly by immigration and establishment, species interactions (e.g., herbivory), and site amelioration (e.g., weathering of inorganic substrates). Primary succession refers to cases with no legacies of the pre-disturbance ecosystem; secondary succession refers to cases where some biota from the pre-disturbance ecosystem persists.

Tephrafall A rain of volcanic particles to the ground following ejection into the atmosphere by an explosive eruption. Tephra is a collective term for particles of any size, shape, or composition ejected in an explosive eruption. (Sources: Pierson and Major 2014; Sigurdsson et al. 2015).

References

Adams, A.B., and V.H. Dale. 1987. Comparisons of vegetative succession following glacial and volcanic disturbances. In *Mount St. Helens 1980: Botanical consequences of the explosive eruptions*, ed. D.E. Bilderback, 70–147. Los Angeles: University of California Press.

Adams, A.B., V.H. Dale, A.R. Kruckeberg, and E. Smith. 1987. Plant survival, growth form and regeneration following the May 18, 1980, eruption of Mount St. Helens, Washington. *Northwest Science* 61: 160–170.

Alfano, F., C. Bonadonna, A.C.M. Volentik, C.B. Connor, S.F.L. Watt, D.M. Pyle, and L.J. Connor. 2011. Tephra stratigraphy and eruptive volume of the May, 2008, Chaiten eruption, Chile. *Bulletin of Volcanology* 73: 613–630.

Andersen, D.C., and J.A. MacMahon. 1985. Plant succession following the Mount St. Helens volcanic eruption: Facilitation by a burrowing rodent, *Thomomys talpoides*. *American Midland Naturalist* 114: 62–69.

Antos, J.A., and D.B. Zobel. 2005. Plant responses of forests in the tephra-fall zone. In *Ecological responses to the 1980 eruption of Mount St. Helens*, ed. V.H. Dale, F.J. Swanson, and C.M. Crisafulli, 47–58. New York: Springer.

Ayris, P.M., and P. Delmelle. 2012. The immediate environmental effects of tephra emission. *Bulletin of Volcanology* 74: 1905–1936.

Banks, N.G., and R.P. Hoblitt. 1981. Summary of temperature studies of 1980 deposits. In *The 1980 eruptions of Mount St. Helens, Washington*, Professional Paper 1250, ed. P.W. Lipman and D.R. Mullineaux, 295–313. Washington, DC: U.S. Geological Survey.

Brattstrom, B.H. 1963. Barcena volcano, 1952: Its effect on the fauna and flora of San Benedicto Island, Mexico. In *Pacific basin biogeography*, ed. L. Gressitt, 499–524. Honolulu: Bishop Museum Press.

Buntin, S.B. 2010. Dirty words on Mount St. Helens. Terrain.org. Issue 26. http://terrain.org/columns/26/buntin.htm.

Che-Castaldo, C. 2014. The attack dynamics and ecosystem consequences of stem-borer herbivory on Sitka willow at Mount St. Helens. PhD dissertation. College Park: University of Maryland.

Crisafulli, C.M., F.J. Swanson, and V.H. Dale. 2005. Overview of ecological response to the eruption of Mount St. Helens: 1980–2005. In *Ecological responses to the 1980 eruption of Mount St. Helens*, ed. V.H. Dale, F.J. Swanson, and C.M. Crisafulli, 287–299. New York: Springer.

Crisafulli, C.M., F.J. Swanson, J.J. Halvorson, and B. Clarkson. 2015. Volcano ecology: Disturbance characteristics and assembly of biological communities. In *Encyclopedia of volcanoes*, ed. H. Sigurdsson, B. Houghton, S.R. McNutt, H. Rymer, and J. Stix, 2nd ed., 1265–1284. New York: Elsevier.

Dale, V.H., A. Lugo, J. MacMahon, and S. Pickett. 1998. Ecosystem management in the context of large, infrequent disturbances. *Ecosystems* 1: 546–557.

Dale, V.H., J. Delgado-Acevedo, and J. MacMahon. 2005a. Effects of modern volcanic eruptions on vegetation. In *Volcanoes and the environment*, ed. J. Marti and G.G.J. Ernst, 227–249. New York: Cambridge University Press.

Dale, V.H., F.J. Swanson, and C.M. Crisafulli, eds. 2005b. *Ecological responses to the 1980 eruption of Mount St. Helens*. New York: Springer.

Dale, V.H., F.J. Swanson, and C.M. Crisafulli. 2005c. Ecological perspectives on management of the Mount St. Helens landscape. In *Ecological responses to the 1980 eruption of Mount St. Helens*, ed. V.H. Dale, F.J. Swanson, and C.M. Crisafulli, 277–286. New York: Springer.

Dammerman, K.W. 1922. The fauna of Krakatau, Verlaten Island, and Sebesy. *Treubia* 3: 61–121.

———. 1948. The fauna of Krakatau, 1883–1933. *Verhandelingen Koniklijke Nederlansche Akademie van Wetenschappen, Afdeling Natuurkunde II* 44: 1–594.

DeGange, A.R., G.V. Byrd, L.R. Walker, and C.F. Waythomas. 2010. Introduction—The impacts of the 2008 eruption of Kasatochi volcano on terrestrial and marine ecosystems in the Aleutian Islands, Alaska. *Arctic, Antarctic, and Alpine Research (Special Section)* 42: 245–341.

del Moral, R., and S.Y. Grishin. 1999. Volcanic disturbances and ecosystem recovery. In *Ecosystems of disturbed ground*, ed. L.R. Walker, 137–169. Amsterdam: Elsevier Sciences.

del Moral, R., and B. Magnusson. 2014. Surtsey and Mount St. Helens: A comparison of early succession rates. *Biogeosciences* 11: 2099–2111. https://doi.org/10.5194/bg.11-2099.2014.

Docters van Leeuwen, W.M. 1936. Krakatau 1883–1933. *Annales du Jardin Botanique de Buitenzorg* 46–47: 1–506.

Edwards, J.S. 2005. Animals and volcanoes: Survival and revival. In *Volcanoes and the environment*, ed. J. Marti and G.G.J. Ernst, 250–272. New York: Cambridge University Press. New York.

Edwards, J.S., and I.W.B. Thornton. 2001. Colonization of an island volcano, Long Island, Papua New Guinea, an emergent island, Motmot, in its caldera lake. VI. The pioneer arthropod community of Motmot. *Journal of Biogeography* 28: 1379–1388.

Eggler, W.A. 1941. Primary succession on volcanic deposits in southern Idaho. *Ecological Monographs* 11: 277–298.

———. 1948. Plant communities in the vicinity of the volcano El Paracutin, Mexico, after two and a half years of eruption. *Ecology* 29: 415–436.

———. 1971. Quantitative studies of vegetation on sixteen young lava flows on the island of Hawaii. *Tropical Ecology* 12: 66–100.

Franklin, J.F., P.M. Frenzen, and F.J. Swanson. 1988. Re-creation of ecosystems at Mount St. Helens: Contrasts in artificial and natural approaches. In *Rehabilitating damaged ecosystems*, ed. J. Cairns, vol. 2, 288–333. Boca Raton: CRC Press.

Frenzen, P., K.S. Hadley, J.J. Major, M.H. Weber, J.F. Franklin, J.H. Hardison III, and S.M. Stanton. 2005. Geomorphic change and vegetation development on the Muddy River mudflow deposit. In *Ecological responses to the 1980 eruption of Mount St. Helens*, ed. V.H. Dale, F.J. Swanson, and C.M. Crisafulli, 75–91. New York: Springer.

Fridriksson, S. 1975. *Surtsey: Evolution of life on a volcanic island*. London: Buttersworth.

Fridriksson, S., and B. Magnusson. 1992. Development of the ecosystem on Surtsey with reference to Anak Krakatau. *GeoJournal* 28: 287–291.

Goodrich, C., K.D. Moore, and F.J. Swanson, eds. 2008. *In the blast zone: Catastrophe and renewal*. Corvallis: Oregon State University Press.

Griggs, R.F. 1918. The beginnings of revegetation of Katmai Valley. *The Ohio Journal of Science* 19: 318–342.

———. 1922. *The valley of ten thousand smokes*. Washington, DC: National Geographic Society.

Grishin, S.Y., R. del Moral, P.V. Krestov, and V.P. Verkholat. 1996. Succession following the catastrophic eruption of Ksudach volcano (Kamchatka, 1907). *Vegetatio* 127: 129–153.

Hoblitt, R.P., C.D. Miller, and J.W. Valance. 1981. Origin and stratigraphy of the deposit produced by the May 18 directed blast. In *The 1980 eruptions of Mount St. Helens, Washington*, Professional Paper 1250, ed. P.W. Lipman and D.R. Mullineaux, 401–419. Washington, DC: Government Printing Office.

Janda, R.J., K.M. Scott, K.M. Nolan, and H.A. Martinson. 1981. Lahar movement, effects, and deposits. In *The 1980 eruptions of Mount St. Helens, Washington*, Professional Paper 1250, ed. P.W. Lipman and D.R. Mullineaux, 461–478. Washington, DC: U.S. Geological Survey.

Kieffer, S.W. 1981. Fluid dynamics of the May 18 blast at Mount St. Helens. In *The 1980 eruptions of Mount St. Helens, Washington*, Professional Paper 1250, ed. P.W. Lipman and D.R. Mullineaux, 379–400. Washington, DC: U.S. Geological Survey.

Lawrence, R. 2005. Remote sensing of vegetation responses during the first 20 years following the 1980 eruption of Mount St. Helens: A spatially and temporally stratified analysis. In *Ecological responses to the 1980 eruption of Mount St. Helens*, ed. V.H. Dale, F.J. Swanson, and C.M. Crisafulli, 111–123. New York: Springer.

Lipman, P.W., and D.R. Mullineaux, eds. 1981. *The 1980 eruptions of Mount St. Helens, Washington*, Professional Paper 1250. Washington, DC: U.S. Geological Survey.

MacArthur, R.H., and E.O. Wilson. 1967. *The theory of island biogeography*. Princeton: Princeton University Press.

Moore, J.G., and T.W. Sisson. 1981. Deposits and effects of the May 18 pyroclastic surge. In *The 1980 eruptions of Mount St. Helens, Washington*, Professional Paper 1250, ed. P.W. Lipman and D.R. Mullineaux, 412–438. Washington, DC: U.S. Geological Survey.

Mullineaux, D.R., and D.R. Crandell. 1981. The eruptive history of Mount St. Helens. In *The 1980 eruptions of Mount St. Helens, Washington*. Professional Paper 1250, ed. P.W. Lipman and D.R. Mullineaux, 3–15. Washington, DC: U.S. Geological Survey.

Newhall, C.G., and S. Self. 1982. The volcanic explosivity index (VEI): An estimate of explosive magnitude for historical volcanism. *Journal of Geophysical Research* 87 (C2): 1231–1238. https://doi.org/10.1029/JC087iC02p01231.

Paine, R.T., M.J. Tegner, and A.E. Johnson. 1998. Compounded perturbations yield ecological surprises: Everything else is business as usual. *Ecosystems* 1: 535–546.

Pallister, J.S., J.J. Major, T.C. Pierson, R.P. Hoblitt, J.B. Lowenstern, J.C. Eichelberger, L. Lara, H. Moreno, J. Muñoz, J.M. Castro, A. Iroumé, A. Andreoli, J. Jones, F. Swanson, and C. Crisafulli. 2010. Interdisciplinary studies of eruption at Chaiten Volcano, Chile. *EOS, Transactions, American Geophysical Union* 91: 381–382.

Peters, D.P.C., A.E. Lugo, F.S. Chapin, S.T.A. Pickett, M. Duniway, A.V. Rocha, F.J. Swanson, C. Laney, and J. Jones. 2011. Cross-system comparisons elucidate disturbance complexities and generalities. *Ecosphere* 2 (7): art81. https://doi.org/10.1890/ES11-00115.1.

Pickett, S.T.A. 1989. Space-for-time substitutions as an alternative to long-term studies. In *Long-term studies in ecology*, ed. G.E. Likens, 110–135. New York: Springer-Verlag.

Pickett, S.T.A., and P.S. White. 1985. *The ecology of natural disturbances and patch dynamics*. New York: Academic.

Pierson, T.C., and J.J. Major. 2014. Hydrogeomorphic effects of explosive volcanic eruptions on drainage basins. *Annual Reviews of Earth and Planetary Sciences* 42: 469–507.

Ruggiero, A., and T. Kitzberger. 2014. Special section: Ecological responses of arthropods to volcanism. *Ecologia Austral* 24(1). http://www.ecologiaaustral.com.ar/secciones/seccionespecial-24(1).pdf.

Siebert, L., E. Cotrell, E. Venzke, and B. Andrews. 2015. Earth's volcanoes and their eruptions: An overview. In *Encyclopedia of volcanoes*, ed. H. Sigurdsson, B. Houghton, S.R. McNutt, S.H. Rymer, and J. Stix, 2nd ed., 239–256. New York: Academic.

Sigurdsson, H.B. 2015. Volcanoes in art. In *Encyclopedia of volcanoes*, ed. H.B. Sigurdsson, B.F. Houghton, S.R. McNutt, H. Rymer, and J. Stix, 2nd ed., 1321–1344. New York: Academic.

Sigurdsson, H.B., and R. Lopes. 2015. Volcanoes in literature and film. In *Encyclopedia of volcanoes*, ed. H. Sigurdsson, B.F. Houghton, S.R. McNutt, H. Rymer, and J. Stix, 2nd ed., 1345–1362. New York: Academic.

Sigurdsson, B.D., and B. Magnusson. 2010. Effects of seagulls on ecosystem respiration, soil nitrogen and vegetation cover on pristine volcanic island, Surtsey, Iceland. *Biogeosciences* 7: 883–891.

Sigurdsson, H., B.F. Houghton, S.R. McNutt, H. Rymer, and J. Stix, eds. 2015. *Encyclopedia of volcanoes*. 2nd ed. New York: Academic.

Smathers, G.A., and D. Mueller-Dombois. 1974. *Invasion and recovery of vegetation after a volcanic eruption in Hawaii*. National Park Service Scientific Monography Series, No. 5. Washington, DC: Government Printing Office. https://www.nps.gov/parkhistory/online_books/science/5/chap2.htm. Accessed 21 September 2017.

Snyder, G. 2004. *Danger on peaks*. Washington, DC: Shoemaker Hoard.

Swanson, F.J. 2015. Confluence of ecology, the arts, and humanities at sites of long-term ecological inquiry. *Ecosphere* 6: 132. https://doi.org/10.1890/ES15-00139.1. http://onlinelibrary.wiley.com/doi/10.1890/ES15-00139.1/full.

Swanson, F.J., J.A. Jones, C.M. Crisafulli, and A. Lara. 2013. Effects of volcanic and hydrologic processes on forest vegetation: Chaitén Volcano, Chile. *Andean Geology* 40: 359–391.

Swanson, F.J., J.A. Jones, C. Crisafulli, M.E. Gonzalez, and A. Lara. 2016. Puyehue-Cordon Caulle eruption of 2011: Tephra fall and initial forest responses in the Chilean Andes. *Bosque* 37: 85–96.

Terrain.org. 2013. Ruin + Renewal, Part 2. Terrain.org. http://www.terrain.org/archives/archives-issue-31/.

Thornton, I.W.B. 1996. *Krakatau. The destruction and reassembly of an island ecosystem*. Cambridge, MA: Harvard University Press.

———. 2000. The ecology of volcanoes: Recovery and reassembly of living communities. In *Encyclopedia of volcanoes*, ed. H. Sigurdsson, B. Houghton, H. Rymer, J. Stix, and S.R. McNutt, 1st ed., 1057–1081. New York: Academic.

———. 2007. In *Island colonization: The origin and development of island communities*, ed. T. New. New York: Cambridge University Press.

Titus, J.H., and E. Householder. 2007. Salvage logging and replanting reduce understory cover and richness compared to unsalvaged-unplanted sites at Mount St. Helens, Washington. *Western North American Naturalist* 67: 219–231.

Turner, M.G., ed. 1987. *Landscape heterogeneity and disturbance*. New York: Springer Verlag.

Turner, M.G., and V.H. Dale. 1998. What have we learned from large, infrequent disturbances? *Ecosystems* 1: 493–496.

Turner, M.G., V.H. Dale, and E.H. Everham. 1997. Crown fires, hurricanes and volcanoes: A comparison among large-scale disturbances. *Bioscience* 47: 758–768.

Turner, M.G., W.I. Baker, C.J. Peterson, and R.K. Peet. 1998. Factors influencing succession: Lessons from large, infrequent natural disturbances. *Ecosystems* 1: 511–523.

Veblen, T.T., M.E. Gonzalez, G.H. Steward, T. Kitzberger, and J. Brunet. 2016. Tectonic ecology of the temperate forests of South America and New Zealand. *New Zealand Journal of Botany* 54: 223–246.

Venzke, E.. 2013. *Global volcanism program. Volcanoes of the World, v. 4.4.3*. Smithsonian Institution. https://doi.org/10.5479/si.GVP.VOTW4-2013. Downloaded 20 Apr 2016.

Voight, B., H. Glicken, R.J. Janda, and P.M. Douglass. 1981. Catastrophic rockslide avalanche of May 18. In *The 1980 Eruptions of Mount St. Helens, Washington*, Professional Paper 1250, ed. P.W. Lipman and D.R. Mullineaux, 347–377. Washington, DC: U.S. Geological Survey.

Waitt, R.B. 1981. Devastating pyroclastic density flow and attendant air fall of May 18—stratigraphy and sedimentology of deposits. In *The 1980 eruptions of Mount St. Helens, Washington*, Professional Paper 1250, ed. P.W. Lipman and D.R. Mullineaux, 439–458. Washington, DC: U.S. Geological Survey.

Winner, W.E., and T.J. Casadevall. 1981. Fir leaves as thermometers during the May 18 eruption. In *The 1980 eruptions of Mount St. Helens, Washington*, Professional Paper 1250, ed. P.W. Lipman and D.R. Mullineaux, 315–320. Washington, DC: U.S. Geological Survey.

Coda. "Another Weather: Mount St. Helens"

Ursula K. Le Guin

Weightless clouds and airy rain drift over
a lower, slower weather in the world
where lava turns in vast typhoon pavanes,
thick fire beneath a ponderous earthen sky:
storms brew a thousand years before they break
in quaking thunder of tectonic shift
to hurl their hot bright hail straight up, send forth
the monstrous overwhelming wave, or still
a city into feathery clouds of glass.

I watch you, my volcano, through the depths
of sunlit air, and see you snowy-flanked
breathing your lazy steam-plume south, yourself
a vapor drifting, a bright veil of stone.

(from *In the Blast Zone*, Oregon State University Press, p. 8–9)

C.M. Crisafulli, V.H. Dale (eds.), *Ecological Responses at Mount St. Helens: Revisited 35 years after the 1980 Eruption*,
https://doi.org/10.1007/978-1-4939-7451-1

Coda. "Pearly Everlasting"

Gary Snyder

Walk a trail down to the lake
mountain ash and elderberries red
old-growth log bodies blown about,
whacked down, tumbled in the new ash *wadis*.
Root-mats tipped up, veiled in tall straight fireweed,
fields of prone logs laid by blast
In-line north-south down and silvery
limbless barkless poles—
clear to the alpine ridgetop all you see
is toothpicks of dead trees
thousands of summers
at detritus-cycle rest
—hard and dry in the sun—the long life of the down tree yet to go
bedded in bushes of pearly everlasting
dense white flowers
saplings of bushy vibrant silver fir
the creek here once was "Harmony Falls."
The pristine mountain
just a little battered now
the smooth dome gone
ragged crown

the lake was shady *yin*—
now blinding water mirror of the sky
remembering days of fir and hemlock—
no blame to the magma or the mountain
& sit on a clean down log at the lake's edge,
the water dark as tea.

I had asked Mt. St. Helens for help
the day I climbed it,   so seems she did.
The trees all lying flat like,   after that big party.
Siddhartha went to on the night he left the house for good,
crowd of young friends whipped from sexy dancing
dozens crashed out on the floor

angelic boys and girls, sleeping it off.
A palace orgy of the gods but what
"we" see as "Blast Zone" sprinkled with
clustered white flowers

"Do not be tricked by human-centered views," says Dogen,
And Siddhartha looks it over, slips away—for another forest—
—to really get right down on life and death.

If you ask for help it comes.
But not in any way you'd ever know
—thank you Loowit, lawilayt-lá, *Smoky Mâ*
Gracias   xiexie   grace

(from *In the Blast Zone*, Oregon State University Press, p. 114–115)

C.M. Crisafulli, V.H. Dale (eds.), *Ecological Responses at Mount St. Helens: Revisited 35 years after the 1980 Eruption*,
https://doi.org/10.1007/978-1-4939-7451-1

Index

C.M. Crisafulli, V.H. Dale (eds.), *Ecological Responses at Mount St. Helens: Revisited 35 years after the 1980 Eruption*,
https://doi.org/10.1007/978-1-4939-7451-1

T

U

MIX
Papier aus verantwortungsvollen Quellen
Paper from responsible sources
FSC® C105338

If you have any concerns about our products,
you can contact us on
ProductSafety@springernature.com

In case Publisher is established outside the EU,
the EU authorized representative is:
Springer Nature Customer Service Center GmbH
Europaplatz 3, 69115 Heidelberg, Germany

Printed by Libri Plureos GmbH
in Hamburg, Germany